4th International Conference on Mechatronics and Mechanical Engineering (ICMME 2017)

MATEC Web of Conferences Volume 153 (2018)

Kuala Lumpur, Malaysia
28-30 November 2017

Editors:

P. Tandon
W. Wisnoe
M. Z. Bin Abdullah

ISBN: 978-1-5108-5875-6

Printed from e-media with permission by:

Curran Associates, Inc.
57 Morehouse Lane
Red Hook, NY 12571

Some format issues inherent in the e-media version may also appear in this print version.

Printed by Curran Associates, Inc. (2018)

For additional information, please contact EDP Sciences – Web of Conferences at the address below.

EDP Sciences – Web of Conferences
17, Avenue du Hoggar
Parc d'Activité de Courtabœuf
BP 112
F-91944 Les Ulis Cedex A
France

Phone: +33 (0) 1 69 18 75 75
Fax: +33 (0) 1 69 28 84 91

contact-edps@webofconferences.org

Additional copies of this publication are available from:

Curran Associates, Inc.
57 Morehouse Lane
Red Hook, NY 12571 USA
Phone: 845-758-0400
Fax: 845-758-2633
Email: curran@proceedings.com
Web: www.proceedings.com

TABLE OF CONTENTS

Preface

2017 The 4th International Conference on Mechatronics and Mechanical Engineering (ICMME 2017) will be held during November 28-30, 2017 in University of Malaya, Kuala Lumpur, Malaysia this year. ICMME is dedicated to issues related to mechanical engineering, aerodynamics, aerospace systems and technology, applied mechanics, automation, biomechanics, computational fluid dynamics, design and manufacturing, energy management, fluid dynamics etc.

The objective of the conference is to bring academic scientists, engineers, industry researchers together to exchange and share their experiences and research results, and discuss the practical challenges encountered and the solutions adopted. We hope you will have a delightful and rewarding experience, and use this occasion to meet old friends and make many new ones in your academic circle.

ICMME 2017 proceedings contain 60 selected papers from the conference which are presented orally at the conference or via poster sessions, and provide up-to-date, comprehensive and worldwide state-of-the art knowledge in this field. Each contributed paper was peer-reviewed by international reviewers who were drawn from the technical committee members as well as other external reviewers invited from all over the world. The proceedings are divided into 9 chapters based on the different topics of the papers, namely, novel materials and properties, robot design and development technology, power machinery and transmission, vehicle design and control, modeling and simulation of mechanical system, researching and designing of machines and mechanisms, control theory and monitoring technology, industrial engineering and manufacturing systems, urban infrastructure construction and planning.

On behalf of the organizing committees, we would like to thank the organization staff, the members of the program committees and reviewers. They have worked very hard reviewing papers and making valuable suggestions for the authors to improve their work. We also would like to express our gratitude to the external reviewers, for providing extra help in the review process, and the authors for contributing their research results to the conference.

We look forward to seeing all of you next year at ICMME.

Conference General Co-chair

Prof. Dr. Mohd Hamdi Abdul Shukor,
University of Malaya, Malaysia
October 28, 2017.

Statement of Peer review

In submitting conference proceedings to MATEC Web of Conferences, I certify to the Publisher that I adhere to the **Policy on Publishing Integrity** of the journal in order to safeguard good scientific practice in publishing.

1. All articles have been subjected to peer review administered by the proceedings editors.
2. Reviews have been conducted by expert referees, who have been requested to provide unbiased and constructive comments aimed, whenever possible, at improving the work.
3. Proceedings editors have taken all reasonable steps to ensure the quality of the materials they publish and their decision to accept or reject a paper for publication has been based only on the merits of the work and the relevance to the journal.

Title, date and place of the conference

The 4th International Conference on Mechatronics and Mechanical Engineering
November 28-30, 2017
Kuala Lumpur, Malaysia

Proceedings editor(s):

Prof. Puneet Tandon
Prof. Wirachman Wisnoe
Prof. Ir. Dr. Mohd Zulkifly bin Abdullah

12/2/2018

Date and editor's signature

 https://doi.org/10.1051/matecconf/201815301001

Effect of deformation and annealing temperature on the mechanical properties and microstructure of Al-4.5Zn-1.5Mg-0.9Cr (wt. %) alloy fabricated by squeeze casting

Maya Putri Agustianingrum, Nuzulian Akbar Arandana, Risly Wijanarko and Bondan Tiara Sofyan[1]

[1]*Department of Metallurgy and Materials Engineering, Faculty of Engineering, Universitas Indonesia, Kampus UI Depok 16424, Indonesia*

Abstract. In order to produce structural products, Al-Zn-Mg alloys undergo various forming processes. Problems that are usually found in the forming process include peripheral coarse grain (PCG) and hot tearing which decrease mechanical properties and corrosion resistance of the alloys. Addition of microalloying element such as chromium (Cr) is an alternative to overcome these problems. The presence of Cr in Al-Zn-Mg alloys supresses the grain growth by preventing excess recrystallization. In this research 0.9 wt. % Cr was added to Al-4.5Zn-1.5Mg alloy and the deformation behaviour as well as subsequent recrystallization was observed. The alloy was fabricated by squeeze casting followed by homogenization at 400 °C for 4 h. The samples were cold rolled for 5, 10, and 20 %. The 20 % deformed samples were then annealed at 300, 400, and 500 °C for 2 h. Material characterization consisted of microstructure analysis using optical microscope and Scanning Electron Microscope (SEM) – Energy Dispersive Spectroscopy (EDS), hardness testing using Micro Vicker methods. The results showed that the deformed grain ratio was 1.6, 2.84, and 2.99 in the 5, 10, and 20 % deformed samples, respectively. The elongated dendrites were effective to increase the hardness of the alloy. Recrystallization was not detected during annealing at 300 and 400 °C, but was observed at 500 °C. Whereas, for the samples without Cr addition, recrystallization occurred at 400 °C. It means that the addition of Cr increased the recrystallization temperature of the alloy. It occured because (Al, Zn)$_7$Cr dispersoids with size less than 1 μm impeded the dislocation motion during annealing, so that recrystallization was retarded. On the other hand (Al, Zn)$_7$Cr dispersoids with size more than 1 μm promoted the formation of new grains around them by Particle Stimulated Nucleation (PSN) mechanism. In this case, the fine (Al, Zn)$_7$Cr dominated so that recrystallization was slower.

1 Introduction

Aluminium 7XXX series are alloys with zinc and magnesium as alloying elements. With 83.1 wt. % solubility in aluminium, addition of zinc enhances the strength of the alloy through solid solution strengthening [1]. While addition of magnesium promotes the $MgZn_2$ precipitate that increases the strength and hardness of the alloy after ageing processes [2]. Thus, these alloys have good mechanical properties and are commonly used in the aircraft industry as small and large components.

Precipitation strengthening in aluminium alloys are achieved through solution treatment and ageing processes [3]. The strength of the alloy can also be further improved by deformation processes, such as extrusion, cold rolling, and drawing. Problems which usually occur during deformation are Peripheral Coarse Grain (PCG) and hot tearing [4]. To overcome the problems, alloying elements such as Cr are added to the alloys to promote the formation of $CrAl_7$ intermetallic phase [5]. These Cr intermetallic phases were found to inhibit grain growth and to retard recrystallization during annealing process.

Deformation during cold rolling changes the microstructure and mechanical properties. For aluminium that possesses FCC crystal structure, the deformation mechanism is interesting because of the dependency on the SFE (Stacking Fault Energy). Aluminium alloys have higher SFE than any other FCC metals such as brass and copper, which results in cross slip as common deformation mechanism. Annealing process after cold rolling will lead to recovery, recrystallization and grain growth. Lee & Kim [6] cold rolled Al7075 by 20, 40 and 60 % and continued with annealing at 300, 400 and 500 °C for 1 h. The results showed that higher deformation led to the increase in nucleation rate during recrystallization so that finer grains were produced and higher strength and hardness were obtained.

This research studied the change in microstructures and hardness of Al-4.5Zn-1.5Mg (wt. %) alloy with addition of 0.9 wt. % Cr after various degree of cold rolling followed by annealing process at different temperatures. It was intended to learn how the level of cold rolling reduction affected the deformation mechanism in aluminium alloy and how the annealing temperature affected the recrystallization process. The role of Cr during deformation and recrystallization process was also investigated.

2 Experimental procedure

Pure Al, Zn, and Mg ingots as well as Al-80Cr master alloy (Figure 1) were melted in an electric furnace at 850 °C. The molten metal was degassed by argon at 850 °C at 5000 rpm for 30 s for proper mixing and then poured into a 300 °C preheated metal mould with the dimension of 17x17x1.5 mm^3. The semi solid metal was squeezed at the pressure of 76 MPa for 10 min. The nominal composition of the as-cast alloy was shown in Table 1. All composition in this paper is in wt. %, unless otherwise stated. The as-cast alloy then being homogenized in a muffle furnace at the temperature of 400 °C for 4 h followed by water quench. Cold rolling was conducted with the degree of deformation of 5, 10 and 20 % in multiple passes. The 20 % deformed samples were then annealed at 300, 400 and 500 °C for 2 h.

Figure 1. Material used in this research (a) aluminium ingot (b) zinc ingot (c) magnesium ingot and (d) Al-80Cr master alloy.

Table 1. The nominal composition of the Al-4.5Zn-1.5Mg-0.9Cr (wt. %) alloy.

Zn	Mg	Cr	Fe	Si	Cu	Ni	Al
4.56	**1.51**	**0.9**	0.21	0.13	0.01	0.02	Bal.

Microstructures were observed by using optical microscope and Scanning Electron Microscope (SEM). Standard metallographic preparation was performed with Keller's etching of 2.5 ml HNO_3 + 1 ml HCl + 1.5 ml HF + 95 ml distilled water within time span of 60-120 s. The as-homogenized microstructures were further quantitatively analyzed using ImagePro to measure the size of Secondary Dendrite Arm Spacing (SDAS). While Matscope was used to calculate the grain deformation ratio (L/d) and grain roundness ratio ($4\pi A/P^2$) on deformed and annealed samples, respectively. Then, Vickers hardness test was performed according to ASTM E384. Five indentations were made for each measurement.

3 Results and discussion

3.1 Deformation Process

The microstructures of the alloy in as-cast and as-homogenized condition are shown in Figure 2. The as-cast microstructure has typical dendritic structure (arrow 1) and tend to be globular because of the applied pressure during squeeze casting process. Bright Cr intermetallic phase (arrow 3) was found to be sparsely spreaded within the interdendritic structure (arrow 2). The homogenization led to the diffusion of interdendritic phases into the aluminium matrix, which was shown by the decrease in the SDAS from 29.71 µm in the as cast condition to 19.86 µm in the as-homogenized condition. It also decreased the hardness of the alloy from 121 HV to 104 HV (Figure 2).

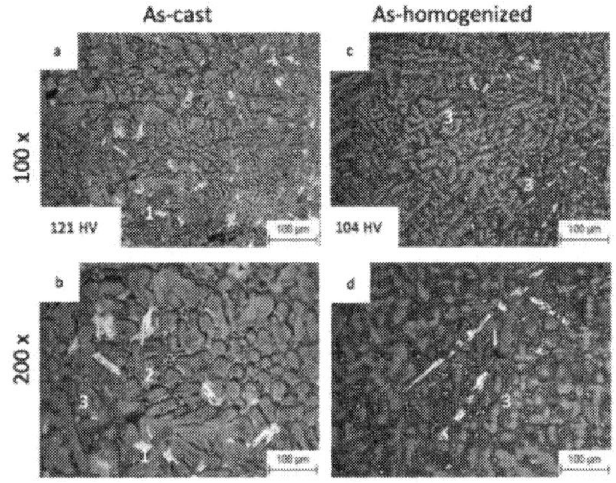

Figure 2. Microstructures of Al-4.5Zn-1.5Mg-0.9Cr alloy in (a, b) as-cast, and (d,e) as-homogenized condition. Arrows 1, 2, and 3 point to dendrites, interdendritic and Cr intermetallic phase, respectively.

Figure 3 shows the comparison between the microstructure of Al-4.5Zn-1.5Mg-0.9Cr and Al-4.7Zn-1.8Mg [7] alloys in as-homogenized condition. It is clear that the addition of 0.9 % Cr reduced the SDAS, therefore increased the hardness from 69.5 HV to 104 HV. Refining of SDAS due to addition of Cr refers to two mechanisms. The first mechanism is Growth Restriction Factor (GRF) due to the presence of Cr atoms within the aluminium dendrites. The solid solubility of Cr in Al is 0.77 wt. % at 661 °C [8]. The amount of Cr, which dissolves in Al matrix, increases the GRF values, as calculated by the following formula [9]:

$$GRF = m.Co.(k-1) \quad (1)$$

where m is liquidus slope, Co is alloy composition, and k is equilibrium partition coefficient. The calculation showed that the GRF for 0.9 % Cr-added alloy and the base alloy are 9.02 and 6.8, respectively. Therefore Cr-containing alloy might have smaller SDAS.

The second mechanism is heteregenous nucleation by Cr intermetallic phases (arrow 3 in Figure 1). This is in line with the results by Kumar et al. [10] which found that the addition of Cr in Al-Si-Mg alloys promoted the Cr intermetallic phases, which then act as heteregenous nucleation sites during solidification and decrease the grain size.

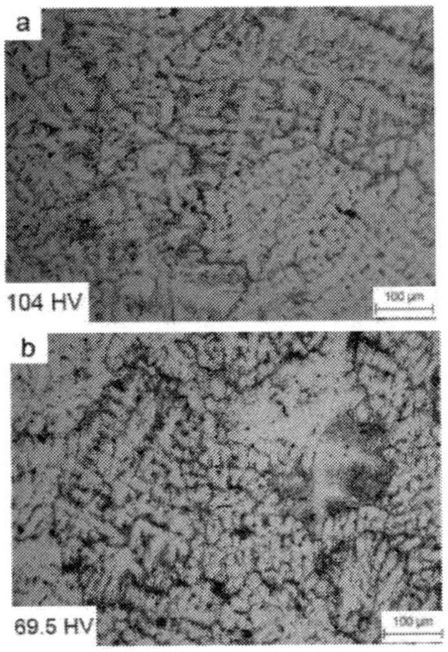

Figure 3. Microstructures of (a) Al-4.5Zn-1.5Mg-0.9Cr and (b) Al-4.7Zn-1.8Mg [7] after homogenization. Red arrow points to Cr intermetallic phase.

Figure 4 shows the microstructures of Cr-containing alloy and the corresponding base alloy after different degrees of cold rolling and the measured grain deformation ratio (L/d). After 5 % of deformation, no significant change on the microstructure was observed. The dendrite structures were not much elongated as shown by the L/d ratios of 1.60 and 2.19 in Cr-containing and the base alloys, respectively. No slip lines were observed, indicating that 5 % deformation did not move the dislocations. Significant elongation of dendrites started to occur at 10 % deformation, followed by cracking of Cr intermetallic phase (Red arrow in Figure 5 (a)). After 20 % deformation, severe elongation of dendrite occured, resulted in L/d ratio of 2.99 and 4.59 in Cr-containing and

the base alloys, respectively. It is noteworthy that the alloy without Cr shows a series of slip lines (yellow arrow in Figure 4 (h)), while on the other hand they were not detected in the Cr-containing alloy (Figure 4 (d)). This is thought to be due to the presence of incoherent Cr intermetallics that hinder dislocation movement.

3.2 Annealing Process

Figures 4 and 5 show the microstructures of the studied alloys after 20 % of rolling and subsequent annealing at 300, 400, and 500 °C. As illustrated in Figure 5 and 6 (a), 20 % of deformation resulted in elongated dendrites and followed by cracking of Cr intermetallic. After annealing at 300 °C for 2 h (Figure 5 and 6 (b)), the dendrite structure can still be observed but not in elongated shape as before the annealing process. This is in accordance with the calculation of grain roundness ratio, as shown in Table 2. The grain roundness ratio increased from 0.010 to 0.079 after annealed at 300 °C. Increasing of grain roundness ratio indicated the occurrence of recovery process after having extreme deformations due to the previous cold rolling process. However, recrystallization was not occur at this temperature. Meanwhile, the grain roundness ratio increased up to 0.092 after annealing at 400 °C. It can be seen in Figure 5 (c) that the dendrite tips are more rounded than that of before the annealing process. However, the SEM image, as presented in Figure 6 (c), has not shown any recrystallization phenomenon. Furthermore, annealing process at 500 °C resulted in more rounded dendrites and Cr intermetallic tips with grain roundness ratio of 0.095. According to Lee & Kim [11], stored energy will decrease along with increasing annealing temperature. However, recrystallization was still undetected after annealed at this temperature, as shown in Figure 6 (d). This was due to the addition of Cr into Al-4.5Zn-1.5Mg wt. % alloy.

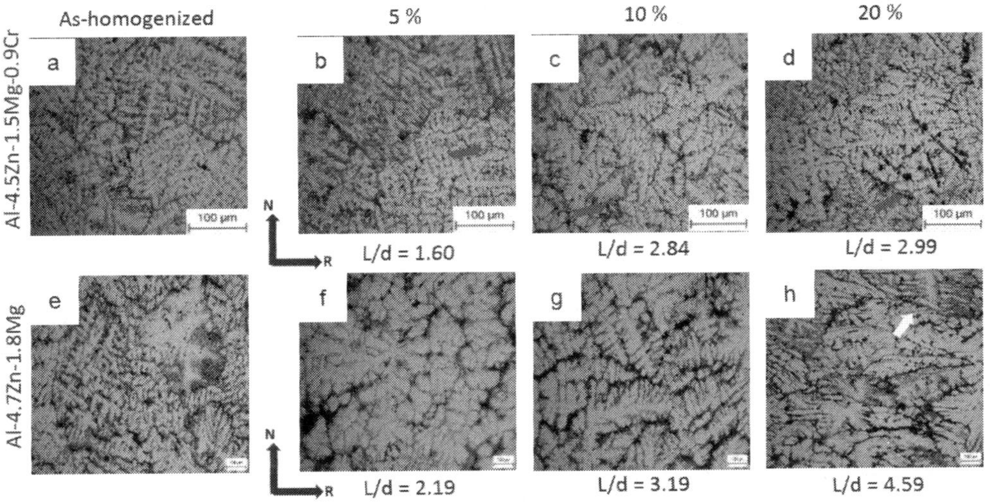

Figure 4. Microstructures of (a-d) Al-4.5Zn-1.5Mg-0.9Cr and (e-h) Al-4.7Zn-1.8Mg [7] alloys in as-homogenized condition and after cold rolling for 5, 10, and 20 %. Red arrow: Cr intermetallic phases, yellow arrow: slip bands, ND: normal direction, RD: rolling direction.

 https://doi.org/10.1051/matecconf/201815301001

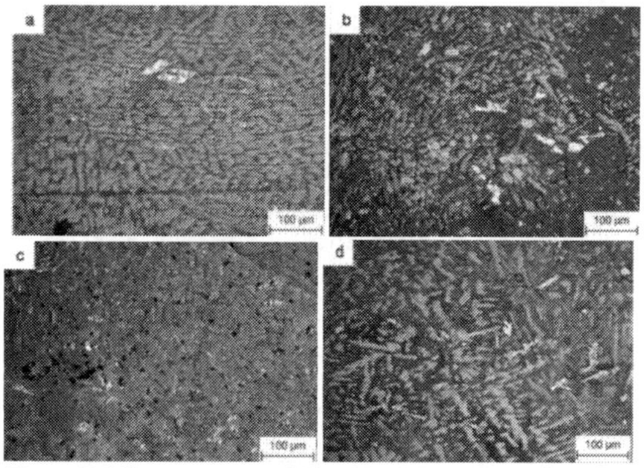

Figure 5. Microstructures of Al-4.5Zn-1.5Mg-0.9Cr alloy after (a) 20 % cold rolling, and subsequent annealing at (b) 300, (c) 400 and (d) 500 °C for 2 h. Red arrow: cracked Cr intermetallic phases.

Table 2. The comparison of grain roundness ratio of the Al-4.5Zn-1.5 Mg-0.9Cr alloy before and after cold rolling.

Treatment	Grain Roundness Ratio ($4\pi A/P^2$)
Cold rolling 20 %	0.010
Cold rolling 20 % + Annealing 300 °C	0.079
Cold rolling 20 % + Annealing 400 °C	0.092
Cold rolling 20 % + Annealing 500 °C	0.095

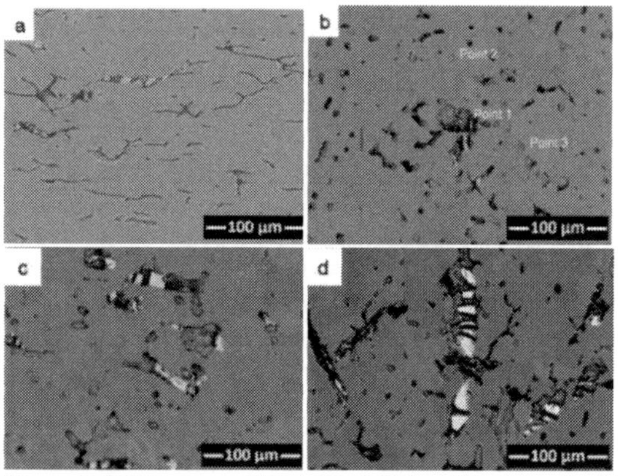

Figure 6. SEM images of Al-4.5Zn-1.5Mg-0.9Cr alloy after (a) 20 % cold rolling, and subsequent annealing at (b) 300, (c) 400 and (d) 500 °C for 2 h.

Table 3. The composition of the Al-4.5Zn-1.5 Mg-0.9Cr alloy after annealing at 300 °C for 2 h at positions as shown in Figure 6 (b).

Point	Composition (at. %)				Possible Phase
	Al	Zn	Mg	Cr	
1	85.1	1.9	-	13	$(Al,Zn)_7Cr$
2	96.2	2.3	1.6	-	α-Al + $MgZn_2$
3	96.6	1.9	1.5	-	α-Al + $MgZn_2$

The SEM-EDS results of the studied alloy, which was annealed at 300 °C, are shown in Figure 6 (b) and Table 3. Point 1 indicated Cr intermetallic in plate-shape, which cracked due to the cold rolling process. The matrix and grain boundary of Al-4.5Zn-1.5Mg-0.9Cr alloy were observed at point 2 and 3, respectively. As listed in Table 3, point 2 contained α-Al with $MgZn_2$ spread in the matrix. Meanwhile, the amount of Zn and Mg were also dominant at point 3, so that $MgZn_2$ existed at the grain boundaries. Point 1 was expected to be the Cr intermetallic due to the composition that was dominated by aluminium, zinc, and chromium and then formed $(Al,Zn)_7Cr$. According to He et al. [12], θ phase (Al_7Cr) contained of 13 at. % Cr, 85 at. % Al, and 1.9 at. % Zn. $(Al,Zn)_7Cr$ phase can strongly pin dislocation and prevents the formation of new grain boundaries during annealing process due to its high melting point and stable composition. Thus, the existence of θ phase in the studied alloy increased the recrystallization temperature.

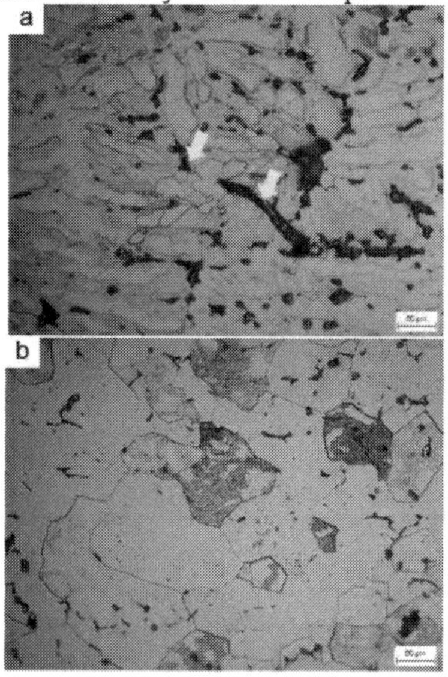

Figure 7. Microstructures of (a) Al-4.5Zn-1.5Mg-0.9Cr and (b) Al-4.7Zn-1.8Mg [7] alloys after annealing at 500 °C for 2 h. Red arrow: recrystallization, yellow arrow: the deformed $(Al,Zn)_7Cr$.

Figure 7 presents microstructures of the studied alloys after annealing at 500 °C for 2 h. As can be seen, $(Al,Zn)_7Cr$ cracked (yellow arrows) and the cracked boundaries tended to be rounded in Al-4.5Zn-1.5 Mg-0.9Cr alloy. Meanwhile, grain growth occurred in alloy without Cr addition. The differences were developed due to the presence of Cr. There are two competing mechanisms work in the Cr-containing alloys due to the formation of $(Al,Zn)_7Cr$ dispersoids. If the $(Al,Zn)_7Cr$ possesses size less than 1 μm, they will pin the dislocations and inhibit grain growth. Liu et al. [13] investigated that the grain growth restriction could effectively occur when Sc and Zr intermetallic were in very small sizes (25 nm).Therefore, the addition of Cr resulted in higher recrystallization temperature than that of alloy without Cr addition. However, when the size of $(Al,Zn)_7Cr$ is ≥ 1 μm, they will promote nucleation of grains during heating, through Particle Stimulated Nucleation (PSN) mechanism. This is clear in Figure 7 (a), where nucleation of new grains was detected around $(Al,Zn)_7Cr$ (red arrow). This phenomenon is in line with Sidor et al. [14], which represented nucleation in AA 5182 alloy by PSN mechanism after cold rolling and annealing at 550 °C. When we compare Figure 7 (a) and (b), of which grain growth was inhibited in the Cr-containing alloy, it is clear that the grain growth retardation by fine $(Al,Zn)_7Cr$ is more dominant than the PSN mechanism by large $(Al,Zn)_7Cr$ dispersoids.

The effect of annealing temperature on the hardness of Al-4.5Zn-1.5 Mg-0.9Cr alloy is shown in Figure 8. Before annealing, the hardness value reached 117.16 HV and significantly decreased to 86.84 HV after annealing at 300 °C due to the recovery of interlocking dislocations. Annealing at 400 °C further increased the hardness to 110.90 HV, which indicated the recrystallization process. However, the microstructure presents the absence of nucleation, as shown in Figure 5 (c). The hardness value continued to increase up to 117.43 HV, which implied recrystallization started to occur (Figure 7 (a)). Furthermore, increased hardness also caused by the pinning effect of $(Al,Zn)_7Cr$ in the studied alloy, which restricted the grain boundaries and dislocation movement during the annealing process. These results correspond to Liu et al. [13], which investigated increasing hardness up to 51.96 % after the addition of 0.10 wt. % Sc and 0.16 wt. % Zr. Meanwhile, the hardness of Al-4.7Zn-1.8Mg wt. % before annealing was lower than that of alloy with Cr addition with the value of 102.90 HV. After annealing at 300 °C, the hardness decreased to 95.7 HV, which was higher compared with alloy with Cr addition. Higher annealing temperature shows a continued decrease down to 94.1 HV. This was due to the absence of grain growth restriction by Cr.

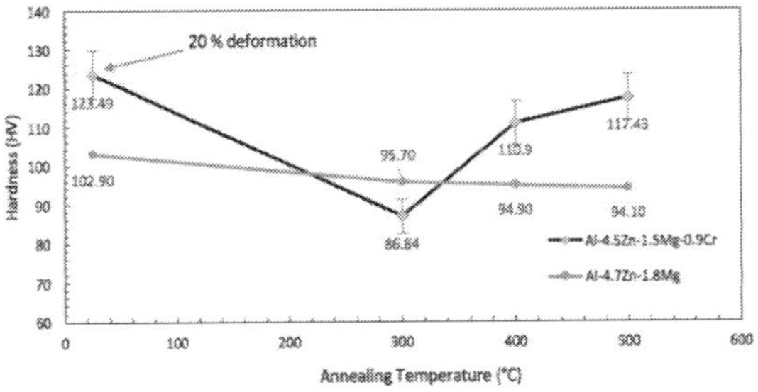

Figure 8. Effect of annealing temperature on the hardness of the 20 % deformed Al-4.5Zn-1.5Mg-0.9Cr compared with Al-4.7Zn-1.8Mg alloys [6].

4 Conclusions

The results of investigation on Al-4.5Zn-1.5Mg-0.9Cr (wt. %) alloy can be concluded as follow:

1. The hardness of as-cast Al-4.5Zn-1.5Mg-0.9Cr alloy was 121 HV and then decreased to 104 HV after homogenization process due to the diffusion of interdendritic phase to the α-Al matrix.
2. The 5 % deformed Al-4.5Zn-1.5Mg-0.9Cr alloy did not indicate any plastic deformation. The dendrite structure was further elongated followed by the cracking of $(Al,Zn)_7Cr$ intermetallic phase after 10 % and 20 % of deformation. The grain deformation ratio after cold rolling for 5, 10, and 20 % are 1.6, 2.84, and 2.99, respectively.
3. The hardness value of Al-4.5Zn-1.5Mg-0.9Cr alloy with variation of 5, 10, and 20 % deformation are 105.95, 118.75, and 117.16 HV, respectively, which hardening mechanism was strain hardening.
4. Annealing process at 300 and 400 °C on Al-4.5Zn-1.5Mg-0.9Cr alloy has not changed the dendritic structure, while the intermetallic tips were more rounded. After annealing at 500 °C, local recrystallization started to occur. The grain roundness ratio after annealing at 300, 400, and 500 °C were 0.079, 0.092, and 0.095, respectively.
5. The presence of Cr promotes the formation of $(Al,Zn)_7Cr$ intermetallic phase.
6. $(Al,Zn)_7Cr$ with size smaller than 1 μm restricted the dislocation movement and grain boundaries migration. Meanwhile, $(Al,Zn)_7Cr$ intermetallic phase with size more than 1 μm initiated nucleation with the mechanism of Particle Stimulated Nucleation (PSN). In the studied alloy, fine $(Al,Zn)_7Cr$ dominated so that recrystallization was slower.
7. Increasing annealing temperature generated increasing hardness values, which are 86.84, 110.9, and 117.43 HV on annealing temperatures of 300, 400, and 500 °C, respectively.

Acknowledgement

The authors are grateful with the provision of research funding through *Hibah Publikasi Internasional Terindeks untuk Tugas Akhir (PITTA)* 2017 from Universitas Indonesia.

References

1. Agrawal, L., Yadav, R., & Sexena, A. *Effect of Magnesium Content on the Mechanical Properties of Al-Zn-Mg Alloys*. Int. J. of Em. Tech., **3**, 137-140 (2012)
2. He, Z., Su, X., Peng, H., Liu, Y., Wu, C., & Wang, J. *600 °C Isothermal Section of the Al-Cr-Zn Ternary Phase Diagram*. J. Alloys and Compounds., **649**, 1239-1245 (2015).
3. Prabhukhot, A. *Effect of Heat Treatment on Hardness and Corrosion Behavior of 6082-T6 Aluminium Alloy in Artificial Sea Water*. Int. J. of Materials Science and Engineering., **3**, 287-294 (2015).
4. Eivani, A. R., Zhou, J., & Duszczyk, J. *Recent Trends in Processing and Degradation of Aluminium Alloys*. Rijeka: InTech. (2011).
5. Rooy, E. L. *Aluminium Alloy Castings: Properties, Processes, and Applications*. Detroit: ASM International Committee. (2004).
6. Lee, Y. S., & Kim, W. K. *Recrystallization behavior of cold rolled Al−Zn−Mg−Cu fabricated by twin roll casting*. Trans. Nonferrous Met. Soc. China., **24**, 2226-2231 (2014).
7. Kurnia, R., & Sofyan, B.T., *Effect of Cold Rolling and Annealing Temperature on The Recrystallization and Mechanical Properties of Al-4.7Zn-1.8Mg (wt. %) Alloy Fabricated by Squeeze Casting*. Proc. Int.Conf. QiR., 1-6 (2017).

8. Rambabu, P., Prasad, N., Kutumbarao, V. & Wanhill, R. ed., *Aluminium Alloys for Aerospace Applications. In: Aerospace Materials and Material Technologies, 1st ed.* Indian Institute of Metals Series, 36 (2017).
9. Zhang, L., Eskin, D., Miroux, A. & Katgerman, L. *Role of Solute and Transition Metals in Grain Refinement of Aluminium Alloys Under Ultrasonic Melt Treatment.* TMS (The Minerals, Metals & Materials Society), 1389-1394 (2012).
10. Kumar, A., Sharma, G., & Sasikumar, A. *Effect of Cr on grain refinement and mechanical properties of Al-Si-Mg alloys.* Applied Mechanics and Materials., **789-790**, 95-99 (2015).
11. Lee, Y. S., & Kim, W. K. *Recrystallization behavior of cold rolled Al−Zn−Mg−Cu fabricated by twin roll casting.* Trans. Nonferrous Met. Soc. China., **24**, 2226-2231 (2014).
12. He, Z., Su, X., Peng, H., Liu, Y., Wu, C., & Wang, J. *600 °C Isothermal Section of the Al-Cr-Zn Ternary Phase Diagram.* J. of Alloys and Comp., **649**, 1239-1245 (2015).
13. Liu, J., Yao, P., Zhao, N., Shi, C., Li, H., Li, X., Xi, D., & Yang, S. *Effect of minor Sc and Zr on recrystallization behavior and mechanical properties of novel Al-Zn-Mg-Cu alloys.* J. Alloys and Compounds., **657**, 717-725 (2016).
14. Sidor, J.J., Decroos, K., Petrov R.H., & Kestens L.A.I. *Particle Stimulated Nucleation in Severely Deformed Aluminium Alloys.* Materials Science Forum., **706-709**, 389-394 (2012).

Fuel Analysis of Jatropha Methyl Ester and n-Tridecane as an Alternative Fuel for the Future

Annisa Bhikuning[1,2], Eriko Matsumura[1] and Jiro Senda[1]

[1]*Doshisha University, Mechanical Engineering Department, 1-3 tatara Miyakodani, Kyotanabe City Kyoto 610-0394, Japan*
[2]*Trisakti University, Mechanical Engineering Department, Jl. Kyai Tapa No.1, Grogol Jakarta Barat 11440, Indonesia*

Abstract. The authors proposed for new fuel between blending of jatropha methyl ester and n-tridecane. Biodiesel has an advantage in reducing emissions. Nevertheless, it has high viscosity and density and has poor spray characteristics compared to diesel fuel. The blending between n-tridecane would overcome the unwanted fuel properties. The n-tridecane and jatropha methyl ester were blended under three condition; JME25% (Jatropha Methyl Ester 25% and n-tridecane 75%), JME50% (Jatropha Methyl Ester 50% and n-tridecane 50%), and JME75% (Jatropha Methyl Ester 75% and n-tridecane 25%). The fuel properties were analyzed under biodiesel standardization from JIS K and ASTM D. FTIR analyzed also showed the characteristics of carbonyl peak that indicates as methyl ester. In the results, JME50% had met the requirements for fuel properties from biodiesel standardization.

1 Introduction

Nowadays, biodiesel has been a warm issue due to its beneficial and its potential to replace fossil fuel in the future. Many researchers have investigated not only edible oil for biodiesel such as palm oil [1], rapeseed [2] but also non-edible oil like jatropha curcas [3] and karanja oil [4]. However, biodiesel from edible oil would confront the competition of feedstock between human's food and fuel. Nevertheless, non-edible oil would be a good alternative since no competition in feedstock between food and fuel.

Biodiesel has advantages for reducing emissions such as HC as well as SOx in combustion through the fuel properties. Biodiesel may reduce some organic wastes, it can also reduce some pollutants to the environment [5].

In this study, the authors investigated new jatropha methyl ester that fulfills with the standard of biodiesel without blended with diesel fuel. Biodiesel has high viscosity, high density and has poor in spray atomization compared to diesel fuel. In order to overcome the undesirable fuel properties, new material was added. In this study, Jatropha Methyl Ester was blended with n-tridecane. Fuel properties and FTIR (Fourier Transform Infrared Spectroscopy) analysis of jatropha methyl ester and n-tridecane were evaluated. The aims of this study were to find the right blending between JME25, JME50, and JME75 that meet the requirements for the new fuel by analyzing some properties and also FTIR.

2 Methodology

2.1 Material and method of the biodiesel blend

The Jatropha Methyl Ester (JME) was purchased from Revo International, Japan. The n-Tridecane was purchased from Nakalai tesque, Japan. JME were blended with n-tridecane under three conditions, JME 25%+n-tridecane 75% (JME25); JME 50% + n-tridecane 50% (JME50); JME 75% + n-tridecane 25% (JME 75).

2.2 Fuel properties analysis

The fuel parameter properties were analyzed in Shimadzu Kyoto-Japan used Japan International standard - JIS K based on the method. The density (15°) was determined according to JIS K official method 2249. The kinematic viscosity (40°) was determined under JIS K 2283. The flash point (PMCC) was analyzed based on JIS K 2265. The Cetane number was analyzed on JIS K 2280. The copper strip corrosion was determined under JIS K 2513. The distillation was analyzed based on JIS K 2254. Nevertheless, n-tridecane chemical properties were taken from PubChem data based [6]. The fuel parameter analysis also compared to the ASTM D standard which based on the American and part of Asian biodiesel standardization.

2.3 FTIR analysis

The FTIR spectra data were taken using Shimadzu, Iraffinity-1, and FTIR 8400 spectrometer, in the range of 4000-400 cm^{-1}.

3. Results and discussion

3.1 Fuel analysis

3.1.1 Density

The density in fuel parameter is important to determine. Density is a chemical property to figure the exact volume of fuel to provide a satisfactory combustion [7]. Density can also impact the efficiency of the combustion system and fuel atomization in engine [8]. Biodiesel usually has high in density compared to diesel fuel. In case, the density is high. It can be influenced by engine output power due to distinctive in fuel mass injection [9]. Figure 1 shows that the density in jatropha blending with n-tridecane have low values compared to standard JIS K 2249. Nevertheless, the value JME 75 has met the requirements of ASTM D 1298 biodiesel standardization.

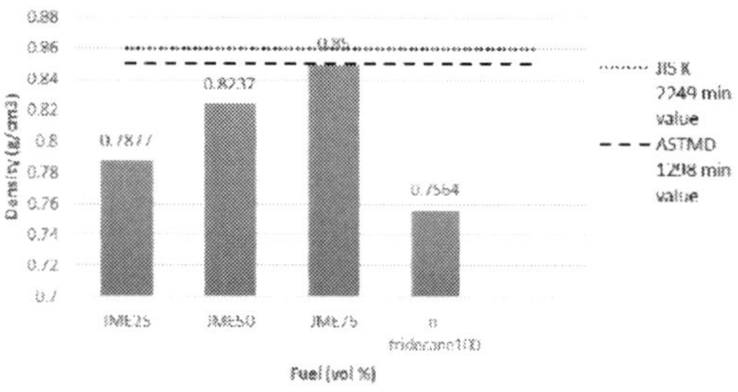

Figure 1. Density in various blends.

3.1.2 Kinematic viscosity

The kinematic viscosity in biodiesel has high value compared to diesel fuel, this condition caused by high fatty acid in biodiesel. High in kinematic viscosity can cause poor in the spray automation and can produce deposits in the engine and it can upgrade the energy to pump the fuel [8]. On the contrary, lower in kinematic viscosity would easier the fuel to pump and attain droplets to injector [10]. Demirbas [10] studied that diesel fuel has a similar viscosity with biodiesel. Transesterification process would decrease the viscosity level than pure biodiesel.

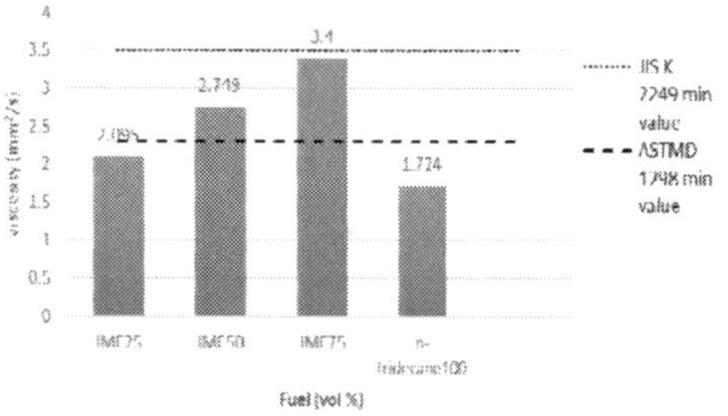

Figure 2. Viscosity in various blends.

From figure 2, the viscosity value of JME50, and JME75 have met the requirement for ASTM D 1298. Nevertheless, JME25, JME50, and JME75 have under value for JIS K 2249 standardization. However, n-tridecane 100 has not met all the standardization for JIS K 2249 nor ASTM D 1298.

3.1.3 Flash point

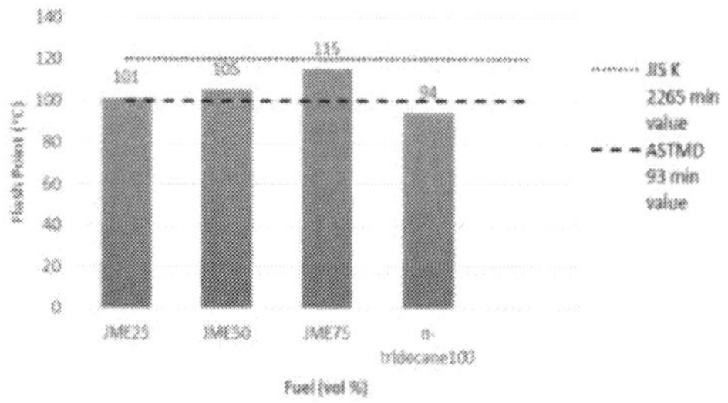

Figure 3. Flash Point in various blends.

The fuel flammability condition is pointed by its flash point. The estimation of flash point diverse and conversely corresponding to the instability of the fuel [11]. The Flash point value of diesel fuel is around 55-66 °C. While biodiesel is around 110-180 °C. Biodiesel has its minimum temperature of flash point. The JIS K 2265 has a minimum standard of a flash point that is 120 oC. Otherwise, ASTM D 93 has a lower standard of a minimum flash point that is 100°C.

From figure 3 shows that all fuels have met the biodiesel standard of ASTMD 93. Nevertheless, all fuels have not met the standard of JIS K2265.

3.1.4 Cetane number

Figure 4 displays the cetane number of various blends. It shows that all the Fuels have met the requirement for cetane number standards.

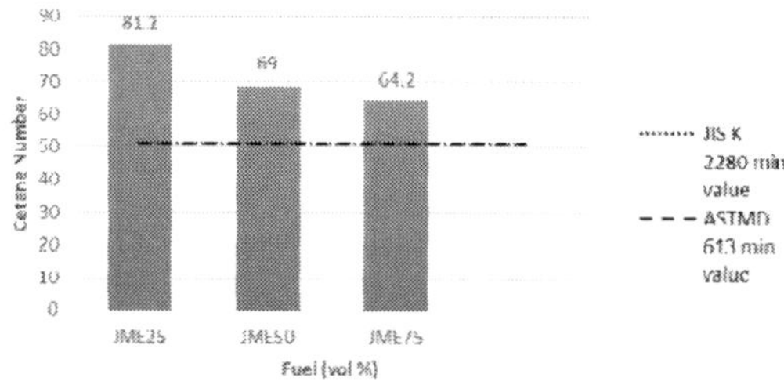

Figure 4. Cetane Number in various blends.

Cetane number is a value that indicates how the fuel will burn completely in the combustion chamber. High cetane number in fuels have several advantages such as the complete combustion will occur in the engine, reduce the knocking and noise in the engine, reduce the warm-up time engine and lower emission during combustion in a safer environment [12].

3.1.5 Copper strip corrosion

Copper strip test is to measure the corrosive of sulfur compounds in the fuel. The ASTM D 130 and JIS K 2513 consist of dipping a strip of copper into the fuel for a specified time and define temperature and observing the corrosive action of the fuel [13].

From figure 5 it shows that JME25, JME50, and JME75 have met the standard requirement of JIS K 213 and ASTMD 130. These indicated that JME 25, JME50, and JME75 are safe from the compound of corrosive sulfur.

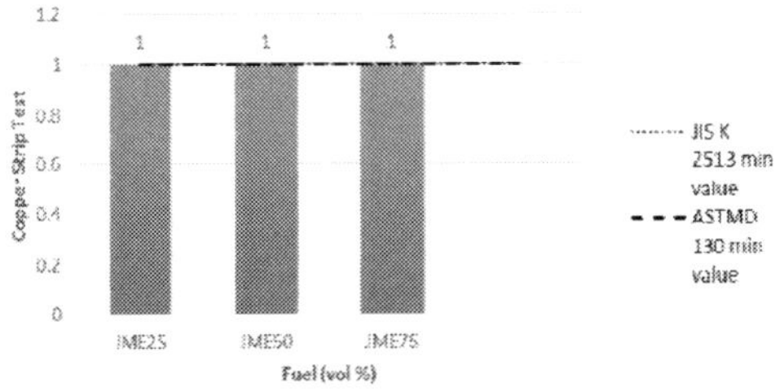

Figure 5. Copper Strip Corrosion from Various Blends.

3.1.6 Distillation

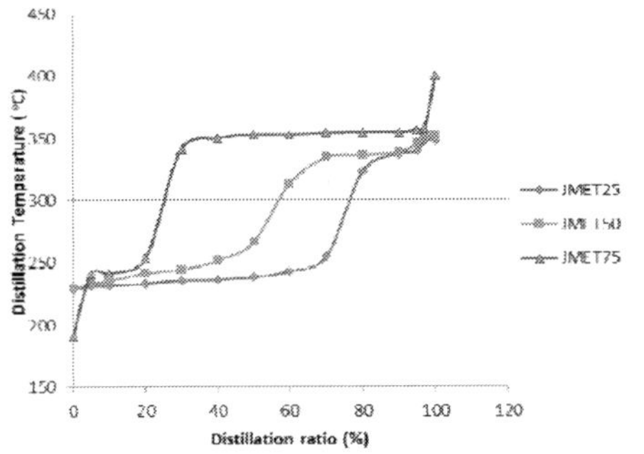

Figure 6. Distillation from various Blends.

The distillation curves can be seen in figure 6. It shows that JME 75% has higher volatility than JME 50% and JME 25%. This indicated that JME 75% has poor in spray characteristics (bigger droplets) than JME 50% and JME 25%.

3.2 FTIR Analysis

FTIR is Fourier Transform Infrared Spectrometry. Presently, FTIR has been characterized to discover the methyl peak positions from the reaction of transesterification of Biodiesel [14]. FTIR spectrometry is a worth equipment to predict wave numbers in biodiesel samples [15]

3.2.1 Tridecane

Tridecane is a combustible fuel. The fuel characteristics from n-tridecane are different from diesel fuel and biodiesel fuel.

The FTIR analysis of n-tridecane is shown in figure 7. In the figure shows there are only three peaks which strong in the FTIR figure. The alkana strong C-H vibration at 2920,23 cm^{-1}, the alkane bending –C-H vibration at 1470 cm^{-1} and the strong alkyl halide C-Cl 720 cm^{-1}. This indicated that n-tridecane is not from vegetable oil since no carbonyl peak between 1740-1750 cm^{-1} [16]. Furthermore, there is no spectral region between 1300-1060 cm^{-1} which is indicated as methyl ester.

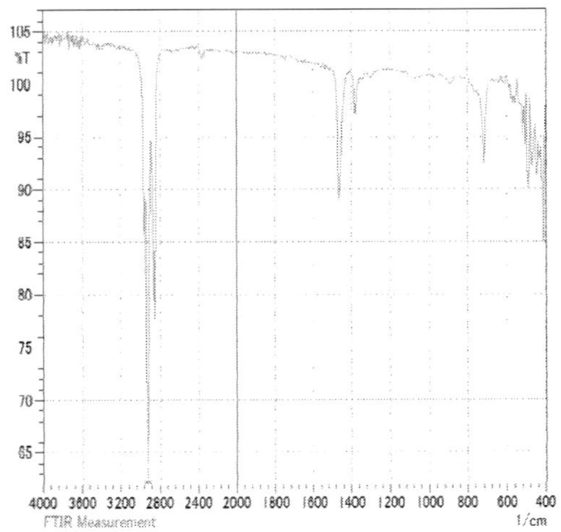

Figure 7. FTIR n-Tridecane.

3.2.2 JME 25% (Jatropha Methyl Ester 25% and n-Tridecane 75%)

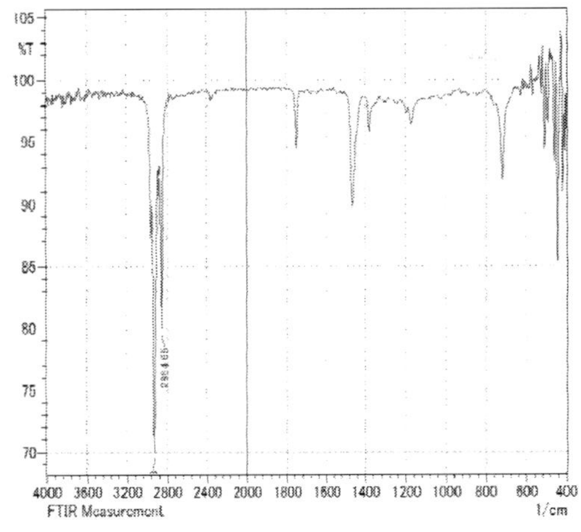

Figure 8. FTIR JME 25%.

The result of FTIR analysis of JME 25% is shown in figure 8. In the figure, there are four peaks from FTIR result. It shows strong band in C-H alkane wave number at 2854.65-2920.23 cm^{-1}. The stretch ether C=O vibration at 1760 cm^{-1}. The alkane bending –C-H vibration at 1460 cm^{-1}. The strong alkyl halide C-Cl vibration at 720 cm^{-1}.

JME25% is pointed that the fuel is made from vegetable oil or methyl ester since the C=O peak at 1760 has appeared although the wave number is not strong. However, it is correlated that the blend between n-tridecane and jatropha methyl ester can be approved from FTIR graph. The strong or weak of wave number have corresponded between the percentage amount of blending between fuel and methyl ester.

3.2.3 JME 50% (Jatropha Methyl Ester 50% and n-Tridecane 50%)

From the figure 9 indicates the FTIR result from JME 50%. From the figure there are five strong peaks in FTIR analysis.

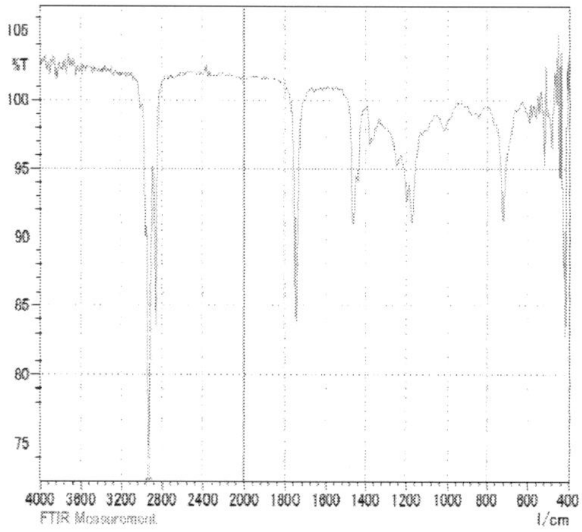

Figure 9. FTIR JME 50%.

The alkane stretch C-H vibration between 2870 – 2920,23 cm^{-1}. The strong stretch ether C=O and the wave number is at 1750 cm^{-1}.The alkane bending –C-H vibration at 1470 cm^{-1}, The ether stretch C-O wave number at 1170 cm^{-1} and the strong alkyl halide vibration at 723 cm^{-1}.

JME 50% shows that the carbonyl peak is in 1750 which is longer than JME 25%. The methyl ester peak is at 1170. This pointed that JME 50% is contained of methyl ester from jatropha, which blended with the higher composition of JME 25% and n-tridecane.

3.2.4 JME 75% (Jatropha Methil Ester 75% and n-Tridecane 25%)

As shown in figure 10, the FTIR analysis of JME 75% is analyzed. There are five strong peaks in the FTIR figure. The alkane stretch C-H vibration at 2870 – 2920,23 cm^{-1}, the ether C=O strong stretch at a wave number of 1750 cm^{-1}, The alkane bending –C-H vibration at 1470 cm^{-1}, the ether stretch C-O at a wave number of 1170 cm^{-1} and the strong alkyl halide C-Cl at a wave number of 723 cm^{-1}.

From the figure it shows that JME 75% has indicated from vegetable oil or methyl ester since it has strong carbonyl peak (1750 cm^{-1}) and strong ester C-O (1168.86). Those lines are longer than JME 25% and JME 50%. This can be proved that JME 75% was contained large blending proportions between jatropha methyl ester and n-tridecane.

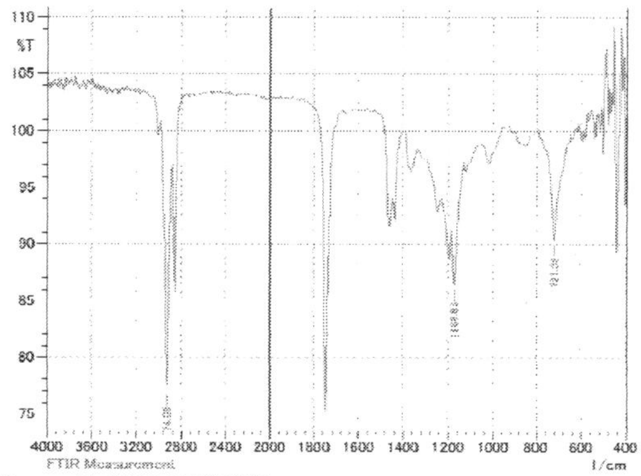

Figure 10. FTIR JME 75%.

4. Conclusion

The conclusions from this study are:

1. Fuel analysis of JME 25%, JME 50% and JME 75% for cetane number, copper strip corrosion point are and flash point are meet the standardization between ASTM D and JIS K.
2. The Viscosity of JME 50% and JME 75% are meet standardization of ASTM D 1298 but not for JIS K 2249.
3. The distillation curves show that JME 75% has higher volatility than JME 50% and JME 25%. This would effect for having a poor spray characteristic than JME 50% and JME 25%.
4. FTIR analysis of n-tridecane is indicated that no vegetable oil or methyl ester is contained in the fuel. This due to no carbonyl peak between 1740-1750 cm^{-1} and ester peak between 1060-1300 cm-^{1}.
5. FTIR analysis of JME 25%, JME 50% and JME 75% showed that there are numbers of carbonyl peak (1740-1750 cm^{-1}) and ester peak (1060-1300 cm^{-1}). This correlated that the longest line of carbonyl peak would have larger blending between jatropha methyl ester. This proved that JME75% has the longer line in carbonyl peak than JME 25% and JME 50%. This can be analyzed that JME 75 is contained 75% jatropha methyl ester and 25% of n-tridecane.
6. From the conclusion of fuel analysis, the JME 50% can be chosen as one of the best alternative fuel for the future.

Acknowledgement

The first author would like to grateful for Directorate General of Higher Education Ministry of Education and Culture (DIKTI) as a financial support, Coordination of Private Universities (KOPERTIS) and Prof. Michiaki Matsumoto Lab. Biochemical Engineering Doshisha University, Kyoto Japan.

References

1. Nongbe MC, Ekou T, Ekou L, Yao KB, Grognec EL, Felpin FX. Biodiesel Production from palm oil using sulfonated graphene catalyst. Renewable Energy. **106**,135-141 (2017)
2. Hu Q, Hua W,Yin Y, Zhang X, Liu L, Shi J, Zhao Y, Qin L, Chen C, Wang H. Rapeseed research and production in China. The Crop Journal. **5**, 127-135 (2017)
3. Rattanaphra D, Srinophakun P. Biodiesel production from crude sunflower oil and crude jatropha oil using immobilized lipase. Journal of Chemical Engineering of Japan. **43**, 104-108 (2010)
4. Meher LC, Kulkarni MG, Dalai AK, Naik SN. Transesterication of karanja (Pongamia pinnata oil by solid basic catalysts. European Journal of Lipid Science and Technology. **108**, 3898-3978 (2006)
5. Ikegami M. The use of fatty acid methyl ester called biodiesel fuel. Meeting for investigation into the actual conditions for diesel exhaust gas emission with alternative fuels. JSAE (2002)
6. https://pubchem.ncbi.nlm.nih.gov/compound/Tridecane
7. Ramirez V, Luis F. Density and viscosity of biodiesel as a function of temperature: empirical models. Renewable and Sustainable Energy Reviews. **19**, 652-665 (2013)
8. Xua J, Grift TE, Hansen AC. Effect of biodiesel on engine performances and emissions. Renewable Energy Reviews. **15**, 1098-1116 (2011)
9. E.Alptekin, M. Canaksi. Determination of the density and the viscosities of biodiesel-diesel fuel blends, Renewable Energy. **33**, 2623-2630 (2008).
10. Demirbas A. Progress and recent trends in biodiesel fuels. Energy Conversion and Management. **50** ,14-34 (2009)
11. A.S. Silitonga et al. Overview properties of biodiesel diesel blends from edible and non-edible feedstock. Renewable and Sustainable Energy Reviews. **22**, 346-360 (2013)
12. http://farmweeknow.com/blogs-benefits-high-cetane-diesel-fuel-3681
13. Knothe G. Analyzing biodiesel: standards and other methods. JAOCS.**83**,823-833 (2006)
14. Barone JR.233rd ACS National meeting. Paper AGRO-153.Chicago.25-29 (2007)
15. Cunha CL, Luna AS, Oliveira RCG, Xavier GM, Paredes MLL, Torres AR. Predicting the properties of biodiesel and its blends using mid-FTIR spectrometry and first-order multivariate calibration. Fuel.**204**,185-194 (2017)
16. Zhang WB. Review on analysis of biodiesel with infrared spectroscopy. Renewable and sustainable energy reviews. **16**, 6048-6058 (2012)

https://doi.org/10.1051/matecconf/201815301003

Simulation Investigation on Application of Powder Metallurgy Damping Effect in Water Hydraulic Pump

Weijie Shi, Xiaohui Luo, Zuti Zhang, Huawei Wang and Yuquan Zhu

School of Mechanical Science and Engineering, Huazhong University of Science and Technology, Wuhan, 430074, China

Abstract. Over the years, the hydrostatic balance slipper is often used to make the hydraulic lubrication film formed between the swash plate and the slipper. In this paper, a slipper containing the powder metallurgy (PM) composite material is proposed. The porous characteristics have an automatic damping effect, which can improve the tribological characteristics of the slipper. Based on the Reynolds equation, porous Darcy law and continuity equation, a lubrication model of PM damper is proposed to analyze the damping effect of PM in this paper. The proposed model is calculated using the finite volume method to capture the water film characteristics. Effects of length, diameter and porosity of the PM damper on the water film thickness are investigated. The results show that the water film can be formed by adopting the PM damper. In order to ensure the slipper with PM damper to form hydrostatic support, the length, diameter and porosity of PM damper is recommended to be 5 mm, 8 mm and 0.7, respectively.

1 Introduction

Water hydraulics system has been of great interest to researchers because of its environmental friendliness, easy disposal and safety. As a crucial component in the hydraulic system, axial piston pumps are widely used for power density, high efficiency and structure compactness. Swash plate-slipper is one of the most important friction pairs in the hydraulic axial piston pumps and it plays a key role in the pump's performance. However, the low viscosity and poor lubrication of water make it difficult to form the water lubrication film, which can produce boundary friction or even unlubrication friction [1]. Therefore, it is essential to study the lubrication characteristics of the swash plate-slipper friction pair.

Hydrostatic slipper bearing with the damping hole is an effective way to maintain a fluid film between swash plate and slipper. Previously, many researches on the damping hole were conducted via simulation and experiment. E. Koç and C. J. Hooke [2] examined the design of hydrostatically balanced bearings as used in the slippers of high pressure axial piston pumps, and outlined a design procedure whereby the slipper behaviour. It is shown that for successful operation the slippers need to have small amounts of non-flatness on the running surfaces. Researchers at the Maha Fluid Power Research Center have developed a simulation software CASPAR to simulate the pump performances. In particular, CASPAR can predict the lubrication performance between the slipper and swash plate [3, 4]. S. Kumar et al. [5] have investigated the characteristics of a piston pump slipper with a groove. The orifice radius and groove width are all 1 mm. A test rig was built to compare experimental and CFD results, finding a good agreement for all of the cases studied. In summary, water film is thinner than oil film between the swash plate and the slipper in piston pump.

On the other hand, the powder metallurgy (PM) material has excellent tribological properties and the most special place of PM is that it contains pores. The porosity can be controlled by adjusting the sintering process. A.G. Kostornovin et al. [6] prepared a series of bronze PM at the pressing pressure of 100-700 MPa, sintering temperature of by 770-860 ℃ and powder diameter of 1-200 μm. They found that the porosity decreases with the increase of pressure and sintering temperature, and the fluctuation range of porosity is 11%-50%. D. Gu [7] used the direct metal laser sintering to consolidate Cu-based alloy powder reinforced with Ni particles and found that a high densification level of 95.2% theoretical density was obtained after sintering. G Sethi et al. [8] studied the sintering behaviour of the Cu-12Sn bronze system in both, a microwave furnace as well as a conventional furnace. They found that the densification degree of a microwave furnace is greatly improved.

In view of the porous properties, the PM is installed into the slipper. The damping effect of PM damper on the lubrication performance is investigated.

2 Simulation methods

2.1 Damping effect model

Figure 1 illustrates the structure and working principle of the damping effect. The PM is installed in the slipper. Similar to the damping hole, the water can flow across the PM into the water chamber to form the water film reaction force. The reaction force lifts the slipper to avoid its contact with the swash plate. When the cylinder pressure is increased, the retain force is raised and the water film thickness is decreased accordingly. On the other hand, the flow rate through the PM damper decreases with decreasing the water film thickness. Further, with decreasing the flow rate through the PM damper, the pressure in water chamber increases, which can balance the retain force.

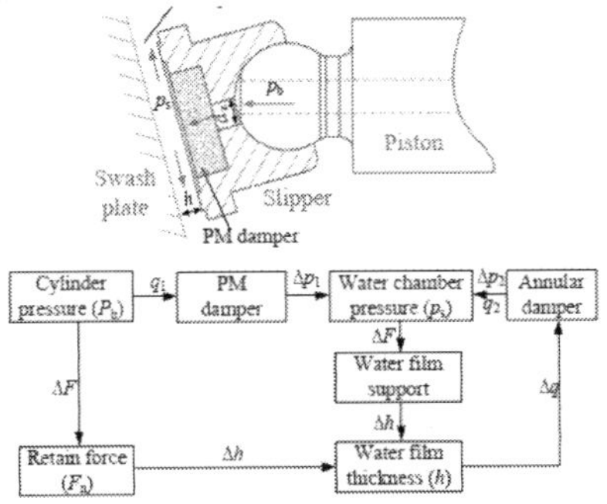

Figure 1. Structure and working principle of damping effect.

Assuming that the porosity of PM distributes uniformly, the flow velocity through the slipper can be calculated by the Darcy law

$$U = \frac{K}{\eta}\nabla p \tag{1}$$

where K is the permeability of the PM and it can be calculated by

$$K = \frac{\varphi^3 d_p^2}{150(1-\varphi)^2} \tag{2}$$

The water in channels between tows is governed by the continuity equation:

$$div(U) = 0 \tag{3}$$

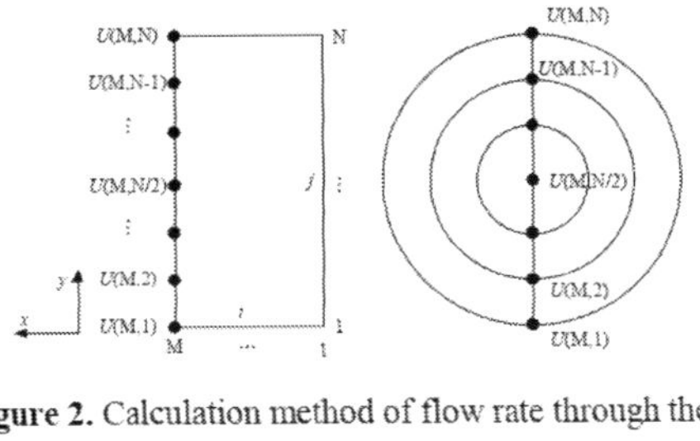

Figure 2. Calculation method of flow rate through the PM.

Figure 2 illustrates the calculation method of flow rate across the PM damper. The area of PM damper in the x and y direction is discretized using the finite difference mesh method with the x step $\Delta x = lx/M$ and y step $\Delta y = ly/N$. Assuming that the Darcy velocity at the outlet of PM damper is equal at the same radius, the flow rate through the slipper can be calculated by

$$q_1 = \sum_{j=1}^{N/2} U(M,j)\times 2\pi j\Delta y^2 \tag{4}$$

The pressure of water chamber between the slipper and the swash plate is calculated by the characteristic equation of annular damper.

$$q_2 = \frac{\pi h^3 (p_s - p_0)}{6\eta \ln(R_0/r_0)} \tag{5}$$

The water through the slipper by the damper effect will flow through the annual damper into the chamber. According to the flow continuity equation, it can be concluded that the flow rate through the PM damper is equal to that through the annual damper.

$$q_1 = q_2 \tag{6}$$

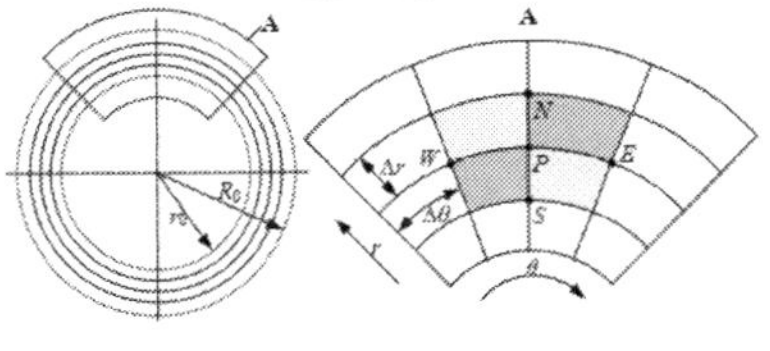

Figure 3. Mesh partition of finite volume method.

2.2 Force characteristic model

In the simulation, the piston force generated by the water pressure is only considered and it can be expressed as follows

$$F_b = \frac{\pi}{4} d^2 P_b / \cos\gamma \tag{7}$$

Assuming that the pressure under the slipper at arbitrary point is $p(r,\theta)$, the water film support force generated by it can be described by

$$F_h = p_s A_s + \iint prdrd\theta \tag{8}$$

According to force equilibrium law, the force generated by water pressure is equal to the force generated by the water film:

$$F_b = F_h \tag{9}$$

2.3 Water film pressure field

To describe the water film pressure field of slipper friction pair, the Reynolds equation [9] is given as shown below

$$\frac{1}{r}\frac{\partial}{\partial r}\left(r\frac{h^3}{\eta}\frac{\partial p}{\partial r}\right)+\frac{1}{r^2}\frac{\partial}{\partial\theta}\left(\frac{h^3}{\eta}\frac{\partial p}{\partial\theta}\right)=6v_{sr}\frac{\partial h}{\partial r}+6v_{s\theta}\frac{\partial h}{\partial\theta}+12\frac{\partial h}{\partial t} \tag{10}$$

The boundary conditions are the important prerequisite for solving Reynolds equation, which is illustrated in Eq. (11).

$$\begin{cases} p(r,0)=p(r,2\pi), & \frac{\partial p}{\partial\theta}\Big|_{(r,0)}=\frac{\partial p}{\partial\theta}\Big|_{(r,2\pi)} \\ p(r_0,\theta)=p_s, & p(R_0,\theta)=p_0 \end{cases} \tag{11}$$

The water film thickness varying with time is ignored ($\partial h/\partial t = 0$). The Reynolds equation is an elliptic partial differential equation, which can be transformed into the linear equations by numerical discretization. According to the discrete theorem, the bottom sealing belt of slipper is conducted the mesh partition, as shown in Figure 3. Adopting the finite volume method with radius and angle steps respectively equal to $\Delta r=(R-r)/U$ and $\Delta\theta=2\pi/V$, a linear equation is acquired

https://doi.org/10.1051/matecconf/201815301003

$$\left(r_N \frac{h_N^3}{\eta} \frac{p_N - p_P}{\Delta r} - r_N \frac{h_N^3}{\eta} \frac{p_P - p_S}{\Delta r}\right) \times 2\Delta\theta$$
$$+\left(\frac{1}{r_E} \frac{h_E^3}{\eta} \frac{p_E - p_P}{\Delta\theta} - \frac{1}{r_W} \frac{h_W^3}{\eta} \frac{p_P - p_W}{\Delta\theta}\right) \times 2\Delta r \tag{12}$$
$$= (6v_{sr} r)_P (h_N - h_S) \times 2\Delta\theta + (6v_{s\theta})_P (h_E - h_W) \times \frac{r_N^2 - r_S^2}{2}$$

Gauss-Seidel over-relaxation iteration method is adopted to solve the linear equation. In order to judge whether the iteration result meets the precision requirement, the relative error of pressure is defined

$$\left| \frac{\sum p_P^{k+1} - \sum p_P^k}{\sum p_P^{k+1}} \right| < \delta \tag{13}$$

where δ the relative error, δ=0.001.

2.4 Water film pressure field

Figure 4 shows the coupling process of the simulation. Initially, the initial water film thickness h_0 and the slipper with the PM damper are given. Adopting the water film thickness as an input, the water film pressure field can be acquired by solving the Reynolds equation and the damping effect model. To avoid the case of negative pressure, a negative pressure is equal to zero. The water film support force is calculated and the force equilibrium equations are solved. When the force imbalance occurs, the water film thickness is adjusted. If the force balance occurs, the simulation terminates. Finally, the water chamber pressure, water film pressure field and water film thickness are acquired.

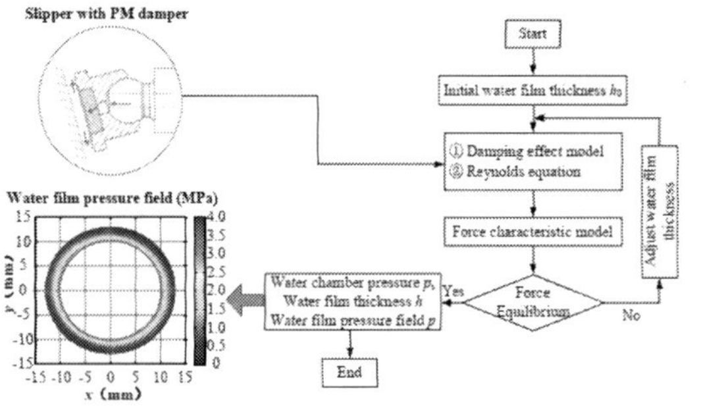

Figure 4. Coupling process of simulation.

Table 1. Parameters used in the simulation.

Category	Parameter	Value
Geometrical parameters	Swash plate angle, γ (°)	15
	Inside radius of the slipper sealing surface, r_0 (m)	0.01
	Outer radius of the slipper sealing surface, R_0 (m)	0.013
	Pitch radius of cylinder, R (m)	0.043
	Diameter of the piston, d (m)	0.023
Operating parameters	Chamber pressure, p_b (MPa)	14
	Rotation speed, n (r/min)	750
PM parameters	Porosity of the PM, φ (-)	0.25

3 Simulation results

3.1 Damping effect of PM

The parameters used in the simulation are shown in Table 1, including geometrical parameters, operating parameters and PM parameters. Damping effect of PM is investigated at piston pressure of 14MPa. When analyzing the pressure variation with porosity, diameter and length respectively equal to 0.7, 8 mm and 10mm, the pressure is shown to be symmetric in angular direction due to the parallelism between swash plate and slipper. The results found by simulation are shown in Figure 5(a). The pressure decreases from inside out, because the angular force can produce movement of flow in angular direction. Figure 5(b) illustrates a two-dimensional cross sectional cut through y=0 position of Figure 5(a). It can be seen that the value of the pressure along the water chamber is approximately constant (13.69MPa) and the pressure loss is 0.31MPa. It indicates that the water can flow through the PM damper which generates the pressure loss. The pressure distributes uniformly along the sealing surface and the growth rate from external to internal is a constant (4.64MPa/mm). In addition, the water film thickness is about 7.41μm. It can be concluded that water film lubrication can be formed with adopting slipper installed the PM damper when the parameter is reasonably designed.

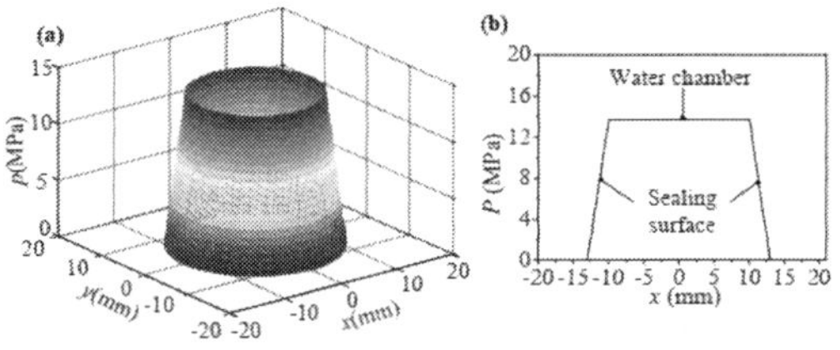

Figure 5. Pressure distribution of sealing surface under piston pressure of 14 MPa.

3.2 Effect of length on the water film thickness

According to the Darcy law, the length of PM damper has great influence on the water velocity into changing the leakage. Figure 6 illustrates the effect of water film thickness at different porosity. As shown in the figures, the water film thickness varies with the length of PM damper. When the length of PM damper increases, the water film thickness will decrease, which indicates that the lubrication becomes poor. In particular, the water film thickness is less than 2μm when the porosity is 0.1. In addition, at different length of PM damper, the attenuation velocity of the water film thickness with increasing length of PM damper is different. Taking Figure 6(a) as an example, the water film thickness will decrease rapidly when the length of PM damper is less than 6mm. But when the length of PM damper is larger than 6mm, the water film thickness begins to decreases slowly. The results indicate that the water film is sensitive to the lower length of PM damper. However, the lower length will decrease its strength of PM damper.

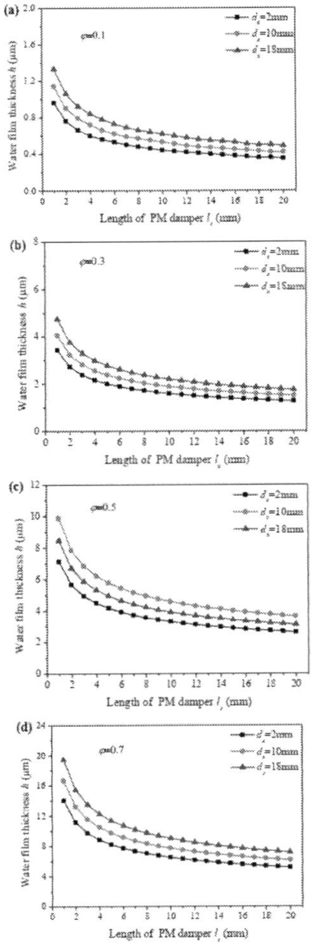

Figure 6. Effect of length on the water film thickness at different porosity: (a) φ=0.1, (b) φ=0.3, (c) φ=0.5, (d) φ=0.7.

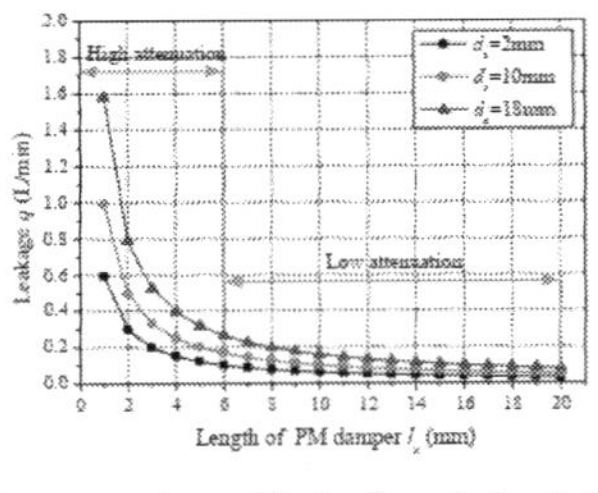

Figure 7. Leakage varying with the length (φ=0.5).

In addition, the leakage will be produced when the slipper with PM adopted. Effect of length of PM damper on the leakage at porosity of 0.5 is illustrated in Figure 6. The leakage is different at different length of PM damper. When the length of PM damper is raised, the leakage will reduce. The attenuation degree is divided into two parts: high attenuation and low attenuation. For example, at the diameter of PM damper of 18 mm, the leakage decreases from 1.58 L/min to 0.26 L/min when the length of PM damper increases from 2 mm to 6 mm. But when the length of PM damper is above 6 mm, the leakage begins to decrease slowly and it maintains at 0.05 L/min.

3.3 Effect of diameter on the water film thickness

Figure 8 shows the relationship between the leakage and the diameter of PM damper. It can be seen that the water film thickness increases linearly with increasing diameter of PM damper, no matter what the porosity. As shown in the Figure 8(c), with the porosity and length of PM damper respectively equal to 0.5 mm and 2 mm, the water film thickness will increase from 7.13 μm to 9.88 μm, when the length of PM damper increases from 1 mm to 18 mm. The growth rate is 0.136 μm/mm and it is basically a constant. In order to ensure the contact area between the slipper and piston, the diameter of PM damper cannot be too large. For the slipper in this study, the diameter is considered to be smaller than 10 mm.

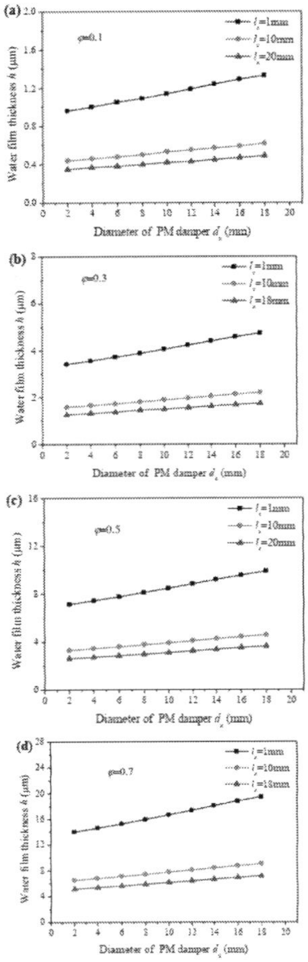

Figure 8. Effect of diameter on the water film thickness at different porosity: (a) φ=0.1, (b) φ=0.3, (c) φ=0.5, (d) φ=0.7.

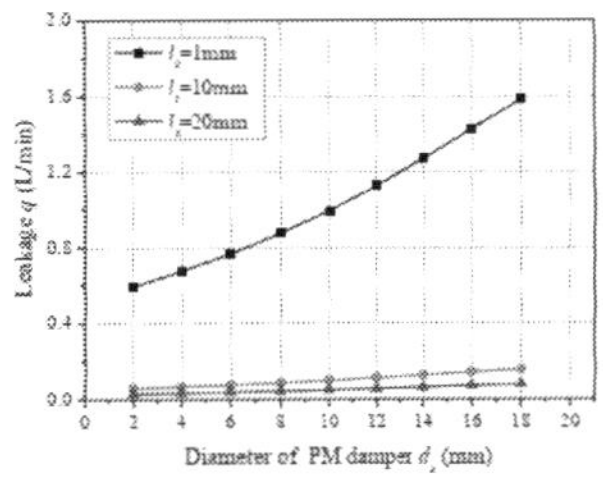

Figure 9. Leakage varying with the diameter (φ=0.5).

Additionally, the slipper leakage at porosity of 0.7 is simulated, which is shown in Figure 9. Different from the relationship between the water film thickness and the diameter of PM damper, the leakage does not increase linearly. The curve at the length of PM damper of 1mm describes well this relationship in Figure 9.

3.4 Effect of porosity on the water film thickness

Porosity is an important parameter of PM damper. Considering the contact area and strength, the diameter of PM damper adopted in this study is 8 mm. Effect of the porosity of PM damper on the water film thickness is described in Figure 10. It can be seen that the water film thickness increases nonlinearly with increasing porosity. At length of 2 mm, the water film thickness increases from 0.87 μm to 12.67μm when the porosity increases from 0.1 to 0.7. Compared with the length and diameter of PM damper, the porosity has the highest effect on the water film thickness Also, the water film thickness will increase rapidly when the porosity is raised.

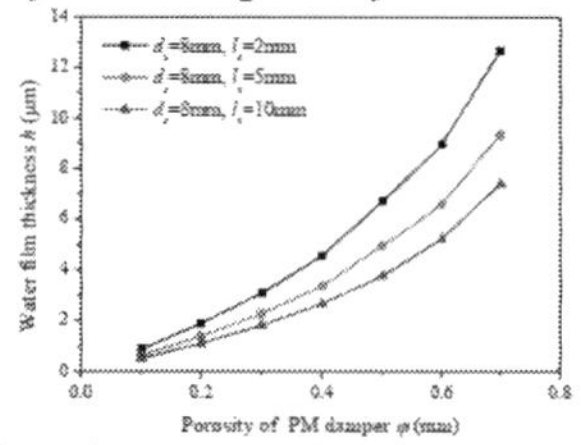

Figure 10. Effect of porosity on the water film thickness.

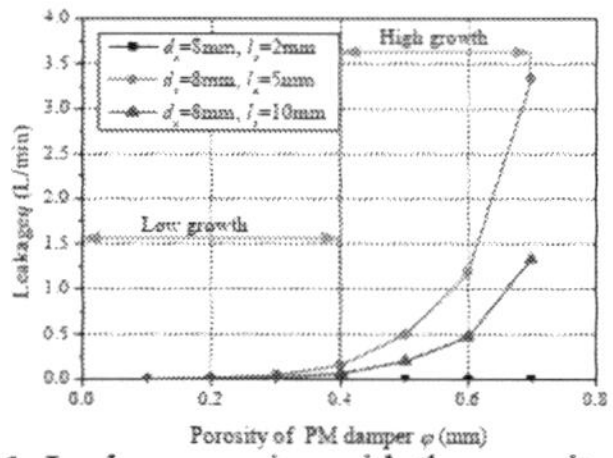

Figure 11. Leakage varying with the porosity (φ=0.5).

In addition, the leakage varying with the porosity of PM damper is shown in Figure 11. As shown in Figure 11, the leakage increases with increasing porosity. But the growth rate is different at different porosity. When the porosity is less than 0.4, the leakage is close to 0. However, the leakage increases rapidly when the porosity is more than 0.4. The growth rate is divided into high growth and low growth, which is illustrated in Figure 11. Considering the strength, lubrication and contact area, the length, diameter and porosity of PM damper is recommended to be l_z=5mm, d_z=8mm and φ=0.7, respectively. The water film thickness is about 9.33μm.

4 Conclusion

A slipper with PM damper is proposed in this paper. Based on the Reynolds equation, continuity equation and Darcy law, a lubrication model is built to analyze the damping effect of PM. The finite volume method is adopted to solve the equation. The results show that the water film can be produced with installing the PM damper. In addition, effects of length, diameter and porosity on the water film thickness and leakage are investigated. The water film and thickness decrease nonlinearly with increasing length of PM damper. On the other hand, the water film thickness increases linearly when the diameter of PM damper is raised. But the growth with increasing diameter is linear. Furthermore, the water film and thickness increase nonlinearly with increasing length of PM damper. As a result, in order to ensure the slipper with PM damper to form hydrostatic support, the parameters of PM damper is recommended to be l_z=5 mm, d_z=8mm and φ=0.7, respectively. Actually, the overturning phenomenon is occurred during the processing of slipper working. As a future work, the dynamic characteristics of slipper with PM damper can be investigated to analyze the damper effect of PM.

Acknowledgment

The authors acknowledge the Youth Science Foundation of National Natural Science Foundation of China (51305146).

References

1. B. Xu, Q. N. Wang, and J. H. Zhang, "Effect of case drain pressure on slipper/swashplate pair within axial piston pump," Journal of Zhejiang Universityence A, vol. 16, pp. 1001-1014, 2015.
2. E. Koç and C. J. Hooke, "Considerations in the design of partially hydrostatic slipper bearings," Tribology International, vol. 30, pp. 815-823, 1997.
3. A. Wondergem, "Piston / cylinder interface of axial piston machines - effect of piston micro-surface shaping," Dissertations & Theses - Gradworks, 2014.
4. A. Schenk, "Predicting lubrication performance between the slipper and swashplate in axial piston hydraulic machines," Dissertations & Theses - Gradworks, 2014.
5. S. Kumar, J. M. Bergada, and J. Watton, "Axial piston pump grooved slipper analysis by CFD simulation of three-dimensional NVS equation in cylindrical coordinates," Computers & Fluids, vol. 38, pp. 648-663, 2009.
6. A. G. Kostornov, "Tribological Characteristics of Materials Based on Bronze for Small Friction Assemblies," Powder Metallurgy and Metal Ceramics, vol. 42, pp. 379-382, 2003.
7. D. Gu, et. al, "Microstructural characteristics and formation mechanism of direct laser-sintered Cu-based alloys reinforced with Ni particles," Materials & Design, vol. 30, pp. 2099-2107, 2009.
8. G. Sethi, A. Upadhyaya, and D. Agrawal, "Microwave and conventional sintering of premixed and prealloyed Cu-12Sn bronze," vol. 35, pp. 264-267, 2003.
9. T. Kazama, "Thermohydrodynamic Lubrication Model Applicable to a Slipper of Swashplate Type Axial Piston Pumps and Motors (Effects of Operating Conditions)," Tribology Online, vol. 5, pp. 250-254, 2010.

Constitutive Parameters Identification of AZ31B-H24 Magnesium Alloy and Energy-Absorption Analysis in Structural Panel Use

Yuedong Yang[1,2], Jiqing Chen[1,2], Fengchong Lan[1,2], Wu Zeng[3] and Zhengwei Ma[4]

[1]School of Mechanical & Automotive Engineering, South China University of Technology, Guangzhou, 510641, China
[2]Guangdong Province Key Laboratory of Automotive Engineering, Guangzhou, 510641, China
[3]School of Mechanical Engineering, Beijing Institute of Technology, Beijing, 100081, China
[4]College of Mechatronics and Control Engineering, Shenzhen University, Shenzhen, 518060, China

Abstract. As a novel lightweight material, AZ31B magnesium alloy is considered as the most potential material to instead baseline steel in some automotive parts. However, their structural use is quite limited and so far proper numerical modeling has not been developed to represent magnesium alloy. In present study, the Split Hopkinson Pressure Bar (SHPB) test is utilized to investigate material dynamic mechanism for AZ31B-H24 over a wide range of strain rates from 1389 s^{-1} to 7296 s^{-1}. Parametric identification for Johnson-Cook (J-C) constitutive model available in the commercial finite element package LS-DYNA is carried out. Proper parameters are obtained by curve fit using genetic algorithm with experimental results. Constitutive model after parametric identification is applied to automotive outer panels for crashworthiness analysis. Energy absorption with magnesium alloy substituted baseline steel under lightweight 51.18% is obtained and the key problem of thin-walled magnesium alloy applied in automotive structure is advanced.

1 Introduction

Stricter regulations on fuel economy and growing concerns on automotive emissions have led to an increased focus on vehicle weight reduction [1]. Automotive lightweight is considered as key strategy to improve fuel consumption and reduce artificial environmental damage emissions [2, 3]. Application of lightweight materials is regard as the most effective way to reduce weight. Magnesium alloys are ideal candidate materials in comparison to baseline steel or aluminum alloys due to their low density, high strength-to-weight ratio, good durability, recyclable, excellent fatigue, potential reduction in part count due to larger castings and the ability of improving structural NVH (Noise, Vibration and Harshness) [4-8]. However, structural use of magnesium alloys is very limited, and even proper material numerical modeling have not been developed to establish crash models as well as forming processes. As we all know, the stress, strain and strain rate of magnesium alloy materials will be changed over the crash and forming process in automotive industry. It always covers high strain rate in these dynamic process. But little knowledge of their dynamic mechanism for magnesium alloys is available [9].

When the dynamic processes with large deformation, highly nonlinear characteristics such as crashworthiness, punch forming and so on are occurred, the constitutive relationships of materials are changed obviously, which result in strain hardening, rate-dependent and thermal softening [10]. Proper material constitutive parameter is the key to connect simulations with experiments. It's an important step to determine the constitutive parameters in simulation. The investigations of constitutive parametric identification for material dynamic mechanism are mainly from two aspects, one is a constitutive model based on physical microstructures and the other is based on empirical phenomenological. The empirical phenomenological constitutive model is widely used as its intuitive and easy identification of material parameters [11]. Necmi Dusunceli et al. [12] carried on an experiment about creep and relaxation behavior of HDPE under loading-unloading, and determined constitutive parameters for viscoplastic materials using genetic algorithm. Yong Zhang et al.[13] utilized successive artificial neural network technique and particle swarm optimization algorithm to predict Deshpande and Fleck model parameters for aluminum foam at high strain rate. M. Anghileri et al. [14] identified model parameters determining damage and failure mechanisms in fibre-composite materials using an inverse technique based on multi-objective optimization. M.Sedighi et al. [15] identified constitutive model parameters of steel 1018 and 4340 at high strain rate using Hopkinson pressure bar test. For the research of AZ31B magnesium alloy materials, D. Ghaffari Tari et al. [16] studied constitutive behavior of AZ31B considering elevated temperature and anisotropy of tensile-compression, and simulated warm forming process. N Peixinho and A C M Pinho [17] presented results concerning the stiffness and denting resistance of 6111-T4 aluminum alloy and AZ31B-H24 magnesium alloy, but there was no data related to AZ31B-H24 magnesium alloy dynamic mechanism response. The 6111-T4 aluminum alloy constitutive model was fit to C-S. So far, little research provides dynamic constitutive, strain hardening, rate dependence and energy absorption for the lightest and potential metallic materials magnesium alloys. Only that Fei Feng et al. [18] established AZ31B constitutive considering strain rate using SHTB and fracture model in tensile state; I.R. Ahmad et al. [19] identified AZ31B constitutive parameters considering strain rate using SHPB. The advantage is study for anisotropy and SEM analysis in [19]. Although many investigators mention the potential of magnesium alloys for automotive and aerospace industries, there is no current research related to high strain rate beyond 3000 s^{-1} to 7000 s^{-1} and the structural simulation use of automotive crashworthiness after AZ31B magnesium alloys constitutive parameters identified, which are necessary for automotive transient collision simulation.

In present work, the constitutive parameters of AZ31B-H24 magnesium alloy are identified using SHPB test to fit J-C constitutive model by genetic algorithm. Constitutive parameters are identified by curve fitting over a wide range of strain rate form quasi-static (about 2.1×10-3 s^{-1}) to high dynamic (about 7296 s^{-1}). Constitutive model after parameters identified is applied to automotive outer panels for crashworthiness simulation. In section 2, unknown parameters are defined in J-C constitutive relation. In section 3, the quasi-static test and the SHPB dynamic test are carried on. In section 4, fit the constitutive parameters. In section 5, apply constitutive model to study automotive crashworthiness and conclusions in section 6.

2 Constitutive model

Obviously, nonlinear processes in engineering such as collision, impact and forming process et al. belong to high strain rate problem. The material mechanism characteristics under high strain rate and quasi-static are different [15]. The mechanism characteristics of materials need to be expressed by constitutive equations. J-C mode [20-22] is common used constitutive equation which is belonged to empirical phenomenological constitutive models based on macro-measurable experimental data. The constitutive model is mainly used for materials dynamic plastic relations, which comprehensive reflect metallic constitutive relationship under loadings of large strain, high strain rate and temperature sensitive. The constitutive model has achieved great applications in actual engineering.

After observing material performances at different strains, strain rates and temperatures, Johnson and Cook [21, 22] established an original model J-C to reveal the relationship of flow stress and strain, strain rate and temperature. J-C model is a power-exponent-hardening function. In this equation, Von Mises flow stress is a multiplication form of strain hardening, strain rate dependent and thermal softening.

$$\sigma = \left[A + B\varepsilon_p^N\right]\left[1 + C\ln\left(\frac{\dot{\varepsilon}}{\dot{\varepsilon}_0}\right)\right]\left[1 - \left(\frac{T_t - T_0}{T_m - T_0}\right)^M\right] \quad (1)$$

where σ is Von Mises flow stress, A is yield stress at the reference temperature and reference strain rate, B is the coefficient of strain hardening, N is strain hardening exponent, ε_p is equivalent plastic strain and $\varepsilon_p = \varepsilon - \frac{\sigma}{E}$, E is material elastic modulus, C is the sensitive coefficient of strain rate, M is thermal softening exponent, $\dot{\varepsilon}_0$ is the reference strain rate, $\dot{\varepsilon}$ is the real strain rate, T_t is the test temperature, T_0 and T_m are the reference temperature and the melting temperature, respectively.

In Eq.(1), A, B, C, N, M are the material parametric constants needed to be identified, among A, B, N are obtained by quasi-static tension-compression test, C is by dynamic response and M is by heat resource temperature test. The process such as transit automotive collision in the engineering field is always finished in hundreds milliseconds, so we don't consider the influence of temperature, and the constitutive equation is simplified to Eq.(2).

$$\sigma = \left[A + B\varepsilon_p^N\right]\left[1 + C\ln\left(\frac{\dot{\varepsilon}}{\dot{\varepsilon}_0}\right)\right] \quad (2)$$

3 Quasi-static and dynamic response test

In order to realize structural use, the tests of quasi-static and dynamic over a wide range of strain rates are carried out to identify constitutive parameters and then to develop the proper numerical modeling under nonlinear dynamic mechanism for AZ31B-H24 magnesium alloy. The AZ31B-H24 magnesium alloy materials in this work are from a commercial magnesium alloy company of China. The chemical compositions are illustrated in Tab.1.

Table 1. Chemical composition of AZ31B-H24

elements	Al	Zn	Mn	Si	Fe	Cu	Ni	Ca	Mg
wt.%	2.87	1.0	0.27	< 0.08	< 0.0018	< 0.01	< 0.001	< 0.04	Bal.

3.1 Quasi-static test and results

The quasi-static test is performed on material universal testing machine of SANS (now it is merged by MTS from USA) in Mechanics Experiment Center of South China University of Technology at room temperature. AZ31B-H24 magnesium alloy materials are of anisotropy[23-25], so in present study we carry out tensile and compressive tests under quasi-static.

Specimens for quasi-static tensile test are manufactured by ASTM E8/E8M-15a[26], specimens diagram is shown in Fig.1. In order to eliminate accidental error, 6 group tests are carried out for the same tensile specimens. The average test strain rate is $7\times10^{-4}s^{-1}$. The average true stress versus strain result of quasi-static tensile test is illustrated in Fig.2(a). AZ31B-H24 magnesium alloy is of no obvious yield step material and the $R_p0.2$ presents its yield strength. The $R_p0.2$ is 147.54 N/mm^2, the tensile strength is 194.63 N/mm^2, and the elongation percentage is 1.25%. Specimens elastic modulus is 45.0GPa.

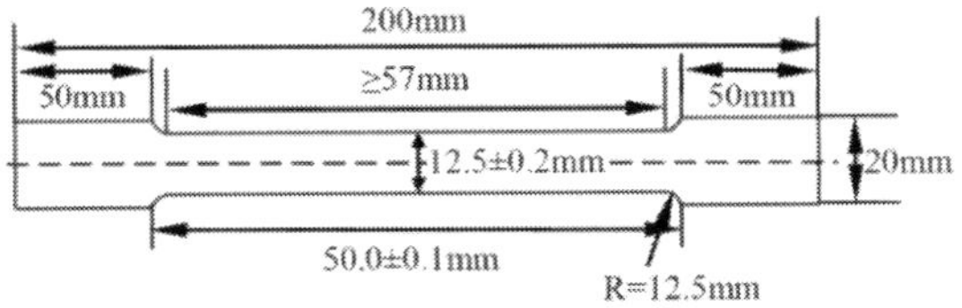

Figure 1. Specimens diagram.

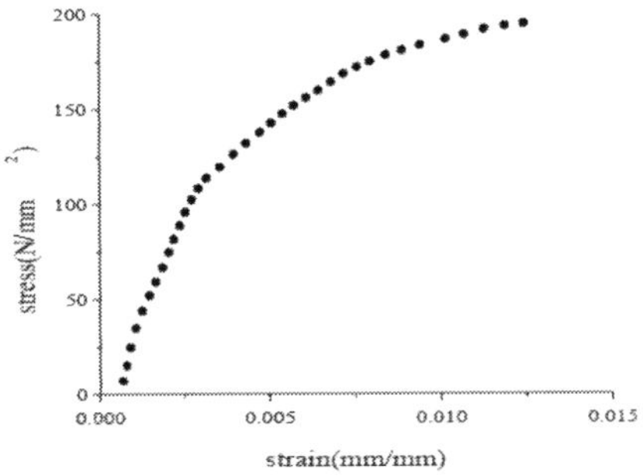

(a) quasi-static tensile

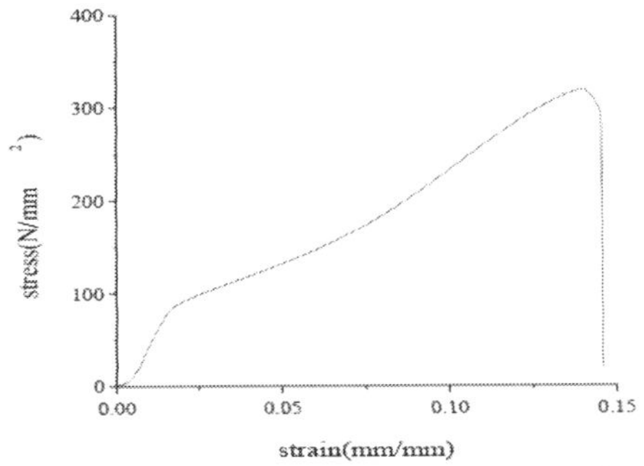

(b) quasi-static compression

Figure 2. Results of quasi-static tensile and compression test for AZ31B-H24.

Specimens for quasi-static compression test are manufactured by ASTM E9-89a (Reapproved 2000)[27], the size for quasi-static compression test is Φ13×25, 4 specimens are used to compression experiment. The average test strain rate is $2.1\times10^{-3}s^{-1}$, which is reference compression strain rate $\dot{\varepsilon}_0$ in Eq.(2). The average true stress versus strain results of quasi-static compression test is illustrated in Fig.2(b). The peak compressive strength (PCS) in quasi-static compression is 318.76 N/mm^2 and the elongation percentage is 14.56%.

3.2 Dynamic test and results

3.2.1 The SHPB apparatus.

The SHPB test is a widely used and typical technical for identifying material dynamic compression response constants under strain rates 10^2~10^4s^{-1} [28]. The SHPB apparatus is comprised of accelerating chamber, strike bar, incident bar, transmission bar and energy absorption device. The apparatus is shown in Fig.3. The test is carried on in Materials Laboratory of Beijing Institute of Technology. The strike bar is accelerated in helium accelerating chamber, and impact the incident bar. A rectangular elastic pulse is produced in incident bar and spread to specimens to put a dynamic loading. The wave impedances are difference in specimens and incident-transmission bar, a part of incident wave reflects back to incident bar, and the other part enters to transmission bar through specimens. Strain gauges are stuck to incident and transmission bars to measure incident, reflect and transmitted stress wave. Ultra-dynamic strain gauge connected to strain gauge is used to get pulse signal. The stress versus strain relationships are obtained through pulse signal according to one-dimensional stress wave theory.

Figure 3. the SHPB apparatus.

3.2.2 One-dimensional stress wave.

Using the SHPB test to measure material dynamic response for AZ31B-H24 magnesium alloys under high strain rate is based on two assumptions, one is one-dimensional assumption and the other is uniform assumption. According to the two assumptions, one-dimensional stress wave theory can be directly used to determine strain rate $\dot{\varepsilon}(t)$, strain $\varepsilon(t)$, and stress $\sigma(t)$ of materials. The principle diagram of measuring is shown in Fig.4.

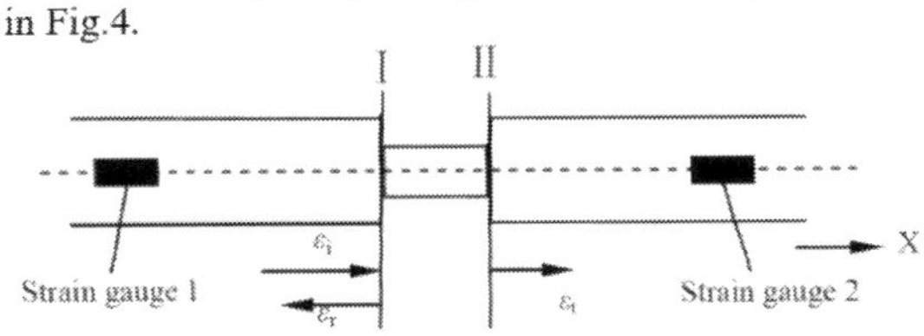

Figure 4. The principle diagram of measuring.

From strain gauge 1 and 2, the incident pulse ε_i, reflect pulse ε_r, and transmission pulse ε_t are obtained. Eq.(3) is the displacement equation according to one-dimensional elastic wave propagation theory.

$$u = C_0 \int_0^t \varepsilon dt \quad (3)$$

Where u is displacement in time t, C_0 is velocity of elastic longitudinal wave, ε is material strain.

In interface I of incident bar and specimens, there are incident strain pulse ε_i in X positive direction and reflect pulse ε_r in X negative direction. The displacement in I is recorded as u_{I},

$$u_{\mathrm{I}} = C_0 \int_0^t (\varepsilon_i - \varepsilon_r) dt \tag{4}$$

Similarly, the displacement u_{II} in interface II is a result by transmission pulse ε_t,

$$u_{\mathrm{II}} = C_0 \int_0^t \varepsilon_t \, dt \tag{5}$$

The average strain $\varepsilon(t)$ in specimens is as follow, l_0 in Eq.(6) is initial length of specimens.

$$\varepsilon(t) = \frac{u_{\mathrm{I}} - u_{\mathrm{II}}}{l_0} = \frac{C_0}{l_0} \int_0^t (\varepsilon_i - \varepsilon_r - \varepsilon_t) dt \tag{6}$$

According to uniform assumption,

$$\varepsilon_r = \varepsilon_t - \varepsilon_i \tag{7}$$

Substitute (7) to (6),

$$\varepsilon(t) = -\frac{2C}{l_0} \int_0^t \varepsilon_r dt \tag{8}$$

$$\dot{\varepsilon}(t) = \frac{2C}{l_0} \varepsilon_r \tag{9}$$

The loadings at both ends of the specimen are F_{I} and F_{II},

$$F_{\mathrm{I}} = EA(\varepsilon_i + \varepsilon_r) \quad F_{\mathrm{II}} = EA\varepsilon_t \tag{10}$$

E is elastic modulus of incident and transmission bars, A is sectional area of bars.

The average stress $\sigma(t)$ can be calculated by Eq.(11),

$$\sigma(t) = \frac{F_{\mathrm{I}} + F_{\mathrm{II}}}{2A_0} = \frac{1}{2} E\left(\frac{A}{A_0}\right)(\varepsilon_i + \varepsilon_r + \varepsilon_t) = E\left(\frac{A}{A_0}\right)\varepsilon_t \tag{11}$$

In Eq.(11), A_0 is specimens section area.

3.2.3 Dynamic response results

In order to study the dynamic mechanism over a wide range of strain rate, the dynamic test is performed on SHPB apparatus in Beijing Institute of Technology. Specimens size is Φ5×5. Specimens of AZ31B-H24 are tested at various strain rates ($1389s^{-1}$, $2475s^{-1}$, $3044s^{-1}$, $3801s^{-1}$, $4019s^{-1}$ and $7296s^{-1}$). To ensure the repeatability, 3-5 tests are performed for each strain rate. The strain corresponding to PCS and energy absorbing in dynamic compression state are taken up to PCS as in [19]. Results of high strain rates tests dealing with 3.2.2 are illustrated in Fig.5.

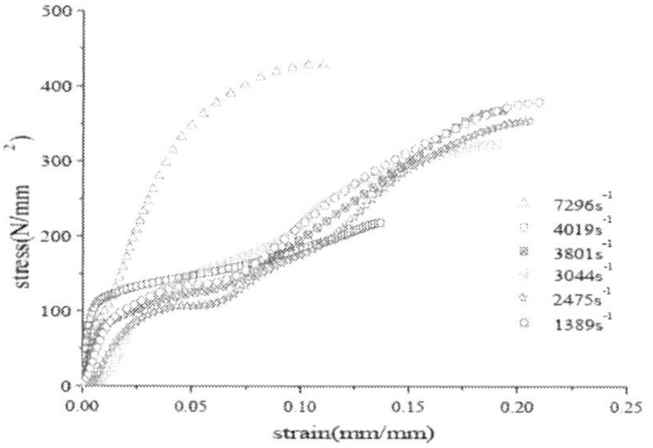

Figure 5. Results of dynamic test.

AZ31B magnesium alloy is a strain-rate-dependent material as described in reference [23-25]. To illustrate the strain rate dependence of the AZ31B-H24 magnesium alloy mechanism at nonlinear stage, the relationship of equivalent plastic strain versus stress is utilized to detailed describe the dynamic response and energy absorption. The first portion of the various strain rates curves is elastic phase, which is linear and of non energy absorption, so it is removed and the rest portion is shown in Fig.6. From Fig.6 we can clearly obtain the strain-rate-dependent characteristic and energy absorption at nonlinear stage. The stress in single strain rate is increasing with increasing equivalent plastic strain. The stress and the initial stress of plastic stage are also increasing with increasing strain rate. The PCS increased with increasing strain rate, however a slight decrease at $3044s^{-1}$, the regularity is also obtained in ref [19]. PCS are 218.24MPa, 354.72MPa, 323.54MPa, 368.51MPa, 379.04MPa and 428.40MPa corresponding to $1389s^{-1}$, $2475s^{-1}$, $3044s^{-1}$, $3801s^{-1}$, $4019s^{-1}$ and $7296s^{-1}$. The equivalent plastic strain corresponding to PCS is increased at first and then decreased at high strain rate $7296s^{-1}$ after $4019s^{-1}$. The value of initial stress of plastic stage (ISPS), PCS and equivalent plastic strain (EQPS) corresponding to PCS at various experimental strain rates are illustrated in Tab.2.

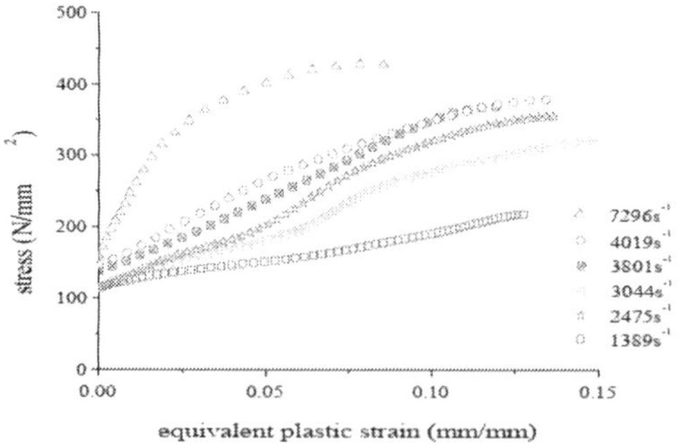

Figure 6. Relationship of equivalent plastic strain versus stress.

Table 2. Detailed value at various strain rates

strain rates(s^{-1})	1389	2475	3044	3801	4019	7296
ISPS (MPa)	112.75	114.50	114.87	138.71	147.30	158.82
PCS(MPa)	218.24	354.72	323.54	368.51	379.04	428.40
EQPS corresponding to PCS(%)	12.76	13.66	15.71	11.94	13.44	8.5

4 Parameters identification

Genetic algorithm (GA) available in Matlab is used to identify parameters in the constitutive equation to meet the results of quasi-static and dynamic mechanism. Evaluation criteria of GA in present work is defined function $f(x)$, minimum of $f(x)$ is the most suitable parameters value.

$$f(x)=\sqrt{\frac{1}{n}\sum_{i=1}^{n}(x-x_i)^2} \tag{12}$$

Where, n is experimental sampling point number, x_i is experimental sampling point value and x is the value of J-C equation with parameters identification.

From the data of Fig.2(b), using GA, we can fit that in J-C Eq.(2) A is 91.487 MPa, B is 4353.112 and N is 1.353. From the data of Fig.6, using GA, we can fit that in J-C Eq.(2) C is 0.033. The contrasts of experimental data versus J-C fit curve are illustrated in Fig.7. The stress versus equivalent plastic strain results of quasi-static compression test is illustrated in Fig.7(a) and high strain rate dynamic compression test is illustrated in Fig.7(b). From Fig.7(a), we can see that the J-C fit curve is completely agreement with the experimental data, and from Fig.7(b) the J-C fit curve is in reasonable agreement with the experimental data except in the last portion of the flow stress curve.

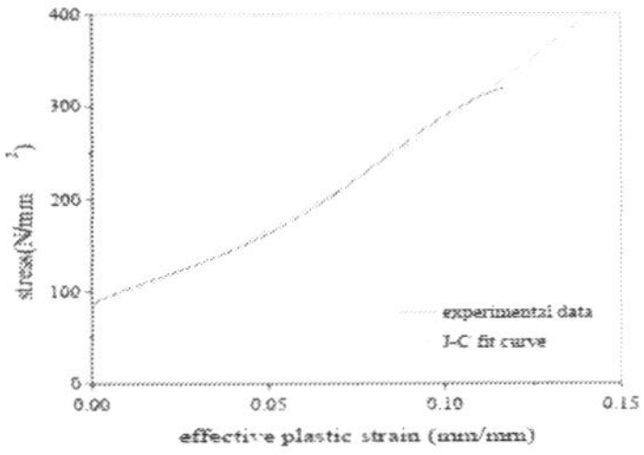

(a) quasi-static compression

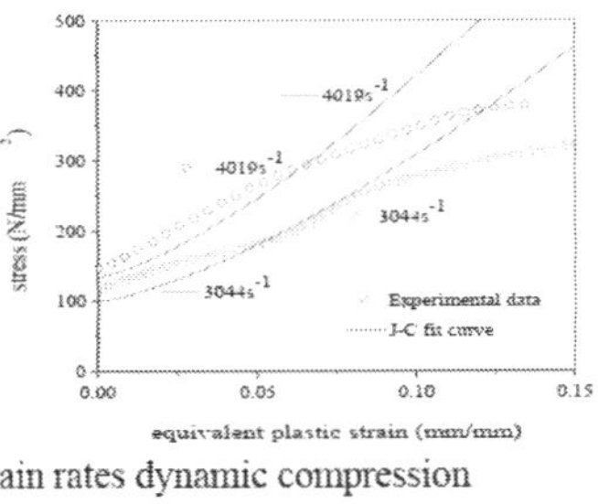

(b) high strain rates dynamic compression

Figure 7. The contrasts of experimental data versus J-C fit curve.

5 Structural panel use simulation

For the use of lightweight magnesium alloy panels into automotive industry, until now only Andrew Parrish, Masoud Rais-Rohani and Ali Najafi [1,8] have replaced the baseline steel parts and performed optimization on the full-scale Dodge Neon model on crashworthiness characteristics. But there is no refer to material mechanism of magnesium alloys, especially dynamic mechanism for nonlinear characteristics such as in this crashworthiness. In present research, identification constitutive parameters are utilized to establish automotive outer panels simulation model for crashworthiness investigation. The contrasts of energy-absorption between AZ31B-H24 magnesium alloy model with J-C constitutive equation and origin steel model are carried on at the premise of 51.18% lightweight.

A full-scale model validated in some automotive group of China is used to compare the energy absorption characteristics of substituting baseline steel with AZ31B-H24 magnesium alloy in current study. A frontal crashworthiness model has been established with Ls-DYNA software, which is shown in Fig.8. Finite element model mass of the whole automotive is 1.588t. The boundary conditions are defined as that initial velocity is 50km/h in the negative direction of X axis, the contact is automatic surface to surface, the friction coefficient between rigid wall and the whole car is 0.3, and the friction coefficient between rigid ground wall and the whole car is 0.7. The AZ31B-H24 magnesium alloy frontal crashworthiness model is established only substitution materials of some panels. The selected automotive panels are illustrated in Fig.9. The change of mass and panel thickness of all selected panels are in Tab.3. The solver time is set as 100ms. Energy curves are calculated and shown in Fig10.

Figure 8. Frontal crashworthiness model

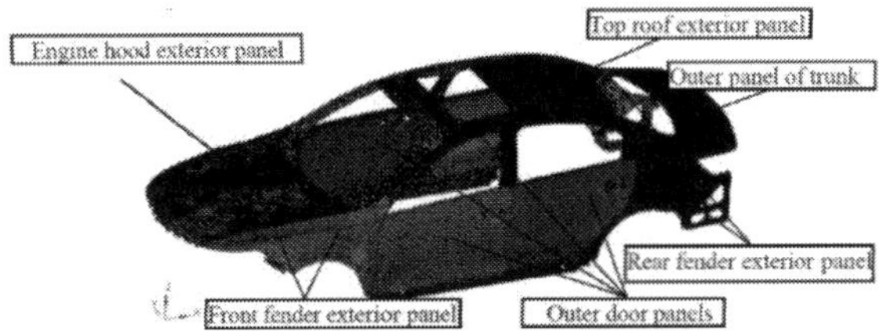

Figure 9. Selected automotive panels

Table 3. The change of materials

automotive parts name	panel thickness(mm)	mass(kg)

	origin model	AZ31B-H24 model	origin model	Z31B-H24 model
engine hood exterior panel	0.7	1.5	8.047	3.932
top roof exterior panel	0.8	1.7	8.767	4.248
outer panel of trunk	0.7	1.5	3.820	1.866
outer door panels	0.7	1.5	15.980	7.806
front fender exterior panel	0.7	1.5	6.497	3.175
rear fender exterior panel	0.7	1.5	23.710	11.59

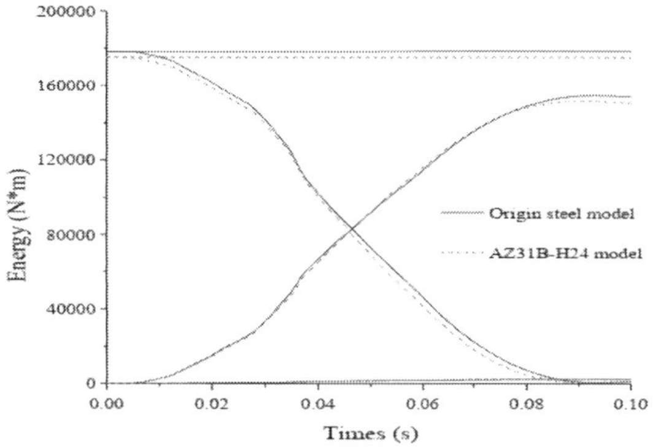

Figure 10. Contrasts of energy-absorption.

Fig.9 shows the automotive panels which we select for structural use investigation. These panels are faced with transient collision and forming impact process. Fig.10 shows the situation of structural energy-absorption, the top of all four curves is total energy, the bottom is hourglass energy, the gradual decline is kinetic energy and the gradual increase is internal energy. Internal energy represents energy absorption capability of component. The mass of selected panels in origin model is 66.82Kg and the energy-absorption is 152.33KJ, while the mass of selected panels in AZ31B-H24 magnesium alloy model is 32.62Kg and the energy-absorption is 149.25KJ. The structural use of AZ31B-H24 magnesium alloy substation of steel can reach 51.18% lightweight and the energy-absorption almost invariable (about 2.0% decrease).

6 Conclusions

The present investigation has provided experimental data for AZ31B-H24 magnesium alloy under quasi-static and high strain rate dynamic situations. In quasi-static mechanism, the material tensile strength is 194.63 N/mm^2 and the elongation percentage is 1.25%, the material compression strength is 318.76 N/mm^2 and the elongation percentage is 14.56%, which reveal anisotropy. In dynamic mechanism, the SHPB test is used to obtain the high strain rate between $1389s^{-1}$~$7296s^{-1}$, which reveal strain-rate-dependent.

Genetic algorithm in Matlab is used to fit J-C constitutive equation. A is 91.487 MPa, B is 4353.112, N is 1.353 and C is 0.033. The J-C fit curve is reasonable agreement with experimental data. These identification parameters are utilized to structural panel use, The selected automotive panels of AZ31B-H24 magnesium alloy substation of steel can reach 51.18% lightweight and the energy-absorption almost invariable.

Acknowledgments

This work is supported by Science&Technology Planning Project of Guangdong Province China (2015B010137002;2016A050503021;2016B090918089) and the Knowledge Innovation Program for Basic Research of Shenzhen City (Fund No. JCYJ20160520164316902).

References

1. Andrew Parrish, Masoud RaisRohani, Ali Najafi, Int. J. Crashworthiness, **17**(3) (2012) 259-281.
2. Pollock TM, Science, **328**(5981) (2010) 986-987.
3. Murray J, King D, Nature, **481**(7382) (2012) 433.
4. W Altenhof, A Raczy, M Laframboise, et al., Int. J. Impact Eng., **30**(2) (2004) 117-142.
5. Xiaoming Chen, Shen-Rong Wu, David A Wagner, et al., Int. J. Crashworthiness, **7**(4) (2002) 429-440.
6. Dørum C, Hopperstad O S, Langseth M, et al., SAE Technical Paper, 2005-01-0724, 2005.
7. Dørum C, Hopperstad O S, Lademo O G, et al., Comput. Struct., **85**(1) (2007) 89-101.
8. A Parrish, M RaisRohani, A Najafi, //52nd AIAA/ASME/ASCE/AHS/ASC Structures, Structural Dynamics and Materials Conference
 19th, 4-7 April 2011, Denver, Colorado.
9. Feng Zhu, Clifford C. Chou, King H. Yang, et al., Int. J. Crashworthiness, **17**(5) (2012) 540-552.
10. DH Liu, HP Yu, CF Li, Mater. Sci. Eng. A, **551** (2012) 280-287.
11. Y Tian, L Huang, H Ma, et al., Mater. Design(1980-2015), **54** (2014) 587-597.
12. N Dusunceli, OU Colak, C Filiz, Mater. Design, **31**(3) (2010) 1250-1255.
13. Y Zhang, G Sun, X Xu, et al., Comp. Mater. Sci., **74** (2013) 65-74.
14. M Anghileri, EC Chirwa, L Lanzi, et al., Compos. Struct., **68**(1) (2005) 65-74.
15. M Sedighi, M Khandaei, H Shokrollahi, Mater. Sci. Eng. A, **527**(15) (2010) 3521-3528.
16. DG Tari, MJ Worswick, J. Mater. Process Tech., **221** (2015) 40-55.
17. N Peixinho, ACM Pinho, P. I. Mech. Eng. D- J. Aut., **220**(9) (2006) 1191-1198.
18. F Feng, S Huang, Z Meng, et al., Mater. Sci. Eng. A, **594** (2014) 334-343.
19. J Xiao, D W Shu, Met. Mater. Int., **21**(5) (2015) 832-831.
20. Johnson G R, Cook W H, *A constitutive model and data for metals subjected to large strains, high strain rates and high temperature* (The Netherlands Publishers, Hague: 1983: 541-547).
21. Johnson GR, Cook WH, Eng. Fract. Mech., **21**(1) (1985) 31-48.
22. G. R. Johnson, W. H. Cook, //Proceeding of 7th International Symposium on Ballistics. The Hague, **21**(1983):541-547.

23. M. T. Tucker, M. F. Horstemeyer, P. M. Gullett, et al., Scr. Metall., **60**(3) (2009) 182-185.
24. G. Wan, B. L. Wu, Y. D. Zhang, et al., Mat. Sci. Eng. A, **527**(12) (2010) 2915-2924.
25. I. R. Ahmad, D. W. Shu, //Applied Mechanics and Materials. Trans Tech Publications, 2012, 151: 726-730.
26. Standard Test Methods for Tension Testing of Metallic Materials.
27. Standard Test Methods of Compression Testing of Metallic Materials at Room Temperature.
28. W. N. Sharpe Jr., *Springer Handbook of Experimental Solid Mechanics*(Springer, 2008).

A finite element method of modelling and designing of an ACCC conductor

Yan Youcan and Duan Yugang

State Key Lab for Manufacturing Systems Engineering, Xi'an Jiaotong University, Xi'an, China

Abstract. The ultimate breaking strength of conductors always plays a vital role in the long-term safe operation of transmission lines and it should be carefully considered in the designing of a conductor. While limited literature has been released regarding the strength calculation of ACCC (Aluminum Conductor Composite Core) conductors. In this study, a finite element method based on real conductor samples is proposed to evaluate the strength of the ACCC conductor, and investigate on how diameters (or sectional areas) of the composite core and aluminum strands influence the conductors' breaking strength, thereby helping engineers design a proper diameter of an ACCC conductor. The finite element analysis shows that increasing the sectional area of the composite core leads to a larger breaking strength than increasing an equal sectional areas of aluminum strands, which is in line with the tensile experimental results. In specific, the relative errors between the simulated breaking strength and the experimental results are as low as less than 3%. Therefore, the finite element method presented in this work, to some extent, fills in the blank of strength calculation for ACCC conductors, and in practical terms, it could serves as guidelines for conductor-design engineers.

1 Introduction

In recent years, the Aluminum Conductor Composite Core (ACCC) conductor has been widely used in overhead transmission lines due to its merits of high specific strength, light weight, and low lag at high temperature compared with the traditional Aluminum Conductor Steel Reinforced (ACSR) conductor [1-5].

As shown in Figure 1, the ACCC conductor consists of two-layer aluminum strands and a composite core based on carbon fibers and glass fibers, which has successfully replaced the high weight steel core in ACSR conductors [5]. The inner strength core is based on continuous carbon fibers while the outer insulation layer is based on glass fiber/epoxy composites to prevent a direct electrical path between the aluminum conductors and the conductive carbon fibers. Both kinds of fibers are impregnated with epoxy and shaped into a composite mandrel bar [6]. Additionally, the out-layer aluminum strands of circular sections are replaced with strands of trapezoidal sections, which can increase the conductive areas, and at the same time, the two-layer aluminum strands are twisted in a reverse spiral to improve the stability of the wire structures. In this paper, the lay ratio (the ratio of screw pitch to the outer diameter) of the ACCC conductor samples is a constant.

Figure 1. Macro-structure of an ACCC wire

It is well known that the ultimate tensile strength of conductors plays a vital role in the long-term safe operation of transmission lines, and it is important to take the breaking strength into consideration when designing a conductor. Prior to this study, plenty of work has been done both analytically and numerically to study the strength calculation methods for the ACSR conductor with circular cross sections [7-14]. Cardou A. et al. [12] developed concise mechanical models of ACSR helical strands, while it cannot be applied to analyze ACCC conductors with more complex sections—trapezoid sections (outer aluminum strands) plus a circular section (composite core). W.G. Jiang et al. [13] presented a theoretical finite element model for simple straight wire rope strand, and MA Jun et al. [14] also established a finite element model using ANSYS to simulate the stress and deformation of ACSR conductors, but limited literature has been released regarding the strength calculation of the ACCC conductor.

In this paper, a finite element method based on real conductor samples is proposed to evaluate the breaking strength of the ACCC conductor, and study how diameters (or cross-section areas) of the composite core and aluminum strands influence the conductors' breaking strength. Therefore, this method can promisingly help engineers design a new type of ACCC conductor, and to some extent, it fills in the blank of strength calculation for ACCC conductors.

2 Method and Assumptions

Prior to modelling, it's necessary to make some assumptions on materials of models so as to simplify the analysis. Firstly, the carbon composite core is regarded as ideal elastic body in axial direction on the premise of a small deformation, and only axial modulus and Poisson's ratio are required to be defined. In addition, as the aluminum strands in out layers have a relatively low axial modulus (around 60GPa compared to 120GPa of the composite core) and a high elongation percentage (more than 20%), we assumed aluminum as bilinear elastic-plastic body, which means that the nonlinear plastic deformation of the aluminum wire after reaching its yield limit $\sigma 0$ (about 80MPa) in the elastomer stage is simplified as a linear deformation with a tiny modulus, i.e. Tangent Modulus $G0$ (around 0.6GPa).

Then the geometrical model of an ACCC conductor is created in ANSYS software based on real wire samples until it is meshed with solid185 eight-node elements. To match the practical situation, contact behaviors between conductor layers are taken into consideration and the flexible three-dimensional contact pairs are applied, in which the target surface adopts TARGE170 units and the contact surface adopts CONTA174 units. After completely constraining one end of the conductor and applying an axial displacement load to the other end, the model can be solved and the calculation results of the stress and deformation are obtained. By integrating the stress results on the cross section of the conductor, we can obtain the corresponding internal force, which is equivalent to the tensile load applied on the end of the wire.

Assuming that if the von Mises stress to any of the components in the ACCC wire material, i.e., the composite core or the aluminum strands reaches or exceeds its ultimate tensile strength, the wire would be considered as been broken (referring to the fourth strength theory or distortion energy theory in material mechanics). Therefore, we can obtain the breaking strength by increasing the axial displacement load stepwise until the maximum von Mises stress to the aluminum or composite core reaches its ultimate tensile strength.

Comparing the simulated breaking strength with the experimental results of tensile tests, we can verify the rationality of this finite element method, and figure out how diameters or cross-section areas of the composite core and aluminum strands influence the ultimate breaking strength of ACCC conductors. This can help engineers choose an optimal diameter of the composite and aluminum strands of ACCC wires on the ground that other mechanical and electrical requirements are satisfied.

3 Finite Element Model

3.1 Geometric Modelling

The finite element model is created on 1:1 based on, for example, the JRLX/T (ACCC) wire samples produced by Far East Composite Technology Co., Ltd., and the wire code in USA is Linnet ACCC/TW431 [15]. In this model, the rod length is set to 43.2mm which approximates to a quarter of the wire pitch, the outer diameter of the wire is 18.29 mm, and the diameter of the inner composite core is 5.97 mm. Specific geometrical parameters of the JRLX/T wire are listed in Table 1 [16], and the specific mechanical properties of the JRLX/T wire are shown in Table 2.

Table 1. Geometrical parameters of the JRLX/T wire

Parameters	Value
Nominal total transverse area, mm²	246.4
Nominal aluminum transverse area, mm²	218.4
Outside diameter of aluminum wires, mm	18.29
Diameter of composite core, mm	5.97
Lay ratio of inner-layer aluminum strands	19
Lay ratio of outer-layer aluminum strands	10
Filling coefficient	0.93

Table 2. Mechanical properties of the JRLX/T wire

Properties	Composite Core	Aluminum strands
Axial Modulus (GPa)	120	60
Poisson's Ratio δ	0.27	0.3
Tangent Modulus G_0 (GPa)	/	0.6
Yield Limit σ_0 (MPa)	/	80
Ultimate Tensile Strength σ_m (MPa)	≥2100	90

In general, the geometrical modelling process of the JRLX/T wire is based on a bottom-top method, which mainly includes four steps as follows:

1. Create the outline and bottom areas by connecting key points and boundary lines respectively (Figure 2a).
2. Create the sub-volume of an aluminum strand by copying, rotating the bottom area and connecting corresponding key points (Figure 2b).
3. Create the circular carbon core by extrusion and single aluminum strands by a similar way in step2 (Figure 2c).
4. Complete the whole model by copying circumferentially (Figure 2d).

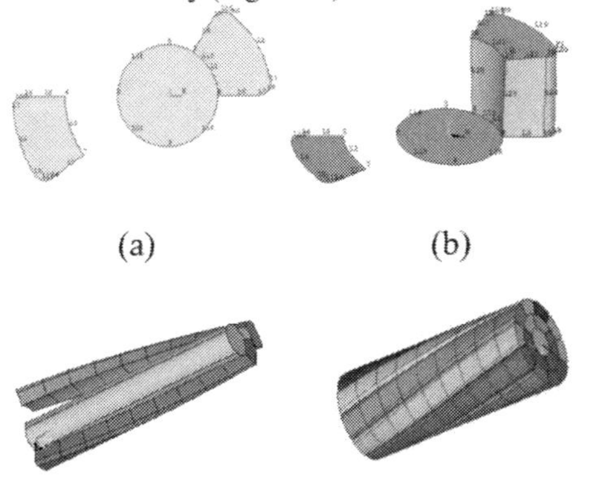

(a) (b) (c) (d)

Figure 2. Procedure of creating a finite element model of the JRLX/T wire: (a) creation of the outline and bottom areas, (b) creation of the sub-volume of an aluminum strand, (c) creation of the circular carbon core and single aluminum strands, (d) creation of the complete wire model

3.2 Contact Pairs Setting

After meshing the finite element model (20,400 8-noded elements are generated), we set contact pairs between the composite mandrel and the inner aluminum strands, as well as contact pairs between the inner and outer layer of aluminum strands [16], which is well matched with the true conditions. As the ACCC rod are meshed with three-dimensional solid185 elements, the types of contact elements are set as flexible three-dimensional TARGE170 and CONTA174. Using the contact wizard in ANSYS and displaying the contact pairs graphically, the contact pairs are shown in Figure 3a and 3b, respectively.

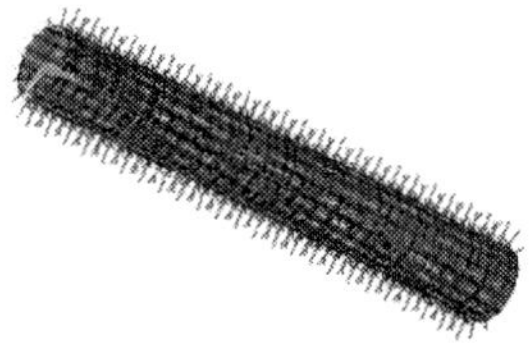

(a)

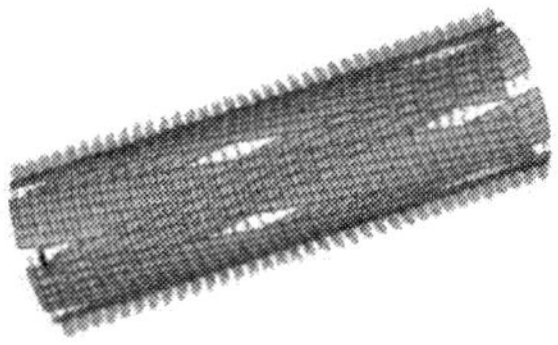

(b)

Figure 3. (a) Contact pair between the composite mandrel and inner aluminum strands, **(b)** Contact pair between the inner and outer layer of aluminum strands.

3.3 Boundary Conditions

Constraining all displacements (X, Y and Z directions) of nodes on the one end (bottom areas) of the wire model and only the radial displacements (X direction in cylindrical coordinates) of nodes on the other end (top areas), then applying axial displacement (Z direction) loads Δl on the top areas.

3.4 Results and Discussions

Increasing the axial displacement load Δl stepwise from 0.05 to 1 mm and setting the number of sub-steps properly, we can obtain the desired stress results of the ACCC wire when the maximum von Mises stress to the aluminum or composite core reaches its ultimate tensile strength.

Figure 4a give the curve of von Mises stresses to the nodes of the wire model over the axial displacement loads up to 0.75mm within 24 sub-steps. To see more clearly, the curves of von Mises stresses to aluminum strands, namely SEQV-2 and SEQV-3 are independently shown in Figure 4b. Figure 5 shows the distribution of axial stresses on the mid-section of the ACCC wire. At the same time, the von Mises stresses to the nodes in Figure 4 are extracted and listed in Table 3 (at Time=1, i.e., Δl=0.75mm).

The von Mises stress to the composite core (SEQV-1) in Figure 4a saw a constant increase and reaches up to 2127.19MPa when Δl =0.75mm, which exceeded its ultimate strength limit—2100MPa (referring to Table 2), indicating that the composite core had broken at that moment. In the meantime, the maximum von Mises stress to the two-layer aluminum strands SEQV-2 and SEQV-3 reaches 88.81MPa and 80.48MPa respectively, which exceeded its yield strength-80MPa while remaining smaller than its ultimate strength limit-90MPa, that is to say, the aluminum strands were deforming plastically while had not yet been broken.

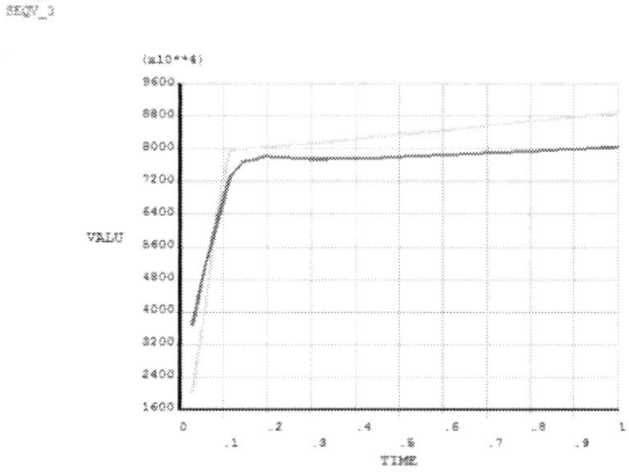

(a)

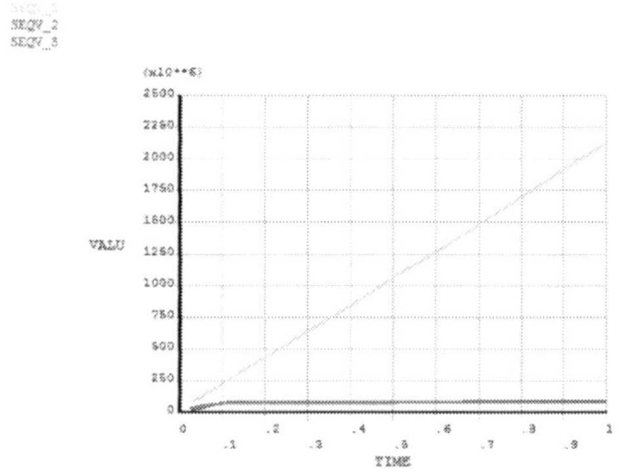

(b)

Figure 4. (a) Curve of von Mises stresses to the wire model over axial displacement loads, **(b)** Curve of von Mises stresses to the two-layer aluminum strands over axial displacement loads. (SEQV-1: the von Mises stress to the composite core, SEQV-2: the von Mises stress to the inner aluminum strands, SEQV-3: the von Mises stress to the outer aluminum strands)

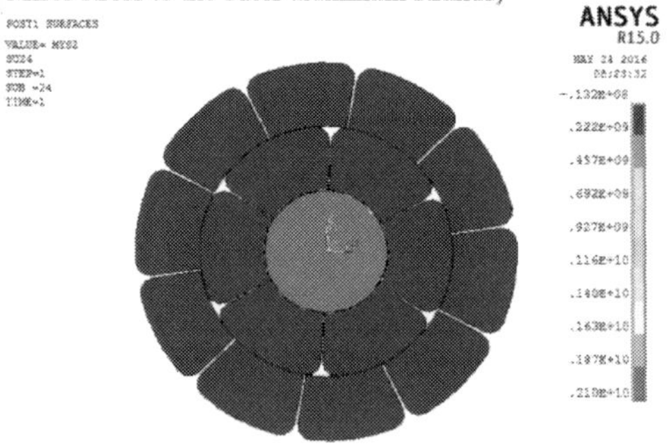

Figure 5. The distribution of axial stresses on the mid-section of the ACCC wire

Table 3. Maximum von Mises stresses to the nodes of components of the ACCC wire model

Components of the ACCC Wire	Maximum Von Mises Stresses (MPa)
Composite core	2127.19
Inner-layer aluminum strands	88.81
Outer-layer aluminum strands	80.48

Consequently, the equivalent axial force, i.e., the simulated tensile breaking force, can be calculated by integrating axial stresses on cross sections of the ACCC wire at last time-step (Time=1). At the same time, we can obtain the simulated breaking strength of ACCC conductors in different sizes using the same method, and the simulated results are listed in Table 4.

4 Experiments

The tensile testing experiments are carried out by Far East Composite Technology Co., Ltd. and guided by the Standards of Q/GDW 1851-2012, which is issued by State Grid Corporation of China. Firstly, the conductors sample are applied with special hardware at two ends so as to connect with the tensile testing machine. In the next step, the conductor is clamped by clamps on the tensile tester and the tensile load is applied on its one end at a loading speed of 20-40kN/min [17]. The tests end when either the composite core or aluminum strands breaks. The experimental results of six conductor samples in various sizes are listed in Table 4.

As shown in Table 4, the simulated breaking load is highly consistent with the experimental results, with a low relative error of no more than 3%, which implies that the finite element method in this work is able to predict the breaking load of an ACCC conductor reasonably in practice.

In addition, it is found in Figure 6 that the breaking strength of conductors increase as the sectional areas of the composite core and aluminum strands become larger, while the increase in sectional area of composite core contributes more to a greater breaking force than that of aluminum strands.

We can draw the conclusion aforementioned particularly by figures for Sample 4 and Sample 5—the experimental breaking strength (F_{NO}) increases by as large as 15.9 kN (from 106.1 to 122.0) as the sectional area of the composite core (S_C) rises by 7 mm^2 and the sectional areas of aluminum strands (S_A) increases by 10 mm^2, while F_{NO} increases by only 2.9 kN (from 103.0 to 106.1) and 2.2 kN (from 122.0 to 124.2) as S_A rises by 40 mm^2 (from 310 to 350) and 20 mm^2 (from 360 to 380) respectively, with S_C stay constant in Sample 3,4 (40mm^2) and Sample 5,6 (47mm^2). It means that increasing the sectional area of the composite core can lead to a larger breaking strength than increasing an equal number of sectional areas of aluminum strands, which is in line with the real condition that the composite core bears most of the tensile load.

Table 4. Comparison between the simulated breaking s and the experimental one based on conductor samples of different sizes [15]

Conductor Samples No.	1	2	3	4	5	6
Sectional Area of the Composite Core S_C (mm^2)	28	28	40	40	47	47
Sectional Area of Aluminum Strands S_A (mm^2)	218	240	310	350	360	380
Simulated Breaking Strength F_{NA} (kN)	70.2	72.9	100.3	103.6	119.3	121.1
Experimental Breaking Strength F_{NO} (kN)	72.1	74.8	103.0	106.1	122.0	124.2
Relative Error η Between F_{NA} and F_{NO} (%)	2.64	2.54	2.62	2.37	2.21	2.50

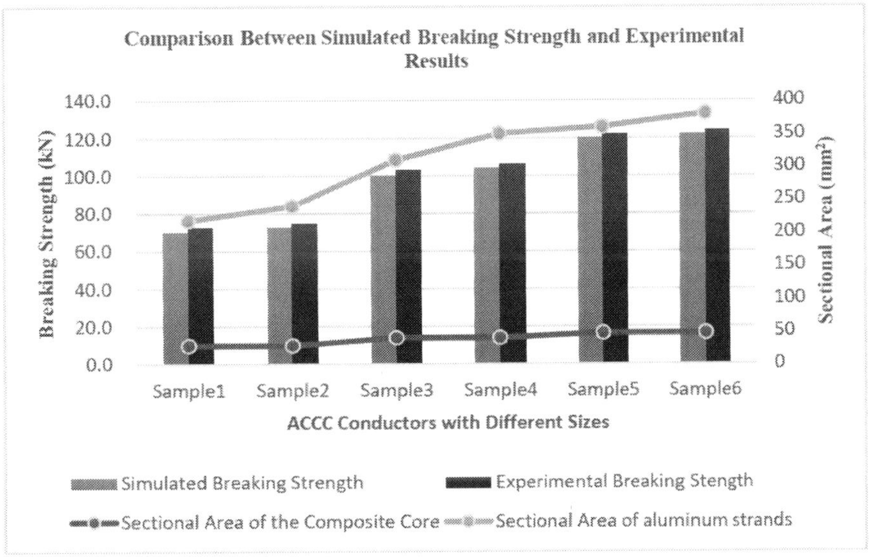

Figure 6. Comparison between simulated breaking strength and experimental results of conductor samples with different sizes

Overall, the increase in sectional area of composite core contributes more to a larger breaking force compared to that of aluminum strands. The calculated tensile breaking force F_{NA} is very close to the experimental result

F_{N0}, and the relative error η between simulations and experiments is less than 3%, which indicates that the simulation result fits well with the actual breaking process.

5 Conclusions

Herein, a finite element method based on real ACCC conductor samples is proposed to investigate on how sectional areas of both the composite core and aluminum strands affects the breaking strength of ACCC conductors. The stress-displacement curves demonstrate that the composite core bears most of tensile load applied on the conductor, compared to the aluminum strands yielding first and deforming plastically afterwards (with no load-bearing capacity).

It is also found that a 7mm2 increase in the sectional area of the composite core and a 10mm2 increase in the sectional area of aluminum simultaneously lead to a rise by 15.9kN of breaking strength, while a 40mm2 increase in the sectional area of aluminum solely produces a rise by only 2.9kN of breaking strength (keeping the sectional area of the composite core unchanged). It means that the breaking strength of conductors increase with a larger sectional area of both the composite core and aluminum strands, while the increase in the sectional area of composite core contributes more to a larger breaking force than that of aluminum strands. In addition, the calculated breaking force in conductors is highly consistent with the experimental results (with a low relative error of less than 3%), which means that the finite element model fits well with the true condition.

The finite element method presented in this work, to some extent, fills in the blank of strength calculation for ACCC conductors, and in practical terms, it could serves as guidelines for conductor-design engineers. For example, they can choose an optimal diameter of the composite and aluminum strands of ACCC wires based on practical requirements (current-carrying capacity, weight, rated breaking strength, etc.), and for conductors operating in extreme conditions (heavy wind, ice or snow), a larger diameter of the composite core is highly recommended to meet the strength requirements.

However, this paper focuses on the strength evaluation of ACCC conductors in room temperature (around 20°C), and in most cases, the mechanical performance of wires under high temperature (around 300 ℃) should be considered. In future work, we will investigate on how temperature affects the breaking strength and the mechanical creep of ACCC conductors as the carbon composite core and aluminum strands have different expansion coefficients and so forth.

References

1. Alawar, E.J. Bosze, and S.R. Nutt, "*A Composite Core Conductor for Low Sag at High Temperatures*," IEEE Trans. Power Delivery, Vol. **20**, pp. 2193-2199, (2005).
2. E.J. Bosze, A. Alawar, Y.I. Tsai, S.R. Nutt, D. Bryant, and G. Bowles,"*Performance of a New Overhead Conductor Design using a Carbon/Glass Fiber Composite Core*", Intern. Conf. Overhead Lines, Fort Collins, Colorado, USA, pp. 121-135, (2006).
3. E.J. Bosze, Y.I. Tsai, E. Barjasteh, S.R. Nutt, and D. Bryant, "*Long-term Durability of the Composite Core Conductor ACCC/TW*", Intern. Conf. Overhead Lines, Fort Collins, Colorado, USA, pp. 516-525, (2008)
4. 3M© Company, "*Aluminum Conductor Composite Reinforced Technical Notebook (477 kcmil family),*" © (March 2003), v.**2.21**.
5. Berjozkina S, Sauhats A, Bargels V, et al. *The technical and economic efficiency of using conductors with composite core in the transmission grid*[C]// International Conference on the European Energy Market. IEEE, (2012):**1-7**.
6. Burks B, Armentrout D L, Kumosa M. *Failure prediction analysis of an ACCC conductor subjected to thermal and mechanical stresses*[J]. IEEE Transactions on Dielectrics and Electrical Insulation, (2010), **17**(2).
7. Páczelt I, Beleznai R. *Nonlinear contact-theory for analysis of wire rope strand using high-order approximation in the FEM*[J]. Computers & Structures, (2011), **89**(11):1004-1025.
8. Stanova E, Fedorko G, Fabian M, et al. *Computer modelling of wire strands and ropes Part I: Theory and computer implementation*[J]. Advances in Engineering Software, (2011), **42**(6):305-315.
9. Frikha A, Cartraud P, Treyssède F. *Mechanical modeling of helical structures accounting for translational invariance.* Part 1: Static behavior[J]. International Journal of Solids & Structures, (2013), **50**(9):1373-1382.
10. Jiang W G, Henshall J L, Walton J M. *A concise finite element model for three-layered straight wire rope strand*[J]. International Journal of Mechanical Sciences, (2000), **42**(1):63-86.
11. Kumar K, Cochran J E. *Closed-Form Analysis for Elastic Deformations of Multilayered Strands*[J]. Journal of Applied Mechanics, (1987), **54**(4):744-744.
12. Cardou A, Jolicoeur C. *Mechanical Models of Helical Strands*[J]. Applied Mechanics Reviews, (1997), **50**(1):1.
13. Jiang W G, Yao M S, Walton J M. *A concise finite element model for simple straight wire rope strand*[J]. International Journal of Mechanical Sciences, (1999), **41**(2): 143-161.
14. Jun M A, Ge S R, Zhang D K. *Distribution of wire deformation within strands of wire ropes*[J]. Journal of China University of Mining and Technology, (2008), **18**(3): 475-478.
15. JRLX/T *(ACCC) Wire products*, Far East Holding Composite Technology Co., Ltd[z], (2009).
16. Gnanavel B K. *Effect of interfacial contact forces in radial contact wire strand*[J]. Archive of Applied Mechanics, (2011), **81**(3):303-317.
17. GB/T 1179—2008, *Round wire concentric lay overhead electrical stranded conductors*[S], (1991).

https://doi.org/10.1051/matecconf/201815301006

Effect of Hybridization on the Mechanical Properties of Chopped Strand Mat/Pineapple Leaf Fibre Reinforced Polyester Composites

Suhas Yeshwant Nayak, Srinivas Shenoy Heckadka, Nishank Minil Amin, Ramakrishna Vikas Sadanand and Linto George Thomas

Department of Mechanical and Manufacturing Engineering, Manipal Institute of Technology, Manipal Academy of Higher Education, Manipal, Karnataka, India-576104

Abstract. Hybridization of synthetic and natural fibres as reinforcement makes the polymer composites environmental friendly and sustainable when compared to synthetic fibres based polymer composites. In this study chopped strand mat/pineapple leaf fibres were hybridized. Four laminates with six layers each, with different stack sequence (GGGGGG, GPPPPG, PGGGGP and PPPPPP) were fabricated using hand layup technique while maintaining a fibre to matrix ratio of 30:70 by weight with polyester resin as matrix. Mechanical properties such as tensile and flexural strength were determined and morphology of fractured specimens was studied. Maximum tensile strength of 180 MPa was obtained for the laminate with six layers of chopped strand mat followed by hybrid laminate with four layers of chopped strand mat at the centre (120 MPa). Tensile strength of hybrid laminate with four layers of pineapple leaf fibres at the centre was in third position at 86 MPa. Least tensile strength of 65 MPa was obtained for the laminate with six layers of pineapple leaf fibres. Similar trend was observed in case of flexural behaviour of the laminates with maximum flexural strength of 255 MPa and minimum flexural strength 107 MPa. Scanning electron microscopy of the fractured specimen reinforced with chopped strand mat only, indicated, fibre pull out, matrix cracking and lack of matrix adhesion to fibres. In case of hybrid composite (GPPPPG and PGGGGP) delamination was observed to be prominent due to improper wetting of the pineapple leaf fibres with the matrix. More significant delamination led to lesser strength in case of pineapple fibres reinforced composites even though the fibre pull out was relatively less.

1 Introduction

With the increased use of polymer composites in our day to day activities comes the need to reduce the cost of manufacturing the polymer composites, thus making it a safer and sustainable material. Polymer Matrix Composite (PMC) generally consist of a reinforcing fibre (natural or synthetic or hybrid) bound together by a polymer based matrix like epoxy, vinyl ester, polyester etc [1,2]. The fibres are the basic load bearing member whereas the matrices surrounding the fibres protect the reinforcement and also help in uniform transfer of the loads [3]. Glass fibre in the form of Chopped Strand Mat (CSM) is used as a reinforcing material in many PMCs as they form relatively strong and lightweight composites popularly known as Glass Fibre Reinforced Polymer (GFRP). Although GFRP offer high mechanical strength, they are expensive to fabricate and excess usage of these is not environmental friendly. On the other hand composites made of natural fibres have a much lower cost, they are more sustainable in comparison to synthetic fibres with lower density values (1.5 g/cm^3 vs 2.6 g/cm^3) for pineapple fibres and chopped strand mats respectively [4,5]. Two major drawbacks in the selection of natural fibre as reinforcement in composites is that, they offer lower strength in comparison to glass fibres and they are prone to moisture absorption either through direct contact or through the environment leading to swelling and the formation of voids in the polymer composite. Therefore by forming a hybrid composite reinforced by glass and natural fibres it is possible to obtain a product that is both cheaper and easy to use with moderate strength requirements [6]. Higher cellulose content of pineapple fibres make them a good choice for reinforcing material as it offers better mechanical properties [7]. George et al. observed that pineapple/polyethylene composites demonstrated stress relaxation behaviour due to better reinforcement caused by an increase in fibre weight fraction [8]. Chemical treatment of the natural fibres helps to reduce the problems associated with water absorption. Alkali treatment helps in cleaning the surface of the fibres and improves its surface properties for better adhesion with the matrix [9]. Pavithran et al. studied the Pineapple leaf fibre (PALF)/polyester composites with unidirectional orientation of the fibres. They observed that an increase in fibre angle resulted in increased composite toughness [10]. Uma Devi et al. characterized PALF/Polyester composite and studied the effects of coupling agents, fibre loading and fibre length. A linear increase in impact, flexural, tensile strengths and Young's modulus was noted when fibre weight fraction was increased. They concluded that the optimum length of fibres was 30mm [7]. Mishra et al. tested the mechanical properties of PALF/glass fibre hybrid polyester composites. They found that there was a noticeable improvement in impact, flexural, and tensile strengths of the composites which indicated a positive effect of hybridization through the addition of a small amount of glass fibres [11].

In the present study chopped strand mat and pineapple leaf fibres were hybridized with different stack sequence. Composites were fabricated using hand layup technique. Mechanical characteristics of the composites such as tensile and flexural strength were determined. Scanning electron microscopy was conducted on fractured specimens. This study intends to identify the possibility of using hybrid composites as alternative to synthetic fibre composites in automotive interiors, light loaded furniture and as temporary structural members.

2 Experimental Details

2.1 Materials

Pineapple fibres were obtained by serrating the pineapple leaves and extracting the fibres using a champ's agro unit. The leaves were procured from local plantations. Fibres were washed, dried and separated according to their diameter by inspecting them visually. They were chopped to 30mm in length and soaked in 5% NaOH solution for half an hour to remove impurities. This is followed by washing in 1% HCl solution and then rinsing with distilled water. Washing with an acid is essential to neutralize effect of any residual alkali [12]. The fibres were then dried for 24 hours at room temperature. Randomly oriented E-glass chopped strand mat (CSM) of 450 gsm was chosen as the synthetic reinforcing material. General purpose polyester resin of density 1.1g/cm^3 was chosen as the matrix material.

2.2 Fabrication of composites

Four laminates of 300mm x 300mm were fabricated using hand layup process with six layers of fabric each, among which two laminates were of hybrid composites. Designation of composites with stacking sequence of plies is presented in Table 1. Fibre weight fraction of 30% was maintained in all laminates. Composites were cured at room temperature for duration of 24h. Cured panels are shown in Figure 1.

Table 1: Designation and stacking sequence of composite laminates

Designation	No. of layers	Stacking sequence
GFRP	6	GGGGGG
HYB1	6	PGGGGP
HYB2	6	GPPPPG
NFRP	6	PPPPPP

P: Pineapple leaf fibres; G: E-glass CSM

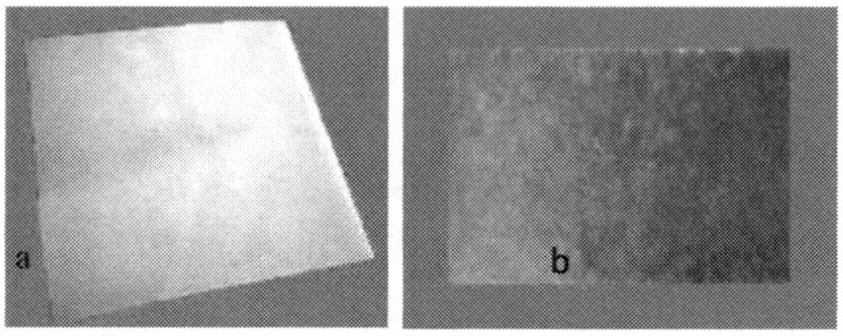

Figure 1. Cured composite panel – (a) GFRP Panel (b) Hybrid panel

2.3 Experimental Methods

Tensile and flexural strengths were determined on the Instron 3366 Universal testing machine having a maximum loading capacity of 10kN. Tensile test specimens of 250mm x 25mm were subjected to a constant cross head speed of 2mm/min as per ASTM D3039 [13].

Flexural testing (ASTM D7264) specimens of 116mm x 13mm for GFRP, 154mm x 13mm for HYB1/ HYB2 and 192mm x13mm for NFRP were subjected to three point bending at a constant cross head speed of 1mm/min [14].

All tests were conducted on conditioned samples at a temperature of 25^{0}C. Test specimens were cut from the panels using water jet cutting process. The results displayed are the averages of five individual tests. Specimens were visually checked for defects before testing.

Fractured surfaces of the laminates were analysed to understand the probable failure mechanism with the help of a Scanning Electron Microscope (SEM). A thin layer of silver coating was deposited on the specimen surface by ion sputtering and a 15kV accelerating voltage was used. The SEM was used in variable pressure mode.

3 Results and Discussions

3.1 Mechanical properties of composites

The comparison of tensile strength of the four laminates is presented in Figure 2. Tensile strength of hybrid laminate HYB1 is lower by 33.3% and that of HYB2 is lower by 52.2% while NFRP laminate is even lower at 63.8% when compared to the tensile strength of GFRP laminate, this can be attributed to fact that pineapple fibres have much lower tensile strength than glass fibres. The possibility of poor interfacial bonding between the layers of the hybrid laminate could be a potential cause as well.

https://doi.org/10.1051/matecconf/201815301006

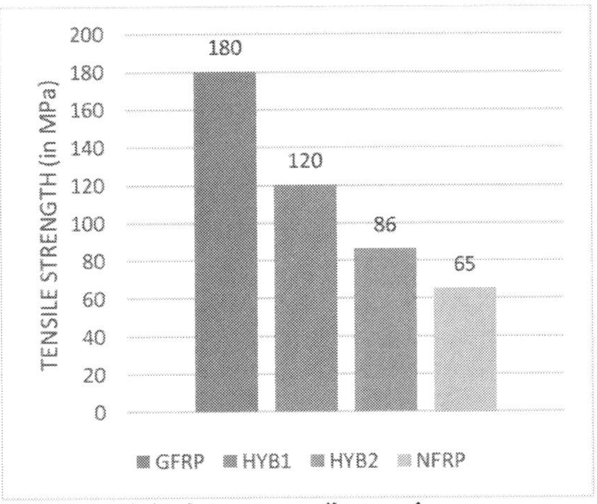

Figure 2. Variation in average tensile strength

From Figure 3 we can understand that the hybrid laminate HYB1 has a lower flexural strength by 24.3%, HYB2 is lower by 50.5% and NFRP laminate is even more weaker at 58.03% when compared to flexural strength of the GFRP laminate. Flexural loading subjects specimens to tension on the outermost layers, compression on the innermost layers and delamination in the intermediate layers. All samples in the above case failed due to tension primarily and the results match the values of the tensile test in which GFRP laminate was stronger. As seen in HYB1, higher number of layers of glass fibre led to a higher strength of the laminate whereas in case of laminate HYB2, after the outer layers of glass failed there is a possibility that delamination occurred in the layers of pineapple fibres leading to faster failure at a lower strength value. Since the pineapple fibres have lower tensile strength, the NFRP laminate displayed the weakest flexural strength among all laminates.

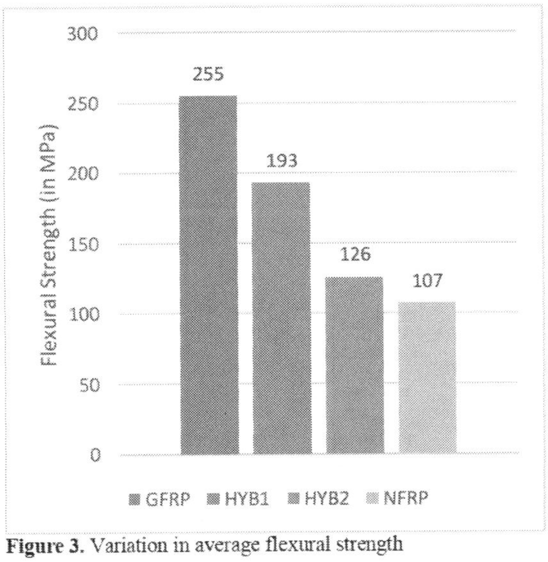

Figure 3. Variation in average flexural strength

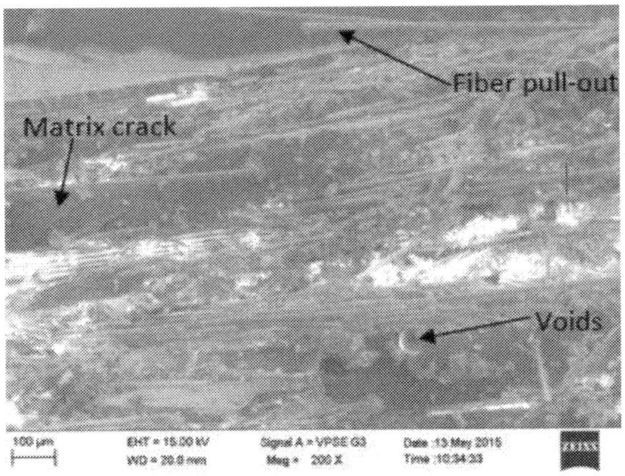

Figure 4. SEM image of GFRP tensile test specimen

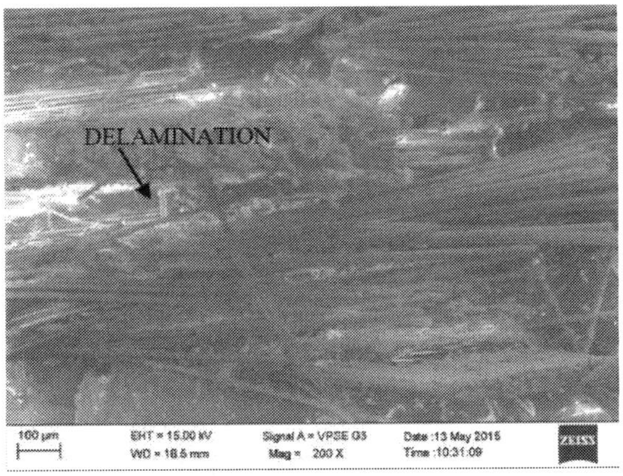

Figure 5. SEM image of GFRP flexural test specimen

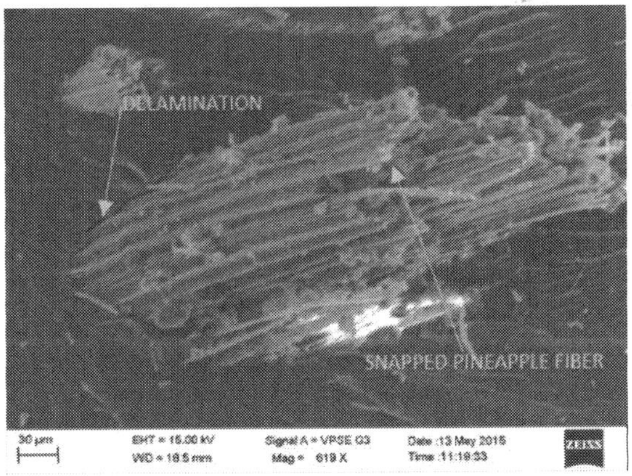

Figure 6. SEM image of NFRP tensile test specimen

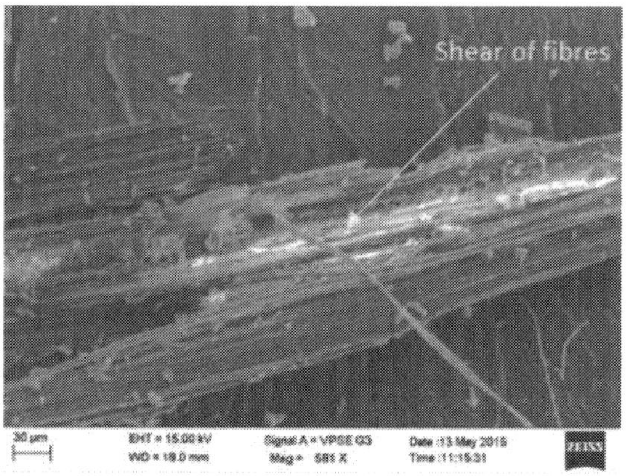

Figure 7. SEM image of NFRP flexural test specimen

3.2 Morphology of fractured surfaces

Micrographs of fractured specimens are presented in Figures 4 to 7. Figure 4 shows the micrograph of GFRP specimen failed through tensile loading. Fibre pull-out is seen at the extremities. Matrix cracking exists between layers of CSM fibres. The presence of voids further contributes to weakening of the laminates. A lack of matrix skin on the fibres displays poor interfacial bonding between fibres and matrix.

Figure 5 shows the micrograph of the GFRP flexural test specimen. A primary failure due to delamination can be observed which was caused by the bending load applied to the specimen. Figure 6 shows the micrograph of NFRP tensile specimen. Delamination can be seen which may be attributed to ineffective treatment of fibres which lead to poor wetting of natural fibres causing poor interfacial bonding with resin. Some fibre pull-out is also visible but it is comparatively lesser than GFRP specimen which is an evidence of lower tensile strength of pineapple leaf fibre. Fibre shearing is evident in the micrograph of NFRP flexural test specimen (Figure 7).

4 Conclusion

The GFRP laminate results obtained show an agreement to the generally accepted values of strength for this type of laminate. Hybrid laminates display lower tensile strength in comparison to GFRP but have a relatively decent flexural strength and hence can be used to reduce costs and increase recycling in objects that primarily undergo flexural loading like surfboards, automotive interiors, light loaded furniture and temporary structural members. Laminates having natural fibres closer to the centre of the laminate show lower strength in both loading conditions. The SEM images show delamination, fibre pull out and matrix cracking as modes of failure. Improving the treatment of natural fibres will help in proper wetting of fibres reducing the chances of delamination based failures, hence increasing the strength of hybrid laminates.

Acknowledgements

The authors are thankful to Dr. K Jagannath, Head of the Department, Mechanical and Manufacturing Engineering for permitting us to make use of the Advanced Material Testing and Research Laboratory. The authors are also indebted to Dr. Satish Shenoy B, Head of the Department and Mr. Padmaraj N. H., Assistant Professor, Department of Aeronautical and Automobile Engineering for allowing us to use their Advanced Composite and Material Testing Laboratory. The authors thank Manipal College of Dental Sciences, Manipal for permitting us to use their material testing facilities. The authors would also like to thank M/s. Konkan Speciality Polyproducts Pvt. Ltd., Mangalore for providing us their testing facility. The authors thank Mr. Anand Moolya for providing the pineapple leaves.

References

1. Ramnath, B. Vijaya, C. Elanchezhian, P. V. Nirmal, G. Prem Kumar, V. Santhosh Kumar, S. Karthick, S. Rajesh, and K. Suresh. *"Experimental investigation of mechanical behavior of jute-flax based glass fibre reinforced composite."* Fibre. Polym. **15**, 6 (2014): 1251-1262.
2. Mahesha, G. T., S. B. Satish, K. M. Vijaya, and K. Subrahmanya Bhat. *"Preparation of unidirectional grewia serrulata fibre-reinforced polyester composites and evaluation of tensile and flexural properties."* J. Nat. Fibres **13,** 5 (2016): 547-554.
3. Deogonda, Patil, and Vijaykumar N. Chalwa. *"Mechanical property of glass fibre reinforcement epoxy composites."* IJSER **1**,4 (2013): 2347-3878.
4. Mohammed, Layth, MOHAMED NM Ansari, Grace Pua, Mohammad Jawaid, and M. Saiful Islam. *"A review on natural fibre reinforced polymer composite and its applications."* Int. J. Polym. Sci. **2015** (2015).
5. Joshi, Satish V., L. T. Drzal, A. K. Mohanty, and S. Arora. *"Are natural fibre composites environmentally superior to glass fibre reinforced composites?."* Compos. Pt. A-Appl. Sci. Manuf. **35**,3 (2004): 371-376.
6. Arib, R. M. N., S. M. Sapuan, M. M. H. M. Ahmad, M. T. Paridah, and HMD Khairul Zaman. "Mechanical properties of pineapple leaf fibre reinforced polypropylene composites." Mater. Des **27**, 5 (2006): 391-396.
7. Devi, L. Uma, S. S. Bhagawan, and Sabu Thomas. *"Mechanical properties of pineapple leaf fibre-reinforced polyester composites."* J. Appl. Polym. Sci **64**, 9 (1997): 1739-1748.
8. George, Jayamol, S. S. Bhagawan, N. Prabhakaran, and Sabu Thomas. *"Short pineapple- leaf- fibre-reinforced low- density polyethylene composites."* J. Appl. Polym. Sci **57**, 7 (1995): 843-854.
9. George, Jayamol, S. S. Bhagawan, and Sabu Thomas. *"Effects of environment on the properties of low-density polyethylene composites reinforced with*

pineapple-leaf fibre." Compos. Sci. Technol. **58**, 9 (1998): 1471-1485.
10. Pavithran, C., P. S. Mukherjee, M. Brahmakumar, and A. D. Damodaran. *"Impact properties of natural fibre composites."* J. Mater. Sci. Lett. **6,** 8 (1987): 882-884.
11. Mishra, S., A. K. Mohanty, L. T. Drzal, M. Misra, S. Parija, S. K. Nayak, and S. S. Tripathy. *"Studies on mechanical performance of biofibre/glass reinforced polyester hybrid composites."* Compos. Sci. Technol. **63**, 10 (2003): 1377-1385.
12. John, Maya Jacob, and Rajesh D. Anandjiwala. *"Recent developments in chemical modification and characterization of natural fibre- reinforced composites."* Polym. Compos. **29**, 2 (2008): 187-207.
13. ASTM International. ASTM D3039/D3039M-14 Standard Test Method for Tensile Properties of Polymer Matrix Composite Materials. West Conshohocken, PA: ASTM International (2014).
14. ASTM International. ASTM D7264/D7264M-15 Standard Test Method for Flexural Properties of Polymer Matrix Composite Materials. West Conshohocken, PA: ASTM International (2015).

Comparative study on porous media combustion characteristics using different discrete materials

Ayub Ahmed Janvekar[1], M. Z. Abdullah[2], Z. A. Ahmad[3], Aizat Abas[1], Musavir Bashir[2], Mazlan Mohamed[1,4]

[1]*School of Mechanical Engineering, Universiti Sains Malaysia, Engineering Campus, 14300 Nibong Tebal, Penang, Malaysia*
[2]*School of Aerospace Engineering, Universiti Sains Malaysia, Engineering Campus, 14300 Nibong Tebal, Penang, Malaysia*
[3]*School of Materials and Mineral Resources Engineering, Engineering Campus, 14300 Nibong Tebal, Penang, Malaysia*
[4]*Advances Material Research Cluster, Faculty of Bioengineering and Technology, Universiti Malaysia Kelantan, Jeli Campus, 17600 Jeli, Kelantan, Malaysia*

Abstract. Occurrence of combustion phenomenon in porous media has always excited researchers to develop various shape and size of burner so that maximum utilization of energy can be taken achieved. Here in this experiential work, dual layer micro burner was exclusively built to carry out porous media combustion characteristic with different type of discrete material in reaction zone. Presently, only alumina and zirconia are compared in discrete form, while preheat layer was made of porcelain ceramic material (foam type). Reaction zone was restricted to thickness of 20mm while preheat zone at 10mm. A concept of equivalence ratio was aided since it involves premixed combustion of air and butane as fuel mixture. Additionally, burner was made to run under lean to ultra-lean modes and finest temperature were recorded. Both surface and submerged flame was generated effectively. Maximum temperatures recorded during surface and submerged flame condition was better by installing alumina rather than zirconia there by reaching a value of 631°C and 470°C respectively. Thus maximum thermal efficiency was calculated and found out to be 84%. Finally, emission parameters like NOx and CO where monitored and found out to be well within acceptable limits.

1 Introduction

Impact of fossil fuel in developing countries will not only affects industrial and domestic sector but greatly endeavour in economic aspects of nation. Thus saving single drop of fuel either in crude or purified state becomes very much necessary [1-3]. Now to identify the large scale consumption of fuel need to snipped surveyed by grassroots level. So by and large burners one of the vital candidate for consuming large amount of fuel. Burners are not only used in domestic sector to fulfil day to day requirements of the peoples, but does act as prime source of energy supply in medium and large scale industries. Tremendous research work has already been carried out in the area of modern burner developments, but there is wide scope to build novel burner with high thermal efficiency [4-6]. Fusing porous media and burner together have created significant results in downgrading the emissions and enhancing the temperature profiles. Addition of porous media in reaction zone shall increase the combustion surface area exponentially their by generating high thermal energy [7-9]. Premixed combustion is popular over non premixed combustion especially with portable, macro and micro burner. Zuo, E [10] worked on micro elliptical tube combustor and compared with regular micro circular tube combustor and focused on the understandings of the combustion characteristics. It is found that the pressure loss and combustion efficiency of the micro elliptical tube combustor is larger than that of the micro circular tube combustor. Panigrahy, Mishra [11] made numerical and experimental analyses of combustion of LPG in a domestic cooking stove with a two-layer porous radiant burner, and conveyed CO emissions of all burners operating at equivalence ratio (ER) 0.4 or below were found to be lower than 5.0 ppm. Mustafa, Abdullah [12] made a combustion driven thermoelectric (TE) and thermophotovoltaic (TPV) power systems using several blends of vegetable cooking oil–kerosene (VCOK) fuels and moreover enabled TE and TPV systems with a porous burner. Restrict their study to combustion characteristics and system performance. Ismail, Abdullah [13] work with porous medium burner (PMB) along micro cogeneration system. The average surface temperature obtained in this study was 544.8°C for a flow rate of 0.25 L/min. Impact of using butane have also shown significant improvement in porous media burners [14, 15]. Present work deal with compassion between utilization of alumina and zirconia porous media in reaction layer. Both surface and submerged flame conditions were taken into account. Burner was made to run from lean to ultra-lean equivalence ratio. Finally, generated emission was fed into chimney to calculated the amount of NOx and CO.

2 Experimental setup

With the help of professional drawing tool burner housing was sketched was as shown in Fig. 1. It helps to understand the major components of the housing. A complete layout of the system is as shown in Fig. 2 whereas actual setup is as shown in Fig. 3.

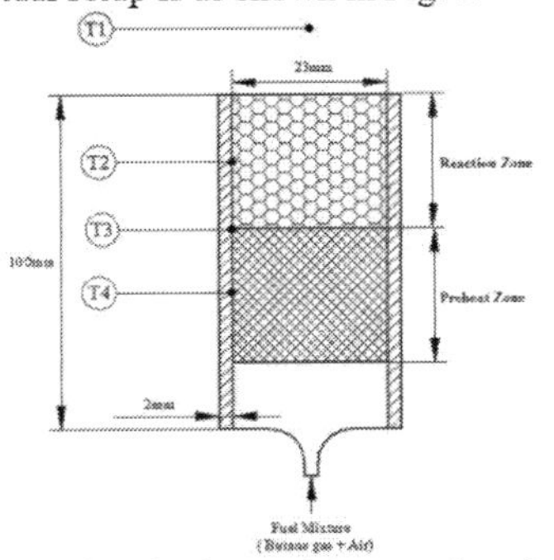

Figure 1. Sectional view of burner housing. (Not to scale).

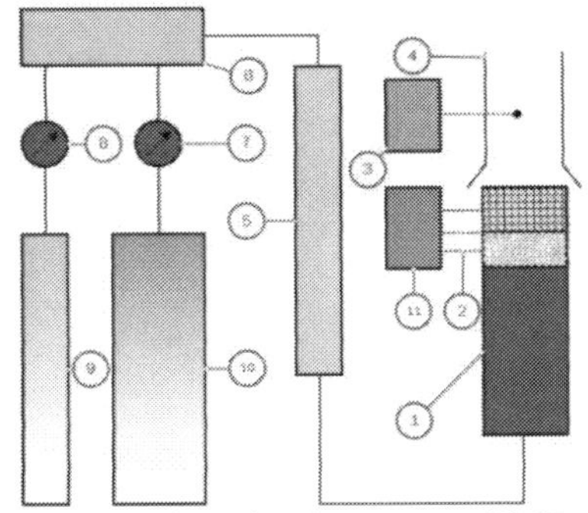

Figure 2. Layout of combustion system. (1) Housing, (2) Thermocouple, (3) Emission gas analyser, (4) Gas conduit, (5) Mixing unit, (6) Pre-mix, (7) Air flow meter, (8) Butane flow meter, (9) Butane supply, (10) Air inject, (11) DAQ (Data acquisition system).

Figure 3. Experimental burner setup.

Burner housing was divided into two major sections namely upper section (Reaction zone) and lower section (Preheat zone). Thickness of reaction zone was freeze to 20mm while the preheat at 10mm. Corresponding experimental trial were taken with both alumina and zirconia in upper section. Whereas porcelain in the foam type was maintained in lower section. Preheat layer was made of porcelain foam with 8 ppcm and a porosity of 84%. Premixing was perfumed using two pre mix units as shown in layout. Butane was supplied from compressed butane cartridge, while air was regulated with domestic natural gas pump. Both fuel and air was controlled using certified digital flow meters. Fuel mixture was introduced from the bottom of the burner housing. Thermocouple (K-type) was used to measure surface temperature (T1) while surface probe was used to for wall temperature at suitable locations (namely T2, T3, and T4). T1 was placed at mid of reaction zone, T2 at the junction between reaction and preheat layer while T3 at the mid of preheat layer. To capture the behaviour of the K type of thermocouple DAQ system was also used, which help to generate the accurate results vai computer. At critical ER, thermal images where taken with FLUKE thermal imager. Flow of air and butane was measure L/min while emissions were captures using KANE-9106 QUINTOX Combustion Analyser in ppm (parts per million).

3 Analysis

Present work deals with natural air, concept of ER was used as shown in Equation (1), where the actual air–fuel ratio (M_a) and while stoichiometric air fuel ratio (M_s) for butane gas, was can be calculated by considering Avogadro's and Dalton's laws [13], which was computed a value of 30.95 with air at room temperature as shown in Table 1.

$$ER = M_s / M_a \quad (1)$$

Table 1. ER for butane and air flow rate.

Butane (Lpm)	0.1				
Air (Lpm)	3.1	3.4	3.8	4.4	5.1
ER	1.0	0.9	0.8	0.7	0.6

Stable flame with minimum fuel consumption was obtained at 0.1 Lpm (litre per min). Trails were performed at room temperature. With alumina in reaction zone, maximum surface temperature was noted as 631°C at 0.7 ER, while zirconia it was around 615°C at 0.7 ER, as shown in Fig 4. and 5. Average temperature profile can be seen in Fig. 6 and 7.

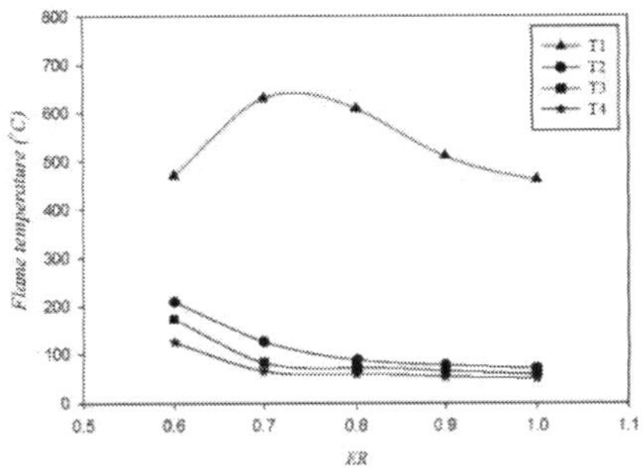

Figure 4. Flame temperature for alumina

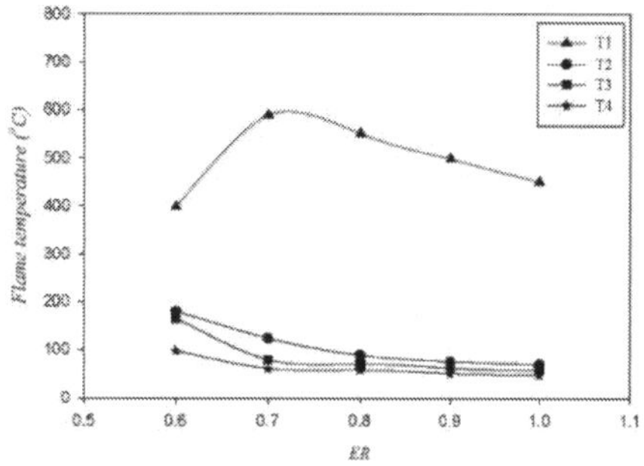

Figure 5. Flame temperature for zirconia

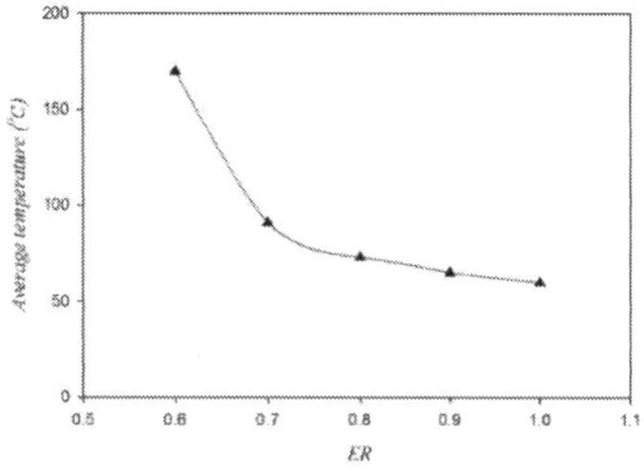

Figure 6. Average temperature for alumina

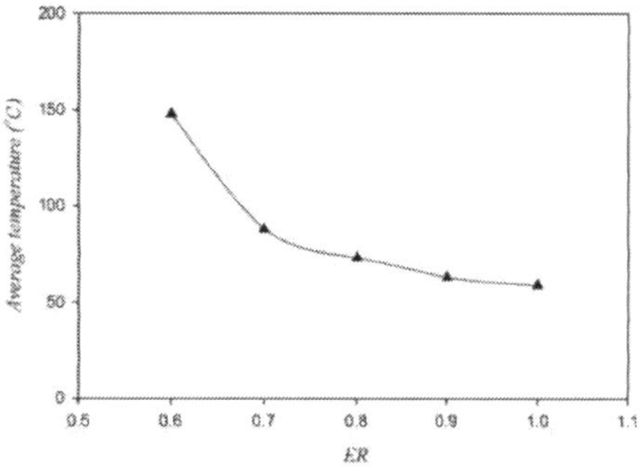

Figure 7. Average temperature for zirconia

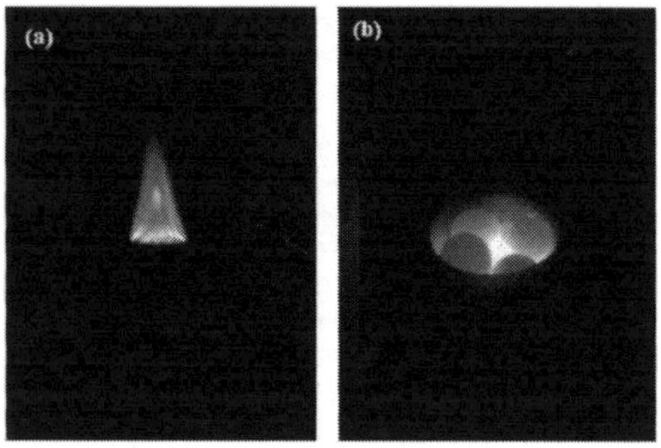

Figure 8. Surface (left) and submerged flame (right)

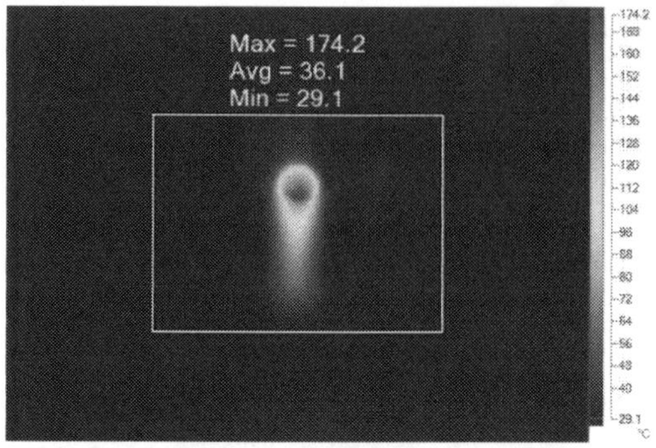

Figure 9. Thermal image with alumina in reaction zone

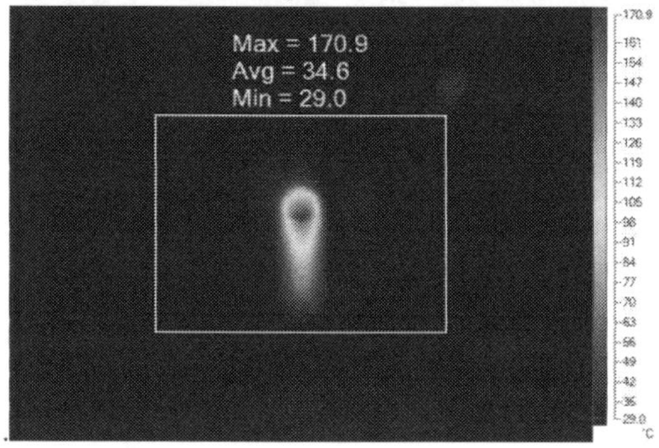

Figure 10. Thermal image with zirconia in reaction zone

The possible reason of burner to undergoes surface or submerged flame mainly depends on availability of oxygen and surface area to perform combustion. Furthermore Fig. 8 indicate the actual photographic images of alumina at 0.7 and 0.6 ER. Thermal images are used to capture the temperature distribution at critical ER. With the help thermal imager generated photographic images indicate the sniped overview of temperature variation on the burner wall for alumina and zirconia as shown in Fig. 9 and 10. Maximum thermal efficiency was calculated during surface flame was recorded as 84% and 77% for alumina and zirconia respectively. Finally, with the help of certified gas analyser, emitted emissions from the burner surface where fed to gas probe to measure NOx and CO. It was found out that emission where less than 1ppm since the minimum amount to detect the ppm for the utilized gas analyser is 1pmm.

4 Conclusions

Presented experimental work deal with comparative study between alumina and zirconia porous media in reaction zone. With Alumina in reaction zone better temperature profiles were obtained. At least 7% more thermal efficiency was noted in comparison with zirconia. Trails were conducted with both surface and submerged flame based on various ER from lean to ultra-lean ER. Predefined thickness of reaction and preheat layer was used to performed combustion. Optimum ER for surface flames and submerged flame are 0.7 and 0.6 respectively. Maximum temperature recorded at best performing ER was 631°C leading to thermal efficiency of 84%.

Emission parameters such NOx and CO where less than 1ppm.

References

1. Udayraj, et al., Heat and mass transfer through thermal protective clothing – A review. International Journal of Thermal Sciences, 2016. 106: p. 32-56.
2. Jiang, X., D. Mira, and D.L. Cluff, The combustion mitigation of methane as a non-CO2 greenhouse gas. Progress in Energy and Combustion Science, 2016.
3. Arya, P.K., et al., DME blended LPG as a cooking fuel option for Indian household: A review. Renewable & Sustainable Energy Reviews, 2016. 53: p. 1591-1601.
4. Daneshvar, H., R. Prinja, and N.P. Kherani, Thermophotovoltaics: Fundamentals, challenges and prospects. Applied Energy, 2015. 159: p. 560-575.
5. Al-Abdeli, Y.M. and A.R. Masri, Review of laboratory swirl burners and experiments for model validation. Experimental Thermal and Fluid Science, 2015. 69: p. 178-196.
6. Zanganeh, J., B. Moghtaderi, and H. Ishida, Combustion and flame spread on fuel-soaked porous solids. Progress in Energy and Combustion Science, 2013. 39(4): p. 320-339.
7. Mujeebu, M.A., et al., Combustion in porous media and its applications--a comprehensive survey. J Environ Manage, 2009. 90(8): p. 2287-312.
8. Mujeebu, M.A., et al., Applications of porous media combustion technology - A review. Applied Energy, 2009. 86(9): p. 1365-1375.
9. Mujeebu, M.A., et al., A review of investigations on liquid fuel combustion in porous inert media. Progress in Energy and Combustion Science, 2009. 35(2): p. 216-230.
10. Zuo, W., et al., Numerical investigations on combustion characteristics of H 2 /air premixed combustion in a micro elliptical tube combustor. Energy, 2017. 126: p. 1-12.
11. Panigrahy, S., et al., Numerical and experimental analyses of LPG (liquefied petroleum gas) combustion in a domestic cooking stove with a porous radiant burner. Energy, 2016. 95: p. 404-414.
12. Mustafa, K.F., et al., Comparative assessment of a porous burner using vegetable cooking oil–kerosene fuel blends for thermoelectric and thermophotovoltaic power generation. Fuel, 2016. 180: p. 137-147.
13. Ismail, A.K., et al., Application of porous medium burner with micro cogeneration system. Energy, 2013. 50: p. 131-142.
14. Janvekar, A.A., et al., Effects of the preheat layer thickness on surface/submerged flame during porous media combustion of micro burner. Energy, 2017. 122: p. 103-110.
15. Janvekar, A.A., et al. Assessment of porous media burner for surface/submerged flame during porous media combustion. in AIP Conference Proceedings. 2017. AIP Publishing.

 https://doi.org/10.1051/matecconf/201815302001

Visual Servoing of Mobile Microrobot with Centralized Camera

Gandjar Kiswanto, Mohamad Safhire, Reza Afrianto and Rifkie Nurcahya

Department of Mechanical Engineering, Faculty of Engineering, Universitas Indonesia, Depok, Indonesia 16424

Abstract. In this paper, a mechanism of visual servoing for mobile microrobot with a centralized camera is developed. Especially for the development of swarm AI applications. In the fields of microrobots the size of robots is minimal and the amount of movement is also small. By replacing various sensors that is needed with a single centralized vision sensor we can eliminate a lot of components and the need for calibration on every robot. A study and design for a visual servoing mobile microrobot has been developed. This system can use multi object tracking and hough transform to identify the positions of the robots. And can control multiple robots at once with an accuracy of 5-6 pixel from the desired target.

1 Introduction

The role of robots in human life, especially in the manufacturing sector continues to increase and more relevant [1]. Today's robot system has reached the point where the system can run automatically and improve performance by interacting with the environment [2].

To do this, robots are equipped with various sensors to be able to understand the environment. But in the applications of microrobot systems that scale from cm to mm the various sensors are difficult to use because of their limited size and the required precision level of the sensor will make it difficult in the calibration process [3]. Visual Servoing or visual control is widely used for various applications of robot manipulator industry and mobile robot, starting from positioning, navigation, object tracking etc. [4] [5] [6]. Microrobot systems thus can benefit from this system as well.

In the case of mobile robots, it is rare to purely use visual control system to get information. Some methods use lasers to obtain distance information with objects around and various combination of sensors [7][8]. With the development of computer vision in general visual control system becomes more cheap, lightweight, compact and practical. Not to mention the amount of information that can be taken from one sensor. This makes the visual control system more attractive.

Visual control is very suitable for use in microrobot to replace the various sensors needed to know the position. This can make the robot system simpler because the inherent unbooted sensor allows us to minimize and simplify the robot design [9]. Visual control also allows us to control more than one robot with only one sensor [10]. With visual controls, we can control the robot large number or swarms of robots. Applications of such systems include micro handling and assembly, micro sensing and one of the most unique of allowing us to simulate the behavior of swarm animals such as fish and bees [11] [12]. Of course, to be applied like this requires a reliable system visual control system. Therefore, it takes a visual control system that can adapt to environmental changes and can process data quickly so that the movement of the robot can be controlled almost real time.

2 System modeling

The system we want to create is where the robot feature will measure with a centralized camera in the form of Logitech HD C270 webcam. Data from this camera will be processed through the controller to be the position data that will be processed by the server to serve as a command to the robot. This command will then be provided via WIFI to each robot.

Control system in use is a closed loop system that compares the features obtained of what we call as (s) with the desired feature (s'). The purpose of this control system is to minimize errors. And **Figure 1** is there to explain this in a better form.

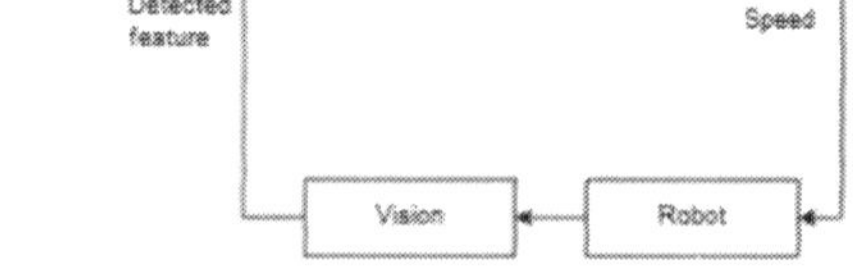

Figure 1. Control Diagram for Microrobot.

Featured in this case is a 2D single point information taken from a color camera picture. Thus, error can be defined as:

$$e(t) = s(t) - s' \quad (1)$$

s(t) defines features detected during ever-changing movements. S' is an input that does not change with time. Here s(t) is the position and orientation data in cartesian coordinates in the image.

$$s(t) = \begin{bmatrix} m(t) \\ n(t) \\ \theta(t) \end{bmatrix} \quad (2)$$

The kinematics model of robots moving in 2D with 2-wheel system can be lowered to.

$$\dot{m} = v\cos(\theta) \quad (3)$$

$$\dot{n} = v\sin(\theta) \quad (4)$$

$$\dot{\theta} = \omega \quad (5)$$

Or can be written as

$$\begin{bmatrix} \dot{m} \\ \dot{n} \\ \dot{q} \end{bmatrix} = \begin{bmatrix} \cos(\theta) & 0 \\ \sin(\theta) & 0 \\ 0 & 1 \end{bmatrix} \begin{bmatrix} v \\ \omega \end{bmatrix} \quad (6)$$

With that in mind if we want to bring the robot towards a position (m', n') with an orientation of ω we can find the speed that is needed as

$$v = K_v\sqrt{(m - m')^2 + (n - n')^2} \quad (7)$$

$$\omega = \tan^{-1}\frac{n' - n}{m' - m} \quad (8)$$

The above equation then describes the movement of the robot against the speed that is given. This equation will then be used to obtain the amount of error that we need to compensate with the vision system.

3 Vision System

This section explains how the vision system detect the robot, assigns the label to the robot, and calculates the position and orientation of the robot. An overview of the entire vision system can be seen in **Figure 2**.

3.1 Robot detection

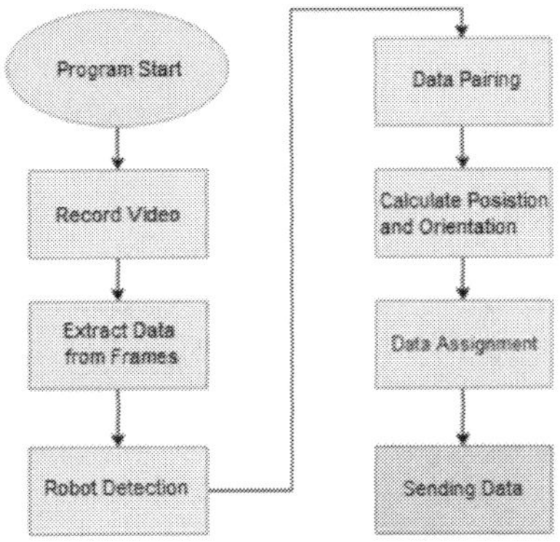

Figure 2. Flowchart for the Vision System that is used.

Robot detection is done using Circular Hough Transform (CHT). CHT is a derivative of the basic form of Hough Transform. This approach is used because of its toughness with noise, occlusion and varying lighting. The foreground pixel with a high gradient is designated as a candidate pixel and allowed to give 'noise' to the accumulator circuit. In the application of classical CHT, candidate pixels vote in patterns around them that form a full circle of fixed radius [11]. The data obtained in the form of coordinate position data to the camera axis. This data is a 2x1 matrix. But the data obtained has not been paired to find the angle. This will be explained in the next section.

3.2 Data pairing, position & orientation calculation

Data pairing is used to pair the data between the center point of the robot circle and the reference center point of the circle data. This needs to be done to calculate the orientation angle of the robot. Data pairing is done by finding the smallest distance between the reference circle to the central point of the robot circle. The data from CHT discussed earlier will be divided into 2 sections. The first part is to circle the robot in the form of a 2xn matrix from the x and y coordinate positions of each circle. The second part is similar but for the coordinate positions of the x and y circles of reference. Data from these two matrices will be paired with the point closest to it. The data in the robot circle matrix is used as a reference.

After the data has been paired we can calculate the orientation of the robot. Orientation data is the result of angle calculation between the robot center point and the reference point. By knowing the distance between the origin point and the center of the robot and the reference point we can know the orientation of the robot as **Figure 3** below illustrates.

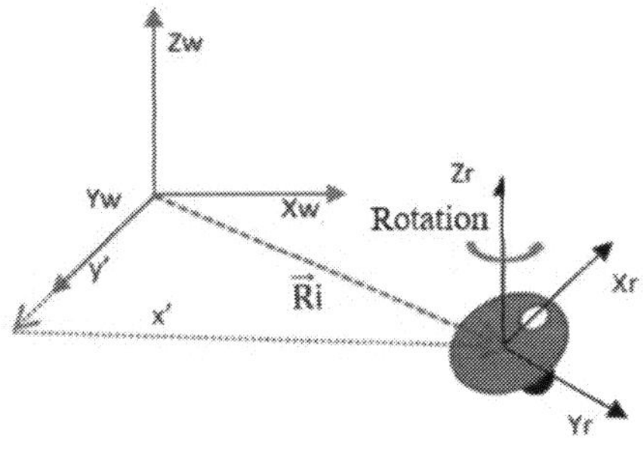

Figure 3. Position and Orientation of the robot.

If we give position notation of robot circle as X and Y and position of reference circle as M and N then orientation can we formulated to be:

$$\emptyset = tan^{-1}\left(\frac{(N-Y)}{(M-X)}\right) \quad (9)$$

Then this process is done for all robots to get the orientation data and position. The final data that is in the form of a 5xn matrix in which n states many robots are detected at that time. Though this data still needs to be labeled to know which robot has this data. For that the part below will explain.

3.3 Data Assignment

The data obtained above is the data that we still cannot identify. Because the data above is not yet indexed and cannot describe which robot has that data. What we need is that the data of robot no 1 to not be confused with the data for robot no 2 and so on. To do so we compare the data that we have now (*T*) with the data in the previous frames (*T-1*). And by using a queue system where the first robot that enters will be labeled robot no 1, the next robot no 2 and so on. We can be sure that each robot data will not be confused with another robot.

New problems will arise when there is a robot that disappeared suddenly or occluded by an object. To solve this problem, we gave a special algorithm when there is an addition or reduction of a robot. When the number of robots detected now (*T*) is less than the previous robot data (*T-1*) the robot location data that has is missing lost will not be updated until the robot appears again. This algorithm adds to the robustness of the system when there is an error. The output of this can be seen in **Figure 4**.

4 Control Algorithm

The control algorithm of the visual servoing will only match the features we want to the ones detected. To achieve this, we need to breakdown into a few simple commands. Due to the robotic construction and also the limitation of the type of movement that can be done we divide it into 4 simple commands that is stop, forward, turn clockwise and counterclockwise. From these 4 commands, we are able to run the robot.

The initial stage is to get the position data from the server the target position that we want. From these two data, we can calculate the amount of angle we need. The robot will then rotate between to the right or to the left to equalize the angle of orientation that it requires. When the robot has equalized its angle, it will be given the orders to move forward. As the robot moves to the position the robot will continue to be updates its location to know the remaining distance that must be taken. When the remaining distance has reached the limit threshold that we have given or what we say as the robot's tolerance it will stop and at this stage the robot position will be the same as the feature we want.

To perform complex movements such as spline circle etc. we can solve the movement to 20 linear motions by way of interpolation. Given the robot can only move linearly then to reduce the error need to use interpolation. Interpolation takes the 20 coordinate points that are in the path we want and include them as target data one by one. With that we can command the robot to move in a complex way without large errors.

The various kinds of motion that we have made is the movement of circle, sinus, waypoint, and point to point. From these various movements, we will test how precise the robots can be moved. By the method described in the next section.

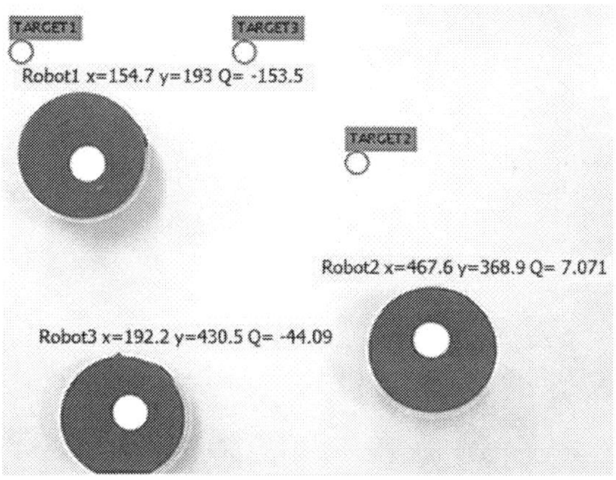

Figure 4. Display of Control System.

5 Experimental testing

We divided our experiments into 4 types of movements in which each will test the motion capabilities that the robot can perform. The four types of movements that are shown in **Figure 5** are squares, "M", circles and spline. Our square shape in **Figure 5** (a) is used to be able to test linear motion of robot's ability. Circle shape in **Figure 5** (b) tests the ability of circular motion. The form "M" in **Figure 5** (c) tests the ability to move in 2 axes at the same time. And spline movement in **Figure 5** (d) to test the fluidity of the movement.

 https://doi.org/10.1051/matecconf/201815302001

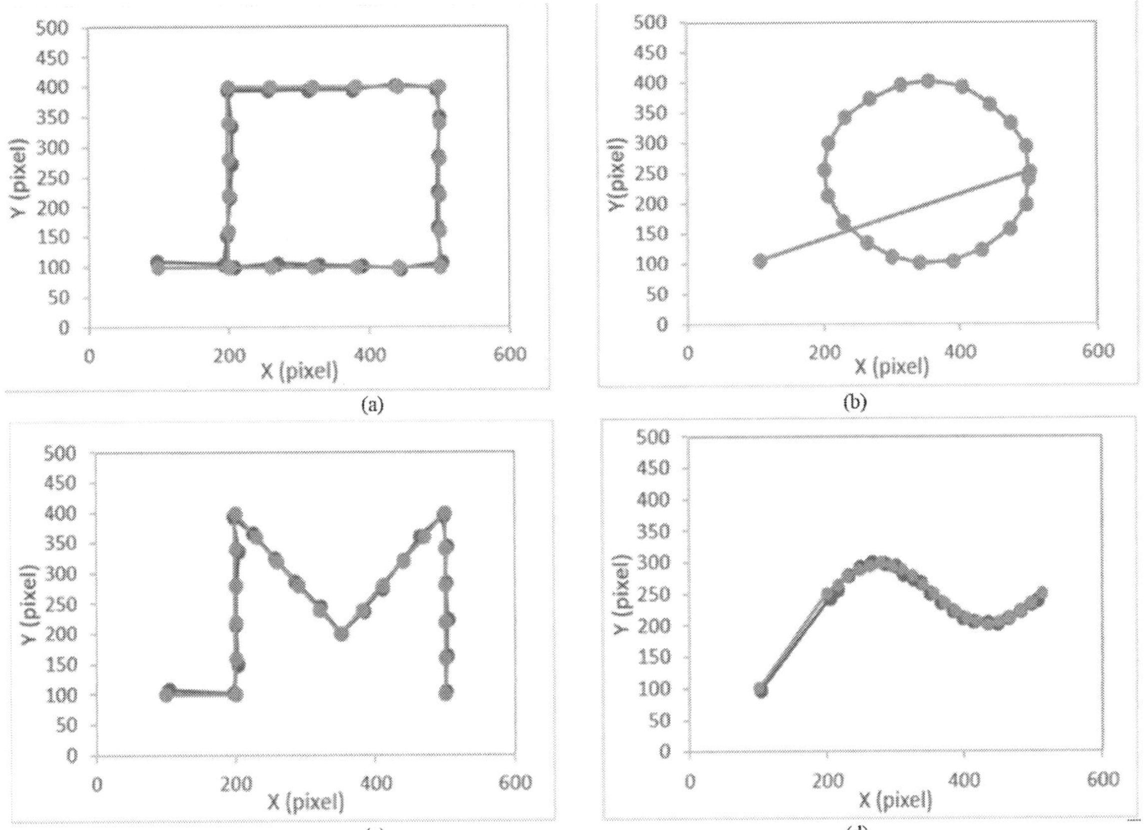

Figure 5. Experimental result (a)Square (b)Circle (c)M (d)Spline.

The result of the robot movement that we analyze is the location of the robot compared to the 20 targets that we command. Retrieval of data on the vision system is done every moment. But because the robot model cannot change direction instantly the data that we will be analyzing is in the moment the robots has achieved its target before it changes its direction to go to the next target. This data is illustrated in **Figure 5.**

As described before we have divided the experiment into 4 groups that is (a)Square (b)Circle (c)M (d)Spline. A total of 22 robot position for each shape are tabulated and graphed. This is done to compare between the data in the form of targets and actual data of the position from the robot. From both data, we calculate the resultant and deviation from the target coordinate in mm. The value gained from the experiment in mm will be shown in **Table 1**. A slight deviation in the starting position like we can see in **Figure 5** is to be expected therefore it is not calculated in the results.

Table 1. Experimental Results in mm

Shape	Maximum	Minimum	Average
Square	9.440	3.092	6.082
Circle	8.975	1.935	5.863
M	8.900	0.795	4.511
Spline	9.791	1.422	4.344

An interesting phenomenon that we found is with the simple shapes like in **Figure 5** (a) there is quite a lot more deviation compared with the more complex shapes. We concluded that this is due to the distance between the targets. With higher distances, it becomes harder to self-correct. Statistical analysis is done to get the precision value from our robot.

From the 4 experiments as many as 84 data in then joined to get the average data value of 5.2 mm and standard deviation of 2 mm. The accuracy of the robot movement in this case can be calculated by knowing the magnitude of the deviation of the movement of the robot. For that we need to know the average size of the robot deviation population. Using the average interval, we can know that the range of the average population is between 4.77-5.63 mm. With that we can say that our robot can move to the destination that we command with an accuracy of 5.63 mm.

6 Conclusion

A Visual control system for mobile microrobot with a centralized camera was created. This vision-based control system is able to control multiple robots at once by comparing the features detected with the desired ones. By using a webcam sensor, we can get the location data and orientation of each robot. Labeling on each robot can be done and there are no errors in labeling such as the miss interpretation of data or others. The vision system is also resistant to external disturbances such as occlusion and light changes. This can be done because it has a robust algorithm in case of interference from the outside. The robot can be command using a control system where it can perform simple movements to create geometric shapes. The control system can move the robot to its destination with an accuracy of 4.77-5.63 mm from the target. All systems such as the robot, server, and control systems have been well integrated to be able to drive the robot. Further study is also needed to determine the optimal distance between targets to achieve better accuracy. Therefore, in a later study the variable of target distance should also be considered.

References

1. K. M. N. T. J. Ogbemhe, "Achieving Sustainability in Manufacturing Using Robotic Methodologies," Procedia Manufacturing, p. 440–446, (2017)
2. M. A. A. K. J. B. K. Hamed Shahbaz, "Demonstration learning of robotic skills using repeated suggestions learning algorithm," Biologically Inspired Cognitive Architectures, p. 21–30, (2017)
3. C. A. ,. N. H. A. C. R. N. Michael Rubenstein, "Kilobot: A low cost robot with scalable operations designed for collective behaviors," Robotics and Autonomous Systems, p. 966–975, (2014)
4. G. C. F. S. F. C. Don Joven Agravante, "Visual Servoing in an Optimization Framework for the Whole-Body Control of Humanoid Robots," *IEEE Robotics and Automation Letters,* pp. 608-615, (2017)
5. S. K. P. P. Gossaye Mekonnena, "Wireless hybrid visual servoing of omnidirectional wheeled mobile robots," Robotics and Autonomous Systems, pp. 450-462, (2016)
6. M.-Y. C. Wei-Che Chang, "Image feature command generation of contour following tasks for SCARA robots employing Image-Based Visual Servoing—A PH-spline," Robotics and Computer-Integrated Manufacturing, pp. 57-66, (2017)
7. D. H.-P. H. Martinez-Barbera, "Autonomous navigation of an automated guided vehicle in industrial environments," Robot. Comput.-Integr. Manufacture., pp. 296-311, (2010)
8. R. O. C. S. C. F. D. Ronzoni, "AGV global localization using indistinguishable artificial landmarks," dalam Proceedings of IEEE International Conference on Robotics and Automation,, Shanghai, (2011)
9. C. D. S. F. Daniel Jasper, "Hardware-based, Trajectory-Controlled," IFAC Symposium on Mechatronic Systems, (2010)
10. C. Diederichs, "Fast Visual Servoing of Multiple Microrobots using an FPGA-Based Smart Camera System," dalam Proceedings of the 18th World Congress The International Federation of Automatic Control, Milano, (2011)
11. Q. L. Y. &. X. N. Xu, "Design, Fabrication, and Visual Servo Control of an XY Parallel Micromanipulator With Piezo-Actuation," IEEE Transactions on Automation Science and Engineering, pp. 710-719, (2009)
12. N. T. H. L. P. &. Q. T. T. Thinh, "Research and Applying Computer Vision for Controlling the School of Fish Robots Using Swarm Model.," dalam 3rd International Conference on Green Technology and Sustainable Development, (2016)
13. E. Davies, Machine Vision: Theory, Algorithms, Practicalities., Morgan Kauffinan Publishers, (2005)

Mechatronics Design of an Autonomous Pipe-Inspection Robot

Mohamed Abdellatif[1], Hazem Mohamed[2], Mohamed Hesham[2], Ahmed Abdelmoneim[2], Amro Kamal[2] and Ahmed Khaled[2]

[1]*Future University in Egypt, Mechanical Eng. Department, Cairo, Egypt*
[2]*Arab Academy for Science, Technology and Maritime Transport, Cairo, Egypt*

Abstract. Pipelines require periodical inspection to detect corrosion, deformation and congestion with obstacles in the network. Autonomous mobile robots are good solutions for this task. Visual information from the pipe interior associated with a location stamp is needed for inspection. In this paper, the previous designs of autonomous robots are reviewed and a new robot is developed to ensure simple design and smooth motion. Images are processed online to detect irregularity in pipe and then start capturing high resolution pictures to conserve the limited memory size. The new robot moves in pipes and provides video stream of pipe interior with location stamp. The visual information can later be processed offline to extract more information of pipeline condition to make maintenance decisions.

1 Introduction

Pipelines are important for economic transport of gases and liquids such as oil, fuels, water and sewer. However, pipelines always suffer from aging and damages, which can cause great waste of resource, environmental pollution, and other accidents. For example, the leak of petroleum pipeline causes financial losses, pollution and ecocatastrophe.

Regular inspections and maintenance of pipelines are essential to keep them functional. Unfortunately, the difficulty and the cost for human inspection can be extremely high, especially with the appearance of increasingly complicated pipelines nowadays [1]. Autonomous inspection and monitoring of pipes is important during several phases including the commissioning stage to ensure that pipes are free from faults of left-over obstacles. Then, monitoring is also needed in the running and routine maintenance phases.

Autonomous robots can be used to inspect the interior of pipes. Sewage pipelines are the most commonly inspected as it is very hard to inspect by humans [2]. Information such as pictures of cracks or any notches in the pipes [3] need to be collected. It may also include videos showing the internal shape of pipes, bends and manholes. Location of the cracks, inside the pipelines, is extremely important for repair. A fully autonomous robot is needed to relax human intervention that is boring and hard to maintain in high quality and attention for longer periods of time. Autonomous robots are needed for inspection because pipeline systems are often hidden underground and are long, thin, and complex. Furthermore, the environment inside pipelines can be dirty and hazardous for humans. Several systems were built in the world to inspect pipe interiors [4, 5]. The classification can be based on several criteria such as autonomous versus manual and tethered versus untethered. The current technology can mainly be classified into three categories according to the degree of robot autonomy: namely; remotely operated, semi-autonomous and fully autonomous. All three categories are tethered in general while the third category can be tethered or not [6, 7].

Although several robots had been introduced in the literature, the mechatronics design is complex and is not easy to operate besides the high cost of the robot [8, 9]. In general the complexity resulted from the requirements to satisfy the projects needs in building such robots, including the ability to climb steep obstacles or to navigate through pipe networks of curved routes and connections. An autonomous robot that can routinely inspect the pipe interior and report deformation will, therefore, be highly desirable for routine maintenance [10-12].

In this paper, we describe the design and development of an autonomous robot capable of inspecting pipes interior of 30 cm diameter which constitute around 60% of pipes used in sewage systems in several countries, including Egypt and Japan [2]. The design objective is to build a simple system and can be made from off the shelf components.

A user interface program for offline visual information processing is also developed in this work to help the users in interpreting the pipe data and customizing their image processing on images and video streams.

This paper is arranged as follows: in the following section, the literature and background will be briefly reviewed showing a number of famous inspection robots and comparing them. In section 3, the design of the new system is considered in view of the system requirements and the motivation for the design parameters and selections are discussed. In section 4, experiment with the real robot is presented and the GUI design for easy operation by users is presented. Finally, conclusions are given in section 5 summarizing the obtained results.

Figure 1. MAKRO a) photograph, b) turning

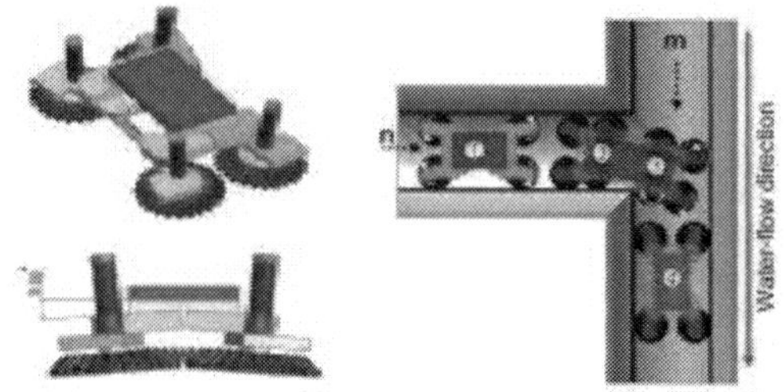

Figure 2. a) Kantaro structure and b) turning in bends

Figure 3. Versatrax system.

2 Background

MAKRO, [4] is an articulated inspection robot of worm-like shape and is designed for autonomous navigation in sewer pipes of 30 to 60 cm width. Its design, as shown in Fig. 1, consists of six segments connected by five motor-driven active joints, allows for simultaneously climbing a step and turning, at a junction consisting of a 600 mm pipe diameter and a branching 300 mm pipe diameter with equal top levels. MAKRO carries all the needed resources on-board and can run for two hours. It has to find and identify manholes and junctions; must be able to turn within manholes up to 90 degrees. The robot can 1) go to a specified point in the sewer system, 2) scan/inspect area and 3) get water sample.

Each joint has three infrared angle decoders and an odometer for the wheels. In each head four infrared sensors are used for distance measurement to the sides, up and down. Ultrasonic sensor detects obstacles in the frontal direction. This robot in our opinion is expensive and complexity is more than needed for the task complexity.

KANTARO, [1] is an autonomous, un-tethered robot, which fits to pipe diameter range of 200-300 mm. Kantaro [12] consists of a bottom box, including a passive-active mechanism and a battery pack, can be presented as a robot platform. The upper box contains the electronic board, the sensors and light connected via the main connector to the bottom box as shown in Fig.2. KANTARO has a passive active mechanism that can move through straight and curved

pipes without sensors or controllers. The design is based on passive adaptation of robot wheels to the bends in the pipe through combined wheels orientation and passive damping of springs.

Kantaro mechanism has four horizontal wheels connected to the four arms and each arm is jointed to the base plate, independently, and they are connected to each other by using four springs.

The laser scanner and fisheye camera are used for inspection and fault-navigation. Two IR sensors are installed in front of robot in two side of the camera to avoid driving the robot in a step down more than 10 cm and protect against obstacles. Rotating laser scanner has been developed and installed in the robot rear to scan radially the inner side of a pipe wall. Kantaro can detect the candidate faulty images on-board, using a fast and simple edge detection program. When robot detects a feature-image, an autonomous control program will start to reduce the robot speed for capturing more images, up to 5 frames, from feature image area. At this time, the laser scanner, that is normally switched off will start to recognize the feature faults as a landmarks. The captured images are transferred to an offline station to find faulty images and estimate their location.

The robot system is complicated because of the need to pass through different successive bends and changing pipe diameters. The case is not with such complexity in several other countries because of high housing density in Japan. It is simpler to develop robots that can work for specific and popular pipe sizes with straight configuration.

The Versatrex robot, shown in Fig.3, has a compact design and uses tracks to move straight in pipes.

3 Mechatronics System Design

The mechanism must enter into the internal pipe diameter 300 mm which is the constraint for robot outer dimensions. According to literature survey, 60% of sewage pipes have diameter of 300 mm.

Sewage pipelines are predicted to be full of sewerage content especially at the bottom. Consequently, the robot base should not touch the pipe bottom. We selected ready-made robot with tracked wheel and split the base of the robot. The two tracks equipped with motors are arranged to form part of two points passing through the arc of pipe with 300 mm diameter. The two tracks were linked by an aluminium base as shown in Fig 4. A casing had been made from acrylic to cover this base and to protect the sensor and computing devices from water or other hazardous materials, as shown in Fig.5. The final mechanical design is shown in Fig.6.

A front ultrasonic sensor is used to detect obstacles. A video camera module is also fitted on robot front to capture the video stream and report it to control module which archive the video stream for retrieval and process on board to detect abnormality. The camera is special to Rasberry PI and have five megapixels sensor array supporting resolution of 1080, 720 and VGA video modes. The camera is attached via a 15 cm ribbon cable to the CSI port on the raspberry module. This camera has high resolution and in same time small and appropriate to our robot. Raspberry pi camera module was found appropriate to our robot, since it can be used to take high-definition video, as well as stills photograph

In case abnormal images are detected then the camera is directed to capture high resolution images for further offline analysis. Encoders are fitted to the DC motors of the tracks to show the travelled distance. The quadrature encoder has resolution of 1000 state change per wheel cycle. Two dc motors with motor torque of 10 kg.cm are used. A lithium battery is also selected for two hours operation.

Raspberry PI was selected since compared to other options; it excels in handling video and multimedia data in general. Raspberry Pi was chosen for several reasons:

- Uses ARM processor with strong computing capability.
- Easy to program in scripting python.
- Low power consumption to increase mission time.
- Has optional memory module to enable storage of video and still images.
- Processing saved data.

The board have SD Card slot for data storage up to 32 G-Byte which is sufficient for expected two hour missions.

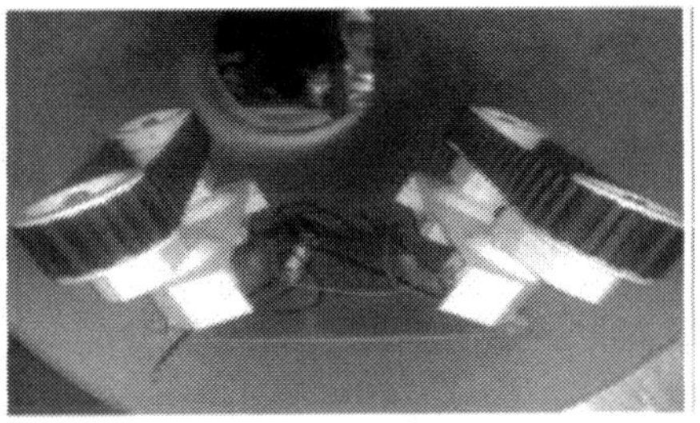

Figure 4. Educational Rover bottom view, it was split into two parts with horizontal cutting section to build the new inspection robot, lower picture shows new arrangement.

Figure 5. The final robot assembly.

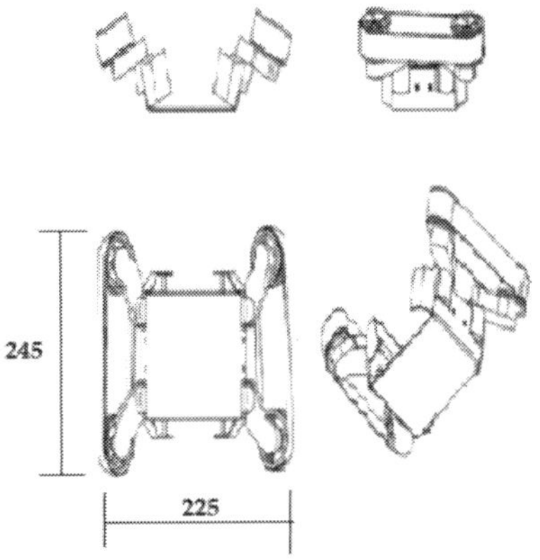

Figure 6. Mechanical Design of the proposed pipe inspection robot.

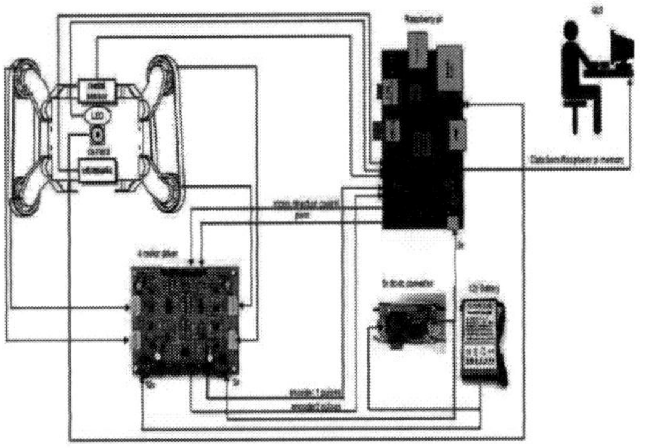

Figure 7. Schematic representation of the pipe inspection robotic system.

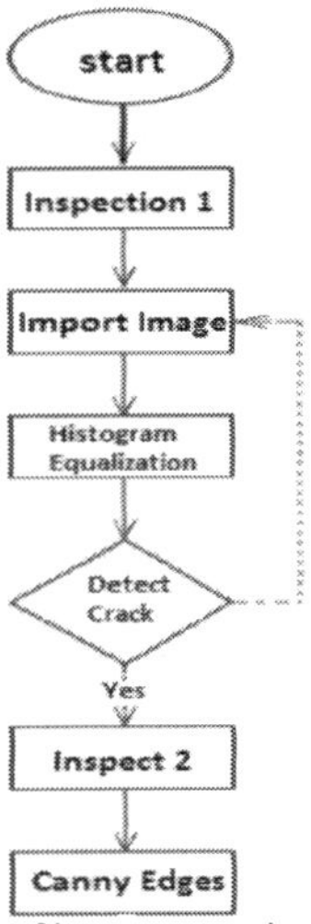

Figure 8. The flow chart of image processing operations.

4 Experiments

The robot is shown while tested to fit in pipe as in Fig.4. The complete system with assembled sensors and computing is shown in Fig.7.

The mechanical chassis of robot is assembled and motors tested to move and fit in pipe of 30 cm internal diameter. The Raspberry acts as the computing brain of robot and responsible for motion commands to track motors through an interface, then it reads signal from the motor encoders. The Raspberry also controls the operation of the camera to capture whether video streams or high resolution still images. The videos or images are stored in memory storage during autonomous operation of robot in pipes. After the robot mission in pipe is over, the robot is retrieved and the memory is used by an offline user with analysis program that can read the images, videos and location stamps of the visual data.

The robot was tested to move inside a test pipe section without defects. The external interface program allows the user to read both still images and videos. It contain image enhancement option by using the Histogram Equalization, HE, techniques. This had the effect of improving the image clarity for user especially for dim lighting. Other color enhancement techniques can be used as well [13] and planned in future.

Objective of robot is to provide the interface for retrieval of visual information collected during the mission of pipe inspection. The location of the robot is written in the title of saved images and videos so that for a specific mission, the robot location inside the pipe relative to pipe entrance location is known. The location information relies now on encoder information only. We intend to use a version of visual SLAM algorithm so that we can have another source for localization inside the pipe. Currently the encoder reading will have accumulated errors which increase with the length of robot mission.

The flow chart of image processing is shown in Fig.8. This program is executed online during robot mission. Program interface is shown in Fig. 9. It is used to enhance image (histogram equalization) and detect edges (Canny algorithm) (inspect 1 and 2).

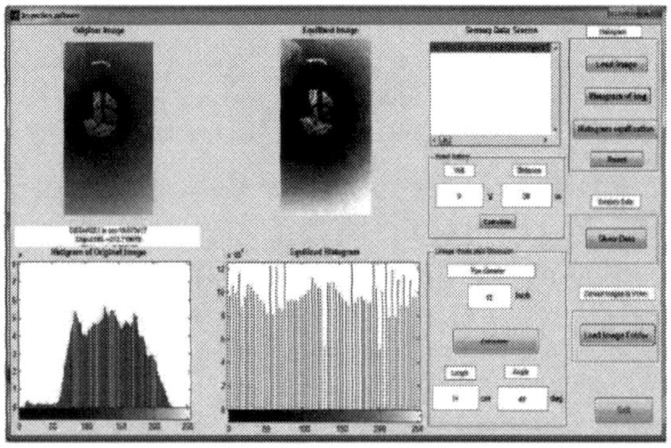

Figure 9. The user interface showing options to display original and enhanced images and showing histogram or other useful statistics (inspect 1 software).

Matlab was used to build a custom GUI. The robot captures live video during its motion in pipe and detects any defect automatically using edge detection algorithm. A flag is raised when detecting abnormality to save still high resolution image in defected areas. When the robot mission is over, the memory card is restored and attached to an offline computer. Paragon software is used to read files from memory and data is imported online by filezilla program.

The offline processing was made by custom developed software with all requirements of user in mind. The GUI shows the images and defect location and offer a variety of useful image processing operation to be customized by the user.

In Fig.10 is shown the original captured image with intentional crack in the pipe shown in a and in b is shown its enhanced version where the crack image have better contrast. The edge contour can be used to check any irregular pipe deformation when compared with a normal circle or ellipse with smooth trajectory. Further processing like this is left for other dedicated image processing software which already exists.

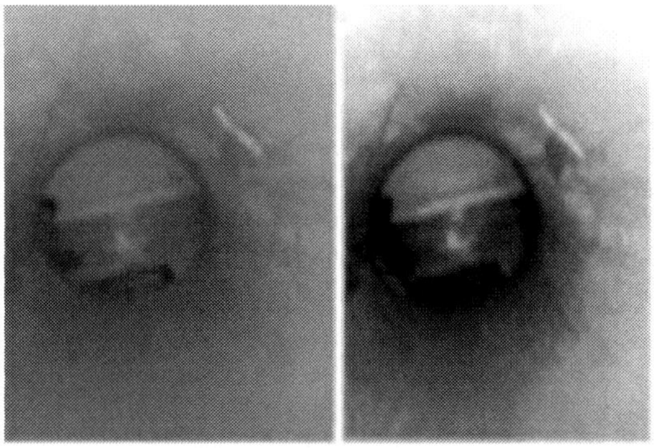

Figure 10. The pipe images showing a) a crack, b) Enhanced image.

5 Conclusions

Since inspection of pipelines is hard for a human inspector, an autonomous robot is needed for this task. We designed a new autonomous robot in consideration of pros and cons of the previous arts. Our mechatronics design is justified through comparison among best available modules and components. The control system is devised to ensure smooth operation of the robot. The new robot captures images and videos of the suspected cracks. It can also detect the location and orientation of the pipe defects between two certain manholes.

References

1. A.F. Nassiraei, Y. Kawamura, A. Ahrary, Y. Mikuriya, and K. Ishii, Concept and design of a fully autonomous sewer pipe inspection mobile robot "KANTARO", Proceedings of the IEEE International Conference on Robotics and Automation (ICRA'07), pp.136-143, 2007.
2. A. Ahrary, Y. Kawamura, and M. Ishikawa, A study of an autonomous mobile robot for a sewer inspection system, Journal of Artificial Life and Robotics, Vol.11, No.1, pp.23-27, January 2007.

3. H. Alireza, G. Hadi, and M. Hossein, Developing a sewer inspection robot through a mechatronics approach, International Conference on Mechanical, Automotive and Materials Engineering (ICMAME'2012), 2012 Dubai, pp.187-191.
4. E. Rome, J. Hertzberg, F. Kirchner, U. Licht, H. Streich, and TH. Christaller, Towards autonomous sewer robots: The MAKRO project, Journal of Urban Water, Vol.1, No.1, pp. 57-70, March 1999.
5. H. R. Choi and S. M. Ryew, Robotics system with active steering capability for internal inspection of urban gas pipelines, Mechatronics, vol.26, no.1, pp.105-112, 2002.
6. L. Hun-ok and O. Taku, Development of pipe inspection robot, ICROS-SICE International Joint Conference 2009.
7. K. Yun-Jong, Y. Kyung-Hyun and P. Young-Woo, Development of In-pipe robot for various sizes, ASME International Conference on Advanced Intelligent Mechatronics Suntec Convention and Exhibition Center Singapore, 2009.
8. K. U. Scholl, V. Kepplin, K. Berns and R. Dillmann, Controlling a multi-joint robot for autonomous sewer inspection, Proceedings of IEEE/RAS International Conference on Robotics and Automation, ICRA 2000, pp. 1701-1706.
9. R. Kirkham, P. D. Kearney and K. J. Rogers, PIRAT - A system for quantitative sewer assessment, The International Journal of Robotics Research, Vol.19, No.11, pp. 1033- 1053, 2000.
10. M. Moghaddam, A. Hadi and T. Abbasian, Design and Manufacturing SEWBOT: a Sewer Inspection Robot, 1st International Conference on Technical Inspection and NDT, October 2007, Tehran, Iran.
11. O. Tatar, D. Mandru and I. Ardelean, Development of mobile minirobots For in pipe inspection tasks, Mechanik, No..6(68) pp. 60-64, 2007.
12. S.G. Roh, S. M. Ryew, J. H. Yang and H. R. Choi, Actively steerable in pipe inspection robots for underground urban gas pipelines, IEEE International Conference on Robotics & Automation, Korea, 2001.
13. M. Abdellatif, Effect of color preprocessing on color based object detection, Proceeding of SICE annual conference, pp:1124-1129, 2008.

Structure Design and Dynamical Modelling of a Spherical Robot Driven by Omnidirectional Wheels

Yonghua Huang, Ganmin Zhu, Changsheng Wang and Hao Huang

School of Mechanical and Electrical Engineering, Guilin University of Electronic Technology, Guilin, China, 541004

Abstract. In this paper, our work focused on designing a novel spherical robot driven by omnidirectional wheels and developing an under-actuated dynamical model for the system. The robot consists of four parts: the running spherical shell (1P), the supporting platform (1P), the driving omnidirectional wheels (3P) and the bearing ball wheels (3P). Considering the noholonomic constrains aroused from the contacts of the shell with the omnidirectional wheels, and the shell with ground, we derived a dynamical model for the system by Chaplygin dynamics. The model illustrated that the robot was an under-actuated system of six DOF (degree of freedom) and three driving-torque inputs. finally, we performed an inverse dynamics simulations of S-curve trajectory to show the effectiveness of the model.

1 Introduction

Spherical robot is a kind of mobile mechanical system typically consisting of a rolling shell and an enveloped driving mechanism. Due to the unique structure, spherical robot has some promise use in the fields of entertainment, wild detection and military patrolling, etc.

Driving method (or actuating mechanism) is a critical problem in developing a spherical robot. In tradition, there are two means to drive a spherical robot.

The first one is the eccentric moment driving mean. In this mean, the actuating mechanism of the spherical robot drives the COG(center of gravity) of the system deviating from the vertical sphere center line, and hence the sphere shell could run on ground. For example, in [1], A.Halme proposed a spherical robot with a single wheel set on the bottom of the spherical shell. When the wheel rolls along the inner surface of the shell, he believed, the eccentric moment would break the system's balance, so the robot shell began running accordingly. In [2], Ranjan Mukhejee presented a spherical robot driven by four moving masses. Mukhejee gave the argument when the four masses move along their individual leadscrews on the sphere diameter, the COM(center of mass) of the robot system should be alternated and consequently the shell would begin to roll. There are similar structures being proposed in [3]-[4]. In [5], J Alves considered a four wheels vehicle to drive his spherical robot. Alves suggested that the running vehicle on the bottom of the shell could give birth of sufficient moment to drive the robot to run. In [6], Zhan exploited an omnidirectional steering wheel and incorporate with a longitudinal driving wheel to drive a spherical robot. Zhan addressed that the cooperating rotations of the two wheels could guarentee the shell to run in all direction. Similar to Zhan, in [7], Wang also discussed a spherical robot of two omnidirectional wheels on two orthogonally axis. In [8], Sun suggested a two DOF pendulum in his spherical robot. Sun argued that the pendulum could be driven to swing on all sides, thus the robot certainly might run in omnidirection. In [9], Zhao presented a spherical robot with two coaxial eccentric masses. Zhao believed when the two masses run with different acceleration and velocity, the robot should move all direction accordingly. The similar structure was also proposed in [10].

The second one is the angular momentum driving mean. In this mean, the spherical robot usually configures with a high speed rotating rotor inside the sphere shell. Because the external friction torque between the running shell and the ground is commonly small, with the principle of the angular momentum conservation, the sphere shell would rotate in the direction reverse to that of the running rotor. Furthermore, by changing the speed and direction of the rotating rotor, the omnidirectional motion of the spherical robot could be easily get. For instance, in [11], Toshiaki Otani developed a spherical robot driven by a three axis mechanical gyroscope. The robot's gyroscope was fixed on a gimbal inside the sphere, and when it rotated with high speed, he believed, there would be a torque that can drive the sphere shell. Likewise, in [12], V.Joshi proposed a spherical robot driven by double rotors which shared the similar principle with the robot in [11].

As mentioned above, there are two kinds of driving means having been used in the existing spherical robot. However, comparing to the angular momentum driving mean(the latter), the spherical robot that use the eccentric moment driving mean(the former) has exhibited a lot of advantages, for instance, more flexibility of the steering, lower speed of the driving mechanism etc,. Specifically, the use of the technique of omnidirectional wheels in the driving mechanism of a spherical robot, just like which in [9] and [10], further highlights this point. As a conclusion, developing a novel structure of the driving mechanism system by utilizing the omnidirectional wheels become an attracting area in the field of the spherical robot research.

This paper deals with an innovating spherical robot driven by three inner omnidirectional wheels. The paper is arranged as follow: at first, we show the mechanical structure of the proposed spherical robot in detail; and next, we develop a dynamical model for the robot system by using Chaplygin equation; and at the end, we consider a S-curve trajectories of the spherical robot system and perform an inverse dynamics simulations to validate the effectiveness of the proposed model.

2 General scheme

In [9]~[10], the omnidirectional wheel were used in the spherical robots successfully. However, their design one wheel contacting with the shell, which might weaken the driving capability of the system, so we use three omni-directional wheels to improve our system's performance.

2.1 Mechanical structure

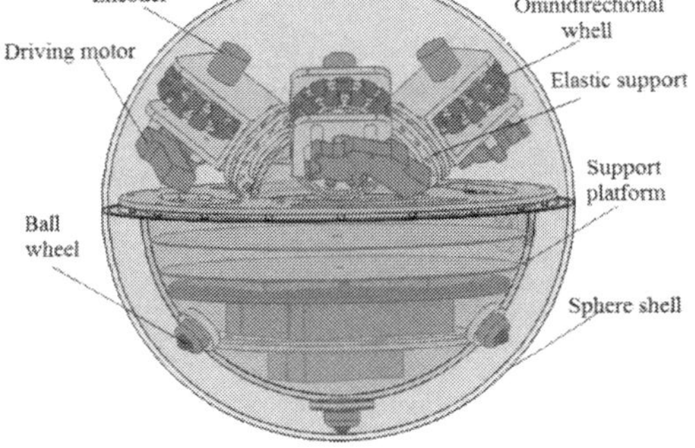

Figure 1. Virtual prototype of our spherical robot.

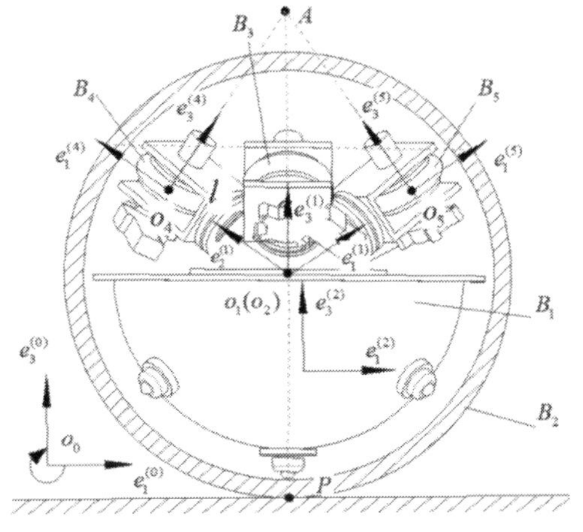

Figure 2. The schematic diagram of our spherical robot.

Our spherical robot approximately consists of two parts: a sphere shell and a support platform. The structure of the system is shown in Fig. 1~Fig. 2.

As it is seen in Fig 1~Fig 2, the three omnidirectional wheels are symmetrically assigned on the upper half part of the inner platform. Additionally, it is noteworthy that, for preventing the three wheels from leaving the internal surface of the sphere shell, we exploit an elastic support for each of the omnidirectional wheel.

Moreover, for ensuring the stability of our system, we exploit three free-running ball wheels on the bottom of the shell to bear the platform. Similarly, the ball wheels are also mounted on the platform via an elastic support.

As for the actuators, three thin motors are used to drive the three omnidirectional wheels to rotate relative to the inner sphere support platform.

2.2 Working principle

The working principle of the proposed spherical robot can be summarized as follow:

The motors drive the three omnidirectional wheels to run with a given law. Simultaneously, the running wheels would synthesize a couple to propel the shell to roll via the contacting friction. And hence, we can conclude if we regulate the rotation speed and direction of the wheels, we could drive the sphere shell in all direction.

After the omnidirectional wheels being driven, because the three wheels are all mounted on the inner support platform, there should be a reaction couple acting on the platform by the omnidirectional wheels. With the bearing action of the three free running ball wheels, the reaction couple would drive the support platform to rotate about the geometric center of the shell.

If we configure the COG of the support platform in different position (that can be done by putting auxiliary weights on the platform), the support platform could work either as a rotor of high speed rotating or a pendulum of low speed swinging.

3 Dynamical model

In this section, we will develop a dynamical model for the robot system by using Chaplygin equation.

3.1 Coordination settings

We represent the supporting platform as B_1, and the sphere shell as B_2, and the omnidirectional wheels as B_i ($i = 3,4,5$). The coordinates of the robot are set as:

- $O - \boldsymbol{e}_1\boldsymbol{e}_2\boldsymbol{e}_3$ $\{0\}$ is the global coordinate system fixed on the ground;
- $O_1 - \boldsymbol{e}_1^{(1)}\boldsymbol{e}_2^{(1)}\boldsymbol{e}_3^{(1)}$ $\{1\}$ is the coordinate system of the inner supporting platform B_1 and its origin is the geometric center of the sphere shell B_2;
- $O_2 - \boldsymbol{e}_1^{(2)}\boldsymbol{e}_2^{(2)}\boldsymbol{e}_3^{(2)}$ $\{2\}$ is the coordinate system of the shell B_2 and its origin is the same as $\{1\}$

- $O_i - e_1^{(i)} e_2^{(i)} e_3^{(i)}$ $\{i\}$ ($i = 3,4,5$) is the coordinate system of the omnidirectional wheel B_i and the origin locates at its geometric center.

3.2 Noholonomic constrains

We suppose the robot runs on a horizontal plane, then the angular velocity of B_1 can be given as:

$$\boldsymbol{\omega}_{B1}^{(1)} = (c_3\dot{q}_2 - c_2 s_3 \dot{q}_1)\boldsymbol{e}_1^{(1)} + (s_2\dot{q}_1 + \dot{q}_3)\boldsymbol{e}_2^{(1)} + (c_2 c_3 \dot{q}_1 + s_3 \dot{q}_2)\boldsymbol{e}_3^{(1)} \quad (1)$$

where $\boldsymbol{e}_i^{(j)}$ ($i = 1,2,3$, $j = 1,2,\cdots$) is the ith base vector of coordinate $\{j\}$. $s_i = \sin(q_i)$, $c_i = \cos(q_i)$ ($i = 1,2,\cdots$); $\dot{q}_i$ ($i = 1,2,3$) is the ith Eular angular rate of B_1.

Similarly, the angular velocity of B_2 can be given as:

$$\boldsymbol{\omega}_{B2}^{(2)} = \dot{q}_5 \boldsymbol{e}_1^{(2)} + c_5 \dot{q}_6 \boldsymbol{e}_2^{(2)} + (\dot{q}_4 + s_5 \dot{q}_6)\boldsymbol{e}_3^{(2)} \quad (2)$$

where $\dot{q}_i$ ($i = 4,5,6$) is the $(i-3)$th angular rate of B_2.

Because B_i ($i = 3,4,5$) rotates about B_1, their angular velocity should be calculated as:

$$\boldsymbol{\omega}_{Bj}^{(j)} = {}^{j}\boldsymbol{R}_1 \bullet \boldsymbol{\omega}_{B1}^{(1)} + \dot{q}_{(j+4)} \boldsymbol{e}_3^{(j)} \quad (j = 3,4,5) \quad (3)\sim(5)$$

where ${}^{j}\boldsymbol{R}_i$ ($i, j = 1,2,\cdots$) denotes the rotation transform matric from the coordinate $\{i\}$ to $\{j\}$, and $\dot{q}_i$ ($i = 7,8,9$) denotes the angular rate of B_i ($i = 3,4,5$), respectively.

We assume the robot runs on a plane without slipping, so the velocity of the point P is straightforward satisfied:

$$\boldsymbol{v}_{o2}^{(0)} + ({}^{0}\boldsymbol{R}_2 \bullet \boldsymbol{\omega}_{B2}^{(2)}) \times \boldsymbol{R}^{(0)} = 0 \quad (6)$$

where $\boldsymbol{R}^{(0)}$ is the position vector of P in coordinate $\{0\}$.

If we assume $\boldsymbol{v}_{o2}^{(0)}$ have the form: $\boldsymbol{v}_{o2}^{(0)} \triangleq \dot{x}\boldsymbol{e}_1^{(0)} + \dot{y}\boldsymbol{e}_2^{(0)}$ and considered the first two items in Equ. (4), we would get two noholonomic constrains as:

$$\begin{cases} \dot{q}_5 = \dfrac{s_4\dot{x} - c_4\dot{y}}{l + r} \\ \dot{q}_6 = \dfrac{c_4\dot{x} + s_4\dot{y}}{(l + r)c_5} \end{cases} \quad (7)\sim(8)$$

where l is the distance between the geometer center of B_2 and the geometer center of B_i ($i = 3,4,5$).

Similar to the presumption of that the shell running on ground, here we also assume B_i ($i = 3,4,5$) all run on the inner surface of B_1 without slipping, then we will get the other noholonomic constrains:

$$\dot{q}_i = g_{(6i-5)}\dot{x} + g_{(6i-4)}\dot{y} + g_{(6i-3)}\dot{q}_4 + \sum_{j=3}^{5} g_{(6i-5+j)}\dot{q}_{(4+j)}$$
$$(i = 1,2,3) \quad (9)\sim(11)$$

In (7)~(9), $\dot{x}$, $\dot{y}$ denote the longitudinal and the lateral velocity of the geometric center of B_2, respectively; g_j ($j = 1,2,3,\cdots,18$) is the function of q_i ($i = 1,2,3,4,5$).

3.3 Velocities of the COM

We set the velocity of the geometric center of B_2 as:

$$\boldsymbol{v}_{C2}^{(0)} = \dot{x}\boldsymbol{e}_1^{(0)} + \dot{y}\boldsymbol{e}_2^{(0)} \quad (12)$$

Under the principle of the relative motion, we will get respectively the velocity of B_i ($i, j = 1,2,\cdots,5$) as:

$$\boldsymbol{v}_{C1}^{(1)} = {}^{1}\boldsymbol{R}_0 \bullet \boldsymbol{v}_{o2}^{(0)} + \boldsymbol{\omega}_{B1}^{(1)} \times \boldsymbol{l}_{C1}^{(1)} \quad (13)$$

$$\boldsymbol{v}_{C2}^{(0)} = \dot{x}\boldsymbol{e}_1^{(0)} + \dot{y}\boldsymbol{e}_2^{(0)} \quad (14)$$

$$\boldsymbol{v}_{Ck}^{(1)} = {}^{1}\boldsymbol{R}_0 \bullet \boldsymbol{v}_{o2}^{(0)} + \boldsymbol{\omega}_{B1}^{(1)} \times \left(\boldsymbol{l}_{ok}^{(1)} + {}^{1}\boldsymbol{R}_i \bullet \boldsymbol{l}_{Ci}^{(k)}\right) \ (k = 3,4,5) \quad (15)\sim(17)$$

where $\boldsymbol{l}_{Ci}^{(i)}$ ($i = 1,3,4,5$) is the position vector in coordinate $\{i\}$ from the center of B_2 to that of B_i; $\boldsymbol{l}_{oi}^{(1)}$ ($i = 3,4,5$) is the position vector in $\{1\}$ from the geometric center of B_2 to the geometric center of B_i ($i = 3,4,5$).

3.4 Kinetic energy

According to the derived $\boldsymbol{\omega}_{Bi}^{(j)}$ ($i, j = 1,2,\cdots,5$) in (1)~(5) and $\boldsymbol{v}_{Ci}^{(k)}$ ($i = 1,2,\cdots,5$; $k = 0,1$) in (10)~(14), we can calculate the system's kinetic energy as:

$$T = \frac{1}{2}\sum_{i=1}^{5}\left((\boldsymbol{\omega}_{Bi}^{(j)})^T \boldsymbol{J}_{Bi}(\boldsymbol{\omega}_{Bi}^{(j)}) + (\boldsymbol{v}_{Ci}^{(k)})^T M_{Bi}(\boldsymbol{v}_{Ci}^{(k)})\right) \quad (18)$$

where $\boldsymbol{J}_{Bi}$ ($\boldsymbol{M}_i$) ($i, j = 1,2,\cdots,5$) represents the inertial matric (mass matrix) of B_i ($i, j = 1,2,\cdots,5$), respectively.

By substituting the noholonomic constrains (5)~(9) into T, we will get another form of the kinetic energy $\tilde{T}$.

3.5 Generalized forces

The external force acting on our spherical robot consist of the driving torque of B_i ($i = 3,4,5$) and the gravity.

We consider the COM of B_1 and B_i ($i = 3,4,5$) in the coordinate $\{1\}$, which is represented as:

$$\boldsymbol{l}_C^{(1)} = \frac{1}{m_1 + \sum_{i=3}^{5} m_i}\left(m_1 \boldsymbol{l}_{C1}^{(1)} + \sum_{i=3}^{5} m_i (\boldsymbol{l}_{oi}^{(1)} + {}^{1}\boldsymbol{R}_i \bullet \boldsymbol{l}_{Ci}^{(i)})\right) \quad (19)$$

If we defined $h \triangleq {}^{1}\boldsymbol{R}_i \bullet \boldsymbol{l}_C^{(1)}[3]$ (where $(*)[3]$ is the 3rd item of the vector $*$), we will get h is the function of q_i ($i = 2,3$).

Moreover, if we investigate the first time derivative of h, then we will get:

$$\dot{h} = f_1\dot{q}_2 + f_2\dot{q}_3 \tag{20}$$

In (18), f_i ($i=1,2$) denotes the explicit function of q_i ($i=2,3$) and m_j ($j=1,3,4,5$).

With (18), we define $\boldsymbol{\tau}_g \triangleq \begin{pmatrix} 0 & \tau_{g2} & \tau_{g3} \end{pmatrix}^T$, in which

$$\tau_{gk} = \left(m_1 + \sum_{i=3}^{5} m_i \right) g \bullet f_{(k-1)} \quad k = 2,3 \tag{21}$$

Again, according to (7)~(9), we define a transform matric as follow:

$$\boldsymbol{R}_\tau \triangleq \begin{pmatrix} g_1 & g_2 & g_3 & g_4 & g_5 & g_6 \\ g_7 & g_8 & g_9 & g_{10} & g_{11} & g_{12} \\ g_{13} & g_{14} & g_{15} & g_{16} & g_{17} & g_{18} \end{pmatrix}^T \tag{22}$$

Consequently, with (19)and(20), the generalized forces of the system can be calculated as

$$\tilde{\boldsymbol{Q}} = \boldsymbol{Q} + \boldsymbol{R}_\tau \bullet \boldsymbol{\tau}_g \tag{23}$$

where $\boldsymbol{Q} = \begin{pmatrix} 0 & 0 & 0 & \tau_7 & \tau_8 & \tau_9 \end{pmatrix}^T$, τ_i ($i=7,8,9$) is the driving torque of B_j ($j=3,4,5$).

3.6 Dynamical model

Considering the following form of Chaplygin equation

$$\frac{d}{dt}\frac{\partial \tilde{T}}{\partial \dot{q}_\sigma} - \frac{\partial \tilde{T}}{\partial q_\sigma} - \sum_{\beta=1}^{\gamma} \frac{\partial T}{\partial \dot{q}_{\varepsilon+\beta}} \sum_{v=1}^{\varepsilon} \left(\frac{\partial B_{\varepsilon+\beta,\sigma}}{\partial q_v} - \frac{\partial B_{\varepsilon+\beta,v}}{\partial q_\sigma} \right) \dot{q}_v = \tilde{Q}_\sigma \tag{24}$$

where T is the kinetic energy and $\tilde{T}$ is the kinetic energy by substituting noholonomic constrains into T; $B_{\varepsilon+\beta,\sigma}$ is the σth coefficient of the βth noholonomic constrain; q_v and q_σ are the generalized coordinates of the system; ε and γ are the numbers of the independent generalized coordinates and the noholonomic constrains; $\tilde{Q}_\sigma$ is the σth generalized force of the system.

We can get the system's dynamics as

$$\boldsymbol{D}(\boldsymbol{q})\ddot{\boldsymbol{q}}_\theta + \boldsymbol{C}(\boldsymbol{q},\dot{\boldsymbol{q}}_\theta)\dot{\boldsymbol{q}}_\theta + \boldsymbol{G}(\boldsymbol{q}) = \boldsymbol{Q} \tag{25}$$

In (23), $\boldsymbol{D}(\boldsymbol{q}) \in \Re^{6\times6}$, $\boldsymbol{C}(\boldsymbol{q},\dot{\boldsymbol{q}}_\theta) \in \Re^{6\times6}$, and $\boldsymbol{G}(\boldsymbol{q}) \in \Re^{6\times1}$ denote the inertia, centripetal-Coriolis, and gravity terms; $\boldsymbol{q}_\theta$ and $\boldsymbol{q}$ are two kinds of generalized coordinates, which are defined as:

$$\boldsymbol{q}_\theta = \begin{pmatrix} x & y & q_4 & q_7 & q_8 & q_9 \end{pmatrix}^T,$$

$$\boldsymbol{q} = \begin{pmatrix} q_1 & q_2 & q_3 & q_4 & q_5 \end{pmatrix}^T.$$

Equation (23) indicates the spherical robot is explicitly an under-actuated system with six independent velocities, and for more detail:

The longitudinal displacementand lateral displacement (x, y) of the geometric center of B_1, and the yaw angle (q_4) of B_1 are all under-actuated;

There are totally three driving inputs in the three joints q_i ($i=7,8,9$), therefore we could regulate control-force inputs τ_i ($i=7,8,9$) of the three joints to control the trajectory (x, y, q_4) of the sphere shell.

4 Inverse dynamics simulations

In order to prove the reliability of the proposed model (see Equ. (25)), here we consider the derived dynamical model, an perform inverse dynamics of S-curve trajectory for the spherical robot.

Table 1 shows the physical parameters which would be used in the numerical simulation. Note that we obtain the parameters from the measurement of a virtual bicycle prototype in Solidworks.

Table 1. Physical parameters.

Physical meanings	Sym	value
Radius of the omnidirectional wheel (m)	r	0.06
Distance between the geometer center of the shell and the omnidirectional wheel (m)	l	0.25
Distance between the geometer center of the shell and the COG of the supporting platform(m)	L_1	0.20
Distance between geometer center of the omnidirectional wheel and the its COG (m)	L_2	0.25
Mass of the supporting platform (kg)	m_1	17.57
Mass of the sphere shell (kg)	m_2	2.87
Mass of the omnidirectional wheel (kg)	m_3	0.45
Inertia matrix of the supporting platform (kg • m²)	J_{B1}	$\begin{pmatrix} 0.78 & 0 & 0 \\ 0 & 0.78 & 0 \\ 0 & 0 & 0.78 \end{pmatrix}$
Inertia matrix of the sphere shell (kg • m²)	J_{B2}	$\begin{pmatrix} 0.13 & 0 & 0 \\ 0 & 0.13 & 0 \\ 0 & 0 & 0.13 \end{pmatrix}$
Inertia matrix of the omnidirectional wheel (kg • m²)	J_{B3}	$\begin{pmatrix} 0.001 & 0 & 0 \\ 0 & 0.001 & 0 \\ 0 & 0 & 0.002 \end{pmatrix}$

The simulation is perform with three steps.

- **Step 1: trajectory planning**

 https://doi.org/10.1051/matecconf/201815302003

We plan the motion trajectory of the geometric center of the sphere shell B_2:

$x(t) = A_1 t$; $\dot{x}(t) = A_1$; $\ddot{x}(t) = 0$;

$y(t) = A_2 e^{-\sigma t}\sin(\omega t + \varphi)$;

$\dot{y}(t) = A_2 e^{-\sigma t}(\omega\cos(\omega t + \varphi) - \sigma\sin(\omega t + \varphi))$;

$\ddot{y}(t) = A_2 e^{-\sigma t}\left((\sigma^2 - \omega^2)\sin(\omega t + \varphi) - 2\sigma\omega\cos(\omega t + \varphi)\right)$

$q_4(t) = 0$; $\dot{q}_4(t) = 0$; $\ddot{q}_4(t) = 0$;

where $A_1 = 0.2$, $A_2 = 0.1$, $\sigma = 0.01$, $\omega = 0.4\pi$, $\varphi = 0$.

■ **Step 2: motion state calculating**

We investigate the first three items of the dynamical model (Equ.(23))

$$\boldsymbol{D}_{u1}(\boldsymbol{q})\ddot{\boldsymbol{q}}_{\theta u} + \boldsymbol{D}_{a1}(\boldsymbol{q})\ddot{\boldsymbol{q}}_{\theta a} + \boldsymbol{F}_1(\boldsymbol{q},\dot{\boldsymbol{q}}_\theta) = \boldsymbol{0} \quad (26)$$

where $\ddot{\boldsymbol{q}}_{\theta u} = (\ddot{x} \quad \ddot{y} \quad \ddot{q}_4)^T$, $\ddot{\boldsymbol{q}}_{\theta a} = (\ddot{q}_7 \quad \ddot{q}_8 \quad \ddot{q}_9)^T$, and $\boldsymbol{F}_1(\boldsymbol{q},\dot{\boldsymbol{q}}_\theta) \in \Re^{3\times 1}$ denotes the first three items of the vector $(\boldsymbol{C}(\boldsymbol{q},\dot{\boldsymbol{q}}_\theta)\dot{\boldsymbol{q}}_\theta + \boldsymbol{G}(\boldsymbol{q}))$, and

$$\boldsymbol{D}_{u1}(\boldsymbol{q}) = \begin{pmatrix} D_{11} & D_{12} & D_{13} \\ D_{21} & D_{22} & D_{23} \\ D_{31} & D_{32} & D_{33} \end{pmatrix}, \boldsymbol{D}_{a1}(\boldsymbol{q}) = \begin{pmatrix} D_{14} & D_{15} & D_{16} \\ D_{24} & D_{25} & D_{26} \\ D_{34} & D_{35} & D_{36} \end{pmatrix},$$

and D_{ij} ($i,j = 1,2,\cdots,6$) is the corresponding component of $\boldsymbol{D}(\boldsymbol{q})$. Then we will get

$$\ddot{\boldsymbol{q}}_{\theta a} = -(\boldsymbol{D}_u(\boldsymbol{q}))^{-1}\boldsymbol{F}_1(\boldsymbol{q},\dot{\boldsymbol{q}}_\theta) \quad (27)$$

From (25), we can calculate angular rotational rate $\dot{\boldsymbol{q}}_{\theta a}$ by the numerical time integration.

Subsequently, with the derived $\dot{\boldsymbol{q}}_{\theta a}$, we further examine the (5)~(9), then we get $\dot{q}_i$ ($i = 1,2,3,5,6$).

Similarly, by integration of $\dot{q}_i$ ($i = 1,2,3,5,6$), we will obtain q_i, definitely.

■ **Step 3: driving torque calculating**

Similar to Step 2, we furtherly investigate the remain three items of the dynamical model(Equ.(23)), and then we will obtain:

$$\boldsymbol{\tau}_a = \boldsymbol{D}_{u2}(\boldsymbol{q})\ddot{\boldsymbol{q}}_{\theta u} + \boldsymbol{D}_{a2}(\boldsymbol{q})\ddot{\boldsymbol{q}}_{\theta a} + \boldsymbol{F}_2(\boldsymbol{q},\dot{\boldsymbol{q}}_\theta) \quad (28)$$

where $\boldsymbol{\tau}_a = (\tau_7 \quad \tau_8 \quad \tau_9)^T$, $\boldsymbol{F}_2(\boldsymbol{q},\dot{\boldsymbol{q}}_\theta) \in \Re^{3\times 1}$ denotes the later three items of the vector $(\boldsymbol{C}(\boldsymbol{q},\dot{\boldsymbol{q}}_\theta)\dot{\boldsymbol{q}}_\theta + \boldsymbol{G}(\boldsymbol{q}))$, and

$$\boldsymbol{D}_{u2}(\boldsymbol{q}) = \begin{pmatrix} D_{41} & D_{42} & D_{43} \\ D_{51} & D_{52} & D_{53} \\ D_{61} & D_{62} & D_{63} \end{pmatrix}, \boldsymbol{D}_{a2}(\boldsymbol{q}) = \begin{pmatrix} D_{44} & D_{45} & D_{46} \\ D_{54} & D_{55} & D_{56} \\ D_{64} & D_{65} & D_{66} \end{pmatrix}.$$

Figure 3 shows the desired trajectory of the inverse dynamics simulation.

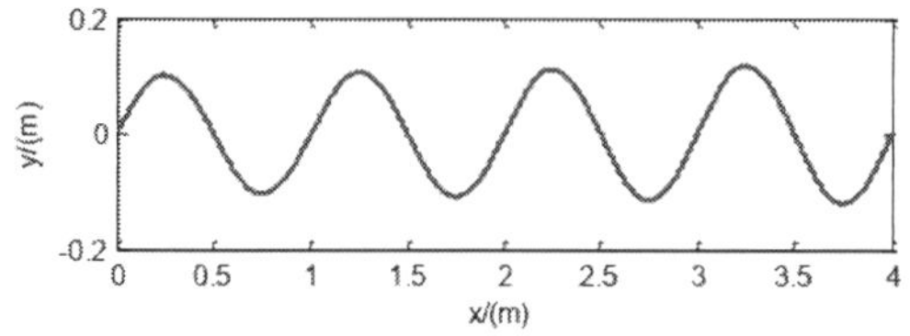

Figure 3. The planning trajectory of the COM of the shell.

Figure 4 demonstrates the driving angular velocity and angular acceleration of the omnidirectional wheels.

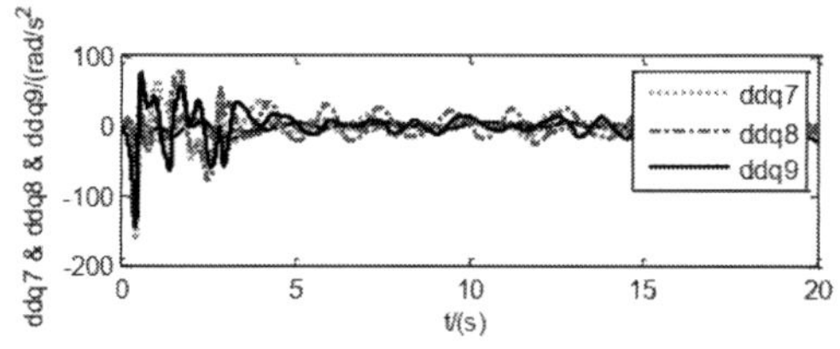

a) Angular acceleration

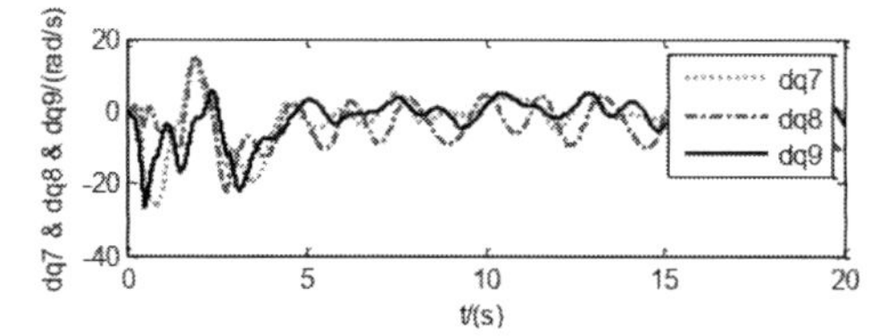

b) Angular velocity

Figure 4. The state of the omnidirectional wheel

Figure 5 shows the velocity constrains of the support platform and the sphere shell.

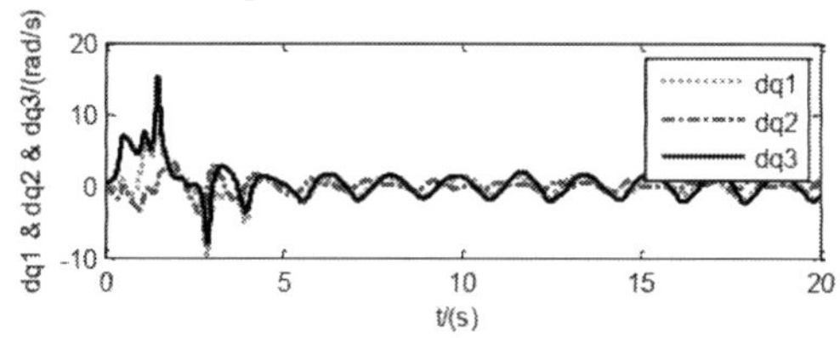

a) Angular rate of the platform

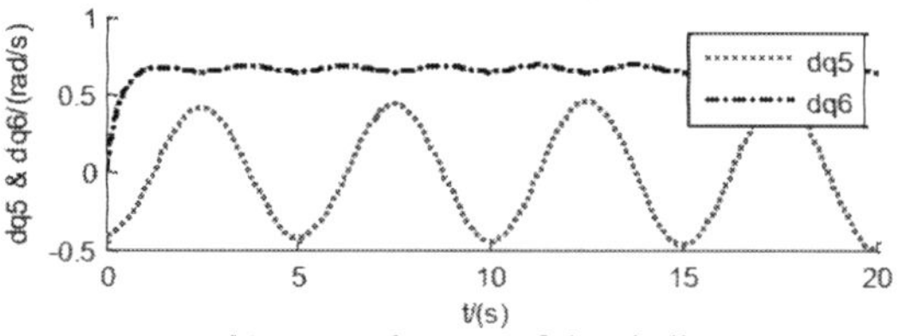

b) Angular rate of the shell

Figure 5. The noholonomic constrains.

Figure 6 shows the driving torques and the powers of the three omnidirectional wheels.

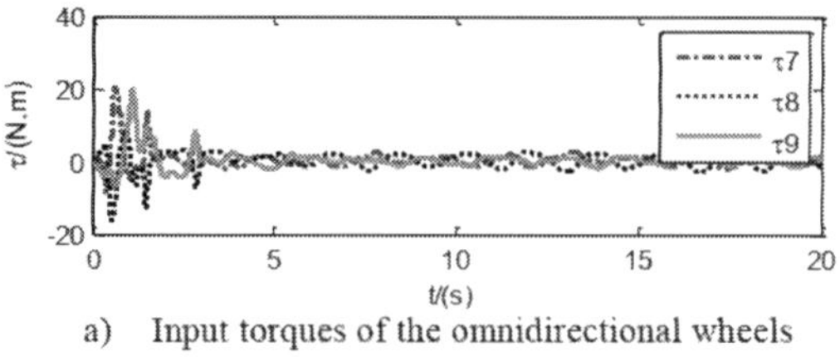

a) Input torques of the omnidirectional wheels

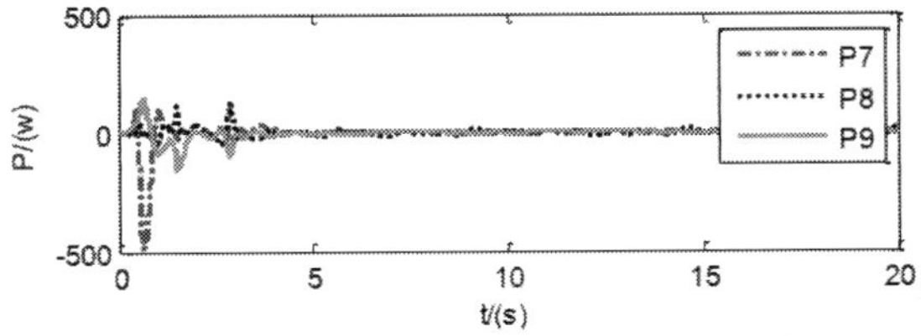

b) Input power of the omnidirectional wheels

Figure 6. The torques and the powers of the omnidirectional wheel.

As a conclusion, the results show that the maximum driving torque of the three omnidirectional wheels is less than 20N.m(see Fig 6(a)), and the maximum input power is less than 500W(see Fig 6(b)). The results are useful in the selection of the driving motor.

Moreover, considering the results in Fig 3~Fig 6, we investigate the comparison between the kinetic energy of the system and the work of the external forces.

Figure 7 examines two kinds of kinetic energy: T1 is get by the current velocity and T2 is by the previous velocity and the elementary work.

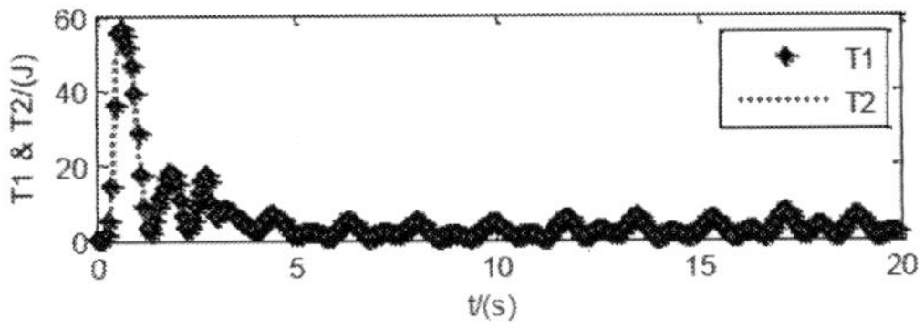

Figure 7. The comparison of two kinds of kinetic energy.

Figure 8 shows the difference between the increment of the kinetic energy and the elementary work.

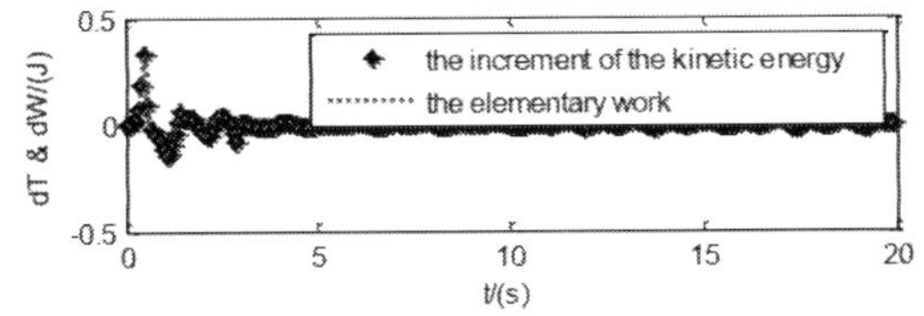

a) The kinetic energy and the elementary work

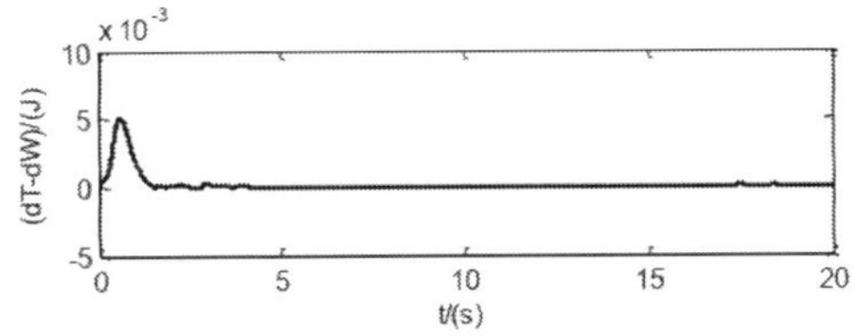

b) The difference of the increment of the kinetic energy and the elementary work

Figure 8. The kinetic energy Vs the elementary work.

As seen in Fig 7~Fig 8, while calculated by use of different variable, the two kinetic energy are coincident, and concretely, their difference is less than 10^{-2} (see Fig 8(a)). The results show the fact that our dynamical model (Equ. (23)) completely obey the law of conservation of energy, thus to a certain extent, we verify reliability of our derivation.

5 Conclusions and future work

As a summary, one of the contribution of this research is that we proposed a novel spherical robot that can be driven by three omnidirectional wheels inside a sphere shell. Another contribution of the research might be that we explored the dynamical model for the system. Our derivation illustrates that the spherical robot is explicitly a noholonomic and under-actuated system, which consist of six independent velocities and three control-torque inputs. With the inverse dynamics simulations of an S-curve trajectory, we validated that the feasibility of our structure scheme of the robot system and the reliability of our dynamical model.

However, currently in our research, there definitely is lack of physical porotype experiments to provide further support for our theoretical analysis, therefore our next work should concentrate on the development a principle prototype so as to perform the presented motion trajectory of the robot system.

Acknowledgement

This work was financial supported by National Natural Science Foundation of China (Grant No.61365012 && No.51305087&&No.51765011) and Director Foundation of Guangxi Key Laboratory of Manufacturing Systems and Advanced Manufacturing Technology (Grant No. 14-045-15- 003Z).

References

1. Halme A, Schonberg T, Wang Y. Motion control of a spherical mobile robot[C]. *In Proceedings of the 4th International Workshop on Advanced Motion Control*. 1996: 259-264.
2. Mukherjee R, Minor M A. A simple motion planner for a spherical mobile robot[C]. *Proceedings of the 1999 IEEE International Conference on Advanced Intelligent Mechatronics*. 1999: 896-901.
3. Amir H J A, Mojabi P. Introducing Glory: A novel strategy for an omnidirectional spherical rolling robot[J]. Journal of Dynamic Systems Measurement & Control, 2004, **126**(3): 678-683.
4. LI Tuan-jie, SU Li, ZHANG Yan. Design and Analysis of a Spherical Omnidirectional Rolling Robot Driven by Linear Motors[J]. Machine Design and Research, 2006, **22**(4): 46-48.
5. Alves J, Dias J. Design and control of a spherical mobile robot[J]. *Proceedings of the Institution of Mechanical Engineers Part I Journal of Systems & Control Engineering*, 2003, **217**(6): 457-467.
6. Zhan Q, Cai Y, Yan C. Design, analysis and experiments of an omin-directional spherical robot [C]. *IEEE International Conference on Robotics and Autmation*. 2011: 4921-4926.
7. Wang Kejun. Mechanism Analysis and Research of Spherical Robot Driven by Omnidirectional Wheel [D]. Beijing Jiaotong University, 2017.
8. SUN Hanxu, WANG Liangqing, JIA Qingxuan, LIU Daliang. Dynamic Model of the BYQ-3 Spherical

Robot [J]. Journal of Mechanical Engineering, 2009, **45** (10): 8-14.

9. ZHAO Bo, LI Man-tian, SUN Li-nin. Turning in place motion control of two pendulums driven spherical robot[J]. Journal of Harbin Institute of Technology, 2011, **43**(11): 49-53.
10. Yoon J C, Ahn S S, Lee Y J. Spherical robot with new type of two-pendulum driving mechanics[C]. *IEEE International Conference on Intelligent Engineering Systems*. 2011: 275-279.
11. Otani T, Urakubo T, Maekawa S, et el. Positional and attitude control of a spherical rolling robot equipped with a gyro[C]. *IEEE International Workshop on Advance Motion Control*. 2006: 416-421.
12. Joshi V A, Banavar R N, Hippalgaonkar R. Design and analysis of a spherical mobile robot[J]. Mechanism & Machine Theory, 2010, **45**(5): 130-136.

 https://doi.org/10.1051/matecconf/201815302004

A review on humanoid robotics in healthcare

Azeta Joseph, Bolu Christian, Abioye A. Abiodun and Festus Oyawale

[1]*Department of Mechanical Engineering, Covenant University, Ota, Nigeria.*

Abstract. Humanoid robots have evolved over the years and today it is in many different areas of applications, from homecare to social care and healthcare robotics. This paper deals with a brief overview of the current and potential applications of humanoid robotics in healthcare settings. We present a comprehensive contextualization of humanoid robots in healthcare by identifying and characterizing active research activities on humanoid robot that can work interactively and effectively with humans so as to fill some identified gaps in current healthcare deficiency.

1 Introduction

The use of robots for surgery has given rise to a large number of new methods which has led to the emergence of a wide variety of robots for use in the medical domain. [1] Points out numerous research areas in the application of robotics in healthcare settings. Such as robotic surgical Systems [2, 3], laparoscopy surgery and tele-rounding robots [4], robot-assisted rehabilitation [5-7], caregiver and patient's assistants [8-10], robotic applications in dentistry, bio-prosthetic [11]. The list continues as the technology advances and even more applications of robot in healthcare could be envisaged.

Humanoid robots that could be used to remotely carry out tasks are also very important. Such robots are endowed with human capabilities to assist caregivers and patients particularly in contaminated environments. These remotely operated robots are to possess certain characteristics for human robot interaction. However, tele-operated semi-autonomous robots can be used to perform assisted healthcare tasks during outbreaks which would reduce the time personnel need to spend in dangerous contaminated areas while putting on their personal protective equipment in high temperature and humid conditions particularly in the West African region. Furthermore [12] highlighted some healthcare robots that have been developed for similar fields of application [13, 12]. The robot Cody is able to wash human limbs autonomously and an evaluation of the patient's system showed that acceptance strongly depends on the interpretation of robot intention [14]. Other robots serve as rehabilitation [15], the nurse assisting lifting of patients [16] or even blood sampling [17].

While research on humanoid robots for general or supportive patients care particularly for disease containment has hit the ground running in other developed countries due to the recent outbreak of Ebola virus disease that shook the world it is also important for those affected directly to find a solution to their problem. We propose a broad overview of some of the current and potential applications of robotics in health care settings and we carefully selected some robots for patients care.

2 Literature Review

Table 1. Summary of different application of robot for general and supportive care [13].

Author	Robot	Description
[18]	ARMAR III	Developed to support tasks in 'human-centered' environments to include households. Upper torso is humanoid in design. Designed to interact with humans and manipulate objects in the environment.
[19]	Care-O-Bot 3	Mobile robot assistant designed to help humans in everyday environments. Includes an arm and gripper for manipulation of objects, a tray for carrying and transferring objects, and a flexible torso enabling butler-like gestures such as bowing and nodding.
[20]	Cody	Robot assistant developed to help caregivers with patient hygiene, specifically bed baths. Uses a compliant arm and gentle force to perform 'wiping motions' similar to those used during bed baths.
[21]	PR2	Mobile humanoid robot design to support tasks in human environments. Capable

 https://doi.org/10.1051/matecconf/201815302004

		of manipulating objects and performing tasks such as setting a table and emptying a dishwasher.
[9]	RIBA	High load capacity robot designed to lift and transfer patients.
[22]	RIBA	Designed in appearance similar to a giant teddy bear. Carries patients in 'human-type' arms. Equipped with tactile sensors to detect patient's contact position, allowing motion adjustment to be made accordingly.
[16]	Robotic Nursing Assistant	Designed to help caregivers with physically demanding tasks such as lifting and transferring patients. Upper torso is humanoid in design.
[23]	Hair-Washing Robot	Robot assistant designed to help caregivers wash patient's hair.
[24]	ASIMO	Human-size humanoid robot. Capabilities include walking, running, and climbing stairs, carrying objects, opening doors, and pushing carts.
[25]	ROSE	Tele-operated robot for home care applications. Capable of performing tasks such as grabbing and placing objects, opening doors, and cleaning.

3 Characteristics of Humanoid robot

3.1 Vision system

Visual perception is fundamental to most robotics systems working in human environments, it is also essential to a wide variety of tasks such as manipulation, tracking, human–robot interaction [26]. [27] Developed a simple and affordable vision-based robotic system for the identification of the Euclidean position of red spheres that emulate ripe tomatoes. This is done by using a RGB-D sensor in a fixed position, together with a 5 DOF manipulator. Robotics should follow the condition and perceive the surroundings to achieve the goal. Due to perception, robots check their state of joints by using encoders or sensors. Robots developed nowadays are covered with force-sensitive skins. Many humanoid robots use senses like laser rangefinders or ultrasonic distance sensors. They also use cameras, to enable them to focus their attention towards specific objects and screens to interpret the images.

3.2 Manipulation tasks

Humans have the ability to move their hands almost 30 DOF [28]. Some of the humanoid robots that were built cannot hold the unknown objects from the environment like humans. The reason behind this is lack of learning ability to perceive and sense the new objects. Improvements were needed so that they could hold the things and sense them by touching [28]. Also learning from demonstration can be used to train soft robotic hands to perform dexterous manipulation tasks. The ReMeDi robot led to the design of a light cable-driven manipulator [29] which then advanced to the construction of a large SCARA type manipulator with 6 DOF which is used for palpation [12].

3.3 Sensing behavior

There are different types of sensors used in humanoid robots. But [28] focused on specific on sensors used to grasp the objects means arm's sensors. Sometimes arms are small so according to them, sensors must be small according to the joints. For loop gesture of hand, there should be a measurement of every joint. [28] Highlighted that from the many approaches that were used for this [30] gave concept of using plastic potentiometer at ever finger's joints. Another technique by [31] gave that linear potentiometers at wrist joint to calculate the angles of joints. But this idea was flopped due to lack of precision cause of friction of tendons. Most useful idea was given by [32].

3.4 Mobile Platform

Humanoid robots have to be flexible enough for easy maneuvering. While fixed robots will always find a place in manufacturing, humanoid robots with mobile base promises additional flexibility to end-user in new applications. These applications include caregiver and patients assistance, medical and surgical uses as well as security. [12] Propose two types of motion for the mobile base, the short distance motion which cater for the robots positioning next to the patient to obtain an appropriate position area for the robot arm and a long distance motion which involves movement between the robot parking lot and the examination room.

4 Applications of Humanoid Robotics in Healthcare

4.1 Tele-Healthcare

A Tele-operated Service Robot is a robot that is controlled by a human being from a distance and performs tasks typically in uncontrolled environments [25]. Tele-operation enables an operator, to act remotely as if the operator was on the spot, by for instance copying the manipulations of the operator at a distance. An example is the Da Vinci Robot used for medical surgery [25].

The use of tele-operated humanoid robots in healthcare represents an exciting opportunity to help doctors, nurses and patients to ease the high risk of infectious disease transfer [33]. [34] Investigated the use of teleoperated robots to allow healthcare workers to perform some of their duties at a safe distance from the infected patients. The result suggest that patients will trust the robot less when the operation is unseen than when the operation is visible. More investigation needs to be carried out to know the extent of the effect. However, the design of the Tele-operated Service Robot [25] can be adapted for the development of a healthcare robot for the general task of supportive patients care.

4.2 Humanoid Robot for Pain Relief

Humanoid robots are used to implement techno-psychological distraction for children in order to reduce their pain as a result of stress and anxiety during a medical procedure. Studies carried out by [35-37] all point to the fact that children are more likely to smile when encountering medical procedures with a robot. The aim is to turn children's attention away from the pain of the needle toward an amusing activity. [38, 37]. States the principles of attentional capacity theory, that the distraction stimulus must be stronger than the pain stimulus to gain the child's attention. Although, music and cartoons have shown effectiveness in reducing pain and anxiety among children undergoing a variety of medical procedures [39, 40, 36]. It would appear that these distractions are not always strong enough to turn children's attention away from the pain. It is now believed that multisensory strategies, which combine visual, auditory, and tactile senses, may have a greater impact on pain than single-sensory strategies [39, 36]. Given the mixed results mentioned above, it stands to reason that stronger and more engaging forms of distraction, which invite the child to engage in an activity, may be necessary for medical procedures.

In addition to other interventions, we suggests that techno-psychological distraction for children for management of pediatric pain be considered. When programmed with humanistic characteristics and to execute psychological strategies, a humanoid robot shows promise of reducing procedural pain and distress in children [37],

4.3 Humanoid Robot for Aging Population

In a society where there is a rise in the disabled and aging population, there is a strong demand for robotics to tackle problems that arise from their inabilities to relate effectively with their environment. Robots for executing patient-transfer tasks are needed in nursing care facilities and hospitals, [9] developed a new prototype robot named RIBA with human-type arms which was designed to perform heavy physical tasks requiring human contact, the robot was able to transfer a human from a bed to a wheelchair and back. The caregiver can intuitively give instructions to RIBA through tactile sensors using a newly proposed method named tactile guidance. RIBA was developed to cope the difficulties encountered previously by RI-MAN [41, 42].

5 Conclusion

More interest are now being directed towords robotics in healthcare because of the successes robots can bring as it has been in other domain. Still, we have not seemed to have been able to appreciably utilize this technology in developing countries for effective healthcare delivery. Healthcare needs a pathfinder solution in which the investment and risk of robotics applications are small. However, There are a number of exciting advances in robotics in recent years, which points to a fruitful future. Humanoid robotics in healthcare settings is rapidly evolving. We examined the state-of-the-art of the emerging field of the current and potential applications of robotics in healthcare.

References

1. Krzysztof Arent, Mateusz Cholewiński, Łukasz Chojnacki, Wojciech Domski, Micha łDrwięga, Janusz Jakubiak, Mariusz Janiak, Bogdan Kreczmer, Adam Kurnicki, Bartłomiej Stańczyk, D. S.-S. Selected Topics in Design And Application of a Robot For Remote medical Examination With use Of Ultrasonography And Auscultation from The Perspective of The Remedi Project. *Journal of Automation, Mobile Robotics & Intelligent Systems*, VOLUME**11**(N◦ 2). (2017).
2. Z. Nawrat. *Postępy robotyki. Przemysłowe i medyczne systemy robotyczne*, chapter Polski robot kardiochirurgiczny, 275–284. Wydawnictwa Komunikacji i Łączności, (2005).
3. "The da Vinci Surgical System, the manufacturer website". http://www.intuitivesurgical.
4. M. Iftikhar, M. J. Majid, M. Muralindran, G. Thayabaren, R. Vigneswaran, and T. T. K. Brendan, "Otorob: Robot for orthopaedic surgeon roboscope: Non-interventional medical robot for telerounding". In: *Proceedings of the 5th International Conference on Bioinformatics and Biomedical Engineering*, 1 – 5. (2011).
5. S. Balasubramanian, J. Klein, and E. Burdet. Robot-assisted rehabilitation of hand function. Curr Opin Neurol, (2010).
6. T. Nef, M. Mihelj, and R. Riener, "Armin: a robot for patient-cooperative arm therapy", *Medical and Biological Engineering and Computing*, vol. **45**, 887–900. (2007).
7. M. Kaczmarski and G. Granosik, "Rehabilitation robot rrh1", *The Archive of mechanical Engineering*, vol. **58**, no. 1, (2011).
8. M. Mariappan, T. Ganesan, V. Ramu, and M. Iftikhar. Intelligent *Robotics and Applications*, chapter Safety System and Navigation for Orthopaedic Robot (OTOROB), 358 – 367. Springer, (2011).
9. T. Mukai, S. Hirano, H. Nakashima, Y. Kato, Y. Sakaida, S. Guo, and S. Hosoe, "Development of a nursing-care assistant robot riba that can lift a human in its arms". In: *Proceedings of the IEEE/RSJ International Conference on Intelligent Robots and Systems*, (2010).

10. C. Urdiales, M. Fernandez-Carmona, J. M. Peula, R. Annicchiaricco, F. Sandoval, & C. Caltagirone. *Problemy robotyki*, chapter A collaborative control scheme for haptics-based blind wheelchair driving. Oficyna Wydawnicza Politechniki Warszawskiej, (2010).
11. J. T. Belter, J. L. Segil, A. M. Dollar, and R. F. Weir, "Mechanical design and performance specifications of anthropomorphic prosthetic hands: A review", *Journal of Rehabilitation Research & Development*, vol. **50**, no. 5, 599–618. (2013).
12. Panzirsch, M., Weber, B., Rubio, L., Coloma, S., Ferre, M., & Artigas, J. Tele-healthcare with humanoid robots: A user study on the evaluation of force feedback effects. *2017 IEEE World Haptics Conference (WHC)*, (June), 245–250. (2017). https://doi.org/10.1109/WHC.2017.7989909
13. K. Doelling, J. Shin, and D. O. Popa, "Service robotics for the home: a state of the art review,"in Int. Conf. on Pervasive Technologies Related to Assistive Environments. ACM, p. 35. (2014).
14. T. L. Chen, C.-H. A. King, A. L. Thomaz, and C. C. Kemp, "An investigation of responses to robot-initiated touch in a nursing context, " Int. Journal of Social Robotics, vol. **6**, no. 1, pp. 141–161, (2014).
15. Barrett Medical, "Proficio," http://www.barrettmedical.com/, accessed: 2016-02-19.
16. J. Hu, A. Edsinger, Y.-J. Lim, N. Donaldson, M. Solano, A. Solochek, and R. Marchessault, "An advanced medical robotic system augmenting healthcare capabilities-robotic nursing assistant," in Int. Conf. on Robotics and Automation. IEEE, pp. 6264–6269. (2011).
17. A. I. Chen, M. L. Balter, T. J. Maguire, and M. L. Yarmush, "Realtime needle steering in response to rolling vein deformation by a 9dof image-guided autonomous venipuncture robot," in Int. Conf. on Intelligent Robots and Systems. IEEE, pp. 2633–2638. (2015).
18. Asfour, T., et al. ARMAR-III: An integrated humanoid platform for sensory-motor control. *IEEE-RAS International Conference on Humanoid Robots.* (Genova, Italy, Dec. 4–6), 169–175. (2006).
19. Reiser, U., et al. Care-O-bot® 3 - creating a product vision for service robot applications by integrating design and technology. *IEEE/RSJ International Conference on Intelligent Robots and Systems.* (St. Louis, USA, Oct. 11-15), 1992-1998. (2009).
20. King, C.-H., et al. Towards an assistive robot that autonomously performs bed baths for patient hygiene. *IEEE/RSJ International Conference on Intelligent Robots and Systems.* (Taipei, Taiwan, Oct. 18-22), 319–324. (2010).
21. Cousins, S., ROS on the PR2. *IEEE Robotics & Automation Magazine*. **1**, 3 23-25. (Sept 2010).
22. Mukai, T., et al.. Tactile-based motion adjustment for the nursing-care assistant robot RIBA. *IEEE International Conference on Robotics and Automation.* (Shanghai, China, May 9-13), 5435 – 5441. (2011).
23. Hirose, T., et al. Development of hair-washing robot equipped with scrubbing fingers. *IEEE International Conference on Robotics and Automation.* (Saint Paul, USA, May 14-18, 2012), 1970-1975. (2012).
24. American Honda Motor Co. Inc. History of ASIMO. Retrieved Jan. 10, (2014) from http://asimo.honda.com/asimo-history/.
25. Osch, M. V., Bera, D., Hee, K. V., Koks, Y., & Zeegers, H. Tele-operated service robots:ROSE. *Automation in Construction* **39**, 152–160. (2014).
26. Sean Ryan Fanello, Carlo Ciliberto, Nicoletta Noceti, Giorgio Metta, F. O. Visual recognition for humanoid robots. *Robotics and Autonomous Systems*, *91*, 151–168. (2017). https://doi.org/10.1016/j.robot.2016.10.001
27. García-Luna, F., & Morales-Díaz, A. Towards an artificial vision-robotic system for tomato identification. *IFAC-PapersOnLine*, **49**(16), 365–370. (2016). https://doi.org/10.1016/j.ifacol.2016.10.067
28. Rabbia Mahum *, Faisal Shafique Butt, Kashif Ayyub, Seema Islam, Marriam Nawaz, D. A. International Journal of Advanced and Applied Sciences. *International Journal of Advanced and Applied Sciences*, **4**(2), 83–90. (2017). Retrieved from http://science-gate.com/IJAAS/Articles/2016-3-12/14 2016-3-12-pp.106-112.pdf
29. A. Kurnicki, B. Stanczyk, and B. Kania, "Manipulator development for telediagnostics". In: *Proceedings of the International Conference on Mechatronics and Robotics, Structural Analysis (MEROSTA2014)*, 214 – 218. (2014).
30. Butterfass J, Grebenstein M, Liu H, and Hirzinger G. DLR-Hand II: Next generation of a dextrous robot hand. In the IEEE International Conference on Robotics and Automation (ICRA), IEEE, Seoul, South Korea, **1**: 109-114. (2001). https://doi.org/10.1109/ROBOT.2001.932538
31. Chua PY, Caldwell DG, Bezdicek M, Gray JO, and Davis S. Tele-operated high speed anthropomorphic dextrous hands with object shape and texture identification. In the IEEE/RSJ International Conference on Intelligent Robots and Systems, IEEE, Beijing, China: 4018-4023. (2006). https://doi.org/10.1109/ IROS.2006.281861
32. Behnke, S. Humanoid robots-from fiction to reality?. In the Künstliche Intelligenz Heft, **4**, 5–9. (2008). Available online at: https://www.ais.uni-bonn.de/papers/KI08_Behnke.pdf
33. Riek, L. D. Healthcare Robotics. (2017). Retrieved from http://arxiv.org/abs/1704.03931
34. Kraft, K., & Smart, W. D. Seeing is comforting: Effects of teleoperator visibility in robot-mediated health care. *ACM/IEEE International Conference on Human-Robot Interaction*, *2016–April*, 11–18. (2016). https://doi.org/10.1109/HRI.2016.7451728
35. Yasemin, M., & Kasımo, Y. Management of dental anxiety in children using robots Reduction of Dental Anxiety and Pain in Children using Robots, (May), 327–332. (2016). https://doi.org/10.1109/SIU.2016.7495721

36. Beran, T. N., Ramirez-Serrano, A., Vanderkooi, O. G., & Kuhn, S. Humanoid robotics in health care: An exploration of children's and parents' emotional reactions. *Journal of Health Psychology*, **20**(7), 984–989. (2015). https://doi.org/10.1177/1359105313504794
37. Beran, T. N., Ramirez-Serrano, A., Vanderkooi, O. G., & Kuhn, S. Reducing children's pain and distress towards flu vaccinations: A novel and effective application of humanoid robotics. *Vaccine*, **31**(25), 2772–2777. (2013). https://doi.org/10.1016/j.vaccine.2013.03.056
38. [38] McCaul KD, Malott JM. Distraction and coping with pain. Psychol Bull; **95**(3):516. (1984).
39. DeMore M, Cohen L. Distraction for pediatric immunization pain: a critical review. J Clin Psychol Med Settings; **12**(4):281–91. (2005)
40. Klassen, JA., Liang, Y., Tjosvold, L., Klassen, TP., Hartling, L. "Music for pain and anxiety in children undergoing medical procedures: a systematic review of randomized controlled trials", Ambulatory Pediatrics, **8**, 117-128. (2008).
41. M. Onishi, Z. W. Luo, T. Odashima, S. Hirano, K. Tahara, and T. Mukai, "Generation of Human Care Behaviors by Human-Interactive Robot RI-MAN," in Proc. IEEE International Conference on Robotics and Automation (ICRA), pp. 3128–3129, (2007).
42. T. Mukai, M. Onishi, T. Odashima, S. Hirano, and Z. W. Luo, "Development of the Tactile Sensor System of a Human- Interactive Robot 'RI-MAN'," IEEE Trans. on Robotics, Vol.**24**, No.2, pp.505– 512, (2008

Application Value of Slider-Crank Mechanism in Pick-and-Place Operation of Delta Robot

QIN Zhe[1,2], LIU Xiao-chu[1,2], ZHAO Zhuan[1,2], XIAO Jin-rui[1,2]

[1]Guangzhou University, School of Mechanical and Electric Engineering, Guangzhou 510006;
[2]Key Laboratory of High-Performance Metal Materials Reinforced Grinding Machining, Guangzhou 510006

Abstract: By absorbing the advantages of the rotary-driven Delta robot and linear-driven Delta robot, a Delta robot for pick-and-place operation that forms a crank-slider at the drive joint is designed.To take the most common gate shaped curve in Cartesian space as the motion trail of robotic pick-and-place operation, according to the kinematics inverse solution theory of Delta robot, this thesis mainly solves the output angular velocity of robot-driven joint. Establishing the static transfer mathematical model and solving the forced condition of driving joint. The simulation analysis show that after the upper slider-crank mechanism is connected to the driving joint, the angular velocity of the driving joint changes suddenly, which caused a rigid impact on the robot in the picking and releasing operation, though the force of the driving joint can be made smaller.

1 Introduction

Delta robots have been proposed by Dr. CLAVEL since 1985[1] and many prototypes for different purposes have been made. Most of these robot widely apply used in electronics, food, medicine and other fields with the advantages of high bearing capacity, high positioning accuracy, simple structure, easy to achieve high-speed movement, etc[2-6]. Many research works have been developed the Delta series of robots include: Pierrot team delvelop H4 I4, Heli4 and Par4 and so on[7-9] in which industrial acceleration reaches 12g and experimental acceleration reaches 20g of Par4. Adept company developed a four degree of freedom high-speed SCARA parallel robot based on the prototype[10] of Par4 robot. And the robot claims to be the fastest in the world. Huang Tian[11] simplified Delta mechanism as Planar mechanism Diamond and invented TJU-Delta manipulator.

At present, to further improve the accuracy, stability and stiffness of the high-speed, high-frequency and pick-and-place operation, the present ways cover two aspects[12-13]: reducing the inertia of moving parts by lightweight design, improving the dynamic performance index by scale integrated; reducing the deviation between the actual pose and the pose of the command using high speed and high precision control system. Improve the precision, stability and rigidity of Delta robot's high-speed, high-frequency and pick-and-place operation play an important role in the design stage and running process.

This article by absorbing the advantages of the rotary-driven Delta robot and linear-driven Delta robot[14-15], a Delta robot for pick-and-place operation that forms a crank-slider at the drive joint is designed.To take the most common gate shaped curve in Cartesian space as the motion trail of robotic pick-and-place operation, according to the kinematics inverse solution theory of Delta robot, this thesis mainly solves the output angular velocity of robot-driven joint, the output force and torque of robot-driven joint, the output linear displacement of the slider that linked at the drive joints. Finally, analyze the value of the Slider-Crank Mechanism used for Pick-and-Place Operation of Delta Robot.

2 Solve the angular velocity of the drive joint

2.1 Trajectory planning for pick-and-place

The pick-and-place operation consists of three parts, including vertical, horizontal and vertical, as shown in Fig.1. The robot's ending part moves from point A at the time of t = 0, completes the entire pick-and-place operation at point D, and divides the whole motion process from A to D into N parts according to time shown in equation (1) Sequence, which is analyzing the position coordinates, velocity and acceleration about the robot's ending part relative to the time variable t[16].

$$t_i = T \cdot \frac{i}{N} (i = 0,1,\cdots,N) \quad (1)$$

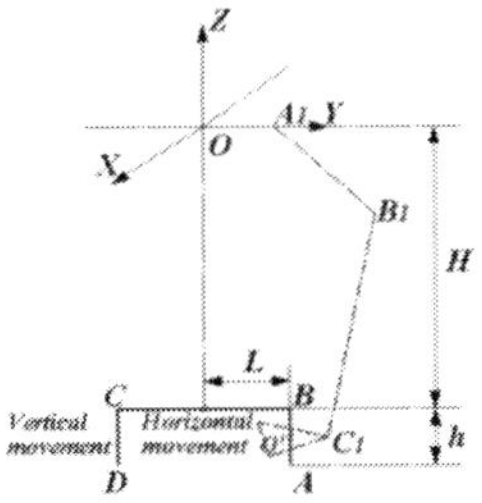

Figure 1. Pick and place operation track

① The robot moves along the vertical axis from point A to point B when $0 \le S_{AB}(t_i) \le h$, during this process, the position coordinates, velocity and acceleration about t_i are as follows (2)

$$\begin{cases} \boldsymbol{S_{AB}} = (0 \quad L/2 \quad -S_{AB}(t_i) - H)^T \\ \boldsymbol{V_{AB}} = \left(0 \quad 0 \quad -S'_{AB}(t_i)\right)^T \\ \boldsymbol{A_{AB}} = \left(0 \quad 0 \quad -S''_{AB}(t_i)\right)^T \end{cases} \tag{2}$$

② When $-L/2 \le S_{BC}(t_i) \le L/2$, the robot's ending part moves from point B to point C in the horizontal direction. At this time, the motion path height z remains unchanged, the position coordinates, velocity and acceleration about t_i are as follows (3)

$$\begin{cases} \boldsymbol{S_{BC}} = (0 \quad S_{BC}(t_i) \quad -H)^T \\ \boldsymbol{V_{BC}} = (0 \quad S'_{BC}(t_i) \quad 0)^T \\ \boldsymbol{A_{BC}} = (0 \quad S''_{BC}(t_i) \quad 0)^T \end{cases} \tag{3}$$

③ Finally, the robot's ending part moves from point C down to point D along the vertical axis to complete the entire pick-up process when $0 \le S_{CD}(t_i) \le h$. At this time, the position coordinates, velocity and acceleraion about t_i are as follows (4)

$$\begin{cases} \boldsymbol{S_{CD}} = (0 \quad -L/2 \quad -S_{CD}(t_i) - H)^T \\ \boldsymbol{V_{CD}} = \left(0 \quad 0 \quad -S'_{CD}(t_i)\right)^T \\ \boldsymbol{A_{CD}} = \left(0 \quad 0 \quad -S''_{CD}(t_i)\right)^T \end{cases} \tag{4}$$

Through the above equations (2) ~ (4), you can ensure the position coordinates $\boldsymbol{S}$, speed $\boldsymbol{V}$, acceleration $\boldsymbol{A}$ that are relative to the changing of t about the robot's ending part movement in the entire pick-up cycle.

The sketch diagram of Delta mechanism, as shown in Fig.2.

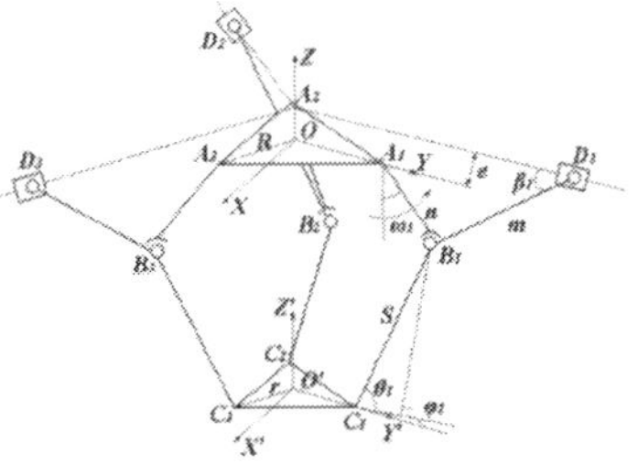

Figure 2. Sketch diagram of Delta mechanism

After solving the motion position of coordinate $\boldsymbol{S}$ about the robot's ending part in the base, the angular velocity of the drive joint can be solved by using the Delta parallel mechanism inverse theory, as the following equation (5) ~ (8)

$$(r + y - n\sin\omega_1 - R)^2 + (z + n\cos\omega_1)^2 = S^2 \tag{5}$$

$$\left(\frac{\sqrt{3}}{2}R - \frac{\sqrt{3}}{2}r + \frac{\sqrt{3}}{2}n\sin\omega_2\right)^2 + (z + n\cos\omega_2)^2 + \left(y - \frac{1}{2}r + \frac{1}{2}n\sin\omega_2 + \frac{1}{2}R\right)^2 = S^2 \tag{6}$$

$$\left(x + \frac{\sqrt{3}}{2}r - \frac{\sqrt{3}}{2}n\sin\omega_3 - \frac{\sqrt{3}}{2}R\right)^2 + (z + n\cos\omega_3)^2 + \left(y - \frac{1}{2}r + \frac{1}{2}n\sin\omega_3 + \frac{1}{2}R\right)^2 = S^2 \tag{7}$$

$$\begin{bmatrix} x_{AB} & y_{AB} & z_{AB} \\ x_{BC} & y_{BC} & z_{BC} \\ x_{CD} & y_{CD} & z_{CD} \end{bmatrix} = \begin{bmatrix} 0 & L/2 & -S_{AB}(t_i) - H \\ 0 & S_{BC}(t_i) & -H \\ 0 & -L/2 & -S_{CD}(t_i) - H \end{bmatrix} \tag{8}$$

2.2 Solution the angular velocity of during the pick-and-place operation

The corresponding parameters of Delta parallel mechanism was shown in Tab.1.

Table1 Delta mechanism parameters

characteristic	*parameter*
R/mm	$200\sqrt{3}/3$
r/mm	*40*
n/mm	*120*
S/mm	*240*
m/mm	*200*
e/mm	*40*
h/mm	*40*
L/mm	*75*
point A- Coordinate/mm	*0, 75, -317*
G/N	*10*

By the equation (5)~(8) can be solved ω_i, $i = 1,2,3$.

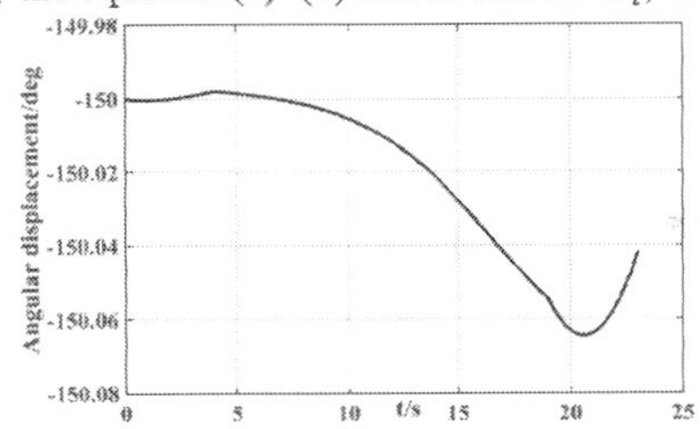

Figure 3. The angular displacement of the first drive joint

General model Fourier4: ω_1 = a0+a1*cos(t*w)+ b1*sin(t*w)+a2*cos(2*t*w)+b2*sin(2*t*w)+a3*cos(3*t*w)+ b3*sin(3*t*w) + a4*cos(4*t*w) + b4*sin(4*t*w)

Coefficients (with 95% confidence bounds):
a0=-150; a1=0.03526; b1=-0.11; a2=-0.08551; b2= -0.01198; a3 =-0.003307; b3 =0.0354; a4=0.007745; b4 =0.001049; w =0.1545.

Goodness of fit: SSE: 0.0001122, R-square: 0.9996, Adjusted R-square: 0.9996, RMSE: 0.0004453.

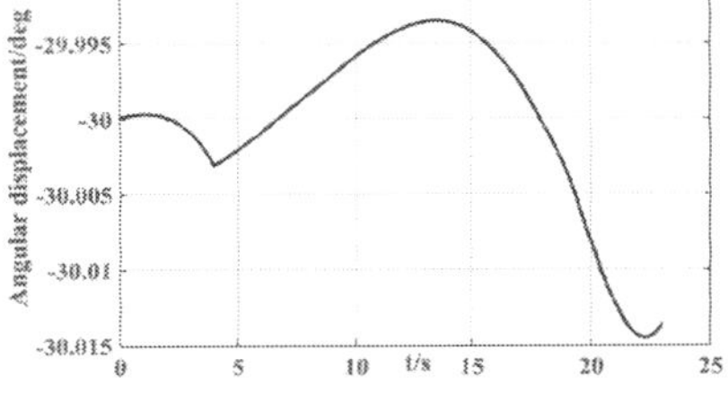

Figure 4. The angular displacement of the second drive joint

General model Fourier4: ω_2 = a0+a1*cos(t*w)+ b1*sin(t*w)+a2*cos(2*t*w)+b2*sin(2*t*w)+a3*cos(3*t*w)+ b3*sin(3*t*w) + a4*cos(4*t*w) + b4*sin(4*t*w)

Coefficients (with 95% confidence bounds):

a0 =-2351; a1=3230; b1=2015; a2 =-927.6; b2 =-1819; a3=-28.86; b3=646.6; a4=46.57; b4=-78.94; w= 0.03169.

Goodness of fit: SSE: 3.142e-05, R-square: 0.9984, Adjusted R-square: 0.9984, RMSE: 0.0002356.

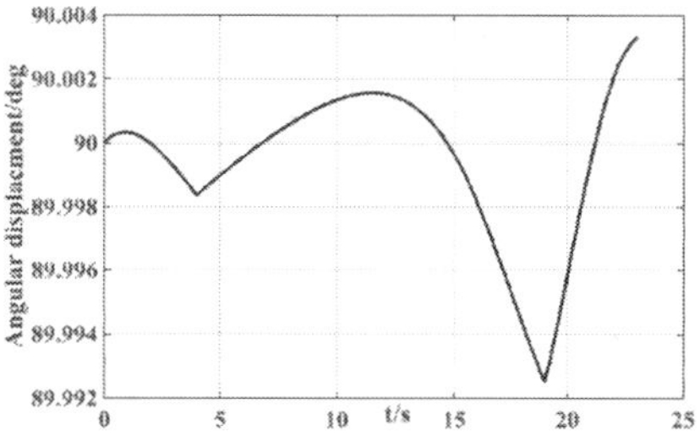

Figure 5 The angular displacement of the third drive joint

General model Fourier4: ω_3 = a0+a1*cos(t*w)+ b1*sin(t*w)+a2*cos(2*t*w)+b2*sin(2*t*w)+a3*cos(3*t*w)+ b3*sin(3*t*w) + a4*cos(4*t*w) + b4*sin(4*t*w).

Coefficients (with 95% confidence bounds):

a0=87.41; a1=2.052; b1=3.798; a2=1.332; b2=-2.052; a3=-0.8618;b3=-0.05547;a4=0.06522;b4=0.128;w=0.09655.

Goodness of fit: SSE: 6.305e-05, R-square: 0.9806, Adjusted R-square: 0.9803, RMSE: 0.0003338.

The angular displacement-time curves of Fig.3 to Fig.5 as the control are added to the robot drive joint. The angular velocity of the robot-driven joint can be solved as follows Fig.6 to Fig.8

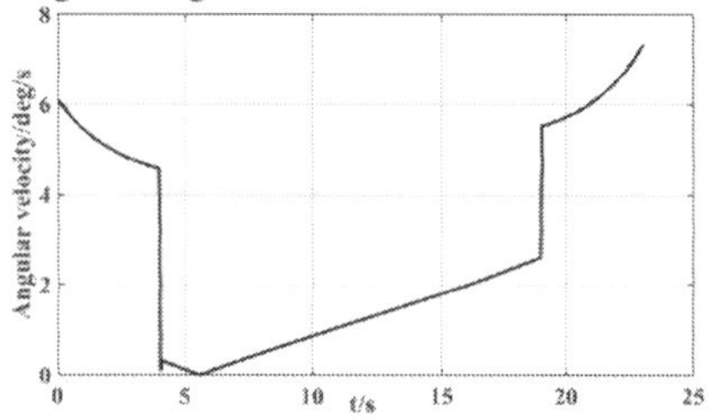

Figure 6. The angular velocity of the first drive joint

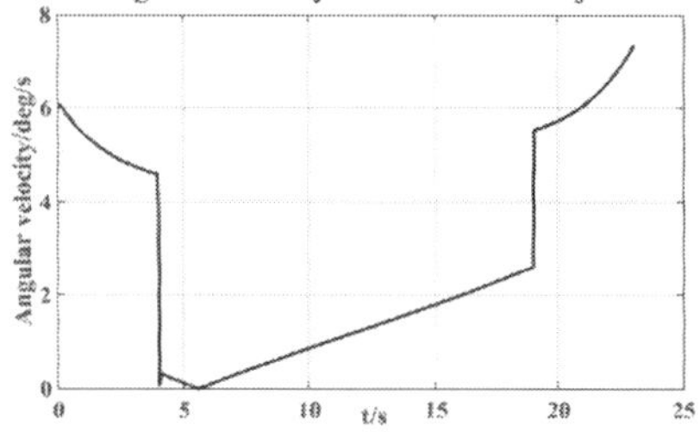

Figure 7. The angular velocity of the second drive joint

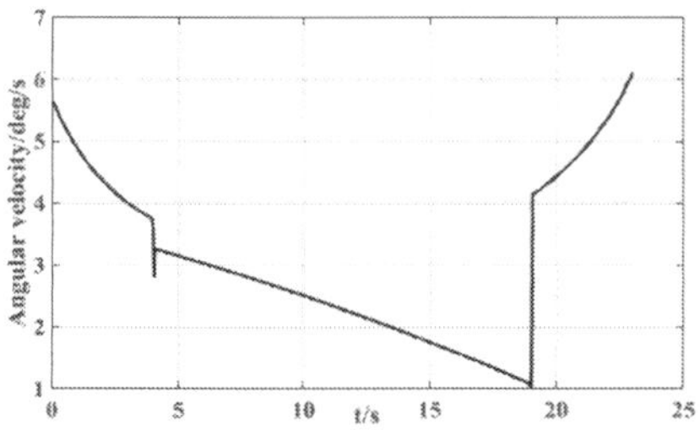

Figure 8. The angular velocity of the third drive joint

The door-shaped curve as the trajectory of the Delta robot pick-and –place operation will give the drive joints to bring the rigid impact.

From Fig. 6 to Fig. 8, it can be seen that when the crank-slider mechanism is applied to the Delta robot pick-and-place operation, the angular velocities of the three driving joints suddenly change at 3.55s and 19.05s. Although the rigid impact under this working condition of the pick-and-place curve of the door-shaped has not been eliminated yet.

3 Solve the driving joint's force and torque

3.1 Forms a crank-slider at the drive joint

Assuming that the load of the moving platform in the Delta parallel robot pick-up operation is G, the force of the three branches on the moving platform is shown in Fig.9, the force of the fixed platform is shown in Fig.10.

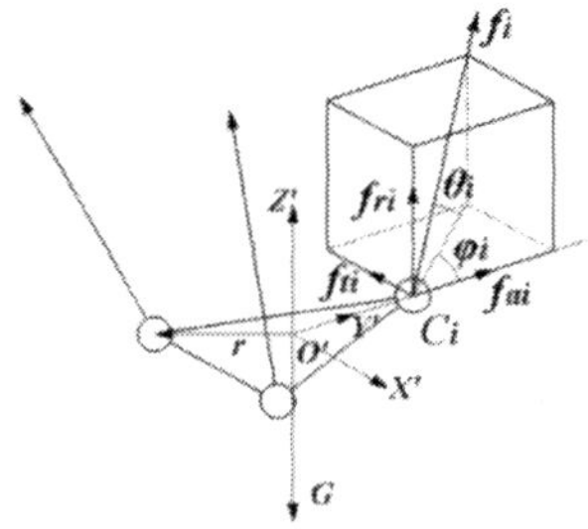

Figure 9. Force diagram of moving platform

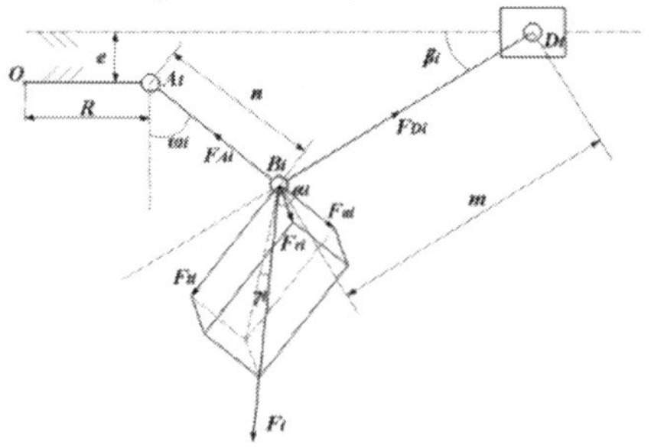

Figure 10. Force diagram of fixed platform

The force and the torque of the drive joints can be solved by using the Fig. 9 and Fig. 10, as the following equation (9) ~ (11).

$$F_{Ai} = F_i \cos\gamma_i \cos\alpha_i + \frac{F_i \cos\gamma_i \sin\alpha_i}{\cos(\omega_i - \beta_i)} \sin(\omega_i - \beta_i) \quad (9)$$

$$\beta_i = \tan^{-1}\frac{e + m\cos\omega_i}{n} \quad (10)$$

$$M_i = F_{Ai} m \sin\gamma_i \quad (11)$$

3.2 Don't form a crank-slider at the drive joint

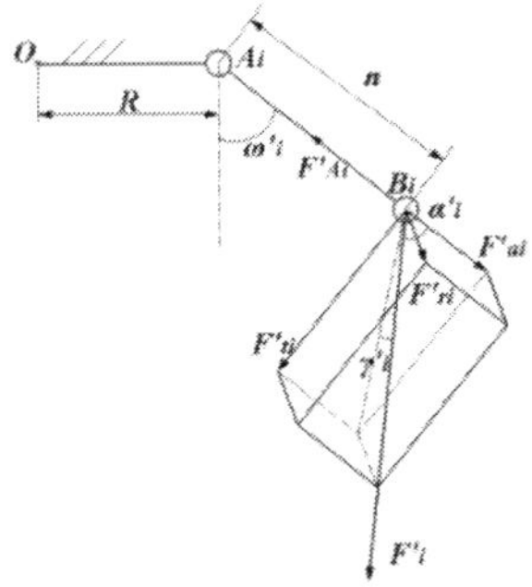

Figure 11. Force diagram of fixed platform

The force and the torque of the drive joints can be solved by using the Fig. 9 and Fig. 11, as the following equation (12) ~ (13).

$$F'_{Ai} = F'_i \cos\gamma'_i \cos\alpha'_i \quad (12)$$

$$M'_i = \sqrt{(mF'_i \sin\gamma'_i)^2 + (mF'_i \cos\gamma'_i \sin\gamma'_i)^2} \quad (13)$$

3.3 The effect of crank-slider on the driving

① On the impact of the force on the drive joints

$$\Delta F_{Ai} = F_i \sin\gamma_i \cdot \frac{\tan\omega_i - \frac{e + m\cos\omega_i}{n}}{1 + \frac{\tan\omega_i\,(e + m\cos\omega_i)}{n}} \quad (14)$$

$|\omega_i|$, $|\gamma_i| \in (0, \pi/2)$.

If $\Delta F_{Ai} < 0$ is less than zero, the Delta robot that drive joints forms a crank-slider mechanism will have less force on the drive joint during the pick-and-place operation.

② On the impact of the output coupling torque of the joint

$$\Delta M_i = M_i - M'_i \quad (15)$$

Through the formulas (11) and (13) we can draw the conclusion: the Delta robot that drive joints forms a crank-slider mechanism will have smaller output torqut on the drive joint during the pick-and-place operation.

4 Solve the displacement changes of sliders

According to the kinematics inverse solution theory of Delta robot, solve the linear displacement of the slider linked at the drive joints, as follows Fig.12 to Fig.14.

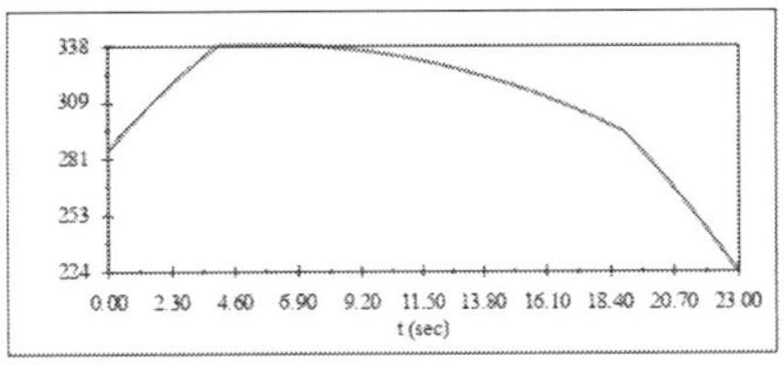

Figure 12. The linear displacement of the first slider

General model Fourier2: S1=a0+a1*cos(t*w)+ b1*sin(t*w) + a2*cos(2*t*w) + b2*sin(2*t*w).

Coefficients (with 95% confidence bounds):

a0 =182.3; a1=54.41; b1 =190.1; a2=46.32; b2= -15.93; w=0.1264.

Goodness of fit: SSE: 924.6, R-square:0.998; Adjusted R-square: 0.998, RMSE: 1.274.

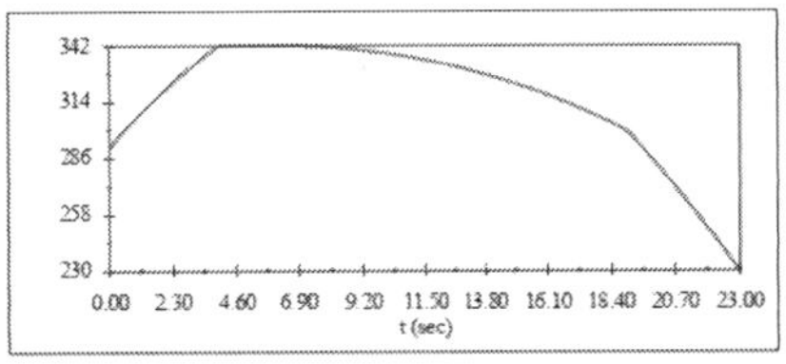

Figure 13. The linear displacement of the second slider

General model Fourier2: S2=a0+a1*cos(t*w)+ b1*sin(t*w) +a2*cos(2*t*w) + b2*sin(2*t*w).

Coefficients (with 95% confidence bounds):

a0 =168.4; a1 =70.78; b1=210.9; a2=49.67; b2= -23.73; w =0.1214.

Goodness of fit: SSE: 870.6, R-square: 0.9981, Adjusted R-square: 0.9981, RMSE: 1.236.

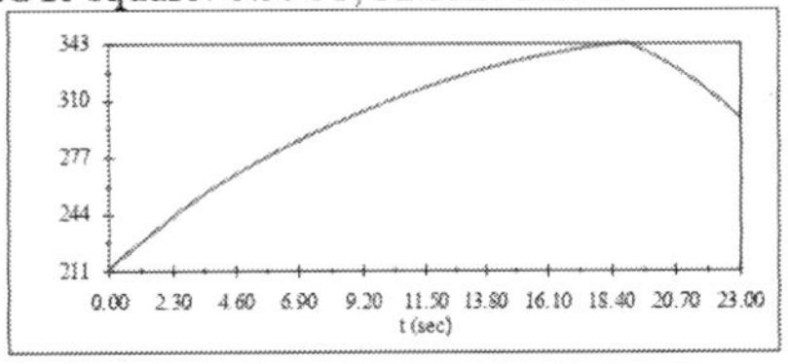

Figure 14. The linear displacement of the third slider

General model Fourier2: S3=a0+a1*cos(t*w)+ b1*sin(t*w) +a2*cos(2*t*w) + b2*sin(2*t*w).

Coefficients (with 95% confidence bounds):

a0 =270.3; a1 =-63.72; b1 =37.54; a2=6.176; b2= 14.68; w =0.172.

Goodness of fit: SSE: 971.6; R-square: 0.9987; Adjusted R-square: 0.9987; RMSE: 1.306.

Solving the linear displacement of the slider that linked at the drive joints, it facilitates use to linear robot and rotary motor to coordinated control the Delta robot that performs the pick-and-place operations.

5 Conclusion

By absorbing the advantages of the rotary-driven Delta robot and linear-driven Delta robot, a Delta robot for pick-and-place operation that forms a crank-slider at the drive joint is designed. To take the most common gate shaped curve in Cartesian space as the motion trail of robotic pick-and-place operation, from many aspects of analysis that application value of slider-crank mechanism in pick-and-place operation of Delta robot.

(1) The angular velocity of during the pickand place operation has been solved. The crank-slider mechanism is applied to the Delta robot for pick-and-place operation, although the rigid impact under this working condition of the pick-and-place curve of the door-shaped has not been eliminated yet.

(2) The Delta robot that drive joints forms a crank-slider mechanism will have smaller force and output torqut on the drive joint during the pick-and-place operation.

(3) According to the kinematics inverse solution theory of Delta robot, solve the linear displacement of the slider linked at the drive joints, it provides a reference for Delta robot which uses rotary motor and linear motor to collaborative control the operation of pick-and-place.

References

1. CLAVELR. Device for the movement and positioning of an element in space, Europe: EP0250470 B1[P]. 1991-07-17.
2. Brogardh T. Present and future robot control development –An industrial perspective[J]. Annual Reviews in Control, 2006, 31(1): 69-79.
3. Poppeova V, Uricek J, Bulej V, et al. Delta robots – Robots for high speed manipulation[J]. Tehnicki Vjesnik, 2011, 18(3):435-445.
4. Bouri M, Clavel R. The linear Delta: Development and application[C] // The 41st International Symposium on Robotics. Frankfurt, Germany: VDE, 2010: 1198-1205.
5. Milutinovic D, Slavkovic N, Kokotovic B, et al. Kinematic modeling of reconfigurable parallel robots based on Delta concept[J]. Journal of Production Engineering, 2012, 15(2): 71-74.
6. Bonev I. Delta paraller robot- The story of success[EB/OL]. (2001-05-06) [2013-06-29]. http://www.parallemic.org/ Review/Review002. html.
7. Pierrot F, Nabat V, Company O, et al. Optimal design of a 4-DOF parallel manipulator: From academia to industry[J]. IEEE Transactions on Robotics, 2009, 25(2): 213-224.
8. Pierrot F, Marquet F, Company O, et al. H4 parallel robot: modeling, design and preliminary experiments[C]//Robotics and Automation, 2001. Proceedings 2001 ICRA. IEEE International Conference on. IEEE, 2001, 4: 3256-3261.
9. Nabat V, de la O RODRIGUEZ M, Company O, et al. Par4: Very high speed parallel robot for pick-and-place[C]//Intelligent Robots and Systems, 2005.(IROS 2005). 2005 IEEE/RSJ International Conference on. IEEE, 2005: 553-558.
10. Liu S, Huang T, Mei J, et al. Optimal design of a 4-DOF SCARA type parallel robot using dynamic performance indices and angular constraints[J]. Journal of Mechanisms and Robotics, 2012, 4(3): 031005.
11. ZHANG Limin, MEI jiangping, HUANG Tian. Dynamic Dimensional Synthesis of Delta Robot[J]. Chinese Journal of Mechanical Engineering，2010，46(3): 2-3.
12. Codourey A. Dynamic modeling of parallel robots for computed-torque control implementation[J]. The International Journal of Robotics Research, 1998, 17(12): 1325-1336.
13. WANG P, MEI J, HUANG T. Time Optimal Path Planning of a 2-DOF High-Speed Parallel Robot [J]. Journal of Tianjin University, 2007, 10: 000.
14. Bouri M, Clavel R. The linear Delta: Developments and applications[C] // The 41st International Symposium on Robotics. Frankfurt, Germany: VDE, 2010: 1198-1205.
15. Milutinovic D, Slavkovic N, Kokotovic B, et al. Kinematic modeling of reconfigurable parallel robots based on Delta concept[J]. Journal of Production Engineering, 2012, 15(2): 71-74.
16. Xie zexiao, Shang Dawei, Ren Ping. Optimization and Experimental Verification of Pick-and-place Trajectory for a Delta Parallel Robot Based on Lame Curves[J]. Chinese Journal of Mechanical Engineering, 2015, 51(1): 52-55.

Parameters Affecting the Erosive Burning of Solid Rocket Motor

Almostafa Abdelaziz[1], Liang Guozhu[1] and Anwer Elsayed[2]

[1]Beihang University (BUAA), Aerospace Department, 100191 Beijing, China
[2]Militery Technical College (MTC), Rocket Department, Cairo, Egypt

Abstract. Increasing the velocity of gases inside solid rocket motors with low port-to-throat area ratios, leading to increased occurrence and severity of burning rate augmentation due to flow of propellant products across burning propellant surfaces (erosive burning), erosive burning of high energy composite propellant was investigated to supply rocket motor design criteria and to supplement knowledge of combustion phenomena, pressure, burning rate and high velocity of gases all of these are parameters affect on erosive burning. Investigate the phenomena of the erosive burning by using the 2'inch rocket motor and modified one. Different tests applied to fulfil all the parameters that calculated out from the experiments and by studying the pressure time curve and erosive burning phenomena.

1 Introduction

The erosive burning refers to the increase in the propellant burning rate caused by the high-velocity flow of combustion gases over the burning propellant surface[1]. It can seriously affect the performance of solid propellant motors. It occurs primarily in the port passages or perforations of the grain as the combustion gases flow toward the nozzle [2]. It is more probable to occur when the port passage cross-section area is low relative to the throat area with a port-to-throat field [3].

Erosive burning increases the gas mass flow and thus also the chamber pressure and thrust during the early portion of the burning, for a particular motor, as soon as the burning enlarges the flow passage (without a major increase in burning area), the port area flow velocity is reduced and erosive burning diminishes until normal burning will again occur. Erosive burning causes also early burn out of the web usually at the nozzle end and lets out the insulation and ceasing to hot combustion gas for a large period of the fourth dimension. This usually needs more insulation layer thickness (and more inert mass) to prevent local thermal failure[4, 5].

A relatively simple model for erosive burning based on heat transfer was first developed in 1956 by Lenoir-Robillard [6] and has since been improved and used in motor performance calculations. The model has the simplified form.

In this paper we present some factors which affect the erosive burning phenomena, and by coding a MATLAB code to investigate the internal ballistic of the rocket motor. Some experiments will take part for describe the phenomena of erosive burying which takes part in small 2 inch rocket motor and then by modified solid rocket motor which shows clearly the phenomena of the erosive burning.

2 Factors Affecting Erosive Burning

Cross-flow velocity, pressure, propellant initial temperature, free stream gas temperature, normal burning rate, metal addition, compressibility, oxidizer particle size, rocket motor size, propellant characteristic, mass flux and Reynolds number. All of this parameters can affect the erosive burning [7, 8], we will describe the effect of total length to port area diameter to describe the phenomena by MATLAB code for investigating the internal ballistic and by some sets of experiments to show the phenomena clearly by different designs of solid rocket motor. A relatively simple model for erosive burning based on heat transfer was first developed by Lenoir-Robillard. The model has the simplified form.

$$r_t = r_o + r_e \quad (1)$$

Where r_t is the total burning rate, r_o is the base burning rate and r_e is the burning due to erosive burning rate.

$$r_t = ap^n + \alpha G^{0.8} L^{-0.2} e^{(\frac{-\beta * r_t * \rho * p}{G})} \quad (2)$$

$$G = \frac{m}{A_p} \quad (3)$$

$$r_t = ap^n + \alpha G^{0.8} d^{-0.2} e^{(\frac{-\beta * r_t * \rho * p}{G})} \quad (4)$$

Where equation (2) used when depending on the length of the propellant and equation (3) when use the inner diameter of the propellant. (Fig. 1) shows the effect of erosive burning. Which show the deviation of the course of pressure variation with time due to erosive burning. where α is the erosive burning coefficient, G is the mass flux, L is the initial grain length, β is the erosive burning pressure coefficient, and ρ_p is the propellant density, Ap is the port area and, d is the inner diameter of the grain.

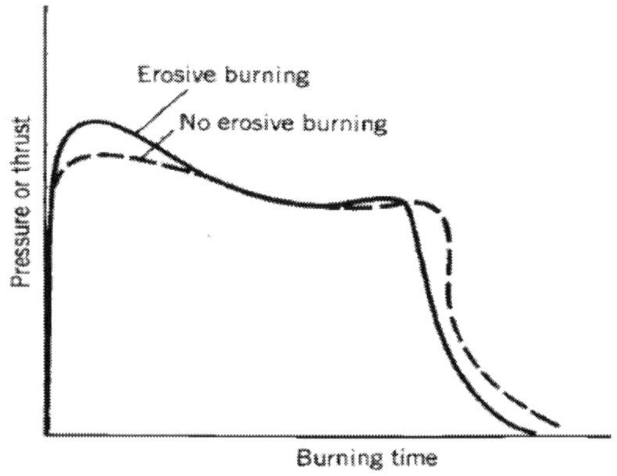

Figure 1. Erosive burning phenomena.

3 Experiments of 2 inch motors

The main task of the test motor is to determine the burning rate of the propellant, combustion index and burning rate coefficient calculated by a fitting curve to determine the burning law. A tubular grain from composite propellant is cast inside a special casing so that the burning is allowed only from inside and from the two ends. The main output of the test is the pressure time-curve.

Figure 2. 2 inch motor.

From the main parameters that affect the erosive burning, decrease the port diameter of the grain by making a modified grain with different port diameter 34, 15, 12.5 and 10 mm respectively to show the effect of the decreased diameter with the same length 108 mm. By decreasing the port diameter the velocity of gases parallel to the burning surface increased, which causes erosive burning in the propellant [9, 10], by using composite propellant consist of ammonium Perchlorate (68-70 %) aluminum powder (16-18 %) and binder (13-15 %).

The results of the tests

Table 1. Output data with different port diameters.

D	mm	34	15	12.5	10
Temp	c	21	21	21	21
Throat	mm	7	7	7	7
WT	Kg	.375	.512	.503	.514
TD	Msec	20	12	10	387
TB	Msec	2.24	4.06	4.31	4.43
T50%	Sec	2.13	3.9	4.1	4.32
Pmax	Bar	71	86	55	74
IPT	Bar.s	151	199	183	197
Pav	Bar	67	48	42	43
P50%	Bar	71	50	43	45
R50%	mm/s	6.1	5.7	5.6	5.7
Cstar	m/sec	1553	1498	1401	1475

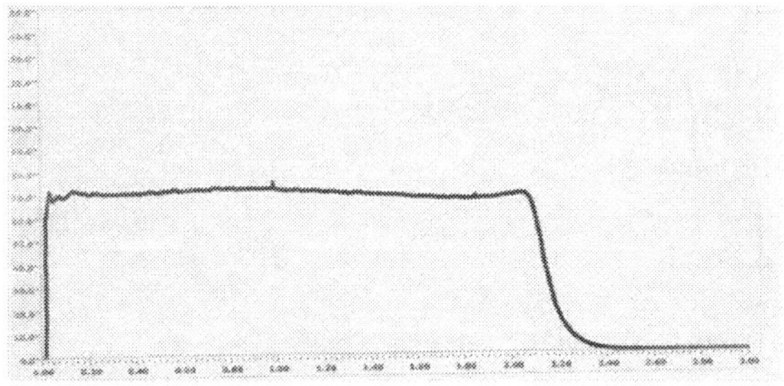

Figure 3. pressure time curve at port diameter 34 mm.

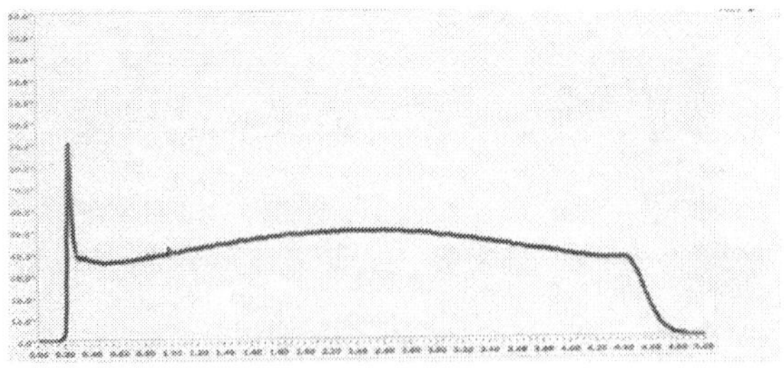

Figure 4. pressure time curve at port diameter 10 mm.

This increase in the beginning of the propellant burning is due to using the greater mass of the black powder.

From the calculation of the parameters of the test and Thermochemical calculation when decreaseing the port diameter and make the critical diameter constant the velocity of gases increase from 37 to 300 m/s, and the erosive ratio from 1 to 1.36.

This information is not enough to investigate the phenomena of erosive burning. In other tests shown later, the length of the rocket motor will be increased to investigate the phenomena of erosive burning clearly.

4 Computer program

The computer program is capable of calculating performance prediction of tabular grain and draw pressure time curve, the program computes in SI units which are used to calculate the main parameters and performance of solid propellant rocket motor, the program may be used for rocket of all size, the main task of the program is to draw the P-T curve and to calculate and the phenomena of the erosive burning.

We will divide the web thickness into equal segments for easy calculation and determine the output with minimum error.

The internal ballistic analysis follow zero dimensional frictionless compressible flow theory with the flow taken as isentropic in the nozzle, equilibrium pressures are established by iteration to find the correct chamber pressure which balance the continuity equation. By dividing the web thickness into segments, solution of the governing equation for each segment will calculated, and make a comparison of the total amount of mass generated with the mass discharge through the nozzle, if the check fails a new assumption is required to obtain a solution for one instant of time. The average burning rate is taken to be liner between the head and nozzle end.

Determination of head and nozzle end burning rate take into account erosive burning effect the burning surface at tail off is reduced in proportion to the length of the propellant that has experienced burn out according to the linear model of the regression surface. The computation are performed in two sections at the head and nozzle end of the grain, it is useful to for computing pressure, temperature, and velocity of gases of the nozzle end of the grain, the web thickness divided into equal segments to determine the output with minimum error.

We assume that the burning rate and pressure at head and nozzle end is linear along combustion

5 Manufactures and Experiment of Modified SRM

Modified two-inch motor consists of 3 segmented grains each one 108 mm by separating them by a small ring similar to the outer case each ring of width 10 mm and assembled together . The results will be shown later. The only modification of the modified 2'inch motor is to vary the length of the motor case; all the other dimensions are the same as the standard 2'inch motor. Also by changing the critical part of the nozzle we can get different pressure and different erosive ratio.

Figure 5. Parts of modified SRM.

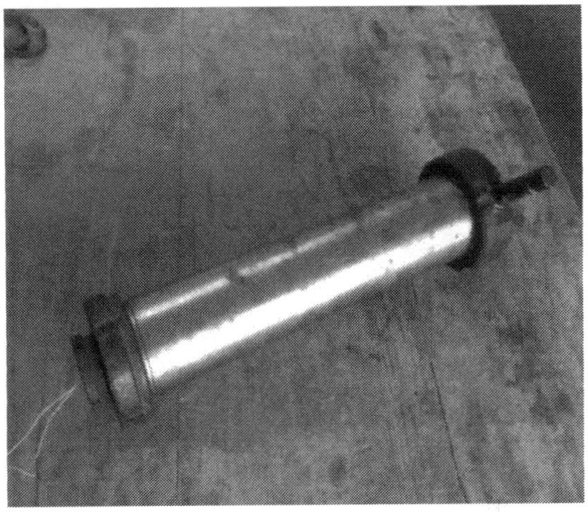

Figure 6. Assembled new motor.

5.1 Experiment results

The length of grain = 324 mm divided into 3 equal parts of solid propellant each one 108 mm separated by 2 steel rings each one of width 10 mm, the outer diameter = 60 mm and the inner diameter = 10 mm, nozzle critical section = 9 mm.

The pressure at the head end of the grain is measured experimentally while pressure at the nozzle end of the grain is predicted from the theoretical model.

Table 2. Results of modified two-inch motor

D	mm	10
Temp	c	2
Throat	mm	9
WT	Kg	1.573
TD	Msec	28
TB	Msec	2.992
T50%	Sec	2.557
Pmax	Bar	187
IPT	Bar.s	334
Pav	Bar	111
P50%	Bar	130
R50%	mm/s	9.7
C star	m/sec	1352

 https://doi.org/10.1051/matecconf/201815303001

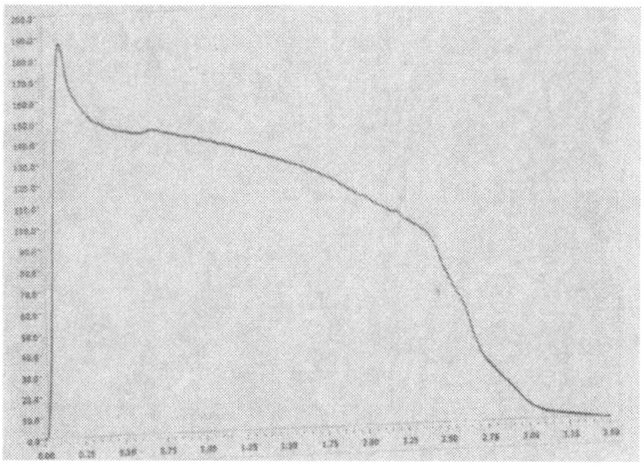

Figure 7. Pressure time curve of modified SRM.

All these curves are from the experiment sheet. In the beginning of the ignition the pressure increases in the small port area and the velocity of gases increased which causes erosive burning. This proves our Prediction and idea. The Very important conclusion could be drawn concerning the effect of very high pressure, very high velocity and the erosive burning. The critical section of the nozzle increases with respect to time, by measuring the critical section of the nozzle before and after burning, the critical section increase from 9 to 10.5 mm. It is observed that increase of the critical area of about 25 %, which result in a drop in the pressure at the end of burning.

So there is a factor taken during calculation of the critical area which is the increase in the critical section with respect to time.

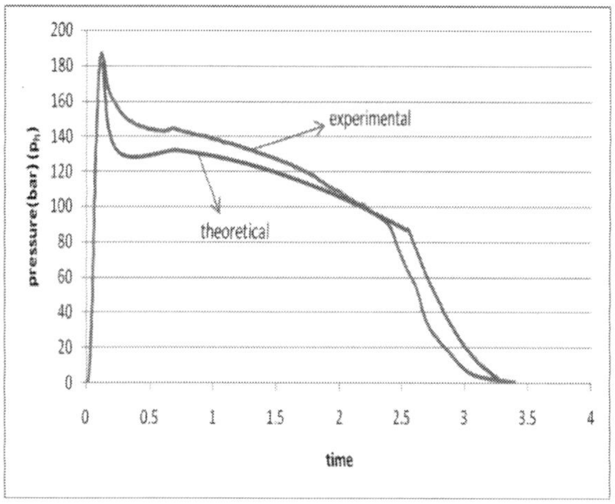

Figure 8. Experimental and theoretical pressure-time curves.

The assumption of the burning rate is taken to be linear along burning, so there is an error between the experimental and theoretical values.

The difference between the pressures at the head end and nozzle end of the grain is due to erosive burning, where at erosive the mass discharge is increased which causes this difference.

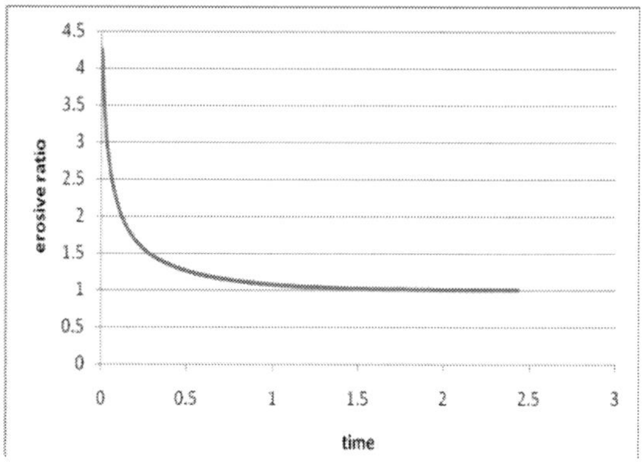

Figure 9. Erosive ratio versus time.

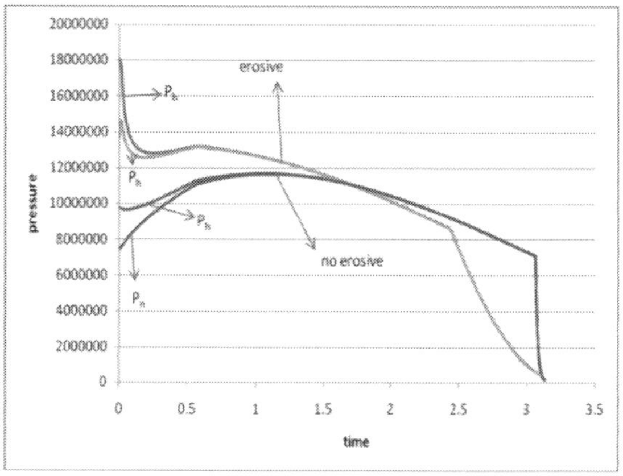

Figure 10. Pressure time cure with and without erosive burning.

6 Conclusion

The flow of combustion gaseous product at high velocity parallel to the surface of solid propellant is often found to lead to a significant increase in burning rate. This phenomenon is referred to as erosive burning and the increase in burning rate is known as the erosive burning rate. Studying the erosive burning is very important for better understanding of effects on the performance of the rocket motor. Good design of solid propellant rocket motor should account for possible erosive burning in order to thoroughly predict the pressure-time history and act accordingly for safe and best performance.

During the present study test rocket motors of standard and modified configurations were designed, manufactured and tested in different working conditions to investigate effects on the burning rate and their impact on the performance. In parallel, a theoretical model of the burning rate was developed and the proposed numerical solution was implemented as a computer code capable of predicting the behavior of the combustion pressure. In particular, the code treats the case of erosive burning.

The comparison between theoretical and experimental results showed fair agreements, which proves the validity of the code. The following results could be drawn from the present investigations:

By varying the grain length-to-diameter ratio or the critical diameter, in order to change the mass flux parallel to the surface, it was possible to generate experimentally and theoretically the condition of erosive burning. It was possible to predict the threshold velocity above which the erosive burning occurs. For a given type of propellant, the erosive ratio and threshold velocity are dependent on the combustion pressure. In the present study, it was found that with the higher pressure, lower erosive ratio, lower threshold velocity and higher threshold flux were encountered.

With smaller values of (a, n) in the burning law, higher erosive ratios are encountered.

Some kind of oscillation occurs in the theoretical model. This may be attributed to the admitted values of the error considered in Mach number, mass balance and burning rate. More runs have been done for smaller values of permitted error between successive iterations and finally smooth curve could be reached. Clever choice of error values should be considered after several trials to get a smooth variation of producing curve.

The erosion at the nozzle critical section was measured and studied in the theoretical prediction. It has a significant effect on the combustion chamber pressure and consequently on the burning rate.

The code can be trusted for used in novel designs, and further investigations, saving numerous firing tests which are costly, time-consuming and risky

References

1. Barrere, "Rocket Propulsion," 1960.
2. O. B. Sutton P, "Rocket Propulsion Elements," *7 th edition,* 2001.
3. J. W. KREIDLER, "Erosive Burning New Experimental Techniques and Methods of Analysis," *solid propellant rocket conference,* 1964.
4. M. K.King, "Erosive Burning of Composite Propellant," *atlantic reserch coperation,* 1976.
5. M. K. King, "Erosive Burning of Solid Propellants," *JOURNAL OF PROPULSION AND POWER,* vol. vol 9, 1993.
6. M. L. a. G.Robillard, "A Matimatical Method to Predict the Effects of Erosive Burning in Solid Rocket Propellant," 1957.
7. C. E. Rogers, "Erosive Burning Design Criteria For Solid Rocket Motor," *NASSA article,* 2002.
8. C. E. Rogers, "Solid Propellant Grain Design and Internal Ballistics,," *NASSA report,* vol. vol 33, 2004.
9. K. Srinivasan, S. Narayanan, and O. P. Sharma, "Numerical Studies on Erosive Burning in Cylindrical Solid Propellant Grain," *Heat and Mass Transfer,* vol. 44, pp. 579-585, 2007.
10. M. K. King, "Erosive Burning of Solid Propellants," *journal of propullsion and power,* vol. vol 9, 1993.

https://doi.org/10.1051/matecconf/201815303002

Optimization on the Law of Variable-Pitch Vertical-Axis Tidal Current Turbine

Kong Fankai, Wang Liying, Zhang Di

Mechanical Engineering, Harbin Engineering University, Harbin 150001, China

Abstract. In order to improve the hydrodynamic performance of the vertical-axis variable-pitch current turbine, with analyzing on the existing design mentality of blade control mechanism, an optimized law of variable-pitch vertical-axis tidal current turbine was given in terms of the instantaneous moment coefficient of the blade. A method for solving the lift - drag coefficient of Blade by wind - hole test data is given. The optimized kinematic model has a significant reference value for the further development of vertical axis turbine model test.

1 Introduction

Tide energy is a kind of clean, stable, and rich renewable energy. As the characteristic of simple, non - directional selective, easy-to-install and easy-to-maintain power generation device, the energy capture efficiency of the vertical-axis turbine is always the focus of nowadays research [1]. Experiments and studies have shown that the study of variable pitch system is of great significance to improve the energy capture efficiency of vertical-axis turbine [2, 3]. And the variable-pitch law is the key part of the variable-pitch system.

In this paper, analysis of hydrodynamic is made and subsequently an optimization model of vertical-axis turbine performance is established, which use the blade instantaneous moment coefficient as an optimization target and blade angle of attack as optimization variable . The range of variation of the design variable is determined by analyzing the lift-drag coefficient and dynamic stall characteristic. Solving the problem by Matlab, the best pitch law is given in terms of the instantaneous moment coefficient of the blade, which provide a new idea on the study of efficiency capture of vertical-axis turbine.

2 Preliminaries

2.1 Hydrodynamic analysis of turbine

The vertical axis turbine with lift-straight blade is selected as the object of study, and the hydrodynamic study is carried out under the fixed velocity environment [4].

Figure 1 shows the analysis of the force of the impeller. During the constant rotation of the turbine, due to the relative movement between the water flow and the blade, the water flow will produce a resistance D_r to the blade opposite to the relative velocity VR and the resistance D_r is perpendicular to the lift L.

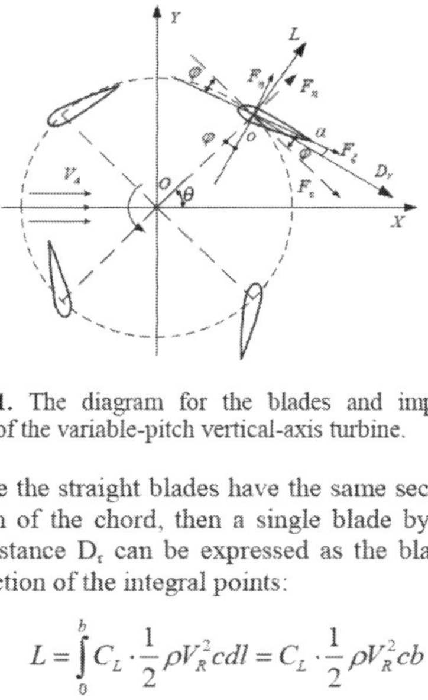

Figure 1. The diagram for the blades and impeller force analysis of the variable-pitch vertical-axis turbine.

Since the straight blades have the same section in the direction of the chord, then a single blade by the lift L and resistance D_r can be expressed as the blade unit of the direction of the integral points:

$$L = \int_0^b C_L \cdot \frac{1}{2}\rho V_R^2 c dl = C_L \cdot \frac{1}{2}\rho V_R^2 cb$$

$$D_r = \int_0^b C_D \cdot \frac{1}{2}\rho V_R^2 c dl = C_D \cdot \frac{1}{2}\rho V_R^2 cb$$

Instantaneous energy utilization factor for a single blade:

$$C_{tp} = \frac{q(\theta)\cdot\omega}{0.5\rho V_A^3 Db} = \lambda C_{tq}$$

In the process of periodic rotation of the blades, the force of both the impeller and the blade are all functions

 https://doi.org/10.1051/matecconf/201815303002

of the azimuth angle θ. The average torque coefficient C_q and the average energy utilization factor C_p expression of the turbine impeller can be expressed as follow:

$$C_q = \frac{\sigma}{2}\int_{-\frac{\pi}{2}}^{\frac{3\pi}{2}} \bar{V}_R^2 \cdot \left[C_L \sin(\alpha+\varphi) - C_D \cos(\alpha+\varphi)\right] d\theta$$

$$C_p = \lambda\frac{\sigma}{2}\int_{-\frac{\pi}{2}}^{\frac{3\pi}{2}} \bar{V}_R \cdot \left[C_L \cdot (1+\lambda_1)\cos\theta - C_D\left(\lambda + (1+\lambda_1)\sin\theta\right)\right] d\theta$$

Where: σ is the blade density,

$$\sigma = \frac{zC}{\pi D}$$

It can be seen from the above equation that the parameters such as the tip ratio λ, the induction ratio λ_1, the relative velocity coefficient VR, the blade lift coefficient C_L and the drag coefficient C_D and the blade angle of attack α are directly affected the overall average energy utilization coefficient C_p of the impeller. For the solution of the induced ratio, there are many solving models in the previous research, and the determination of the airfoil drag coefficient is often dependent on the fluid simulation results, which is different from the actual airfoil parameters. A method for solving the lift-drag coefficient by wind-hole test data of lift-drag coefficient is given in this paper.

2.2 Analysis on lift - drag coefficient of Blade

Taking the NACA0018 airfoil as an example, the corresponding wind tunnel experimental data points for the lift-drag coefficient and the range of $0 \sim \pi$ the angle of attack under different Reynolds number was shown in Figure 2:

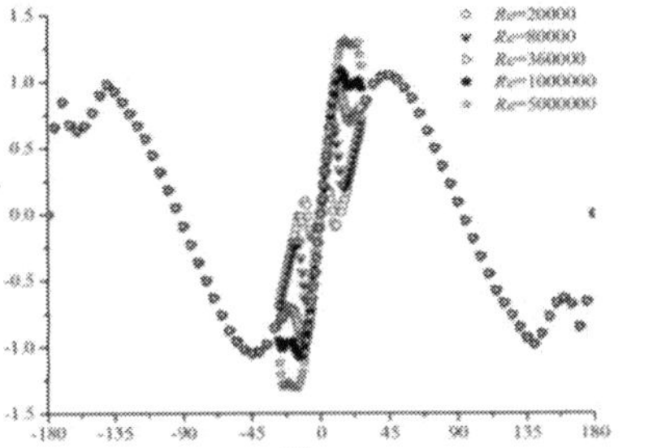

(a) The lift coefficient data points for blade angle of attack

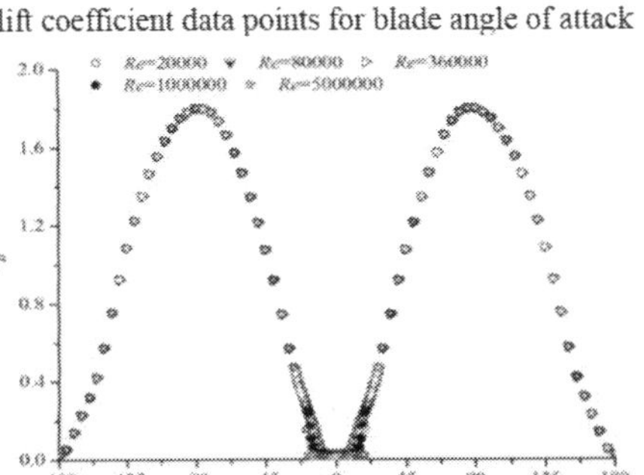

(b) The drag coefficient data points for blade angle of attack

Figure 2. Lift-Drag coefficient blade angle of attack under different Reynolds number

From the above data points, we can see that the Reynolds number and the angle of attack on the airfoil surface are the two main influencing factors of the airfoil. The Reynolds number here is determined by the relative velocity VR between the blade and the fluid, and the relative velocity here is often referred to as the water flow VA. The three-dimensional matrix was established by using the test data and the limited Reynolds number series and the data points under the blade angle of attack. The experimental data matrix is interpolated by cubic spline function in Matlab to solve the blade lift - drag coefficient under the Reynolds number and blade angle of attack.

3 Optimization model of law of variable-pitch

Based on the hydrodynamic analysis of the vertical axis turbine, the performance parameters of the vertical axis turbine and the expression of the parameters of the vertical axis turbine are obtained:

$$C_q = \frac{\sigma}{2}\int_{-\frac{\pi}{2}}^{\frac{3\pi}{2}} \bar{V}_R^2 \cdot \left[C_L \sin(\alpha+\varphi) - C_D \cos(\alpha+\varphi)\right] d\theta$$

The key factor that determines the output performance of the vertical axis turbine is the pitch law. By changing the law of variable pitch, in order to obtain the optimal value under the fixed tip ratio or the average optimum value in the constant speed ratio to solve the performance optimization problem of the turbine. In order to figure out the optimal pitch law of the vertical-axis turbine, the optimal value of the performance at the fixed blade tip ratio is chosen as the optimization target of the performance optimization.

Thus, the above optimization problem is described in function form as follows:

$$\begin{cases} S(\varphi(\theta)) = C_P(\lambda) = \lambda C_Q(\lambda) = \lambda \int_{-\pi/2}^{3\pi/2} C_{tq}(\lambda,\theta) d\theta \\ \max S(\varphi(\theta)) \end{cases}$$

The integral process of the average energy utilization coefficient value of the turbine is the integral process of the discrete energy utilization coefficient. The blade position angle may be discretized in one rotation period of the impeller:

$$\begin{cases} \varphi_i = \varphi(\theta_i) \\ \theta_i = -85^\circ + \dfrac{360^\circ}{n} * i \quad (i = 0,1,2...n) \end{cases}$$

Then the above optimization problem can be expressed as:

$$S(\varphi(\theta)) = C_P(\lambda) = \lambda C_Q(\lambda) = \int_{\theta_0}^{\theta_n} C_{tq}(\theta_i) d\theta$$

$$\max S(\varphi(\theta))$$

It is known from the above equation that if the optimal target is maximized, a series of declination of the integral value in one cycle should be the maximum, thus the optimization problem is turn to kind of multi-variable

 https://doi.org/10.1051/matecconf/201815303002

nonlinear programming problem. At a specific ratio speed, the optimization target is only the integral of a series of discrete values $C_{tq}(\theta_i)$ in a position of the angular period, so the optimization target the integral value transferred to the discrete value. If the arbitrary position of the instantaneous torque coefficient $C_{tq}(\theta_i)$ comes to maximum, undoubtedly torque coefficient value and the average energy utilization coefficient value of the turbine are fundamentally increased. The designed optimization variable is the variable pitch at any position angle, thus the objective function is also a plurality of discrete values, expressed as follow:

$$\begin{cases} \max S(\varphi_i) = C_{tq}(\theta_i) = \overline{V}_R^2 \overline{C} \cdot \left[C_L \sin(\alpha_i + \varphi_i) - C_D \cos(\alpha_i + \varphi_i) \right] \\ \varphi_i = \varphi(\theta_i) \\ \theta_i = -85° + \dfrac{360°}{n} * i \qquad (i = 0,1,2...n) \end{cases}$$

and:

$$\begin{cases} \max S(C_{Li}, C_{Di}) = C_{tq}(\theta_i) \\ \qquad = \overline{V}_{Ri} \overline{C} \cdot [C_L (1+\lambda_1)\cos(\theta_i) - C_D(\lambda + (1+\lambda_1))\sin(\theta_i)] \\ \qquad \overline{V_{Ri}} = \sqrt{(1+\lambda_1)^2 + \lambda^2 + 2\lambda(1+\lambda_1)\sin\theta_i} \\ \qquad C_{Li} = C_L(\alpha_i), C_{Di} = C_D(\alpha_i) \\ \qquad \alpha_i = \alpha(\theta_i) \\ \qquad \theta_i = -85° + \dfrac{360°}{n} * i \qquad (i = 0,1,2...n) \end{cases}$$

It can be seen that the factors that directly affect the optimization target are the position angle θ_i, the lift-drag coefficient C_{Li}, C_{Di} and the relative flow velocity VR. According to the above analysis, it can be seen that in these four factors, except for the position angle θ_i, the second influencing factor is the blade angle of attack α_i.

The blade angle of attack at any position is used as the optimization design variable, so that the relationship between the optimization target and the design variable can be expressed more directly.

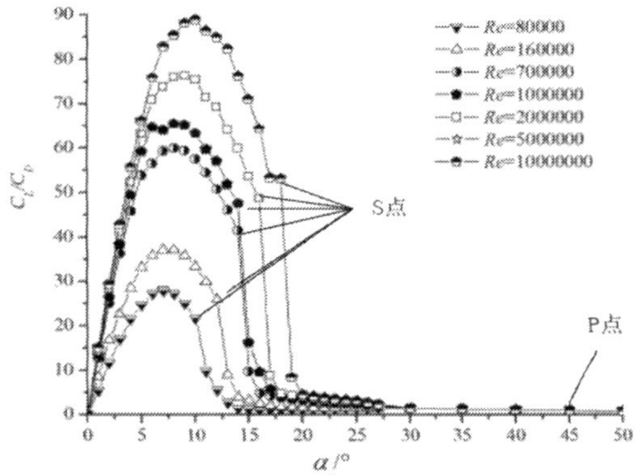

Figure 3. The relationship between airfoil lift - drag coefficient and the angle of attack.

In the design process of the turbine, it is desirable that the airfoil is in good working condition in other words is subject to a large lift and a small drag force. Combined with the airfoil lift - drag coefficient solution given above, the curve of the airfoil lift - drag coefficient with the variation of blade angle of attack is shown in Figure 3. The airfoil lift increases with the increase of the angle α , however the airfoil does not produce a large lift in any case. If the blade angle of attack α is large enough to a certain extent, the airfoil coefficient ratio will appear a sharp decline in the phenomenon, called dynamic stall, which was marked point S in Figure 3. It can be seen that, the slope of the airfoil increases with the increase of the Reynolds number, and the slope of the lift - drag ratio increases, and the stall point of the airfoil gradually increases after the process of rising and then slowly descending before the point S, while they are all below 16 degrees.

4 Simulation result

The objective function of the above-mentioned vertical axis turbine performance optimization problem is a complex nonlinear function with high nonlinearity, multi-peak and univariate. The objective function can not be derived. The design variables are discrete and no fixed population. Traditional research algorithms (such as Newton method, quasi-Newton method, conjugate gradient method, rotation direction method) can not be used. Taking into account the design of this variable range is relatively small, the use of simple enumeration method is adopted in this paper. The enumeration method enumerates all the solutions of the objective function in the feasible solution set, and obtains the optimal solution. This method is very suitable for the variable which is relatively small. And the algorithm is simple, easy to solve.

By Using matlab, when at the specified tip ratio, the blade angle of attack - position data points are figured out at the designed speed.

In the calculation of the performance of the vertical axis turbine optimization model, taking into account of working conditions of the turbine blades, run the set angle range (-16 °,16 °) of the matlab program,

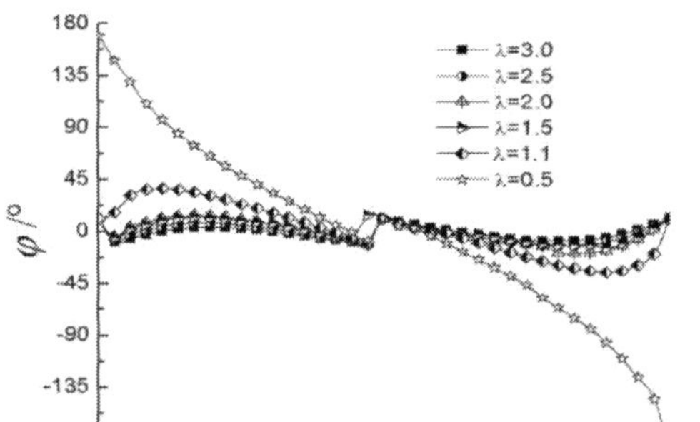

Figure 4. Result of the blade angle of attack and position angle by optimized performance model.

It can be seen from Fig. 4 that the angle of attack of the blade with optimized by the performance optimization model is limited to the set range of the stall angle, and the angle of attack angle of 90 ° and -90 ° changed dramatically, the variation of the two intervals is about the point (90°, 0) symmetry.

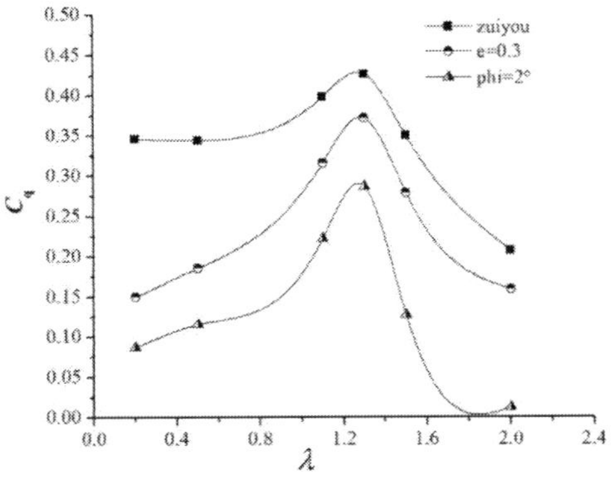

Figure 5. Comparison of Cq-λ curves under different pitch law.

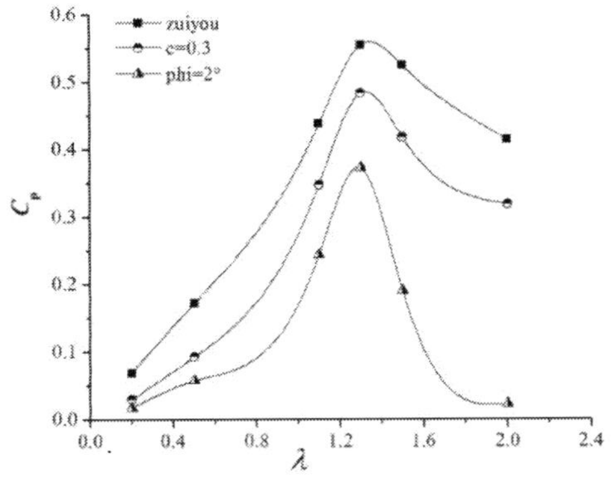

Figure 6. Comparison of C_p-λ curves under different pitch law.

By comparison, when the tip ratio is within the range of 0~2, it can be seen that the solved pitch law is higher than the starting torque under the other two pitch laws, the maximum angle of the average energy capture rate is up to 0.554, which is better than that of the reference whose theoretical analysis is 0.47. In the result, the correctness of the performance optimization model of the vertical-axis turbine is verified.

5 Conclusion

Based on the analysis of the hydrodynamic performance of the vertical axis tidal power turbine, the instantaneous moment coefficient of the blade is used as the optimization target while the blade angle of attack is the design variable. The model is used to obtain the blade speed. The instantaneous angle of attack of the blade with the best moment coefficient and the variation curve of the energy utilization coefficient under the fixed angle relatively to the conventional cycloidal turbine and the fixed angle turbine.

The energy capture efficiency of the vertical-axis turbine largely depends on the variation pitch law. In this paper, the turbine performance optimization model is established and the optimized pitch law is finally given, which improve the average energy capture efficiency of the turbine to the max 0.554. It undoubtedly provides the necessary theoretical basis for further research on the hydraulic performance of turbine.

Acknowledgment

The research is supported by Harbin Applied Technology Research and Development Project:
(NO.2015RQXXJ016);
Harbin Engineering University of basic university basic scientific research business special funds: (HEUCFP201606)

References

1. Z. Liang, Journal of Hydroelectric Engineering, **35**,1-15(2016)
2. P. Koschorrek, C. Siebert, A. Haghani, T. Jeinsch, IFAC-Papers ,**48,**187-183(2015)
3. J. Yuan, HEU Press, **1**,3-4,(2008)
4. H.Wang, I.S., Y.H. Lee, and S.J. Kim. Applied Energy,**86(9),**1532-1540,(2009)
5. U.Schubert, P. Lampe, IFAC-Papers OnLine ,**49**, 557 - 561(2016)

Power Loss and Heat Dissipation Characteristics of High-Power Electromechanical Actuators

Li Peng, Gao Zhi-gang and Zhou Jun

Institute of Precision Guidance and Control, School of Astronautics, Northwestern Polytechnical University, 710072, Xi'an, China

Abstract. Due to serious power loss and heating caused by compact volume and high power density of high power electromechanical actuator (EMA), thermal analysis is becoming the key aspects in its design optimization. Numerical calculation method is proposed to calculate the electrical loss, iron core loss and drive chain mechanical loss in EMA. An equivalent thermal network model based on lumped parameters is established, with the acquisition of heat transfer matrix equation of all network nodes. Temperature rise of each part of the EMA is obtained by solving the equation. By comparison of simulated temperature rise in rated operating condition with test results, the accuracy of the model is verified. Then temperature rise simulation of EMA under simulated actual flight condition is carried out, with the results showing positive guidelines in the optimization design of EMA systems.

1 Introduction

In aerospace field, electromechanical actuator (EMA) is widely used as flight control system to change the attitude of flying vehicles by manipulating the deflection of control surfaces. Due to harsh space limitations and highly dynamic behavior requirements, dimension of EMA is reduced continuously, while the power level is elevating significantly. Therefore, power density and working speed of the servo motor are increasingly high, while its dimension and inertia are cut down to a greatest degree. One problem brought out by this is that power loss and heating of the EMA motor is severe, which has become one of the important factors that limit further improvement of EMA systems [1-4].

Currently, studies on power loss and heat dissipation characteristics of electromechanical servo system focus on the servo motor itself in most cases, while those performing systematic thermal analysis on an actuator can be rarely seen. Furthermore, the influences of transmission mechanism and the actual working condition of actuator system are also not taken into account. Therefore, this paper focuses on the power loss and temperature rise characteristics of EMA from a system perspective, laying a foundation for optimization design and power level improvement of actuator systems.

Research of this work is based on a 4KW EMA for a primary flight control surface in more electric aircraft. The EMA is composed by a permanent magnet synchronous motor (PMSM), a gear reducer and a high precision potentiometer. A numerical calculation method is used to estimate electrical loss, core loss and mechanical loss of the actuator. Then, temperature field modeling is performed for the system according to an equivalent thermal network method using lumped parameters. In addition, a temperature rise matrix equation is also set up for all nodes in the temperature field. Computational complexity is reduced while the accuracy of the model is guaranteed by this way. At last, system temperature rise simulation and comparative analysis were carried out respectively to the EMA in rated load working condition and simulated flight condition. Measures beneficial to further improvement to the actuator are proposed according to the simulation results.

2 Analysis on Power Loss of EMA

Power loss of an EMA is mainly derived from the servo motor and the transmission gear. Specifically, losses incurred by the servo motor include the winding loss and the core loss, and losses of the transmission gear are mainly constituted by the frictional loss of bearings. The power and loss flow of an EMA is presented in Figure 1.

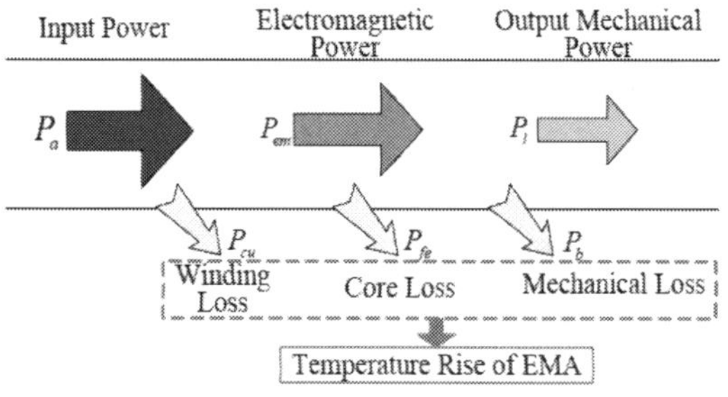

Figure 1. The power flow of EMA.

Power losses generated in the EMA dissipate in a form of heat, which makes temperature of motor and other mechanism parts rise. Furthermore, temperature rise affects the lifetime and reliability of EMA. Thus, EMA loss and heating modelling aims to check temperature rise situations of all heating parts in a state of rated operating to guarantee that temperature rise of a mechanism within service life does not exceed the permitted value.

2.1 Servo Motor Loss

2.1.1 Winding Loss

Winding loss, also known as copper loss, is caused by the resistance characteristics of stator winding in PMSM, which can be written as

$$P_{cu} = NI^2R = \rho NI^2 \frac{L}{A} = \rho NJ^2 LA \tag{1}$$

where N is the phase number of winding, I is the phase current, R is the phase resistance, L is the length of a phase, A is the cross sectional area of the wire, and ρ is the winding resistivity related with temperature and $\rho=\rho_{20}(1+\alpha(T\text{-}20))$, among which T is winding temperature, α is temperature coefficient (α=0.0039 for copper wire) and ρ_{20} is resistivity at 20℃. Thus, winding loss is related to the material, and directly proportional to both the square of current density and the volume of winding.

As is shown in (1), winding loss depends mainly on the phase current decided by the mechanical load, and is independent of revolving speed. Moreover, winding resistance of high-power motor is relatively small, usually at a magnitude of mΩ, so the winding loss is usually not the largest part among the losses. However, when a harsh condition is imposed, the winding loss may not be negligible due to its temperature rise.

2.1.2 Iron Core Loss

Core loss is one of the major losses in high-power motor and it is especially obvious for multi-pole high-speed motors. Due to the variation of magnetic field inside the motor, magnetic hysteresis loss, eddy current loss and added loss are all incurred in the iron core. Magnetic hysteresis loss is irrelevant with waveforms of the magnetic field but has a direct proportion to amplitude and frequency of the magnetic field. As for the eddy-current loss, it is not only associated with alternate frequency and amplitude of the magnetic field, but closely related to its waveform.

Magnetic field changes sinusoidally in PMSM. As a result, based on the classical Bertotti model [5-7], core loss can be simplified as:

$$\begin{aligned} P_{fe} &= P_h + P_{cl} + P_{ex} \\ &= K_h f B_m^{\alpha} + K_{cl} f^2 B_m^2 + K_{ex} f^{1.5} B_m^{1.5} \end{aligned} \tag{2}$$

where P_{fe} is the total core loss of servo motor, P_h, P_{cl} and P_{ex} are magnetic hysteresis loss, eddy-current loss and excess loss respectively, K_h, K_{cl} and K_{ex} are computing coefficients of magnetic hysteresis loss, eddy-current loss and excess loss, related to both the type and the shape of silicon steel sheet, and they can be obtained by tests. In addition, α is an empirical coefficient equal to 1.6, f is the frequency of alternate magnetic field and B_m is the magnitude of the magnetic field.

Alternate frequency of the magnetic field within a servo motor increases with the rise of revolving speed, which further leads to rapid growth of core loss. Consequently, temperature rise occurs in the system which cannot be ignored.

2.2 Mechanical Loss of Transmission Chain

Mechanical loss is mainly composed of bearing friction loss of the transmission chain. Bearing friction loss is directly proportional to the revolving speed, which takes only a small percent, but it cannot be neglected as far as the high-speed heavy-load mechanism is concerned. Bearing gives rise to friction power loss, which can be expressed as

$$P_b = \sum_i T_i \omega_i \tag{3}$$

where P_b is the total loss of all drive bearings, T_i is torque of bearing friction related to bearing structure, lubrication and sealing, and ω_i is the revolving speed of bearing. Heat generated by bearing loss has a direct influence on its service life. Therefore, requirements of bearing load, bearing roller service life and reliability should be taken into the comprehensive consideration at the time of type selection.

3 Heat dissipation model of EMA

Temperature field of the actuator was simplified into a thermal network structure with the lumped parameter to perform heat dissipation analysis. The main purpose is to obtain temperature rises of motor winding, core, permanent magnet and the shell. All these parts are heat-transfer media with heat sources. The following assumptions are made before calculation [8-11]: (a) all parts inside the mechanism were isothermal bodies of uniform temperature rises, (b) all parts are cooled down naturally, and (c) air temperature of the external environment is constant.

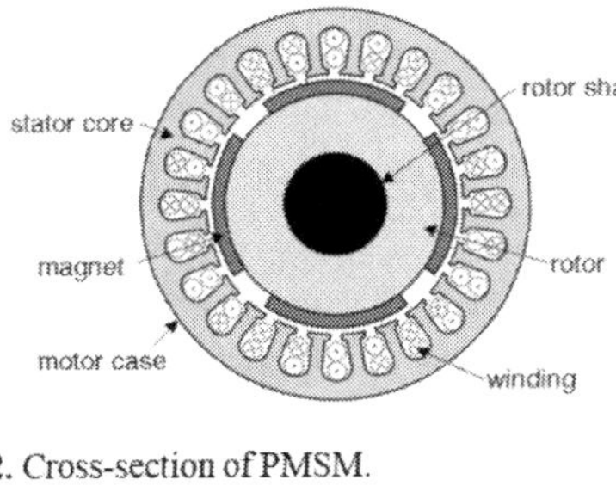

Figure 2. Cross-section of PMSM.

Cross-section structure of the PMSM is shown in Figure 2. Considering the major heat source and loss distribution in the motor as well as the connection between the motor and the reducer, a temperature network model containing 15 nodes is established, as

https://doi.org/10.1051/matecconf/201815303003

shown in Figure 3. Each node represents a main component involved in the heat transfer of the EMA.

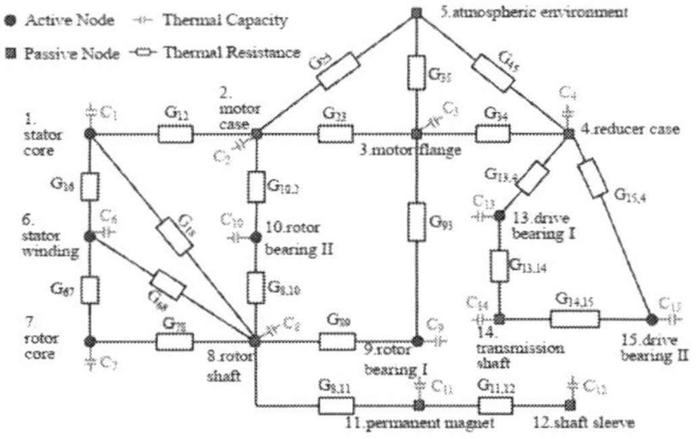

Figure 3. EMA Lumped parameter thermal network model.

Nodes in the temperature network above are adopted according to different characteristics of the main parts of actuator and determined by factors such as temperature distribution, mechanical complexity and material behaviour, etc. All nodes are linked together by thermal conductance (G_{ij}, where i and j refer to two nodes adjacent to each other respectively) depending on a thermal conductivity relation[12-15]. Furthermore, the thermal conductivity coefficient is principally related to the material of each node, as is shown in Table 1.

Table 1. Material properties in the EMA thermal model

Parts	Material	Density (kg/m^3)	Heat conductivity coeff.(W/mK)	Heat capacity (J/(kg·K))
core	silicon steel	7680	42.5	480
winding	copper	8900	382	405.7
rotor shaft	steel	7800	50	460
PM	NdFeB	7500	40	480
rotor sleeve	aluminum	2700	205	880
atmos phere	air	1.23	0.028	990
insulation	epoxy	980	0.2	1200

Based on heat conduction principle, temperature equation of all nodes in the model can be written as

$$GT + C\frac{dT}{dt} = Q \quad (4)$$

where, T is a n-dimensional temperature matrix of each node, G is a $n \times n$ dimensional thermal conductivity coefficient matrix of each node, C is a $n \times n$ dimensional heat capacity matrix of each node, Q is a n-dimensional heat source matrix, and n is the number of nodes in the thermal network and n=15 in this paper. By solving the above equation, temperature curves for all nodes in the thermal network model can be obtained.

4 Simulation and Analysis

Table 2. gives the main parameters of the subject used in simulation. The EMA works in a short-time duty mode with a continuous working time of 5min in a rated load condition.

Table 2. Parameters of the EMA model

Servo Motor		Reducer	
mass(kg)	2.6	mass(kg)	3.0
size(mm)	Φ60×110	size(mm)	80×100×150
rated voltage(V)	220	drive ratio	21
rated power(W)	4000	type of drive	gear
rated speed(rpm)	13000	material	steel
pole pairs	6	position feedback	potentiometer
phase resistance(Ω)	0.22	cooling pattern	passive

4.1 Rated Working Condition Simulation

Based on the established loss calculation model, loss simulation results for main parts of the EMA in a rated working condition (continuously rotating in rated load) are presented in Figure 4. Core loss of the stator is 358W taking the greatest proportion (61%), followed by the winding loss (34%) and the mechanical loss takes only 1% at the minimum level.

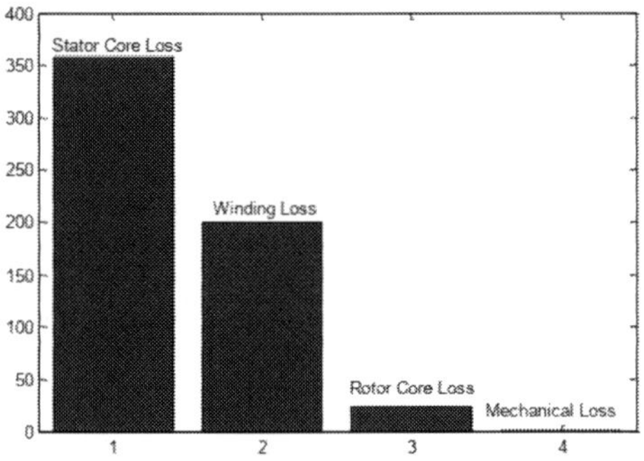

Figure 4. Power loss in rated working condition.

Figure 5 shows temperature rise simulation curves for all major parts inside the actuator, where the ambient temperature is set at 300K or 26.85℃. At 300s, sequencing of temperature rise values are winding > stator core > shell > rotor. Specifically, temperature rises of the winding and the shell are 81℃ and 59℃ respectively.

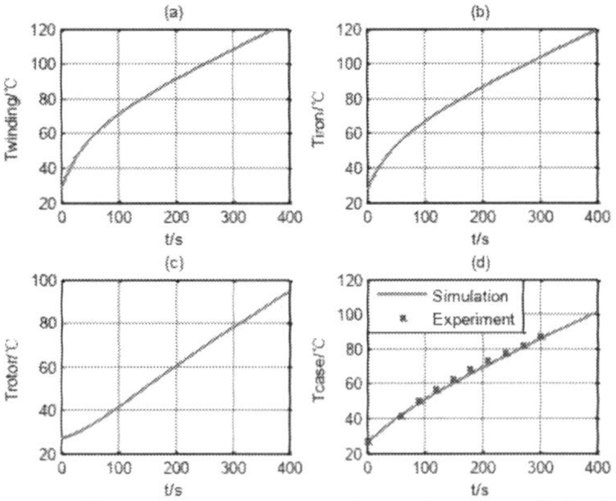

Figure 5. Simulated temperature rise and test result in rated working condition.

 https://doi.org/10.1051/matecconf/201815303003

In Figure 5(d), the measured temperature curve of shell is compared with the simulation curve, which clearly signifies that both curves highly coincide with each other. Initial temperature of the test is 26.6℃ and temperature rise reaches 60.2℃ at 300s. Thus the accuracy of the heat dissipation model is verified. Figure 6 gives the thermograph of the mechanism when the test ended at 300s, with the hottest spot of 86.8℃.

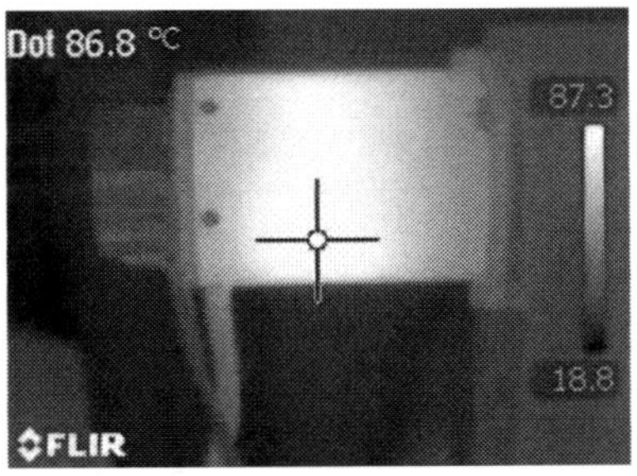

Figure 6. Thermograph of EMA in rated working condition.

Figure 7 displays the comparison of temperature rise curves between an actuator (with reducer) and a single motor (without reducer) under identical rated load circumstances. It can be found that temperature rising in the actuator is slower than a single motor due to heat capacity and heat dissipation effects of the transmission devices in the actuator. At 300s, average temperature rise of each part of the actuator is about 30℃ lower than that of the motor.

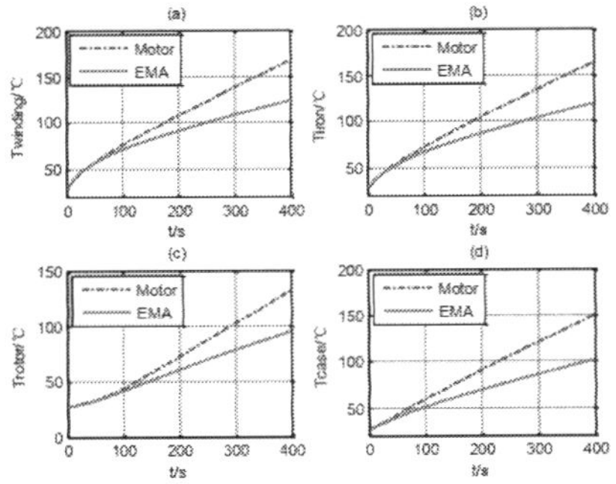

Figure 7. Motor and EMA temperature rise comparison in rated working condition.

4.2 Flight condition simulation

During actual flight mission, the EMA does not entirely work in rated load state. So it is necessary to perform simulation analysis on practical situations. A "three-stage process" approach is proposed in this paper to simulate general flight characteristics. In detail, it consisted of low-frequency sinusoidal motion, high-frequency sinusoidal motion and step motion; and each of them occupies a third segment of the whole flight time. Position command signals sent to the actuator are as follows: u=10sin(0.4πt) for 0~100s, u=sin(20πt) for 100~200s, u=10square(0.4πt) for 200~300s and u=0 for 300~400s. Besides, unit of the command is V and the corresponding rudder coefficient is 2°/V. Position command and actuator position feedback curves are given in Figure 8(a).

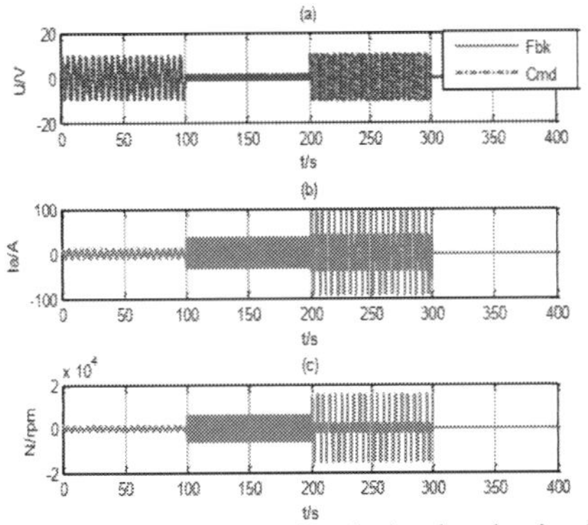

Figure 8. Motor current and velocity in simulated flight condition

Figure 8(b) and 8(c) represent the bus current of the actuator and the revolving speed of the motor under the given command respectively. It can be seen that at a low-frequency stage of 0~100s, both current and revolving speed of the actuator are very low, at a high-frequency stage of 100~200s, current rises up at a peak of about 40A and the revolving speed also increases, at a step motion stage of 200~300s, current surge approaches 100A in each cycle and the corresponding revolving speed also reaches about 16,000rpm in transient process.

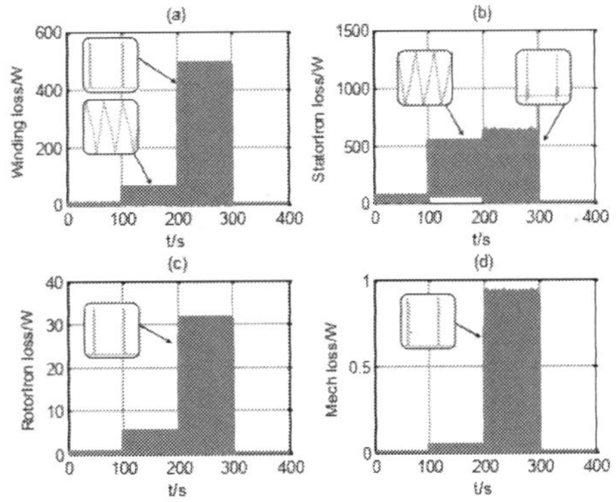

Figure 9. Loss curve of EMA in simulated flight condition

Figure 9 represents loss curves of the main parts in the actuator under the given commands. Clearly, loss curve characteristics are in consistence with the variation features of current and motor's revolving speed shown in Figure 8.

Temperature rise curves of the main parts of EMA in a simulated flight condition are shown in Figure 10. At the low-frequency stage of 0~100s, temperature rises of all parts in the actuator are very low. At the high-frequency stage of 100~200s, rapid temperature rise is incurred by large current, high speed and great average loss, and this is also a primary stage of temperature rise. At the step motion stage of 200~300s, the average loss is not high although it has a high peak power loss. Gradual heat dissipation leads to decreases in temperature of winding and core, while temperature of rotor and shell still rises by 5℃ and 1.5℃ respectively. At 300s, temperature rise of winding parts arrives at the maximum value of 30℃, which is only 37% of that in a rated working condition.

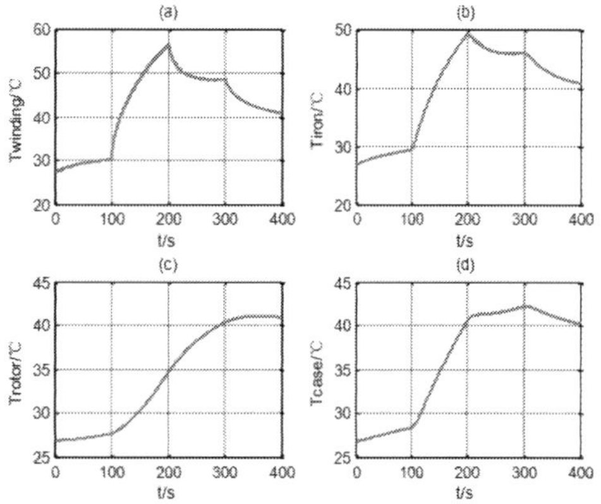

Figure 10. EMA temperature rise in simulated flight condition.

5 Conclusion

Aiming at severe loss and heating problem of high-power electromechanical actuators, a thermal analysis approach based on equivalent thermal network method with lumped parameters is proposed in this paper. An equivalent thermal network model is constructed based on a numerical calculation method to precisely estimate actuator losses. Thermal equilibrium matrix equation is established for all nodes in the thermal network to obtain temperature rise curves of all main parts in the actuator in rated load condition and simulated flight condition.

According to simulation results, it can be concluded that: (a) transmission mechanism of the actuator is able to substantially improve temperature rise of the EMA due to its heat capacity and heat dissipation effects, despite that it incurs a part of mechanical loss, (b) integrated design for servo motor and transmission mechanism shell can further enhance thermal characteristics of the system, and (c) actual operation mode of EMA has a dramatic effect on it temperature rise, thus the thermal analysis should be performed for the actual flight course so as to optimize the design of the actuators.

Acknowledgement

This work was supported by the Fundamental Research Funds for the Central Universities (No.3102017GX14003) of China.

References

1. Chen Yiguang, Zheng Jun, Wei Juan, et al. Design of PMSM for actuator and its temperature field analysis[J]. Transactions of China Electrotechnical Society, **30**(14): 94-99, (2015).
2. Takeda Y, Takahashi Y, Fujiwara K, et al. Iron Loss Estimation Method for Rotating Machines Taking Account of Hysteretic Property[J]. Magnetics IEEE Transactions on, **51**(3):1-4, (2015).
3. Wang W, Zhou Y, Chen Y. Investigation of lumped-parameter thermal model and thermal parameters test for IPMSM[C]. International Conference on Electrical Machines and Systems, IEEE, (2015).
4. Zhang Jianzhong, Jiang Yongjiang. Thermal Analysis of Constant Frequency Double Rotor Permanent Magnet Generator Based on Equivalent Thermal Network Method. Transactions of China Electrotechnical Society. 30(02): 87-97, (2015).
5. Kong Xiaoguang, Wang Fengxiang, Xing Junqiang. Losses calculation and temperature field analysis of high speed permanent magnet machines[J]. Transactions of China Electrotechnical Society, 27(9): 166-172, (2012).
6. Zheng D, Wang D, Li S, et al. Eddy current loss calculation and thermal analysis of axial-flux permanent magnet couplers[J]. Aip Advances, 7(2):24-35, (2017).
7. DONG Jianning, HUANG Yunkai, JIN Long, et al. Review on High Speed Permanent Magnet Machines Including Design and Analysis Technologies[J]. Proceedings of the Chinese Society for Electrical Engineering, (27): 4640-4653, (2014).
8. LI Liyi, ZHANG Jiangpeng, YAN Haiyuan, et al. Study on the Optimization of Thermal Conductivity and 3D Temperature Filed Calculation for the High Power Density Motor[J]. Proceedings of the Chinese Society for Electrical Engineering, 36(13): 3642-3650, (2016).
9. Woodburn D, Wu T F, Zhou L, et al. High-Performance Electromechanical Actuator Dynamic Heat Generation Modeling[J]. IEEE Transactions on Aerospace & Electronic Systems, 50(1):530-541, (2014).
10. Sefkat G. Investigating static and dynamic characteristics of electromechanical actuators (EMA) with MATLAB GUIs[J]. Computer Applications in Engineering Education, 18(2):383-396, (2009).
11. Bahman A S, Ma K, Blaabjerg F. A Lumped Thermal Model Including Thermal Coupling and Thermal Boundary Conditions for High Power IGBT Modules[J]. IEEE Transactions on Power Electronics, **PP**(99):1-14, (2017).
12. Kim K S, Lee B H, Hong J P. Improvement of Thermal Equivalent Circuit Network and Prediction on Heat Characteristic of Motor by Calculation of Convection Heat Transfer Coefficient[C]. The 6th International Conference on Electromagnetic Field Problems and Applications, IEEE, (2012).
13. Boglietti A, Carpaneto E, Cossale M, et al. Stator-Winding Thermal Models for Short-Time Thermal Transients: Definition and Validation[J]. IEEE Transactions on Industrial Electronics, **63**(5): 2713-2721, (2016).
14. Ahmed F, Ghosh E, Kar N C. Transient thermal analysis of a copper rotor induction motor using a lumped parameter temperature network model[C], IEEE Transportation Electrification Conference and Expo, IEEE, (2016).
15. Sciascera C, Giangrande P, Papini L, et al. Analytical Thermal Model for Fast Stator Winding Temperature Prediction[J]. IEEE Transactions on Industrial Electronics, **64**(8): 6116-6126, (2017).

Modeling of the kinematic geometry of spur gears using matlab

[1]Peter I. Anakhu , [1]Abiodun A. Abioye, [1]Christian A. Bolu and [1] JosephAzeta.

[1]Mechanical Engineering Department, Covenant University, Ota, Nigeria

Abstract. The analytical method of gear design is calculation-intensive and it is usually difficult to achieve optimum backlash and interference-free involute profile that are required to generate geometrical compatibility in a pair of meshing gears when design procedure is entirely based on this method. Some amount of backlash is often required in the assembly of gears but excess backlash can lead to increase vibration and wear of the gear assembly. Also, interference-risk profile can result in undercutting of gear tooth. This paper optimized a spur external involute-profile gear by developing an application for the modeling of its geometrical compatibility using Matlab®. The application uses existing models to test for interference and a proposed model to determine effective backlash in a gear. The backlash values resulting from the application are more confined and the model is applicable to a wider range of modules suggested by American Gear Manufacturers Association. Simulation of the gear-set in Solidworks® for kinematic geometry presents an interference-free tooth contour and an effective backlash.

1 Introduction

A gear is a wheel on which toothed members are cut to transmit rotary motion or rotational force (torque) from one shaft to another [1, 2]. Gears are generally the most rugged and durable machine elements for transmitting mechanical power [3, 4]. In particular, involute gears are mission critical and indispensable elements in the design of motion and power transmission applications [5]. Such applications include external gear pump [6], wrist mechanism of a spherical parallel manipulator [7], automobile and aircraft power transmissions and more.

However, gear design is a very complicated process [8, 5]. The complexity is not unconnected with the combined basic design requirements imposed by the fundamental law of gearing, the kinematics of the gear tooth profile, the variable loading of gear tooth, the joint rolling and sliding contact between curved tooth surfaces, and the gear interaction with other machine elements [9]. The law states that a pair of meshing gears must transmit constant velocity ratio throughout the mesh [10]. That is, the ratio must not vary between when a given pair of teeth comes into mesh and when they go out of mesh [3]. Any such variation will produce oscillation in the output velocity and torque, even if the input is constant with time [10]. Therefore, a design objective to conform with the fundamental law of gearing and to obtain gear teeth that will run smoothly through the angle of action must consider three major design tasks, viz: (1) to avoid interference, (2) to achieve a contact ratio greater than one to assure of one tooth pair (preferably more) is in contact at all times and (3) to provide an optimum amount of backlash in the mesh [9]. Interference defines the contact of portions of tooth profiles, around the flanks, that are not conjugate [11]. Contact ratio is a method used to determine how many teeth are in contact at any time and the possibility of load sharing among the teeth [4]. Backlash is the clearance between the width of space and tooth thickness, both measured on the pitch circle, to compensate for unavoidable inaccuracies in gear assembly [12, 4].

Table 1. Nomenclature: Gear terms, symbols and description

Symbols	Description	Symbols	Description
m	module	k	A factor by which the standard addendum of the gear should be multiplied.
α	Pressure angle	u	Gear ratio
d_{p1}	Pitch circle diameter of pinion	z_P	Number of teeth on pinion
d_{p2}	Pitch circle diameter of gear	z_G	Number of teeth on gear

$\frac{d_{pa}}{2}$	Limiting function of interference	R_{actual}	Addendum radius of the gear
j	Backlash		

To avoid violating the law of gearing, the gear tooth contours on mating teeth must be conjugates of one another [12, 10]. The authors argue that one of the possibilities of tooth contours for conjugate pairs that is useful as gear teeth for practical application is the involute contour. Also, involute curve is the most widely used shape for creating the active profiles of spur and helical gears to achieve constant velocity throughout the mesh Figure 1 [8, 1]. This study developed an application for the modeling of involute-profile gears in consideration of interference-free contour and optimum backlash using Matlab® and Solidworks®.

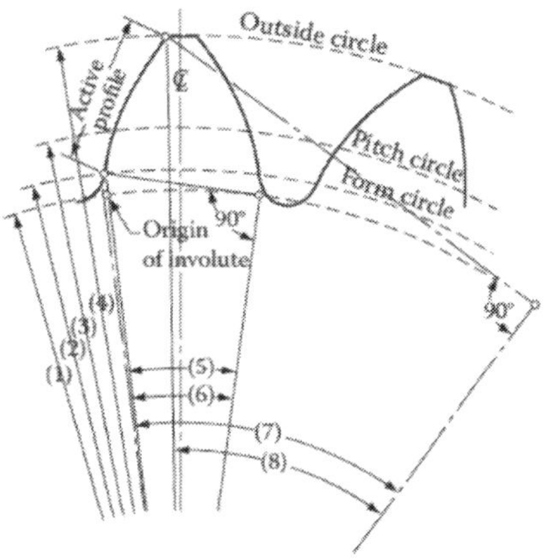

Figure 1. Active profile of an involute curve

2 Review of past work

In designing a mechanism, the real work lies in the determination of the sizes and shapes of its members [13]. Gear, as a type of mechanism, often requires complicated analysis to determine its optimum geometry [9]. Similarly, gear performance is in turn greatly dependent on gear tooth shape [14]. To this effect, the author suggests the optimization of gear tooth geometry as part of the design process to maximise gear drive performance. The tooth geometry optimization is aimed at increasing load capacity, revolution per minute (RPM), life span, and reducing vibrations and noise. The developed application will optimize the tooth shape of a spur gear using involute profile.

Several literatures have proposed different methods for interference-free design. They noted that in a gear design process, when interference is present, it is not a good practice to proceed on the design with the resulting parameters. It may lead to overlapping of the involute profiles of the gear or cut into the mating gear [15] thereby causing undesirable noise and early gear failure. One method involves producing gears through generation process that automatically eliminates interference by removing the interfering portion of the flank through the cutting tool [16]. However, the authors argue that this merely substitutes the original problem with another one called undercutting and it considerably weakens the gear. Undercutting is the removal of a portion from the active profile of the tooth by secondary cutting action [1]. Other methods, based on gear geometry, involve the use of minimum number of teeth on the pinion to check for interference. For a spur gear-set with a gear ratio of unity, the lowest number of teeth on both gears without interference may be determined from equation 1 [17].

$$z_P = \frac{2k}{3\,sin^2\alpha}\left(1 + \sqrt{1 + 3\,sin^2\alpha}\right) \qquad (1)$$

When the gear-set is for a speed reduction application in which the gear ratio is more than unity, the lowest number of teeth on the pinion without interference is given in equation 2 [11]

$$z_P = \frac{2k}{(1+2u)\,sin^2\alpha}\left(u + \sqrt{u^2 + (1 + 2u)\,sin^2\alpha}\right) \qquad (2)$$

These models are based on using more teeth on the pinion to obtain interference-free gear. However, if the pinion is to transmit a given amount of power, increasing the number of teeth on the pinion requires a simultaneous increase in the pitch diameter [16]. On the other hand, undercut-free design usually requires lower number of teeth [1]. Analytical approach to this trade-off may be intensive and error-prone. Therefore, an application for testing interference-free geometry is developed to reduce the drudgery. The application provides better flexibility in the optimization of interference-free gear tooth. It is based on a more general model of checking interference in involute gear geometry for most types of gears including spur, helical, bevel and more [15]. The objective function is to determine that the actual addendum radius is less than or equal to the maximum value of addendum radius of the gear to avoid interference. There is a maximum possible gear addendum that should avoid interference [3]. Therefore, the module can be reduced to design a gear geometry with interference-free contour based on the models below.

$$R_{actual} = \frac{d_{p2}}{2} + m \qquad (3)$$

$$\frac{d_{pa}}{2} = \sqrt{\left(\frac{d_{p2}}{2}\right)^2 cos^2\alpha + \left(\frac{d_{p2}}{2}\,sin\alpha + \frac{d_{p1}}{2}\,sin\alpha\right)^2} \qquad (4)$$

$$R_{actual} \leq \frac{d_{pa}}{2} \qquad (5)$$

In addition, using the right cutter number to generate a tooth profile is important in achieving interference-free geometry. Cutter number usually depends on the number of teeth on a gear [18]. The developed application will determine the cutter number based on the number of teeth on a gear.

The unavoidable inaccuracies in gear assembly has often impose the requirement of making the tooth thickness slightly thinner than the width of space between teeth, both measured on the pitch circle, to provide some clearance and this always result in backlash, Figure 2, [12,

4]. Consequently, in many applications, gears exhibit various backlashes that occur because of the loose fit essentially required between two meshed gears. While generous backlash is suitable for power gearing, narrow backlash is preferred for control gearing [1]. Optimum backlash is very critical in gear assembly and in providing a clearance for the lubrication of mating gear pair. Thus, the amount of backlash for each application is often indicated for that particular application and discretion is often required in utilizing such values for design [8]. For instance, a table of backlash was suggested as a guide for gear design for spur, helical, bevel, and spiral bevel gears [8]. But the author equally cautioned against using it in every design application. As useful as backlash could be for gear assembly, extra backlash could lead to loss of motion and causes undesirable gear performance, especially for reversing gear drive [1]. It usually increases stress and wear of the mating gears and also causes undesirable positional error in some applications [13]. The proposed model used in developing the application will generate a tooth geometry that has a more effective backlash and may be used for any module.

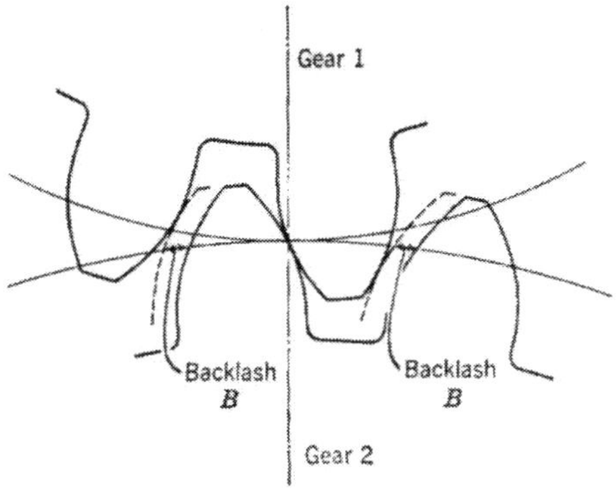

Figure 2. Backlash in gears

3 Materials and methods

3.1 Design and Modeling Parameters

The application adopted standard modules and pressure angles as published by American Gear Manufacturers Association (AGMA). The same module and pressure angle were used for both pinion and gear wheels because standardised cutting tools are available for such gear teeth geometry [5].

In determining an interference-free involute profile, a more general model is used in the application. The limiting functions for checking interference in a gear tooth contour are given in equations 3 to 5 and reproduced here as equations 6 to 8.

$$R_{actual} = \frac{d_{p2}}{2} + m \quad (6)$$

$$\frac{d_{pa}}{2} = \sqrt{\left(\frac{d_{p2}}{2}\right)^2 cos^2\alpha + \left(\frac{d_{p2}}{2} sin\alpha + \frac{d_{p1}}{2} sin\alpha\right)^2} \quad (7)$$

$$R_{actual} \leq \frac{d_{pa}}{2} \quad (8)$$

Interference is avoided if the addendum radius of the gear is less than or equal to the square root of the sum of squares of the base radius and the product of centre distance and pressure angle. Decreasing the module and increasing the gear pitch diameter is usually sufficient to avoid interference.

Backlash optimization using the proposed model is based on the hypothesis that backlash is a function of pitch circle diameter, d_p, the inverse of the number of teeth, z, and pressure angle, α. That is

$$j = f(d_p, \frac{1}{z}, \frac{1}{\alpha}) \quad (9)$$

3.2 Simulation analysis

The developed application was used to test a set of design variables including pressure angle, module, and number of teeth on pinion and gear wheels as shown in Table 2.

Table 2. Comparison of suggested backlash with modeled backlash

Pressure angle (θ)	20		
Number of teeth on pinion	18		
Number of teeth on gear	24		
Module	Interference	Suggested backlash* (mm)	Backlash from model (mm)
25	Interference-free	0.63 - 1.02	0.625
20	Interference-free	No matching values	0.500
18	Interference-free	0.46 - 0.69	0.450
12	Interference-free	0.35 - 0.51	0.300
15	Interference-free	No matching values	0.375
10	Interference-free	0.28 - 0.41	0.250
8	Interference-free	0.23 - 0.36	0.200
7	Interference-free	No matching values	0.175
6	Interference-free	0.18 - 0.28	0.150
5	Interference-free	0.15 - 0.23	0.125
4	Interference-free	0.13 - 0.20	0.100
3	Interference-free	0.10 - 0.15	0.075

2	Interference-free	0.08 - 0.13	0.050
1	Interference-free	0.05 - 0.10	0.025

***Source: Radzevick, 2012**

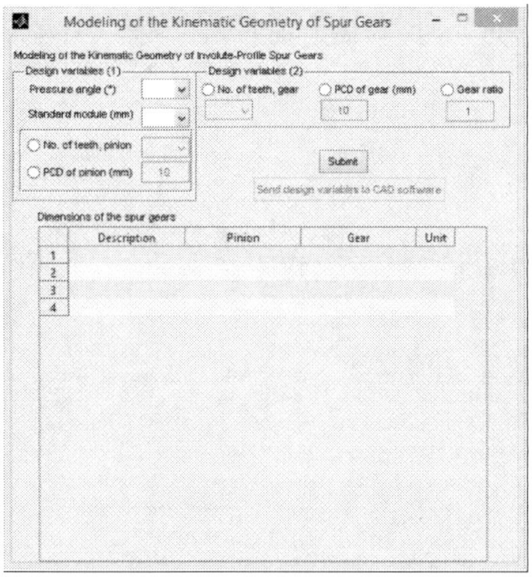

Figure 3. Spur gear geometry modeling GUI

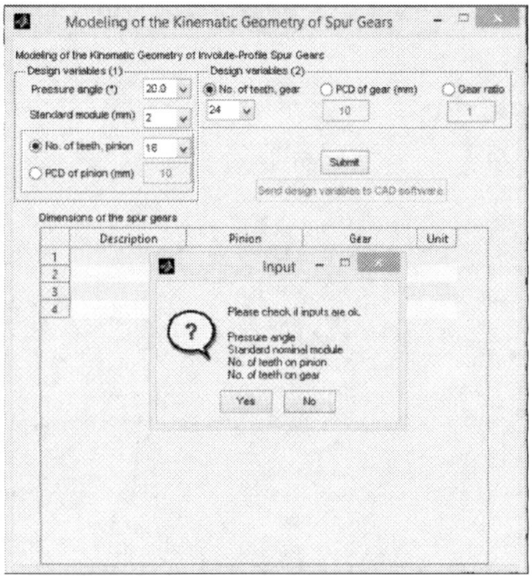

Figure 4. Modeling with gear design variables

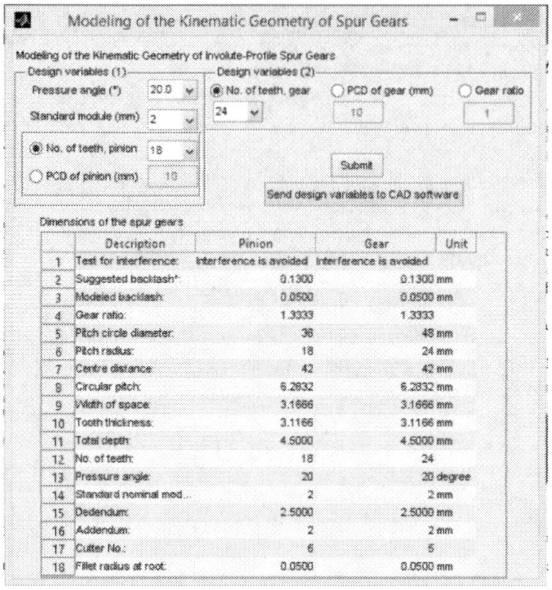

Figure 5. Modeling output

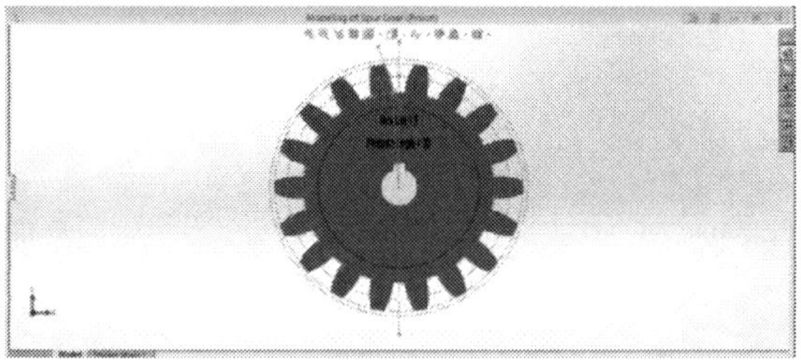

Figure 6. 3D Model of the pinion wheel

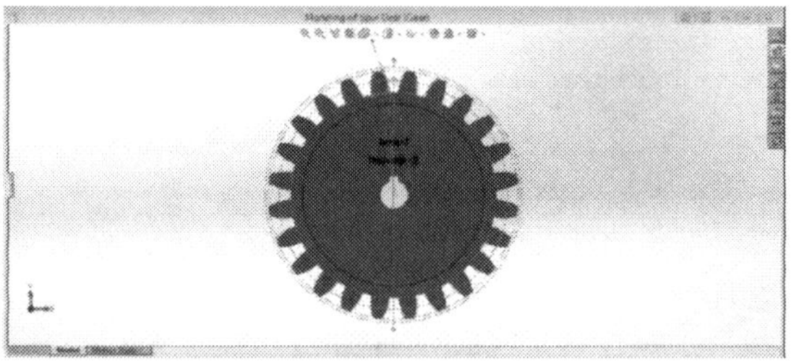

Figure 7. 3D Model of the gear wheel

4 Results

Backlash values suggested by previous authors are applicable to a narrow range of modules. Some authors limited the use of such values in certain design. The proposed model for this application can be used for a combination of any module, pressure angle and number of teeth to determine the optimum backlash required for the given design. Design of spur gear by modeling its kinematic geometry with this application created an involute-profile gear pair in Solidworks®, Figure 8. The application can generate text files for any parametric

CAD software to create a 3D model of an involute-profile gear. The text file synchronises the computation and the 3D modeling of the gear in a CAD software. It also tests for interference-free or interference-risk portion on the involute profile. The backlash model for gear tooth design is dependent on the design variables such as pitch circle diameter, d_p, the inverse of the number of teeth, z, and pressure angle, α. Therefore, it is simultaneously generated as shown in Table 3 rather than look-up tables.

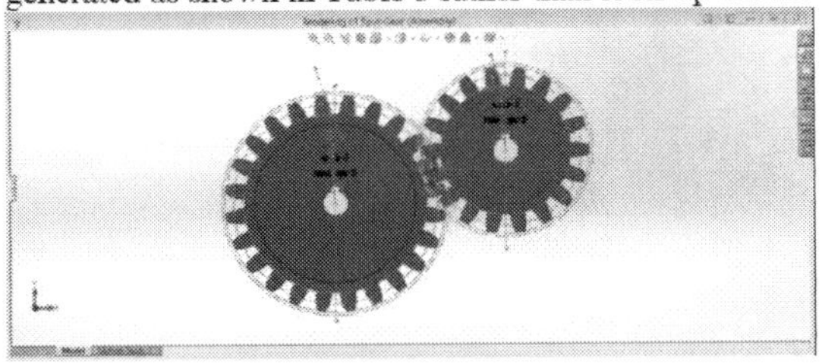

Figure 8. 3D Model of the gear assembly

5 Conclusions

In this study, the tooth shape of a spur gear was modeled using Matlab® and Solidworks®. The developed application was used to determine the effective backlash during the modeling of the gear tooth shape. Backlash can be obtained for any module and/or a combination of other gear design variables including pressure angle, number of teeth on pinion and gear, and gear ratio without the need for a look-up table or the use of thumb rule. Also, the presence of interference can be checked during the modeling process and the designer is prompted to adjust input variables to create an interference-free geometry. Furthermore, the application selects a cutter number depending on the number of teeth on a gear. Finally, in addition to the accuracy of the application, time of computation involved in gear tooth modeling is drastically reduced.

References

1 P.S. Radzevick, Dudley's Handbook of Practical Gear Design and Manufacture. 2nd Edn. CRC Press, Taylor & Francis Group, London, (2012), ISBN 13: 978-1-4398-6602-3

2 R.N.C. Peter, Mechanical Design Engineering Handbook. Butterworth Heinemann, (2013), DOI: https://doi.org/10.1016/B978-0-08-097759-1.00008-3 . Available online www.sciencedirect.com/science/article

3 C.R. Juvinall and M.K. Marshek, Machine component design: 5th Edn. John Wiley & Sons Singapore Pte. Ltd., (2012), ISBN 978-1-118-09226-2

4 C.A. Ugural, Mechanical Design of Machine Components. 2nd Edn. CRC Press, Taylor & Francis Group, (2015), ISBN 13:978-1-4398-8780-6 (Hardback).

5 A. Amani, C. Spitas and V. Spitas, Generalised non-dimensional multi-parametric involute spur gear design model considering manufacturability and geometrical compatibility. Mechanism and Machine Theory 109, 250 – 277, (2017). Available online www.elsevier.com/locate/mechmachtheory

6 X. Zhao and A. Vacca, Formulation and optimization of involute spur gear in external gear pump. Mechanism and Machine Theory 117, 114 – 132, (2017). Available online www.elsevier.com/locate/mechmachtheory

7 L-W. Tsai, Robot analysis: the mechanics of serial and parallel manipulators. John Wiley & Sons, Inc., (1999), ISBN: 0-471-32593-7

8 W.D. Dudley, Handbook of Practical Gear Design. CRC Press LLC, (1994), www.crcpress.com ISBN: 1-56676-218-9

9 A.J. Collins, R.H. Busby and H.G. Staab, Mechanical Design of Machine Elements and Machines. 2nd Edn. John Wiley & Sons, USA, (2010), ISBN 13: 978-0-470-41303-6

10 L.R. Norton, Machine Design: An Integrated Approach. 3rd Edn. Pearson Prentice Hall, (2006), ISBN: 0-13-202012-2

11 E.J. Shigley and R.C. Mischke, Mechanical Engineering Design. 6th Edn. McGraw-Hill Education (Asia), (2001), ISBN: 0-07-365939-8

12 J.B. Hamrock, R.S. Schmid and B. Jacobson, Fundamental of Machine Elements. 2nd Edn. (2005), ISBN 0-07-246532-8

13 R.L. Norton, Design of Machinery: An Introduction to the Synthesis and Analysis of Mechanisms and Machines. 2nd Edn. McGraw-Hill Inc, USA, (1999), ISBN 0-07-048395-7

14 A.L. Kapelevich, Direct Gear Design. CRC Press, Taylor & Francis Group, (2013), ISBN 13: 978-1-4398-7618-3 http://www.taylorandfrancis.com

15 S.A. Hall, R.A. Holowenko and G.H. Laughlin, Theory and Problems of Machine Design. Tata McGraw-Hill Publishing company Ltd, (2002), ISBN 0-07-048333-7

16 R.G. Budynas and J.K. Nisbett, Shigley's Mechanical Engineering Design. McGraw-Hill Education (Asia), (2008), ISBN 978-007-125763-3

17 R. Lipp, "Avoiding Tooth Interference in Gears," Machine Design, Vol. 54, No. 1, pp. 122 – 124, (1982)

18 L. Ivan, Gears and gear cutting, Special Interest Model Books Ltd, England, (2003), ISBN 1-85242-911-8

Nonlinear dynamics modelling of multistage micro-planetary gear transmission

Jianying Li[1], Zhiyong Xu [2], Qingchun Hu[3], Changfu Zong[1] and Tianjun Zhu[1]

[1]*School of Mechanical and Automotive Engineering, ZhaoQing University, Zhaoqing 516260, Guangdong, China,*
[2]*School of Electronics and Electrical Engineering, ZhaoQing University, Zhaoqing 516260, Guangdong, China,*
[3]*School of Mechanical and Automotive Engineering, South China University of Technology, Guangzhou510640, China*

Abstract. The transmission structure of a 2K-H multistage micro-planetary gear transmission reducer is described in detail, and three assumptions are supposed in dynamic modelling. On basis of these assumptions, a three stages 2K-H micro-planetary gear transmission dynamic model is established, in which the relative displacement each meshing gear pairs can be obtained after including the comprehensive transmission error. According to gear kinematics, the friction arms between the sun gear, the ring gear and the nth planet are also obtained, and the friction coefficient in the mixed elastohydrodynamic lubrication is considered, the transmission system motion differential equations are obtained, including above factors and the time-varying meshing stiffness, damping and backlash, inter-stage coupling stiffness, it can be provided an theoretical foundation for further analysing the parameter sensitivity, dynamic stability and designing.

1 Introduction

The micro-planetary gear transmission is widely used in aviation equipment, space technology and medical equipment, small cars and so on [1], because it has the characteristics of planetary transmission and miniature gear. The appearance background, structure, size and function of micro-planetary gear reducer were discussed in the literature [2]. The design calculation and the modular design of micro-planetary gear transmission were discussed in detail [3, 4]. The manufacturing and intelligent assembly processes of micro-planetary gear reducer were analyzed after briefly introducing its advantage and application prospect [5]. The strength of the tubular multistage micro-planetary gear reducer was analyzed in Ref. [6]. The lubrication is a very important problem during the development of micro-planetary gear transmission[7], concerning about the friction field and establishing the nonlinear dynamic model of multistage micro-planetary gear have not been reported, but Zhu EnYong, et al. [8]built the translational-torsional nonlinear dynamic model including the friction, time-varying meshing stiffness, backlash and comprehensive transmission error. Wang Haofei[9] studied the dynamic properties of double helical planetary gear based on the tooth surface friction. Chen Changzheng, et al [10] established multi-degree of freedom translational-torsional nonlinear dynamic model with mixed elastohydrodynamic lubrication friction, time-varying meshing stiffness, and transmission error, backlash. Bao Heyun, et al. [11] established a translational-torsional dynamic model with the friction, time-varying meshing stiffness and damping, comprehensive transmission error and flexible deformable ring gear beyond pitch point meshing. Chen Fu, et al. [12] built two-stage planetary gearbox coupling torsion dynamic model considering the tooth surface friction, time-varying meshing stiffness and inter-stage coupling stiffness.

The non-linear dynamic model of multistage micro-planetary transmission including the tooth surface friction, time-varying meshing stiffness, comprehensive transmission error, backlash and inter-stage coupling stiffness in this paper, it can be provided a theoretical foundation for further dynamic analysis.

2 Dynamic modelling of multistage micro-planetary gear transmission

2.1 Dynamic model of multistage micro-planetary gear transmission

The transmission diagram of a 2K-H multistage micro-planetary gear reducer is shown in Fig. 1. The power inputs through the first stage sun gear (s^{I}), the third stage carrier is output (c^{III}). The connections of the former stage carrier and the rear stage sun gear are regarded as the spring and damping, the connections of the former stage ring gear (r^{i}) and the rear stage ring gear are also regarded as spring and damping, and each stage ring gear connects with the body of a 2K-H multistage micro-planetary gear reducer.

The following assumptions are supposed in dynamic modelling:

(1) All gears are involute spur gears and the tooth meshing force is always in the meshing plane;

(2) The same stage planetary gears distribute evenly with the same quality, size and moment of inertia;

(3) The bearing stiffness of each stage sun gear, carrier and planetary gears are linear stiffness.

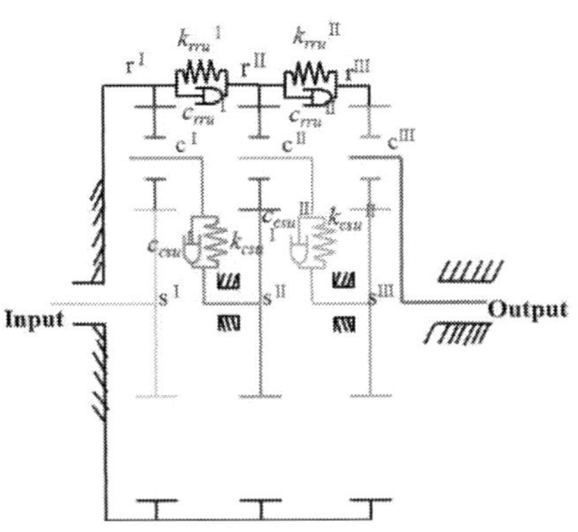

Figure 1. The transmission diagram of a 2K-H multistage micro-planetary gear reducer

Based on the above assumptions, the co-moving coordinates of each stage carrier are modeled. Unless specially indicating, i= Ⅰ, Ⅱ, Ⅲ represent planet set Ⅰ, Ⅱ and Ⅲ, respectively. $n=1,2,\cdots N_i$, where $N_{\text{I}}, N_{\text{II}}, N_{\text{III}}$ represent numbers of planetary gear of planet set Ⅰ Ⅱ and Ⅲ. The three stage micro-planetary gear transmission dynamic model is established through inter-stage coupled method based on the translational-torsional dynamic model (See Fig. 2). $x_h^i, y_h^i, u_h^i, (h=s,c,r)$ are translational and torsional displacements of component h of planet set i (the sun gear, carrier and ring gear), respectively. θ_h^i is angular displacement of component h of planet set i, and $u_h^i = r_h^i\theta_h^i$, where r_h^i is base radius of component h of planet set i (r_c^i is radius of the circle passing through planetary gear centers). x_n^i, y_n^i, u_n^i are translational and torsional displacements of the nth planet of planet set i, respectively. m_h^i, I_h^i are mass and moment of inertia of component h of planet set i; m_n^i, I_n^i are mass and moment of inertia of the nth planet of planet set i; k_{hj}^i, c_{hj}^i ($j=x,y,u$) are bearing stiffness and damping, torsional stiffness and damping of component h of planet set i, $k_{h'n}^i, c_{h'n}^i (h'=\text{s,r})$ are meshing stiffness and damping between the sun gear, ring gear and the nth planet of planet set i; k_{pn}^i, c_{pn}^i are bearing stiffness and damping of the nth planet of planet set i; k_{rru}^i, c_{rru}^i are coupling torsional stiffness and damping of the former stage ring gear and the rear stage ring gear of planet set i; k_{csu}^i, c_{csu}^i are coupling torsional stiffness and damping of the former stage carrier and the rear stage sun gear. $b_{h'n}^i$ is the backlash between the sun gear, ring gear and the nth planet of planet set i.

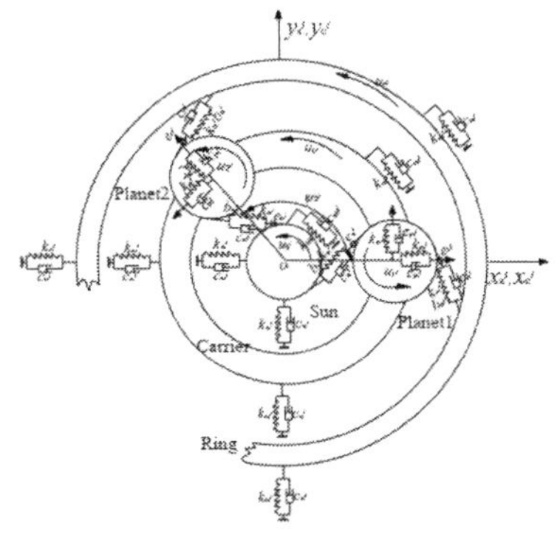

Figure 2. Dynamic model of a 2K-H multistage micro-planetary gear transmission system

2.2 Relative displacement and meshing force

The relative positional relationship of each meshing gear pairs of a 2K-H multistage micro-planetary gear transmission can be obtained after including the comprehensive transmission error.

1) The elastic deformation between the sun gear and the nth planet of planet set i

$$\delta_{sn}^i(t) = (x_n^i - x_s^i)\sin\psi_{sn}^i + (y_s^i - y_n^i)\cos\psi_{sn}^i + u_s^i + u_n^i + e_{sn}^i(t) \quad (1)$$

2) The elastic deformation between the ring gear and the nth planet of planet set i

$$\delta_{rn}^i(t) = (x_n^i - x_r^i)\sin\psi_{rn}^i + (y_r^i - y_n^i)\cos\psi_{rn}^i + u_r^i - u_n^i + e_{rn}^i(t) \quad (2)$$

3) The linear displacement and tangential direction elastic deformation of the carrier relative to the nth planet of planet set i

$$\delta_{cnx}^i(t) = x_c^i - x_n^i - u_c^i \sin\psi_n^i \quad (3)$$

$$\delta_{cny}^i(t) = y_c^i - y_n^i + u_c^i \cos\psi_n^i \quad (4)$$

$$\delta_{cnu}^i(t) = (x_n^i - x_c^i)\sin\psi_n^i + (y_c^i - y_n^i)\cos\psi_n^i + u_c^i \quad (5)$$

where $\psi_n^i(t)$ is the nth planet location of planet set i with x axis positive direction. $\psi_n^i(t) = \omega_c^i t + 2\pi(n-1)/N_i$ · $\psi_{rn}^i(t) = \psi_n^i(t) + \alpha_r^i$, $\psi_{sn}^i(t) = \psi_n^i(t) - \alpha_s^i$, where α_r^i, α_s^i is pressure angle of the ring gear and sun gear of planet set i, respectively. $e_{h'n}^i$ is the comprehensive transmission error between the sun gear, ring gear and the nth planet of planet set i[13].

$$e_{sn}^i(t) = E_{sn}^i \sin[\omega_m^i(t + \gamma_{sn}^i T_m^i)] \quad (6)$$

$$e_{rn}^i(t) = E_{rn}^i \sin\{\omega_m^i[t + (\gamma_{rn}^i + \gamma_{sr}^i)T_m^i]\} \quad (7)$$

where comprehensive transmission error amplitude between the sun gear, ring gear and the nth planet of planet set i E_{sn}^i, E_{rn}^i; meshing frequency and period ω_m^i, T_m^i; internal and external meshing phasing γ_{rn}^i, γ_{sn}^i; phasing between internal and external meshing for a given planet γ_{sr}^i.

The backlash can be described by piecewise function between the sun gear, ring gear and the nth planet of planet set i

$$\delta_{h'n}^{b\ i}(t)=\begin{cases}\delta_{h'n}^i(t)-b_{h'n}^i, & \delta_{h'n}^i(t)\geq b_{h'n}^i\\ 0, & \left|\delta_{h'n}^i(t)\right|<b_{h'n}^i \\ \delta_{h'n}^i(t)+b_{h'n}^i, & \delta_{h'n}^i(t)\leq -b_{h'n}^i\end{cases} \quad (8)$$

The time-varying meshing stiffness with the nth planet of planet set i $k_{h'n}^i(t)$

$$k_{h'n}^i(t)=\bar{k}_{h'n}^i+\sum_{l=1}^{\infty}(a_{h'n}^{i(l)}\sin l\omega_{m}^i t+b_{h'n}^{i(l)}\cos l\omega_{m}^i t) \quad (9)$$

$\bar{k}_{h'n}^i$ is average meshing stiffness with the nth planet of planet set i; $a_{h'n}^{i(l)}$, $b_{h'n}^{i(l)}$ are Fourier series expansion coefficient; l is Fourier series expansion number.

The time-varying meshing force of a 2K-H multistage micro-planetary gear transmission can be expressed as

$$F_{mh'n}^i(t)=k_{h'n}^i(t)\delta_{h'n}^{b\ i}(t)+c_{h'n}^i(t)\dot{\delta}_{h'n}^{b\ i}(t) \quad (10)$$

2.3 Friction arm

The diagram of the outer and internal meshing gear pairs friction arm are shown in Fig. 3, p^i is pitch point in the outer meshing gear pairs friction arm diagram (Seen Fig. 3(a)), $N_1^iN_2^i$ is theoretical line of action, $B_1^iB_2^i$ is actual line of action. r_b^i, r_a^i are radius of the gear base circle and addendum circle, p_{bt}^i is gear normal pitch. The symbolic meaning shown in Fig. 3(b) is similar to Fig. 3 (a).

According to gear kinematics, the friction arm between the sun gear and the nth planet of planet set i[8,12]

$$l_{sn,k}^i=(r_{bs}^i+r_{bn}^i)\tan\alpha_{sn}^i-\sqrt{(r_{an}^i)^2-(r_{bn}^i)^2}+(k-1)p_{bt}^i+M_k^i \quad (11)$$

$$l_{ns,k}^i=\sqrt{(r_{an}^i)^2-(r_{bn}^i)^2}-(k-1)p_{bt}^i-M_k^i \quad (12)$$

Similarly, the friction arm between the ring gear and the nth planet of planet set i[8,12]

$$l_{rn,k}^i=\sqrt{(r_{ar}^i)^2-(r_{br}^i)^2}+(k-1)p_{bt}^i+M_k^i \quad (13)$$

$$l_{nr,k}^i=\sqrt{(r_{ar}^i)^2-(r_{br}^i)^2}-(r_{br}^i-r_{bn}^i)\tan\alpha_{rn}^i+(k-1)p_{bt}^i+M_k^i \quad (14)$$

where $M_k^i=\text{mod}(\omega_n^i r_{bn}^i, p_{bt}^i)$, and "mod" is the modulus function defined as $\text{mod}(x,y)=x-y\cdot\text{floor}(x/y)$, if $y\neq 0$. ω_n^i is gear angular velocity of planet set i.

2.4 Friction coefficient

The lubrication of a 2K-H multistage micro-planetary gear transmission is regarded as in the mixed elastohydrodynamic lubrication and its friction coefficient of meshing gear pairs of planet set i[14,15]

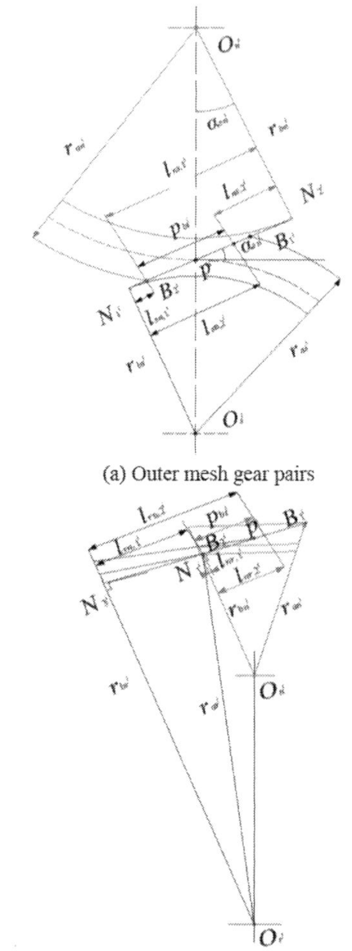

(a) Outer mesh gear pairs

(b) Internal mesh gear pairs

Figure 3. The diagram of the outer and internal meshing gear pairs friction arm

$$\mu^i=\left[0.0127\zeta^{1.2}\left(\frac{50}{50-39.37\sigma}\right)\lg\frac{29.66W}{\rho V_s(\varphi)V_R^2(\varphi)}+0.15(1-\zeta)\right]^i \quad (15)$$

where ρ is viscosity of the lubricating oil; $V_s(\varphi)$, $V_R(\varphi)$ are instantaneous sliding and rolling speed of the meshing point, respectively, as shown in the equations (16) and (17)

$$V_s(\varphi)=\frac{\pi}{30}n_1r_{b1}[(1+\frac{1}{\kappa})\tan\alpha'-(1+\frac{1}{\kappa})\tan\alpha]\times 10^{-3} \quad (16)$$

$$V_R(\varphi)=\frac{\pi}{30}n_1r_{b1}[(1-\frac{1}{\kappa})\tan\alpha'+(1+\frac{1}{k})\tan\alpha]\times 10^{-3} \quad (17)$$

where n_1 is driving gear speed; r_{b1} is base radius of the driving gear; κ is transmission ratio; α' is pressure angle of the circle of the meshing point; α is pressure angle of the pitch circle.

ζ shown in the equation (15) is bearing ratio of full film lubrication in mixed lubrication, so the bearing ratio of the meshing gear pairs full film lubrication area is shown in the equation (18)[16]

$$\zeta^i = \frac{1.21\left(h^i_{\min}\Big/\sqrt{{\sigma_1^i}^2+{\sigma_2^i}^2}\right)^{i\,0.64}}{1+0.37\left(h^i_{\min}\Big/\sqrt{{\sigma_1^i}^2+{\sigma_2^i}^2}\right)^{i\,1.26}} \tag{18}$$

where $h^i_{\min}$ is minimum oil film thickness; σ_1^i, σ_2^i are root-mean-square deviation of the profile of meshing tooth faces of the driving and driven gear, respectively.

The minimum oil film thickness can be calculated by the equation (19)[17]

$$h^i_{\min} = \left[\frac{2.65\alpha^{0.54}(\eta_0 U)^{0.7}R^{0.43}}{E'^{0.03}W^{0.13}}\right]^i \tag{19}$$

where α is lubricating oil viscosity coefficient; η_0 is dynamic viscosity of the lubricating oil at normal temperature; U is average speed of tooth faces; R is radius of the equivalent curvature of the meshing tooth; E' is comprehensive elastic modulus of materials, W is unit normal load of the meshing tooth. They are shown in the equations (20)-(23).

$$U = \frac{\pi}{60}n_1 r_{b1}[(1-\frac{1}{\kappa})\tan\alpha' + (1+\frac{1}{\kappa})\tan\alpha] \tag{20}$$

$$E' = 2[(1-\nu_1^2)/E_1 + (1-\nu_2^2)/E_2]^{-1} \tag{21}$$

$$R = r_{b1}\tan\alpha' - \frac{r_{b1}\tan^2\alpha'}{(1+\kappa)\tan\alpha} \tag{22}$$

$$W = \begin{cases} \dfrac{F_n}{2b} \text{ Double teeth mesh} \\ \dfrac{F_n}{b} \text{ Single tooth mesh} \end{cases} \tag{23}$$

where ν_1, ν_2 are Poisson's ratio of the driving and driven gear, respectively; E_1 , E_2 are elastic modulus of the driving and driven gear; F_n is normal force; b is tooth width.

The meshing tooth friction force of the mixed elastohydrodynamic lubrication

$$F^i_{fh'n}(t) = \lambda^i \mu^i F^i_{mh'n}(t) \tag{24}$$

where λ^i is direction coefficient of the friction force, it changes with the relatively rotation speed of the driving and driven gear, and can be taken as 1, 0, -1.

3 Motion differential equations of multistage micro-planetary gear transmission

According to Newton's second law, the lumped parameter dynamic model of a 2K-H multistage micro-planetary gear transmission system is established, so its motion differential equations of the ring gear, sun gear, carrier and and the nth planet of planet set i are shown in the equations (25)-(28).

$$\begin{cases} m_r^i\ddot{x}_r^i - \sum_{n=1}^{N_i}(F^i_{mrn1}+F^i_{mrn2})\sin\psi^i_{rn} - \sum_{n=1}^{N_i}(F^i_{frn1}+F^i_{frn2})\cos\psi^i_{rn} + k^i_{rx}x^i_r + c^i_{rx}\dot{x}^i_r = 0 \\ m_r^i\ddot{y}_r^i + \sum_{n=1}^{N_i}(F^i_{mrn1}+F^i_{mrn2})\cos\psi^i_{rn} - \sum_{n=1}^{N_i}(F^i_{frn1}+F^i_{frn2})\sin\psi^i_{rn} + k^i_{ry}y^i_r + k^i_{ry}\dot{y}^i_r = 0 \\ \left[I^i_r/\left(r^i_r\right)^2\right]\ddot{u}^i_r + \sum_{n=1}^{N_i}(F^i_{mrn1}+F^i_{mrn2}) - \sum_{n=1}^{N_i}(F^i_{frn1}l^i_{r1,1}+F^i_{frn2}l^i_{r2,2})\frac{1}{r^i_r} + k^i_{ru}u^i_r + c^i_{ru}\dot{u}^i_r + \Delta^i_{ru} = 0 \end{cases} \tag{25}$$

$$\begin{cases} m_s^i\ddot{x}_s^i - \sum_{n=1}^{N_i}(F^i_{msn1}+F^i_{msn2})\sin\psi^i_{sn} - \sum_{n=1}^{N_i}(F^i_{fsn1}+F^i_{fsn2})\cos\psi^i_{sn} + k^i_{sx}x^i_s + c^i_{sx}\dot{x}^i_s = 0 \\ m_s^i\ddot{y}_s^i + \sum_{n=1}^{N_i}(F^i_{msn1}+F^i_{msn2})\cos\psi^i_{sn} - \sum_{n=1}^{N_i}(F^i_{fsn1}+F^i_{fsn2})\sin\psi^i_{sn} + k^i_{sy}y^i_s + c^i_{sy}\dot{y}^i_s = 0 \\ \left[I^i_s/\left(r^i_s\right)^2\right]\ddot{u}^i_s + \sum_{n=1}^{N_i}(F^i_{msn1}+F^i_{msn2}) - \sum_{n=1}^{N_i}(F^i_{fsn1}l^i_{s1,1}+F^i_{fsn2}l^i_{r1,1})\frac{1}{r^i_c} + k^i_{su}u^i_s + c^i_{su}\dot{u}^i_s + \Delta^i_{su} = \frac{T^i_s}{r^i_s} \end{cases} \tag{26}$$

$$\begin{cases} m^i_c\ddot{x}^i_c + \sum_{n=1}^{N_i}(k^i_{pn}\delta^i_{cnx} + c^i_{pn}\dot{\delta}^i_{cnx}) + k^i_{cx}x^i_c + c^i_{cx}\dot{x}^i_c = 0 \\ m^i_c\ddot{y}^i_c + \sum_{n=1}^{N_i}(k^i_{pn}\delta^i_{cny} + c^i_{pn}\dot{\delta}^i_{cny}) + k^i_{cy}y^i_c + c^i_{cy}\dot{y}^i_c = 0 \\ \left[I^i_c/\left(r^i_c\right)^2\right]\ddot{u}^i_c + \sum_{n=1}^{N_i}(k^i_{pn}\delta^i_{cnu} + c^i_{pn}\dot{\delta}^i_{cnu}) + k^i_{cu}u^i_c + c^i_{cu}\dot{u}^i_c + \Delta^i_{cu} = -\frac{T^i_c}{r^i_c} \end{cases} \tag{27}$$

$$\begin{cases} m^i_n\ddot{x}^i_n + (F^i_{mrn1}+F^i_{mrn2})\sin\psi^i_{rn} + (F^i_{frn1}+F^i_{frn2})\cos\psi^i_{rn} \\ +(F^i_{msn1}+F^i_{msn2})\sin\psi^i_{sn} + (F^i_{fsn1}+F^i_{fsn2})\cos\psi^i_{sn} - k^i_{pn}\delta^i_{cnx} - c^i_{pn}\dot{\delta}^i_{cnx} = 0 \\ m^i_n\ddot{y}^i_n - (F^i_{mrn1}+F^i_{mrn2})\cos\psi^i_{rn} + (F^i_{frn1}+F^i_{frn2})\sin\psi^i_{rn} \\ -(F^i_{msn1}+F^i_{msn2})\cos\psi^i_{sn} + (F^i_{fsn1}+F^i_{fsn2})\sin\psi^i_{sn} + k^i_{pn}\delta^i_{cny} + c^i_{pn}\dot{\delta}^i_{cny} = 0 \\ \left[I^i_n/\left(r^i_n\right)^2\right]\ddot{u}^i_n + (F^i_{msn1}+F^i_{msn2}) - (F^i_{mrn1}+F^i_{mrn2}) + (F^i_{frn1}l^i_{r1,1}+F^i_{frn2}l^i_{r2,2})\frac{1}{r^i_r} + (F^i_{fsn1}l^i_{s1,1}+F^i_{fsn2}l^i_{s1,1})\frac{1}{r^i_s} = 0 \end{cases} \tag{28}$$

$$\Delta^i_{ru} = \begin{cases} k^{\mathrm{I}}_{rru}\dfrac{r^{\mathrm{I}}_e}{r^{\mathrm{I}}_r}\left(u^{\mathrm{I}}_r\dfrac{r^{\mathrm{I}}_e}{r^{\mathrm{I}}_r} - u^{\mathrm{II}}_r\dfrac{r^{\mathrm{I}}_e}{r^{\mathrm{II}}_r}\right) + c^{\mathrm{I}}_{rru}\dfrac{r^{\mathrm{I}}_e}{r^{\mathrm{I}}_r}\left(\dot{u}^{\mathrm{I}}_r\dfrac{r^{\mathrm{I}}_e}{r^{\mathrm{I}}_r} - \dot{u}^{\mathrm{II}}_r\dfrac{r^{\mathrm{I}}_e}{r^{\mathrm{II}}_r}\right), i=\mathrm{I} \\ k^{\mathrm{II}}_{rru}\dfrac{r^{\mathrm{II}}_e}{r^{\mathrm{II}}_r}\left(u^{\mathrm{II}}_r\dfrac{r^{\mathrm{II}}_e}{r^{\mathrm{II}}_r} - u^{\mathrm{III}}_r\dfrac{r^{\mathrm{II}}_e}{r^{\mathrm{III}}_r}\right) + c^{\mathrm{II}}_{rru}\dfrac{r^{\mathrm{II}}_e}{r^{\mathrm{II}}_r}\left(\dot{u}^{\mathrm{II}}_r\dfrac{r^{\mathrm{II}}_e}{r^{\mathrm{II}}_r} - \dot{u}^{\mathrm{III}}_r\dfrac{r^{\mathrm{II}}_e}{r^{\mathrm{III}}_r}\right), i=\mathrm{II} \\ k^{\mathrm{III}}_{rru}\dfrac{r^{\mathrm{III}}_e}{r^{\mathrm{III}}_r}\left(u^{\mathrm{III}}_r\dfrac{r^{\mathrm{III}}_e}{r^{\mathrm{III}}_r} - u^{\mathrm{I}}_r\dfrac{r^{\mathrm{III}}_e}{r^{\mathrm{I}}_r}\right) + c^{\mathrm{III}}_{rru}\dfrac{r^{\mathrm{III}}_e}{r^{\mathrm{III}}_r}\left(\dot{u}^{\mathrm{III}}_r\dfrac{r^{\mathrm{III}}_e}{r^{\mathrm{III}}_r} - \dot{u}^{\mathrm{I}}_r\dfrac{r^{\mathrm{III}}_e}{r^{\mathrm{I}}_r}\right), i=\mathrm{III} \end{cases}$$

$$\Delta^i_{su} = \begin{cases} 0, i=\mathrm{I} \\ k^{\mathrm{II}}_{csu}\dfrac{r^{\mathrm{II}}_e}{r^{\mathrm{II}}_s}\left(u^{\mathrm{II}}_s\dfrac{r^{\mathrm{II}}_e}{r^{\mathrm{II}}_s} - u^{\mathrm{I}}_c\dfrac{r^{\mathrm{II}}_e}{r^{\mathrm{I}}_s}\right) + c^{\mathrm{II}}_{csu}\dfrac{r^{\mathrm{II}}_e}{r^{\mathrm{II}}_s}\left(\dot{u}^{\mathrm{II}}_s\dfrac{r^{\mathrm{II}}_e}{r^{\mathrm{II}}_s} - \dot{u}^{\mathrm{I}}_c\dfrac{r^{\mathrm{II}}_e}{r^{\mathrm{I}}_s}\right), i=\mathrm{II} \\ k^{\mathrm{III}}_{csu}\dfrac{r^{\mathrm{III}}_e}{r^{\mathrm{III}}_s}\left(u^{\mathrm{III}}_s\dfrac{r^{\mathrm{III}}_e}{r^{\mathrm{III}}_s} - u^{\mathrm{II}}_c\dfrac{r^{\mathrm{III}}_e}{r^{\mathrm{II}}_c}\right) + k^{\mathrm{III}}_{csu}\dfrac{r^{\mathrm{III}}_e}{r^{\mathrm{III}}_s}\left(\dot{u}^{\mathrm{III}}_s\dfrac{r^{\mathrm{III}}_e}{r^{\mathrm{III}}_s} - \dot{u}^{\mathrm{II}}_c\dfrac{r^{\mathrm{III}}_e}{r^{\mathrm{II}}_c}\right), i=\mathrm{III} \end{cases}$$

$$\Delta^i_{cu} = \begin{cases} k^{\mathrm{I}}_{csu}\dfrac{r^{\mathrm{I}}_e}{r^{\mathrm{I}}_c}\left(u^{\mathrm{I}}_c\dfrac{r^{\mathrm{I}}_e}{r^{\mathrm{I}}_c} - u^{\mathrm{II}}_s\dfrac{r^{\mathrm{I}}_e}{r^{\mathrm{II}}_s}\right) + c^{\mathrm{I}}_{csu}\dfrac{r^{\mathrm{I}}_e}{r^{\mathrm{I}}_c}\left(\dot{u}^{\mathrm{I}}_c\dfrac{r^{\mathrm{I}}_e}{r^{\mathrm{I}}_c} - \dot{u}^{\mathrm{II}}_s\dfrac{r^{\mathrm{I}}_e}{r^{\mathrm{II}}_s}\right), i=\mathrm{I} \\ k^{\mathrm{II}}_{csu}\dfrac{r^{\mathrm{II}}_e}{r^{\mathrm{II}}_c}\left(u^{\mathrm{II}}_c\dfrac{r^{\mathrm{II}}_e}{r^{\mathrm{II}}_c} - u^{\mathrm{III}}_s\dfrac{r^{\mathrm{II}}_e}{r^{\mathrm{III}}_s}\right) + c^{\mathrm{II}}_{csu}\dfrac{r^{\mathrm{II}}_e}{r^{\mathrm{II}}_c}\left(\dot{u}^{\mathrm{II}}_c\dfrac{r^{\mathrm{II}}_e}{r^{\mathrm{II}}_c} - \dot{u}^{\mathrm{III}}_s\dfrac{r^{\mathrm{II}}_e}{r^{\mathrm{III}}_s}\right), i=\mathrm{II} \\ 0, i=\mathrm{III} \end{cases}$$

where Δ^i_{ju} is torsional coupling term of the component h

of planet set *i*.

4 Conclusions

A three stages 2K-H micro-planetary gear transmission mechanism is taken as the research object in this paper, on the basis of modeling assumptions, the lumped parameter dynamic model of a 2K-H multistage micro-planetary gear transmission system is established and its motion differential equations are obtained, including the time-varying meshing stiffness, damping, transmission error and backlash, inter-stage coupling stiffness, and considering tooth face friction and lubrication condition of the meshing gear pairs, it can be provided an theoretical foundation for further analyzing the parameter sensitivity, dynamic stability and dynamic designing.

References

1. Z.G. Rao, Drive System Technique **27**,16-22 (2013) (in Chinese)
2. Z.G. Rao, J.S. Maehine Building & Automation 10-11,15 (1999) (in Chinese)
3. Z.G. Rao, Drive System Technique 18-24,34 (2003) (in Chinese)
4. Z.G. Rao, Drive System Technique **26**,10-17 (2012) (in Chinese)
5. Y.Z. Zhao, Machinery Design & Manufacture 242-243(2010)(in Chinese)
6. C.P.Yang, HeFei *University of Technology* (HeFei, 2012) (in Chinese)
7. Z.G Rao, *Chemical Industry Press* (Beijing, 2014) (in Chinese)
8. E.Y. Zhu, S. J. Wu, X.S. Wang, M.X. Deng, B. Qian, J. Vibration and shock **29**, 217 -221(2010) (in Chinese)
9. H.F. Wang, *Nanjing University of Aeronautics and Astronautics* (Nanjing, 2015) (in Chinese)
10. C.Z. Chen, L. Zhang, J. Liu, Z.Q. Sun, X.T. Bai, Acta Energiae Solaris Sinica **37**, 1695-1701(2016) (in Chinese)
11. H.Y. Bao, X.J. Zhou, R.P. Zhu, F.X. Lu, J. of Central South University (Science and Technology) **47,** 3005-3010 (2016) (in Chinese)
12. F. Chen, Z.M. Xiao, L.R. Huan, Machine Design and Research **33**, 63-71 (2017) (in Chinese)
13. C. Zhang, *Higher Education Press* (Beijing, 2008) (in Chinese)
14. J. Castro, J. Seabra, Proceedings of the Institution of Mechanical Engineers, Part J: Journal of Engineering Tribology (2007)
15. B. Wang, X.B. Chen, J. Tongji University (Natural Science), **42**, 1904-1911 (2014) (in Chinese)
16. Y.Z. Hu, D. Zhu, J. Tribology **122**, 1-9 (2000) (in Chinese)
17. N.X. Wang, X.P. Jiang, J. Shaanxi University of Science & Technology **27**, 84-86 (2009) (in Chinese)

Research on control system of truck-mounted rig for coalbed methane

Hejian Wang, Dong Fan, Yinsheng Weng

Xi'an Research Institute of China Coal Technology & Engineering Group Corp, Xi'an, Shaanxi, China

Abstract. The coal-bed methane (CBM) as a kind energy is clean and efficient, also it can become a security risk in mining process if it could not get out of the coal seam. In view of the current large-scale exploitation of coal-bed methane resources, the development of drilling rig for CBM drilling is needed. The parameters and structures were introduced in the paper. The rig uses a highly integrated approach that integrates the required functions on the chassis of the vehicle to meet the needs of rapid installation and transportation. Drilling control system uses hydraulic control and electro-hydraulic control dual control mode, can achieve short-range and remote control operations. The control system include security circuits and electric control system. Through the field trial, it is shown that the rig can meet the construction of the majority of coalbed methane drilling in the country and the performance is stable and the operation is simple.

1 Introduction

Coal-bed methane (known as gas) is a kind of clean energy stored in coal seam. The development and utilization of coalbed methane resources can effectively solve the geological disasters caused by gas, and can increase the energy supply and reduce its pollution to the environment. Ground coalbed methane extraction is an effective way to solve the problem of coal mine safety by extracting the gas in the coal field through the ground drilling technology, reducing the coal content and providing the safety guarantee for the large-scale coal mining [1-2]. China's coalbed methane resources are enriched in the Qinshui Basin, with more mountainous and hilly areas. Coalbed methane development wells are scattered and the well site space is small, which is difficult to set drilling equipment. Truck-mounted CBM drilling rigs are designed and developed for specific drilling environments. The main working parts of the rig are integrated into the chassis of the vehicle, and the installation and transportation are very easy. This type of rig was mainly used in drilling wells traditionally, and in recent years it has been gradually applied to the construction of coalbed methane drainage [3-7]. These rigs process adaptability, high construction efficiency and small footprint, is widely used in the development of coalbed methane. At present, the development of ground coalbed methane in China is limited by the factors such as geological conditions and mining technology. Because the output of single well is low and the comprehensive economy is poor, the demand for cost-effective vehicle drilling rig is urgent. In recent years, some manufacturers through the introduction, digestion, absorption and re-innovation, have launched a variety of vehicle rigs, to meet the domestic CBM drilling construction. The 100t truck-mounted drilling rig was developed by Xi'an Research Institute of China Coal Science and Technology Group which has the capability of coalbed methane drilling.

2 Drilling parameters and mechanical systems

2.1 Parameters

Ground coalbed methane well is divided into two types of directional well development and straight well [8]. Directional well development comprehensive low cost, large output, which is widely used in coalbed methane development. Mud motor must be needed in directional drilling, so the drilling rig requires brake function. In view of the specific geological conditions and development factors of coalbed methane burial in China, the capacity of drilling rig is 2500m. The main technical parameters of the drilling rig are shown in Table 1:

Table 1. Main Parameters of ZMK5530TZJ100 Truck-mounted Drilling Rig

depth	Torque	speed	Feeding force	Lifting force
2500m	30000Nm	120r/min	180kN	1000kN

2.2 Feeding device

The mast is made of telescopic structure, with a transportation size of only 13.6 m, providing a 15 m stroke. The telescopic mast is mainly composed of a feed cylinder, a primary feeding device and a secondary feeding device. The cylinder barrel is fixed by the pin to the secondary feeding device, the piston rod is fixed with the first stage, and the feeding cylinder can drive the primary feeding device to move along the guide rail on the secondary feeding device. The top head can be moved through the wire rope in the feeding device. The feeding speed is twice the cylinder to improve the efficiency of drilling. Lifting, unloading and drilling from the single and so on the operation process efficient and reliable. With active pressurization function, handling hole accident ability.

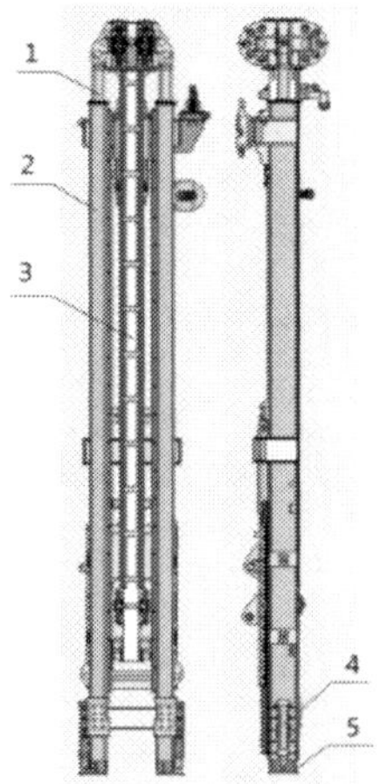

Figure 1. The structure of the feeding system.
1—feed cylinder 2—Secondary feeding device 3—First feeding device 4—Mechanical outrigger 5—Solid device

2.3 Top head

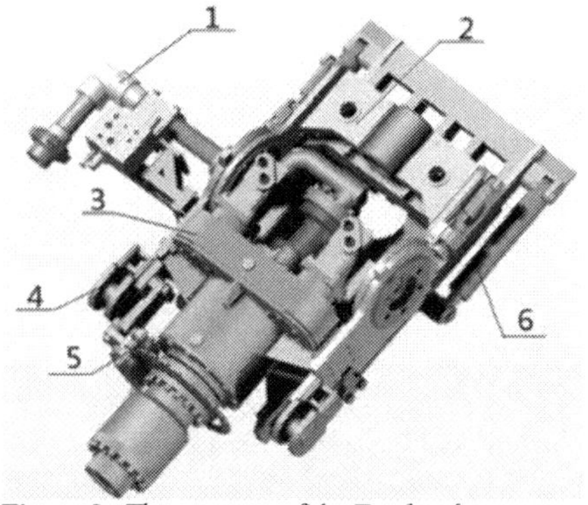

Figure 2. The structure of the Top head.
1—mud pipes 2—pallet 3—rotation box 4—lug 5—brake 6—angle tilting cylinder

Top head, also known as the top drive, is an important executive part of the rig. It is responsible for the hydraulic motor output speed and torque into a suitable parameter range and applied to the drill device. The top head can provide down pressure for the rig, reducing the dependence on the increased drill pipe at the beginning of the drilling. At the same time, in the event of an accident in the hole, you can provide additional pressure to improve the accident handling capacity. Top head (Figure 2) using four two-speed hydraulic motors with four small gears to drive a large gear in a gearbox, with a speed range of 0 ~ 120 r / min., which can meet a variety of drilling needs. Top head through the brake device to achieve the spindle locking function to obtain accurate azimuth and slope, to achieve the bottom of the hole dynamic directional drilling, to meet the directional hole and branch hole construction. The diameter of spindle is up to 150 mm, to meet the mud circulating, air circulation, air reverse circulation and gas lift reverse circulation and other drilling process.

2.4 Power units

Power units are the power source of the rig, made of a diesel oil tank, hydraulic oil tank, diesel engine and pump series. The diesel engine and the main pump are connected by an elastic clutch. In order to guarantee the normal work of the hydraulic system, the power units is equipped with many hydraulic accessories, such as oil filter, air cooler, air filter, oil temperature tester, liquid level indicator. Drill control system using electro-hydraulic proportional control system which can achieve hydraulic control and remote electronic control dual control. The control system is the core of the full hydraulic rig, directly affects the overall performance and maneuverability of the rig. Drilling rigs need to control the action, including: top head rotation, quick feeding and pulling, slow feeding and pulling, clamp control, the main winch to enhance the laying, attitude adjustment, mast setting and landing, etc., .Torque adjustment, pressure control and other functions are also achieved. In order to ensure the safety of drilling, the system designed a variety of safety interlock protection circuit; important circuit are used multi-pump, multi-valve simultaneous control of the redundant design which had high reliability to prevent damage caused by a single hydraulic components. The display can monitor the running status of the diesel engine in real time and the drilling speed of the power head, torque, drilling.

3 Control system

3.1 Hydraulic control system

A drawing of the hydraulic control system is illustrated in Fig 3(a). The hydraulic system adopt open-type system, which provides the hydraulic oil to rotary and feed separately. The hydraulic system consists of 4 main pumps and 3 assistant pumps. This system is easy to control and repair. The speed control system is made up of variable pumps and four two-speed motors. The rotary speed of the motors can be adjusted according to practical conditions when drilling a borehole. The speed of motor can be adjusted within a certain range, so it improves the adaptability of drilling rig for different kinds of drilling technology. The drilling operation system controls the rig to carry out the drilling functions. The hydraulic system is a load-sensitive system and uses a hydraulic load-sensitive valve. The hydraulic control system is equipped with load sensing pumps and constant pressure pumps. The load sensing pump is used for the rotation of the top drive and fast moving of the mask. These controls feature load sensing and maximum pressure compensation. Load sense controls are used to match pump flow and pressure to system demands, thus minimizing losses due to wasted horsepower. The pump automatically adjusts for changes in drive speed and load pressures to match the pump output flow to the load requirement. Since the pump load sense control will maintain a constant pressure drop across the main system throttling valve, the flow rate will remain constant, independent of changes in load pressure and pump shaft speed. The data of load sensing pump is illustrated in Fig.3(b).

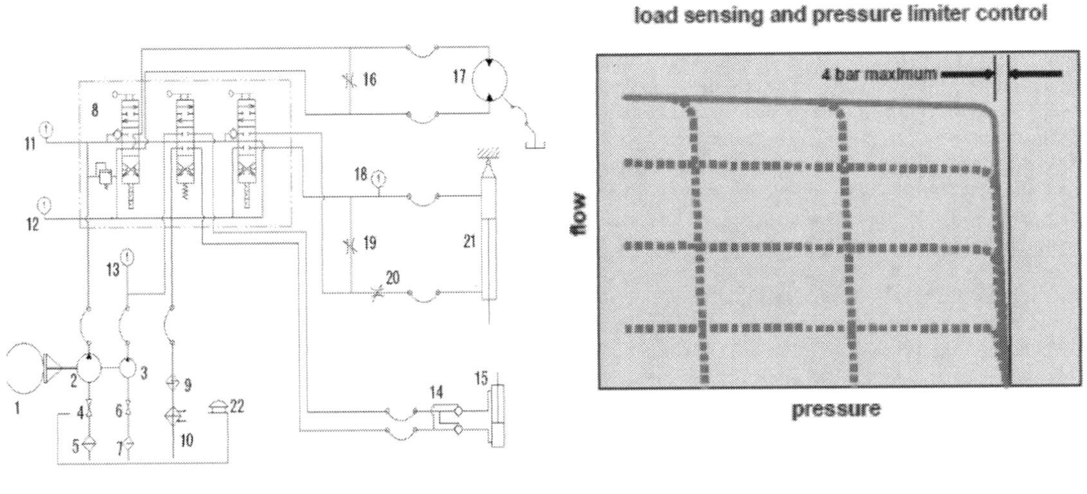

(a) (b)

Figure 3. Hydraulic System Drawing and Load Sensing Pump Data.

3.2 Diesel engine control system

Hydraulic oil and diesel engine cooling control system as shown in Figure 4. The oil-return of hydraulic system can back to the air-cooled tank. Variable pump drive fan motor to cool the oil, radiator cooling capacity depends on motor speed. Hydraulic oil and diesel engine coolant circuit were set up on the temperature sensor, the electronic control system according to the temperature sensor temperature information on the solenoid valve issued a control signal, solenoid valve control variable pump displacement changes to change the fan motor speed. This control system allows the hydraulic system temperature from the environment, the system is always working in the best temperature range.

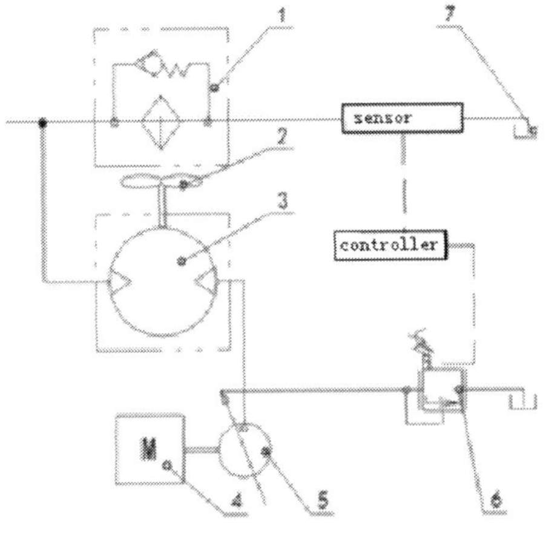

Figure 4. The closed-loops control system of temperature
1—radiator 2—cooling funs 3—motor 4—power unite 5 —variable motor; 6—electric relief valves 7—tank

https://doi.org/10.1051/matecconf/201815304001

3.3 Security circuits

The reliable braking of the spindle is the guarantee of directional drilling. The pressure relief of the brake device will lead to the change of the angle of the mud motor. The pilot oil is used to prevent the pressure relief, and the pressure sensor is used to monitor the pressure change through the pressure sensor. Activated braking device will be damaged when the top head worked. In order to prevent misoperation, a new circuit was designed as shown in Figure 5, which is the spindle brake anti-rotation hydraulic principle circuit. The manual reversing valve controls the clamping and release of the brake, and the pressure sensor monitors the brake condition. When the brake device is clamped, the sensor detects that the pressure signal exceeds the set value, sends a signal from the controller, controls the electromagnetic directional valve commutation, cuts off the oil source of the pilot valve, and deactivates the hydraulic control valve of the operation top head. So as to avoid the braking device in the state of the power head forced rotation caused by the accident, improve the rig safety. The use of control pilot valve oil source on and off the principle of the design of the winch and power head anti-collision, winch arm and mast anti-collision and other security interlock protection circuit to avoid the accident caused by the accident, the system more secure and reliable.

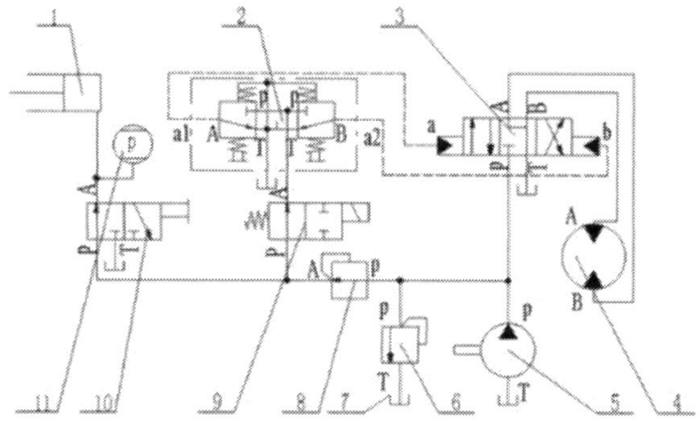

Figure 5. The hydraulic principle diagram with protection function

1—brake 2—pilot valve 3—Hydraulic directional valve 4—motor 5—pump 6—relief valve 7—tank 8—reducing valve 9—electric directional valve 10—manual directional valve 11—pressure sensor 12—controller

3.4 Operating system

Drilling machine operating system using short-range hydraulic control and remote electro-hydraulic control in two ways. Remote operating control panel is shown in figure 6. The main valve body are arranged in the rig body platform, the control valve is arranged in a separate control box. The hydraulic control console is arranged on the rear side of the vehicle body, close to the working table, easy to observe the drilling situation. Remote electro-hydraulic console placed in a separate control room, can be controlled within 50 meters, the use of wireless signal transmission. In the environmental conditions are good when using short-range console operations, harsh environment using remote console operations.

Figure 6. Remote operating control panel.

3.5 Electric Control System

All controls are conveniently located for easy operation. The control system structure is shown in Fig.7. The signal of analogue sensor is treated by signal adjustment circuit, the signal of digital sensor is isolated form ARM by high speed optoelectronic isolator, the switch of the switch joystick is treated directly by ARM, the signal of the joystick with potentiometer is enlarged by amplifier circuit. ARM outputs digit to FET driver circuit to drive switch electromagnetic valve, and outputs PWM to proportional amplifier circuit to drive proportional electromagnetic valve. The communication interface is CAN2.0 by CAN transceiver 82C251[9-20] and the system is communicated with outside by zigbee wireless and exchanges data between zigbee wireless chip and ARM by SPI bus inside. HMI is developed by LCM.

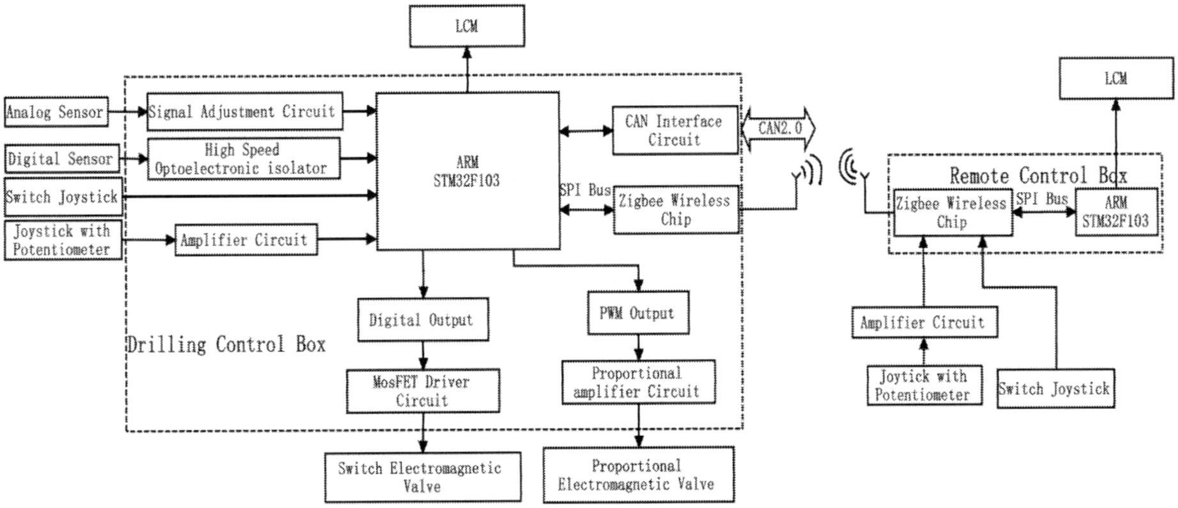

Figure 7. Electric Control System Structure.

4 Field trial

The Verification machine was tested to develop coalbed methane wells in Sihe District, Shanxi Province. The target coal seam was 3 # coal seam. The average thickness of the coal seam was 6.25 m. Ordinary straight wells cannot fully cover the pre-pumping area for the reasons of high dense CBM, affecting the safety of coal production and mining convergence. In order to solve the above problems and achieve rapid reduction of gas content, a horizontal hole was need to be drilled to ensure the safety of coal production purposes, the structure of the horizontal hole is shown in Figure 8 . The test site is located above the hills and the road is narrow. Part of the road is not hardened, the slope of more than 25 °, the rig get through the test and the climbing ability and mobility were verified. The DTH drilling technology, PDC rotary drilling technology, directional drilling technology were tested in the construction. To complete a depth of 1675m horizontal directional hole, the performance and process adaptability were verified.

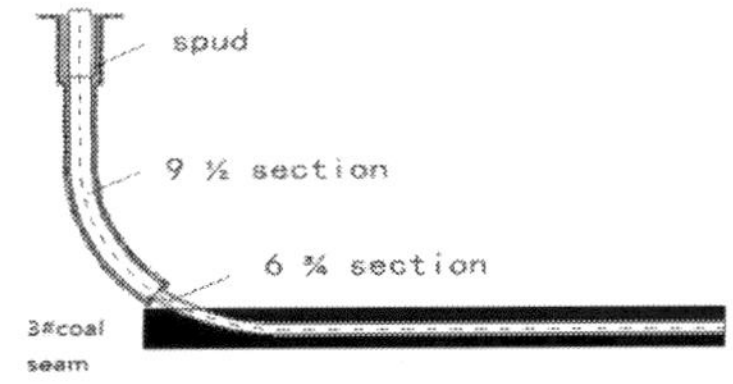

Figure 8. Structure of the horizontal hole.

5 Conclusions

The truck-mounted drilling rig adopts hydraulic drive, hydraulic control and electro-hydraulic control double control mode, to meet a variety of conditions construction requirements. The multiple protection circuit used in the control system effectively protects the safety of the equipment. Experiments show that the rig can meet the 2500 m coalbed methane resource mining requirements.

Acknowledgement

The work presented in this paper was financial supported by the Major Project of the National Science and Technology (No. 2016ZX05045001-002).

References

1. Hu Qianting, Sun Haitao, Du Zijian. Practices and application prospects on development engineering of Coal-bed methane surface well, Coal Science and Technology, 2015(9);59-64
2. LI Guo-fu, LI Gui-hong, LIU Gang. Analysis on the ground extraction effect of coal-bed methane at typical area in Jincheng, China, JOURNAL OF CHINA COAL SOCIETY, 2014,39(9):1932－1937.
3. Gao Hongliang. Application in geological exploration on truck-mounted drilling rig [J]. Equipment for Geotechnical Engineering. 2009(02):38-41
4. XU Liuwan, CAO Fude, GE Hewang. Current Situation of Application of Drilling Technology and Equipment for Hydrological Well, Exploration Engineering (Rock & Soil Drilling and Tunneling) 2007,(01):37-42.
5. TIAN Hongliang, FAN Dong, CHANG Jianghua. Design of ZMK5530TZJ60 type Truck-mounted

Drilling Rig. Technology and application of coal mine safety and high efficiency mining geology, China Coal Industry publishing House, 2014, 435-440

6. Chen Yueqiang. Technology Analysis of The Domestic Rig Used in CBM [D]. Xi'an. Xi`an University of Science and Technology.
7. Yan Xuguang. Design and theoretical research on hoisting system of the slant CBM well rigs [D]. Xi'an. Xi'an Shiyou University
8. ZHANG Peihe, ZHANG Mingshan. Analysis of application status and adapting conditions for different methods of CBM development. COAL GEOLOGY & EXPLORATION, 2010(4):9-13
9. Holger Kinzel. The Designing and Building of an All-Purpose Automatic Torque-Wrenching Machine for Drilling Rigs, IADC/SPE Drilling Conference, 02, 1992.
10. W.A. Hustrulid. Underground Mining Methods Handbook, 1982.
11. YIN Xin-shen. Application of load sensitive technology to mine roadway hydraulic driving head drilling rig, 01, 2008.
12. Parker. PD Series Axial Piston Pumps Variable Displacement, Catalog HY28-2663-01/NA,EU.
13. Yinsheng Weng. CAN Bus Communication System of Tunnel Drilling Rig in Coal Mine. Coal Mine Mechinery. 02, 2013.
14. 14.Tian Hongliang, Wang Hejian, Weng Yingsheng. Design of Track-mounted Drilling Rig Based on Wireless Remote Control System [C], ICMEE, 2013
15. ShueySA.Mining technology for the 21th century[C]:IncodigsdeepinSudbury[J].E&MJChina,1 992,2:7-11
16. KellyM. Developing coalmining technology for the 21th century [C]// Proc Ming Sciand Tech, Rotterdam:AABalkema,1999
17. MichaelJ. Intelligent Drill Pipe Creates the Drilling Network SPE Asia Pacific Oil&Gas Conference, 80454,2003
18. MichaelE,Reeves.IntelligentDrillStringFieldTrialsDe monstrateTechnologyFunctionalitySPE/IADCDrillin gConference,92477,2005
19. TyberP.Controlling Cement Operationfroman Onshore Operation Center SPE/IADC Drilling Conference,92334,2005
20. LI Lin, Nurzat. Adaptive PID control strategy for the drilling-rig rotary system [J]. AC TA PET ROLEI SINICA.2011(1):158-161

Torsional Stiffness Comparison of Different Tube Cross-Sections of a Formula SAE Car Space Frame

Cong Hao Liu[1], Gang Li[1], Ying Hao Ma[1] and Xu Guang Yang[2]

[1]*College of Automobile and Transportation Engineering, Liaoning University of Technology, Jinzhou 121001, China*
[2]*Department of Transportation, Jinzhou School of Electromechanic and Engineering, Jinzhou 121000, China*

Abstract. Since torsional loading and the accompanying deformation of the frame and suspension parts can affect the handling and performance of the car, torsional stiffness is generally thought to be a primary determinant of frame performance for a FSAE car. According to the FSAE Rules, different tube cross-sections are available for some members of a space frame. By finite element simulation, this research compared different tube shapes and thicknesses. Compared with 1.6 mm thickness round tube, square tube with the same wall thickness can improve the torsional stiffness by 23% in test Mode I, and 65% in test Mode II. The 1.2mm thickness square tube also can improve the torsional stiffness by 6% and 39% in test Mode I and Mode II. From these comparisons, it can be found the usage of square tube can improve the frame torsional stiffness efficiently.

1 Introduction

FSAE is a competition sponsored by the Society of Automotive Engineers (SAE). Several kinds of frame styles are available in this competition, space frame, monocoque frame, and ladder frame etc. The most commonly used are monocoque frame and space frame [1, 2]. A space frame is a kind of separate frame construction. The purpose of the frame is to rigidly connect the front and rear suspension while providing attachment points for supporting the different systems of the car [3].

The original frame of this research is from a FSAE electric car, which successfully performed in 2016 competition season.

Based on the original frame design, this research focus on how different tube cross-sections effect on the torsional stiffness of a vehicular frame. Comparisons were simulated by considering different boundary conditions.

2 Modelling

2.1 Finite element model

A frame is mainly consisted by tubes, which can be simplified to beam elements in a finite element (FE) analysis. This research conducted the analysis of the space frame by commercial finite element software ABAQUS.

As shown in Figure 1, the frame is constructed by 86 members, and the cross-sections are defined as beam section. The selected material is Steel 4130, mechanical properties is listed in Table 1. Steel 4130 is a very commonly used material for making a space frame due to its good structural property [4].

The model is meshed by 2-node linear beam elements; element type is B31 in ABAQUS. The approximate global mesh size is 5mm, total number of elements used in this frame model is 7652.

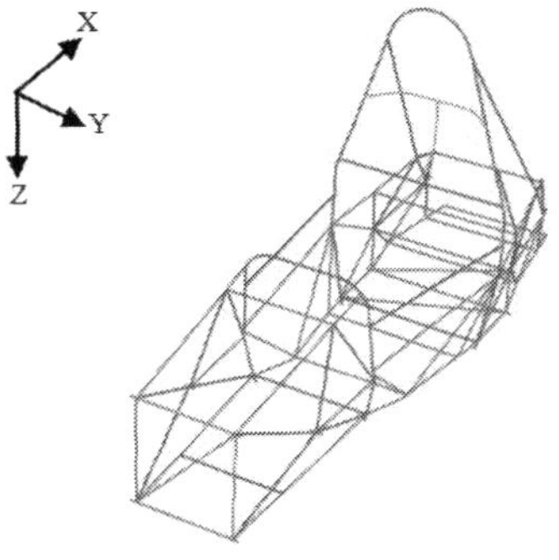

Figure 1. Tubular frame FE model.

Table 1. Material Properties

Properties	4130 Steel
Young's Modulus (GPa)	210
Density (g/cm^3)	7.8
Poisson's Ratio	0.27

2.2 Different tube shapes

The FSAE rules [5] define a minimum size for all the frame members, to avoid adding un-necessary weight, the frame design should make best use of the required members and optimize the size of the members. The rules require different cross-sections of the tubes according to the components location, as shown in Figure 1 and Table 2, different colours represent different tube cross-sections. The parameters in Table 2 were used in the competition of year 2016, which were proofed to be satisfied FSAE rules. However, the rules also provide alternative shape of some members. For example, the green members in Figure 1 can be manufactured by square shape tubing.

Table 2. Member size.

Colour	Outside Dimension(mm) ×Wall Thickness(mm)
Red	Round 25.4×2.4
Green	Round 25.4×1.6
Blue	Round 25.4×1.25
Grey	Round 14×1.5

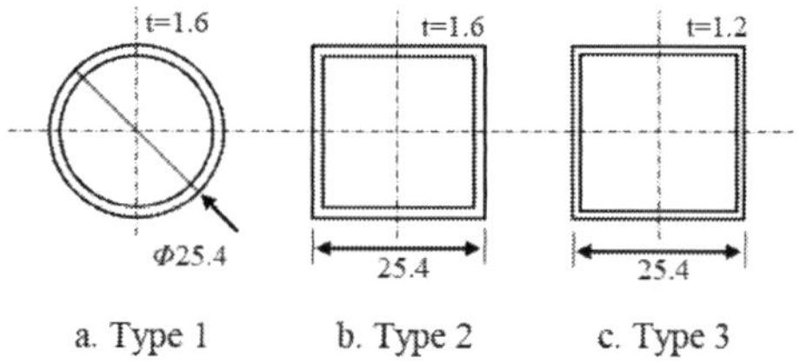

Figure 2. Square and round tubing.

This research studied three different tube shapes of green members, as illustrated in Figure 2, Type 1 is the original cross-section; Type 2 is square shape with outside dimension 25.4mm and wall thickness 1.6mm; Type 3 is square shape with outside dimension 25.4mm and wall thickness 1.2mm. All types are allowed by the rules.

2.3 Boundary Conditions

To make the simulation more accurate, the frame model is connected with MPC control points. This model verifies the condition to be the most related to the real rods and frame connection.

Torsional stiffness is an important influence factor of the racing car riding performance. Since the frame is not equidistant in the longitudinal direction, two modes will be considered and two kinds of boundary conditions are applied to the model.

Mode I is explained by Figure 3, the load F=1000N is perpendicularly applied on the two front sides with opposite directions at point A and B. The rear sides are fixed at point C and D [6].

Similarly, Mode II is explained by Figure 4, the load F is perpendicularly applied on the two rear sides with opposite directions at point C and D. The front sides are fixed at point A and B.

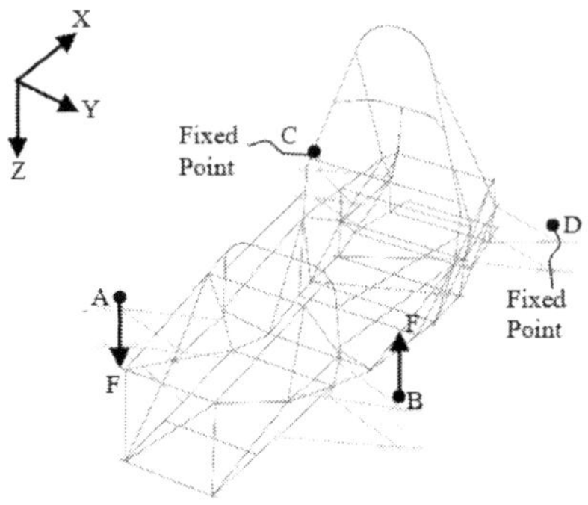

Figure 3. Mode I boundary conditions.

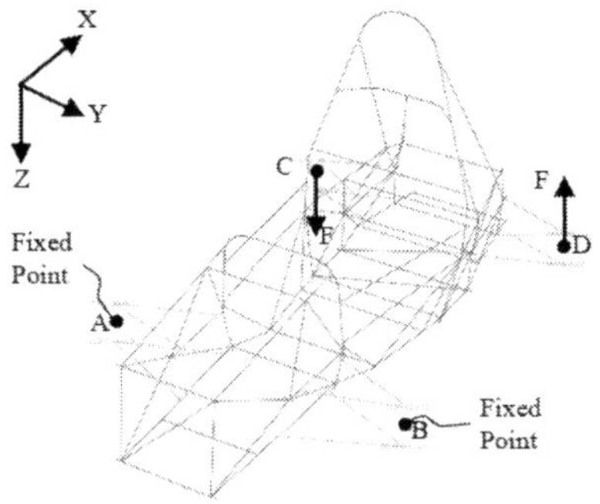

Figure 4. Mode II boundary conditions.

3 Torsional stiffness calculation

The torsional stiffness simulation was performed as shown in Figure 3 and Figure 4. The load is applied and produced the torque to the frame. Then, the torque applied on the frame can be determined by

$$T=FL \tag{1}$$

Where L is the distance between point A and B (Mode I), or C and D (Mode II).

The torsional stiffness can be described by the rotation angle produced by torque load [7, 8], the angle can be calculated from

$$\theta=arctan[(z_1+z_2)/L] \tag{2}$$

Where z1 and z2 are the magnitude of the vertical displacement caused by the torque on point A and B (Mode I), or C and D (Mode II) of the chassis.

And the torsional stiffness, K can be determined from

$$K=T/\theta \tag{3}$$

4 Results and Discussion

4.1 Torsional Stiffness

Three types of tube cross-sections and two test modes, total six combinations were simulated to calculate the torsional stiffness.

 https://doi.org/10.1051/matecconf/201815304002

The compared results are listed in Table 3, it clearly shows that in both Mode 1 and Mode 2, if the green members use Type 2 and Type 3, the space frame have considerable improvement compared with Type 1.

In the condition of Mode 1, the torsional stiffness of Type is 23% higher than Type 1; Type 3 is 6% higher than Type 1.

In the condition of Mode 2, the comparison is more obvious, the torsional stiffness of Type 2 is 65% higher than Type 1; Type 3 is 39% higher than Type 1.

Gross mass value also can be outputted from ABAQUS data file, and the results are compared in Table 4. This table present that if the square tube Type 2 has the same wall thickness with the original round tube Type 1, the frame weight is 18% increased; however, if the square tube reduced the wall thickness to 1.2mm (Type 3), the weight is 2% decreased.

From Table 3 and Table 4, it can be found that the usage of Type 3 in green members is the better alternative choice, it can reduce the weight slightly and increase the torsional stiffness considerably. Also the choice of Type 3 is allowed by the competition rules.

Another phenomenon also should be pointed out, in comparison of Mode I and Mode II, the torsional stiffness of the rear fame is higher than the front frame, especially when the square shape tubing is used. This is a reasonable distribution of torsional stiffness, the rear frame has more important influence on the vehicle handling stability and riding performance; if the rear frame suffers large deflection, it will change the roll stability and rear axle steering characteristic [9].

Table 3. Frame torsional stiffness (Unit: Nm/Deg).

	Type1	Type 2 (Difference)	Type 3 (Difference)
Mode1	1438	1765 (23%)	1522 (6%)
Mode2	1553	2556 (65%)	2153 (39%)

Table 4. Frame mass (Unit: Kg).

	Type1	Type2 (Difference)	Type3 (Difference)
Mass	33.39	39.46 (18%)	32.74 (-2%)

4.2 Torsional Angle Curve

To plot the torsional angle curve, a set of data point points should be selected on both bottom sides of the frame along longitudinal direction, as marked in Figure 5. The displacement magnitude in Z direction of left and right sides points (total 434 points) will be outputted. The distance between each pair of left and right point is known, the torsional angle can be determined by using Equation (2).

The results of torsional angle are plotted in Figure 6 and Figure 7. Horizontal axis is the coordinate of the frame in X direction, the original zero-point means front wheel vertical centre line; vertical axis is the torsional angle of the selected points.

Figure 6 shows the results of Mode I, the angle decreases as the data points change from front to rear. The trend of Figure 7 is on the contrary, the angle increases as the data points change from front to rear.

For different cross-section types, these comparisons show that the maximum torsional angle of Mode I is higher than Mode II, this also confirm that the space frame has higher torsional stiffness at the rear part.

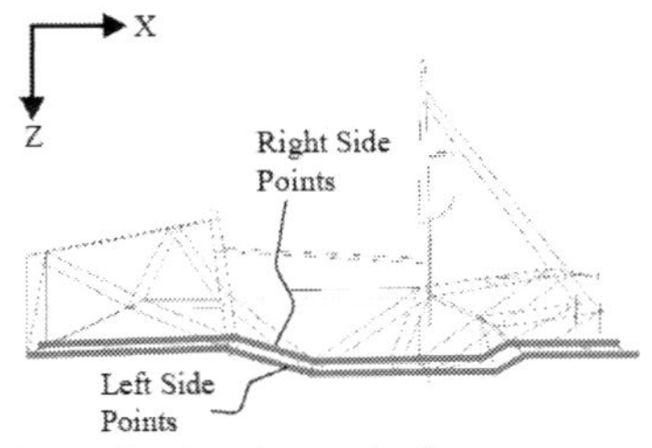

Figure 5. Data points location on the frame.

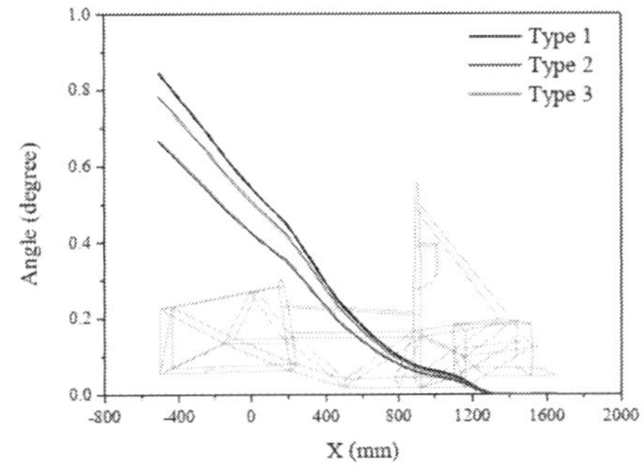

Figure 6. Frame torsional angle curve (Mode I).

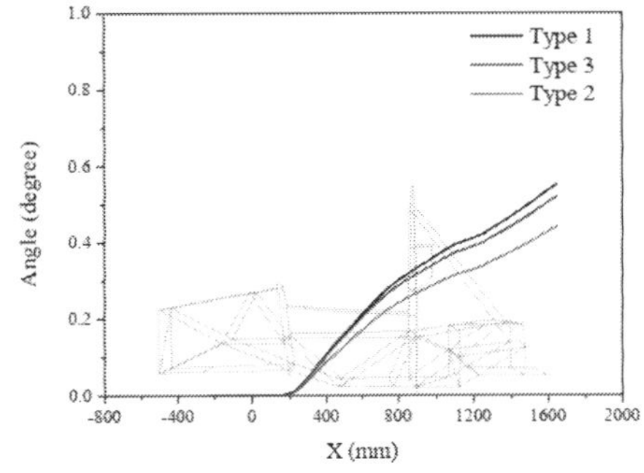

Figure 7. Frame torsional angle curve (Mode II).

Under the condition of satifying FSAE rules, every team intends to make the racing car have lighter weight and better riding performance. During the frame design process, works are usually done by optimizing the geometry structure. Based on the frame model of privious optimization study, this research additionally studied the influence of the different cross-section of the tubes.

Two different torsional stiffness test modes were considered, it can be clearly conclued the rear part has higher torsional stiffnes than front part. As disscussed in

Part 4.1, this trend accords with the ride characteristics of a FSAE car.

There tube types tube were disscussed, Type 1, Type 2 and Type 3. If green members have the same 1.6mm wallthickness (Type 1 and Type 2), square tubing has much better performance in torsional stiffness ; if the square tubing wallthickness is 1.2mm (Type 3), it also performs better than round tubing with 1.6mm wallthickness (Type 1), and even has 2% lighter weight.

This research studied the simulation methods of torsional stiffness test of a FSAE car frame, and disscussed the feasibility of square tubing. It can be seen that square tubing can be used in frame optimization to improve the torsional stiffness and frame weight.

Acknowledgment

This work was supported by the major scientific and technological platform project of Liaoning Province (JP2016020).

References

1. E.F. Gaffney, A.R. Salinas, Introduction to Formula SAE® Suspension and Frame Design, SAE Technical Paper 971584, SAE International (1997)
2. A. Tyagi, Design and Analysis of a Space Frame Tubular Chassis for a Formula Student Car, International Journal of Innovative Research in Science, Engineering and Technology, 5(9), 16451-16456 (2016)
3. C. Sithananun, C. Leelaphongprasut, C. Baitiang, N. Rungpipatphol, N. Noomwongs , T. Singhanart, SAE Student Formula Space Frame Design and Fabrication, The Second TSME International Conference on Mechanical Engineering (2011)
4. B.R. William, R.G. Albert, Design Analysis and Testing of a Formula SAE Car Chassis, SAE Technical Paper 2002-01-3300, SAE International (2002)
5. Formula SAE China Rules, SAE-China (2016)
6. D. Slimarik, F. Bauer, Design of Tubular Space Frame For Formula Student Race Car, Mendelnet 2013, 862-866(2013)
7. W.J. Deng, F.C. Lan, J.Q. Chen, Research on Topology Optimization and Lightweight Design for Frame Structure of FSAE Racing Car, Chin J Auto Eng, **6**(1), 35-42 (2016)
8. D.W. Xue, Y.Y. Zhao, Finite Element Static Analysis of Passenger Car Frame, **07**,1075-1078 (2016)
9. G.F. Yu, H.W. Huang, J.H. Wu, Finite Element Calculation and Analysis of FSAE Race Car's Strenght and Stiffness, **17**(4), 29-32 (2009)

Research and Implementation of Tractor Power Shift Clutch Control System

Chengjun Li, Meicong Ke and Youyu Wu

School of information engineering, Wuhan University of Technology, China

Abstract. Applying power shift control technology on tractor can improve the power, economy and driving comfort of tractor. The separation and engagement control of clutch in tractor power shift transmission system is the key factor to optimize the quality of shift. Through many experiments and tests, an electronic control unit based on MC9S12 was developed, and an effective clutch control strategy was put forward, which improved the jerk of power shift and shortened the gear shift time.

1 Introduction

The power shift tractor adopts electronic control system to simplify the complicated operation process to simple push-button operation, greatly reducing the driver's control difficulty and labor intensity, and improving the quality of work, so as to improve the work efficiency.

The engaging law of clutch is one of the key technologies of tractor power shift. The automatic clutch is based on the traditional mechanical clutch and is equipped with an electronic control system, which uses modern electronic technology to realize the automation of clutch manipulation. When shifting, the driver presses the shift button, and the clutch automatically completes separation and engagement.

2 Design of electronic control system

2.1 System overall design

Electronic control unit is the core of the whole control system. Taking MC9S12 as a microcontroller, the information of high and low gear sensors (digital quantities) and clutch travel sensors (analog signals) is extracted and analyzed, data is processed by software, and clutch travel is monitored accurately. According to the feedback information of several sensors, the electro-hydraulic proportional valve and on-off valve are hydraulically controlled, and the engagement speed of the clutch is precisely controlled with proper clutch engagement law. Combined with the clutch travel, the clutch speed is changed in real time, and the movement law of the clutch is effectively controlled so that the gear shift is steady and fast. The hardware system diagram is shown in figure 1.

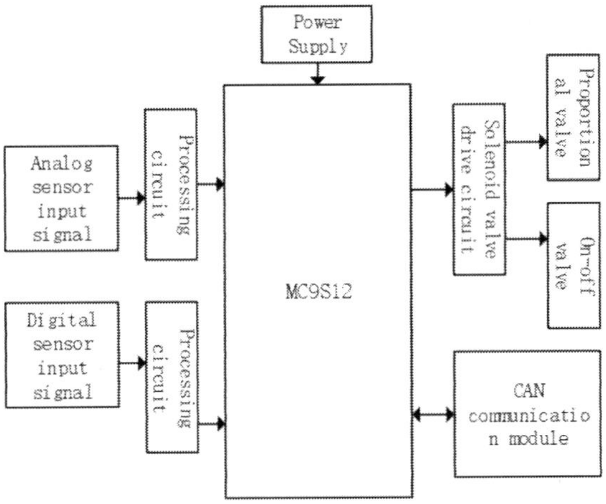

Figure 1. System hardware circuit diagram.

2.2 Clutch control principle

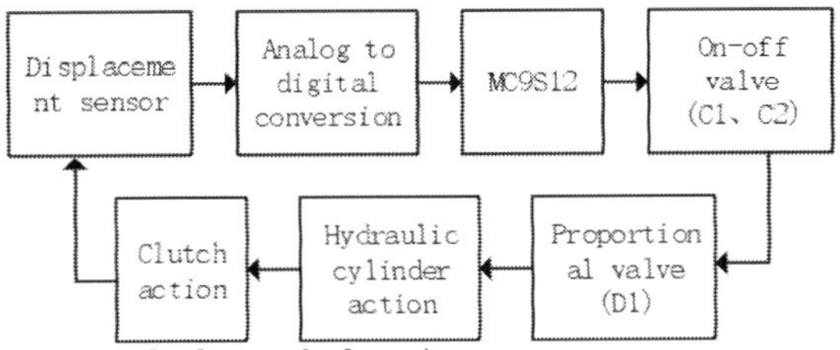

Figure 2. Clutch control schematic.

The action of the clutch is accomplished by the clutch hydraulic cylinder. As shown in figure 2, electronic control system analyzes hydraulic cylinder displacement

https://doi.org/10.1051/matecconf/201815304003

sensor in real-time, through the on-off valve and proportional valve control to achieve the hydraulic cylinder stroke control and speed control, and finally realize the clutch movement control. The physical diagram of the hydraulic cylinder and the displacement sensor is shown in figure 3.

Figure 3. The physical diagram of the hydraulic cylinder and displacement sensor.

3 Clutch control strategy

Clutch control strategy determines the quality of shift, and it is the key and difficult point to realize power shift. Through the clutch separation and joint control, it can shorten the shift time, reduce the shift jerk, and improve the shift-feel.

The speed of the engaging or disengaging of the clutch can be realized by the control of the proportional solenoid valve and the on-off valve by the electronic control unit.

Clutch control strategy is divided into two parts: clutch separation control and clutch engagement control.

3.1 Clutch separation control

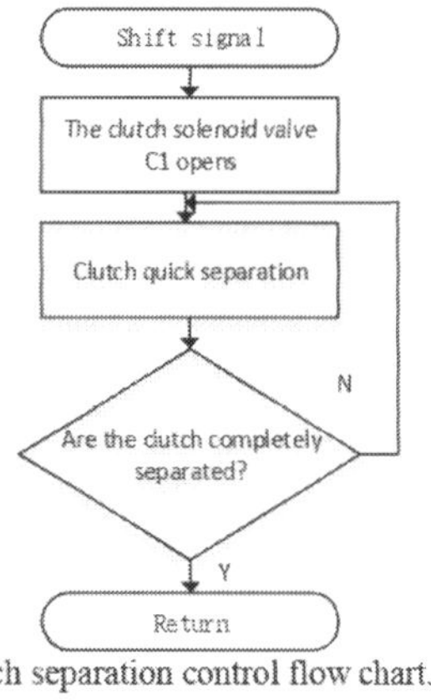

Figure 4. Clutch separation control flow chart.

During the gear shifting process, the power interruption time should be reduced as little as possible in order to avoid excessive power losses which Causes the vehicle speed to fall too much. Therefore. The clutch separation process should be completed quickly.

TCU(Transmission control unit) detects the output signals of each sensor in real time. When the TCU obtains the shift signal, the processor sends out an action instruction to turn on the clutch solenoid valve C1 so that the hydraulic cylinder rapidly drives the clutch to be rapidly separated. When the TCU obtains that the clutch is completely detached, the TCU closes the clutch solenoid valve C1, leaving the clutch in a state of separation. See figure 4.

3.2 Clutch engagement control

The clutch engagement law directly affects the overall shift quality. The "fast - slow - fast" clutch engagement control strategy can effectively improve the shift quality, referred to as "two fast and one slow".

In the clutch engagement process, the clutch engagement control is divided into three modules, namely, half joint front, half joint point and half joint behind.

The so-called "clutch half joint" means the half linkage. The clutch is engaged, but it is not fully engaged. That is to say, the half joint point is that the vehicle body is about to get enough output torque, and the vehicle body is in a critical state that will not move.

The clutch is rapidly approaching the half joint before the semi junction, which is "one fast". After the half joint , the clutch is quickly and completely engaged to shorten the gear shift time, which is "two fast". At the half joint point, the system controls the flow of the proportional valve so that the clutch engages slowly, ensuring the smoothness of the shift, which is "one slow". So precise control of "two fast and one slow" is the key and difficulty of clutch engagement control.

The total clutch travel is 50mm.After many experiments and tests, the clutch half joint point is between 30-40mm.Therefore, in clutch engagement control process, it should be guaranteed that during the travel 30-40mm, clutch is slowly engaged. The engaging time is about 800ms.The diagram of clutch travel and speed is shown in figure 5.

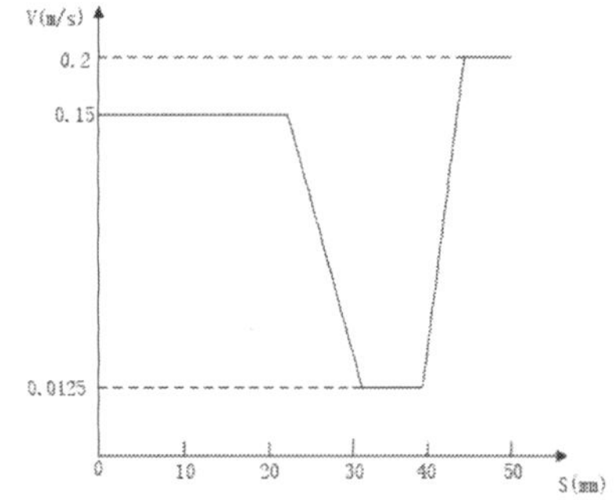

Figure 5. Relationship between clutch engagement speed and stroke.

When the TCU obtains the clutch engagement signal, the TCU sends a control signal to open the clutch solenoid valve C2, and controls the proportional valve D1 to output a larger flow, and the clutch engagement speed v_1 is 0.15m/s. At the same time, the analog sensor is detecting the clutch travel position, and the TCU processes the acquired analog data by means of software (seeking variance and average value) to obtain the real-time travel of the clutch. When the clutch travel is approaching 30mm, the TCU sends a control signal to make the proportional solenoid valve D1 output smaller flow, and the clutch engagement speed v_2 is 0.0125m/s; When the clutch travel is measured to be 40mm, the TCU controls the flow of the proportional valve again, and the clutch engagement speed v_3 is changed to 0.2m/s; When the clutch is detected fully engaged, the TCU closes the solenoid valve C2 and the proportional valve D1 to complete the engagement of the clutch. The process is shown in figure 6.

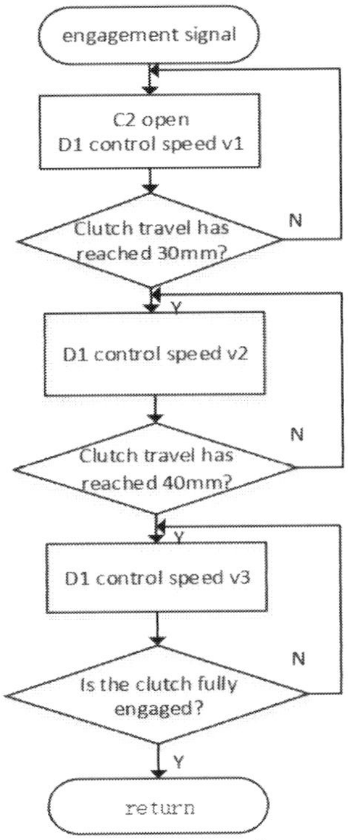

Figure 6. clutch engagement control flow chart.

4 Experiment analysis

4.1 Experimental conditions

It is necessary to ensure the stability of the hydraulic oil and solenoid valve. Their experimental conditions are shown in table 1. The whole tractor performance is in good condition, the experimental road is standard leveling cement ground.

Table 1. Experimental conditions

oil pressure	12.0MPa
oil temperature	25-57℃
solenoid valve C1	25-58℃
solenoid valve C2	25-61℃
proportional valve D1	25-55℃

4.2 Experimental evaluation index

The experiment evaluation index contains two parts: shift jerk j and total shift time t.

Shift jerk j is expressed as follows:

$$j = \frac{da}{dt} = \frac{d^2v}{dt^2} \tag{1}$$

v is tractor speed; a is tractor longitudinal acceleration. The Germany recommended value of jerk is: $j = 10m/s^3$; The China recommended value is: $j = 17.64m/s^3$.

Total shift time t is expressed as follows:

$$t = t_a + t_b + t_c \tag{2}$$

t_a is clutch separation time; t_b is block time; t_c is clutch engagement time;

In order to shorten the total shift time, the clutch separation time can coincide with the block time. Because $t_a > t_b$, total shift time is

$$t = t_a + t_c \tag{3}$$

As the "two fast and one slow" clutch engagement control strategy, clutch engagement time is divided into three parts:

$$t_c = t_{c1} + t_{c2} + t_{c3} \tag{4}$$

t_{c1} is clutch first segment quick engagement time; t_{c2} is clutch second segment slow engagement time; t_{c3} is clutch third segment quick engagement time.

Therefore, the total shift time is

$$t = t_a + t_{c1} + t_{c2} + t_{c3} \tag{5}$$

4.3 Experimental data analysis

For the reliability of the data, the following experimental data are collected from two to three gear.

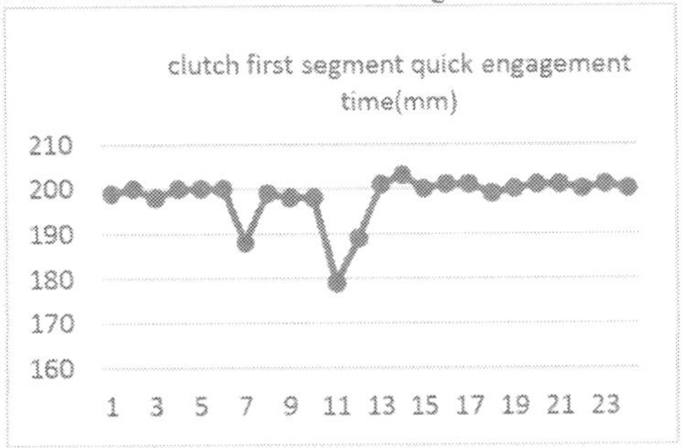

Figure 7. Clutch first segment quick engagement time.

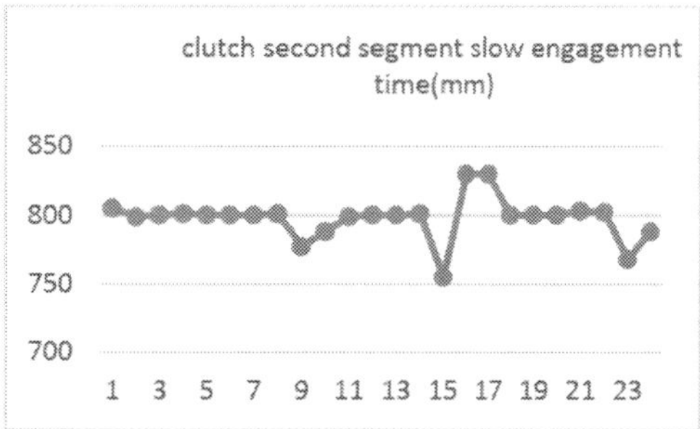

Figure 8. Clutch second segment slow engagement time

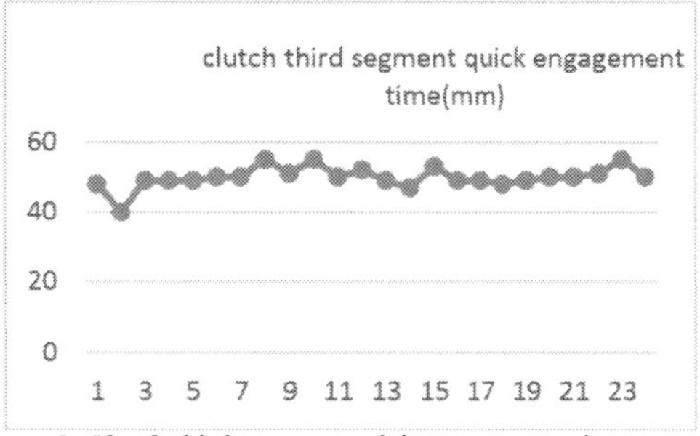

Figure 9. Clutch third segment quick engagement time.

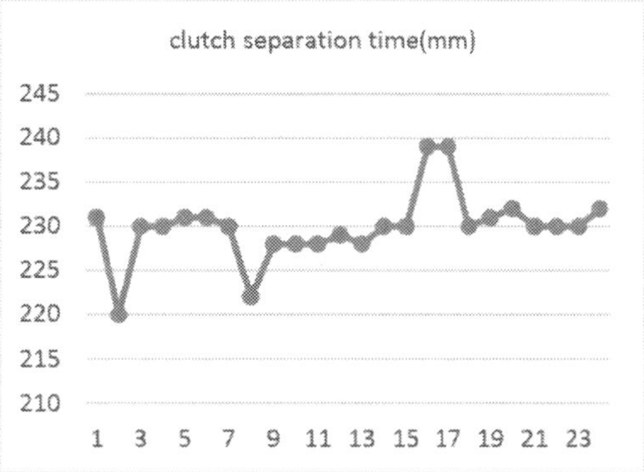

Figure 10. Clutch separation time.

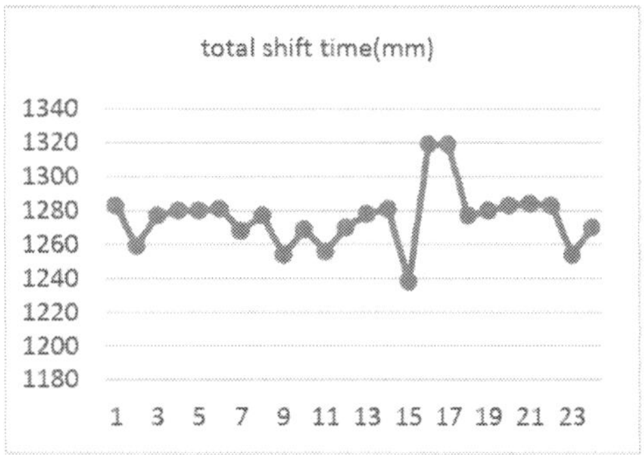

Figure 11. Total shift time.

Fig.7-11 shows the clutch engagement and separation of each part time. The mean value of clutch first segment quick engagement time t_{c1} is 198.16 ms ,the mean value of clutch second segment slow engagement time t_{c2} is 797.79 ms ,the mean value of clutch third segment quick engagement time t_{c3} is 49.91 ms ,the mean value of clutch quick separation time t_a is 229.96 ms .So the total shift time is t 1275.83 ms .

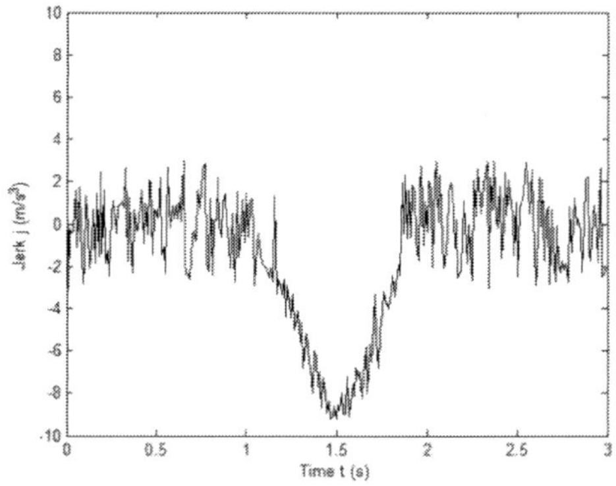

Figure 12. The measured jerk of the gear shift.

Fig.12 is the measured diagram of the jerk of the tractor, and the highest absolute value of the jerk is 9.3 m/s^3 .

The results show that the clutch engagement law adopts the "two fast and one slow" control strategy, which can effectively reduce the shifting sense of frustration and improve the total shift time.

5 Conclusion

1) The key problem of tractor power shift technology is put forward, that is, the separation and engagement of clutch.

2) Through many experiments and tests, an electronic control unit based on MC9S12 is developed, and the "two fast and one slow" clutch engagement control strategy is put forward.

3) Through the loading test, the correctness of the "two fast and one slow" clutch control strategy is verified, the total shift time is shortened to 1275.83 ms, and the shift jerk is improved to 9.3 m/s^3 .

References

1. Liu, Jiguo, et al. "Study on Shifting Control Strategy for Tractor AMT System." Agricultural Equipment & Vehicle Engineering 7(2014).
2. Saitou, Yoshitami, N. Isono, and N. Suzumura. "Electronically controlled clutch engagement speed correcting for an automatic transmission system." US, US5109967. 1992.
3. Ling, Ling, Z. S. Chen, and Y. Y. Luo. "Study on Control Process of Automatic Mechanical Transmission System for Pure Electric Bus."Applied Mechanics & Materials 220-223(2012):1795-1799.
4. Gao, A., Z. Fu, and W. Zhang. "Fuzzy shift schedule of automatic mechanical transmission for tractors." Transactions of the Chinese Society for Agricultural Machinery 37.11(2006):1-4.

5. Ota, Atsuo, S. Narita, and T. Masuda. "Shift control unit of electric-power assist transmission." (2000).
6. Nitsche, Martin, and A. Englisch. "Method and apparatus for gear shifting control with improved shift quality during gear shifting." US, US5946984. 1999.
7. Li, Guozheng, and Y. Tan. "Research of automatic gearshift control technology for tractor mechanical gear-box." Journal of China Agricultural University 9.3(2004):85-87.

Design and Optimization of Automatic Shifting Mechanism for the Tractor

Shiheng Sun, Huiting Shi

Automotive application engineering, Wuhan University of Technology, Wuhan 430070, China

Abstract. The stability of the shifting mechanism is the basis of the reliability of the automatic shift system. The gear shifting process of Taishan TG1804 tractor is hard and long. In order to solve this problem, we design a tractor automatic shift system and apply it to the tractor successfully. This paper designs a shifting mechanism, optimizes the shifting mechanism and dose the dynamic bench test. The test results prove that the optimized shifting mechanism is more reliable and more reasonable.

1 Introduction

With the improvement of mechanization in our country, the traditional agricultural production mode died out gradually, because it's low efficiency. Agricultural land has experienced three stages of development: dispersal, integration and extensive outsourcing. The reform of agricultural and the support of the country have greatly promoted the development of agricultural machinery industry.

The market of agricultural machinery industry continues to expand. With the increasing number of users, new requirements have been put forward for the agricultural machinery industry. The users require not only the products to be affordable, but also easy to operate. At present, most of the agricultural tractors use manual shift method in our country. The shift is time-consuming and the laborious, especially for tractors with more gears. The drivers work at a hard environment, have to work a long time, need to change gears and operate the agricultural machinery frequently. The drivers are easy to get tired.

The application of automatic transmission in the tractor can complete the shift automatically, reduce the operating burden of the drivers greatly, and meet the needs of the users. We can optimize the design of the automatic shift system to control the additional cost within the acceptable degree of the user. Therefore, the application of automatic transmission in tractor is significant.

2 Development of automatic transmission

Automatic transmission of tractor mainly has Automatic Transmission (AT), Automatic Mechanical Transmission (AMT), and Continuously Variable Transmission (CVT).

2.1 Automatic transmission

Automatic Transmission mainly consists of hydraulic torque converter, planetary gear mechanism and automatic shift control system. Hydraulic torque converter changes the torque through the change of liquid moment of momentum, has a good ability to adjust and adapts the external load automatically, contributes to a smoothly starts of the tractor, prevents transmission from overload. However, when the engine is at a low speed, the transmission efficiency is very low, and the torque can not be changed, and the drive ratio can not be controlled accurately. As a result, AT is seldom used on tractors.

The application of AT in tractors firstly appeared in the late 1950s. The American Caterpillar company firstly used AT in its crawler tractors. At the same period, such as the series of crawler tractors of the Case industrial, industrial tractors of the German Fendt company, etc., have also used this transmission.

2.2 Automatic mechanical transmission

Automatic Mechanical Transmission is realized based on the traditional manual transmission by adding shifting mechanism, hydraulic (pneumatic or electric) power system and electronic control system. The structure of AMT is relatively simple, has a good inheritance of production, and has high transmission efficiency. However, due to the power interruption during shift, there is a shift shock, which affects the ride comfort.

The overseas researches on AMT start earlier and have more achievements. Eaton's SAMT system, Scania CAG system and Daimler's EPS system have a representative contribution to the start control of the tractor. Japanese Isuzu Motor Ltd, Germany ZF, Japan

Nissan, France Renault and other companies promote the application of full automation AMT on tractors greatly.

Some universities in China have made great achievements in the research of tractor AMT. Tsinghua University has carried on the simulation experiments in the AMT clutch control aspect, and made a further research, realized the clutch control automatically. YTO Group Corporation and Henan University of Science and Technology develop different gear shifting strategy for different operation conditions together. Jilin University makes a further research on the clutch control strategy when starting and shifting, and formulates the optimum strategy of starting clutch control, and proposes the dynamic three parameter shift schedule.

2.3 Continuously variable transmission

The change of speed ratio of continuously variable transmission is continuous. CVT is different from traditional gearbox which has fixed axle. The transmission of power can be sustained without interruption. There are two types of CVT applied in tractors. One is the hydraulic-mechanical CVT which changes the speed ratio continuously by changing the diameter of the driven pulley. The transmission has a light weight, a small size, few components, a high transmission efficiency, a low fuel consumption. But the transmission ratio has a narrow range, and the belt is easy to damage and can not bear large loads. Another is composed of hydraulic pump and hydraulic motor named Hydro-static Transmission. Its main features are: speed can change continuously, can absorb the shift shock and prevent the tractors from overload, starting fast and soft, can adapt different agricultural machinery, can match with different engine. But the machining accuracy is high and the price is expensive.

Table 1. Advantages and disadvantages of different automatic transmissions

	Advantages	Disadvantage
AT	Mature technology Smooth starting process High shifting quality	Low transmission efficiency.
AMT	Simple structure Low cost, Low fuel consumption.	Shift power interruption Shift shock Complex clutch control.
CVT	Continuous change of speed ratio Good shift quality	complex structure High cost

Due to the high cost of manufacture, the first application of hydraulic-mechanical CVT in 926Vario tractors of Fendt company started in 1996. At present, it's used in Agrotron TTV series of Deutz-Fahr company, Massey Ferguson company's 7400 series and John Deere company's 7R and 8R series abroad. Hydro-static transmission is mainly applied in small tractors, such as South Korea's Machindra 3510, New Holland Boomser 35, John deere3038E. The study on Hydro-static transmission started in 1970s in our country. Hydro-static transmission is mainly used in engineering and military vehicles. At present, the domestic study on hydro-static transmission is still in the prototype stage, has made some achievements. The research institutions are Henan University of Science and Technology, Jiangsu University, etc. The advantages and disadvantages of different automatic transmissions are shown in table 1.

3 The design of shifting mechanism

The test vehicle is TG1804 tractor with 16 gears in the gearbox. The tractor includes forward and reverse gears, 4 gears in the main transmission box, 4 gears in the secondary transmission. The main transmission box adopts synchronizer to change gears, and the secondary transmission uses meshing sleeve to change gears. We designed a new shift mechanism, and designed a clutch separation mechanism to realize the automatic shift in the main transmission. Figure 1 shows the modeling of the main transmission gear of TG1804 tractor.

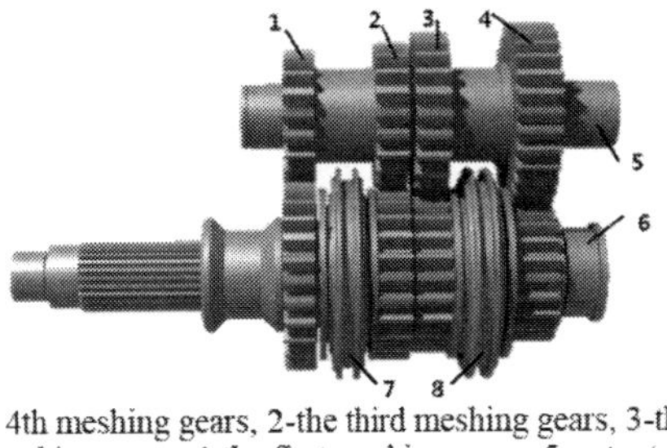

1-the 4th meshing gears, 2-the third meshing gears, 3-the second meshing gears, 4-the first meshing gears, 5-output shaft, 6-input shift, 7-3/4 synchronizer, 8-1/2 synchronizer.
Figure 1. The modeling of the main transmission gear.

3.1 Determination of the gear shifting force

To determine the gear shift force, we should determine the transmission ratio of each gear first. We need to determine the angular speed difference when the gears are shifted at a given engine speed. And we know when the engine speed is 2000rad/s, the engine has a maximum power. When the engine speed is 1500rad/s, it has a maximum torque. And we know the gear ratio of the transmission. The angular speed difference when upshift or downshift is shown in table 2. $\Delta\omega$ means the angular speed difference when shift.

Table 2. The angular apeed difference when shift

	1-2	2-3	3-4	4-3	3-2	2-1
$\Delta\omega$ (rad/s)	28.00	23.23	22.89	19.32	19.67	24.93

In the process of shift, the inner cone of the lock ring contacts the outer cone of the gear to be engaged. Force analysis of the gear to be engaged is shown in figure 2. According to the conservation of energy, the formula 1 should be satisfied when complete the synchronization at a given time [4].

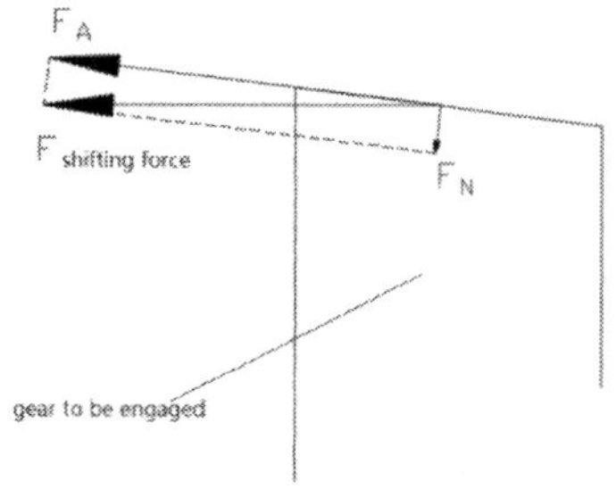

Figure 2. Force analysis of the gear to be engaged.

$$J\Delta\omega \leqslant FR_1\mu t/\sin\alpha \qquad (1)$$

The F is the shift force, and other parameters are shown in table 3. When the given synchronization time is 0.2s, in order to meet all gear shifting requirements, we need to take the maximum angular speed difference as $\Delta\omega$, so the $\Delta\omega$ takes the angular speed difference when up-shift from 1 gear to 2 gear, and the angular speed difference is 28 rad/s.

Table 3. The meaning and value of the parameter

Symbols	Meaning	Value
J	Input moment of inertia	0.263kg.m^2
$\Delta\omega$	Angular velocity difference when shifting	28 rad/s
t	Synchronization time	0.2s
R_1	Average working radius of friction cone	54.63mm
μ	friction coefficient	0.1
α	the angle of friction cone	7°

3.2 Schematic diagram of shift

According to the requirements of the company, we choose pneumatic shift. The maximum pressure that the tractor can provide is 0.8Mpa. The displacement of the synchronizer is 11 millimeters when the shift fork shaft moves from neutral palace to gear palace. A shifting fork shaft has three positions: first (or third) gear, second (or 4th) gear and neutral position.

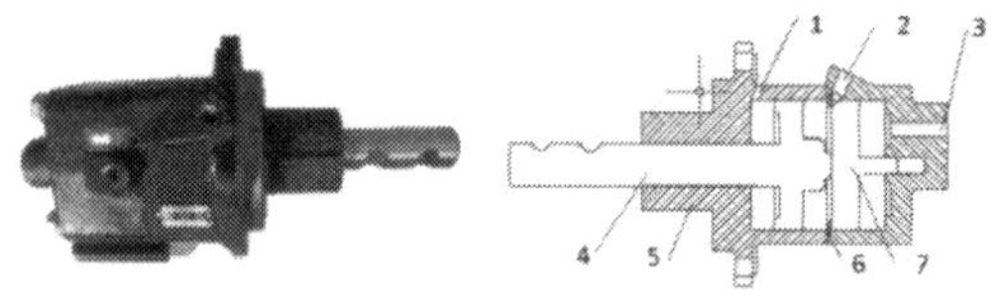

(a) (b)
1-A inlet, 2-B inlet, 3-C inlet, 4- piston rod, 5-cylinder block, 6- piston rod's right limiting ring, 7 limit piston,

Figure 3. Shift cylinder.

We choose the Fastre shifting cylinder, as shown in Figure 3 (a). The piston rod has three fixed positions by controlling the inlet of the cylinder. As shown in Figure 3 (b), there is a limiting piston inside the cylinder. When A inlet intakes air, the piston rod will move to the right limit position, B inlet intakes air, the piston rod moves to the left limit position, A, C inlet intake air at the same time, limiting piston moves to left limiting position, the piston rod moves to the neutral position, the pressure of C inlet is greater than the A inlet.

The diameter of the shift cylinder bore is 76mm, and the diameter of the piston rod is 25mm. After calculation, the theoretical value of the shift gas pressure is 0.2Mpa. And we take 0.22Mpa as the actual value. In order to have a neutral position, the air pressure at the C inlet is greater than the A inlet, so the pressure at the C inlet is 0.3Mpa. Pressure relief valve regulate pressure at 0.35Mpa, the pressure reducing valve 1 regulate pressure at 0.3Mpa, and the pressure reducing valve 2 regulate pressure at 0.22Mpa. We choose two 3 dimensional 5 way solenoid valve and two 2 position 3 way solenoid valve. The schematic diagram of shift is shown in figure 4, the pressure generated by the air pump is stable at 0.35Mpa after overflow pressure relief valve. The air is divided into two directions, one flow to A and B inlet, another flow to C inlet. The pressure at A and B inlet is 0.22Mpa, and the pressure at the C inlet is 0.3Mpa. The control program controls the opening and closing of the electromagnetic valve, and realizes the opening and closing of the air inlet to meet the shift requirements.

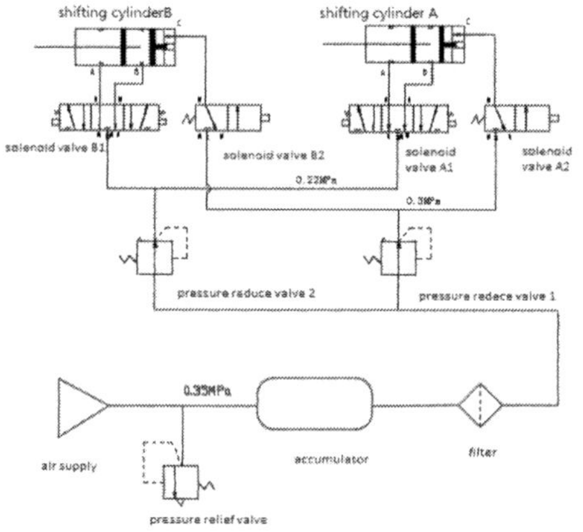

Figure 4. The schematic diagram of shift.

3.3 Shifting mechanism design

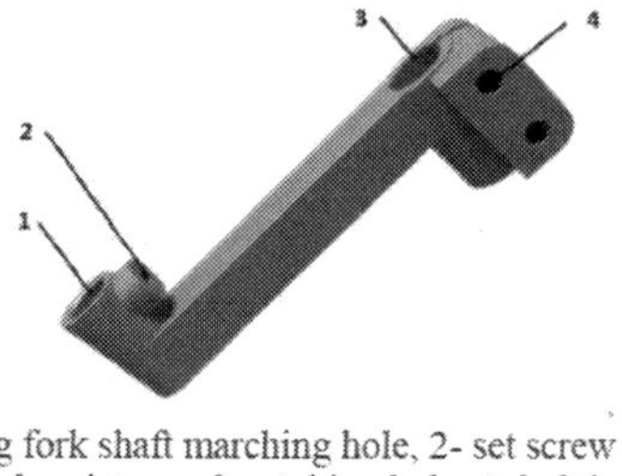

1-Shifting fork shaft marching hole, 2- set screw hole, 3 - cylinder piston rod matching hole, 4- bolt holes;

Figure 5. The model of connecting part.

The three-dimensional model of the connecting part is shown in figure 5. The connecting part is connected with the shifting fork shaft and the cylinder piston rod respectively. The number 1 is the matching hole of the fork shaft, number 2 is the set screw hole, the connecting part is fixed with the shifting fork shaft through a set screw. Number 3 is the matching hole of the cylinder piston rod, through two M12 bolts to achieve the solid connection between the connecting part and the cylinder piston rod.

The assembly diagram of the pneumatic shifting mechanism is shown in figure 6. By controlling the opening and closing of the electromagnetic valve, we can control the opening and closing of the corresponding intake and exhaust ports, and control the movement of the piston rod of the cylinder. Through the connecting part, the piston rod drives the fork shaft to move. Control the cylinder to realize the shift operation.

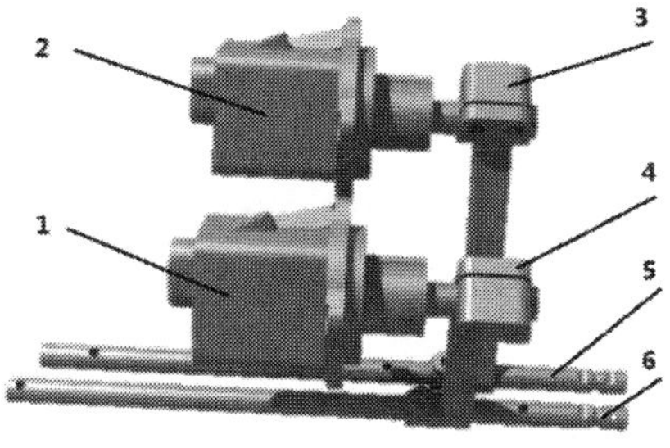

1- 3/4 shift cylinder, 2 -1/2 shift cylinder, 3- 1/2 connecting part, 4- 3/4 connecting part, 5- 1/2 shift fork shaft, 6- 3/4 shift fork shaft.

Figure 6. Assembly of pneumatic shift shifting mechanism.

3.4 The problems

Because the piston rod and the shifting fork move in the same direction, so the parallelism tolerance of the two holes of the connecting part is small, and it leads to higher machining accuracy. If the parallel tolerance does not meet the design requirements, the required shift force increase, and the wear of the gearbox will increase. There will be movement interference between the cylinder piston rod and fork. In extreme condition, we can not shift. Because the upper cover's space is small, it is hard to install. After installation, the physical drawing is shown in figure 7.

Figure 7. The physical drawing.

After the installation, we did the dynamic bench test and debugged the program, and the reliability of the shifting mechanism was tested. Under the engine idling speed and the wheel suspended, we shift 4 hours every day for 5 days, a total of 6000 times shifting, and record the 400 groups of data about the total gear shift time and shift fork shaft movement time. These are several times we failed to upshift from first gear to second gear. When raising the inlet pressure, it does not happen anymore. The analysis result is the connecting parts do not meet the parallel tolerance. There is interference between the shifting fork shaft and the piston rod.

4 Optimization of shifting mechanism

The connecting part of the original design has a high machining accuracy, and it is hard to install. When the machining errors existed it will increase the shift force and the shift time. The connecting part can not meet the requirement to change gears when the machining errors reach a certain degree.

4.1 The design of the shifting mechanism

After optimization, the connecting part is changed into two parts. This design avoids the high machining accuracy. The optimized connecting parts reduce the quality by 0.736kg compared with the original designed part. The cost increases little because of the reduction of mass and machining accuracy. It is convenient to install the new connecting parts. The force condition of the shifting mechanism gets better. The designed 3D model and assembly model of the shifting mechanism are shown in Figure 8

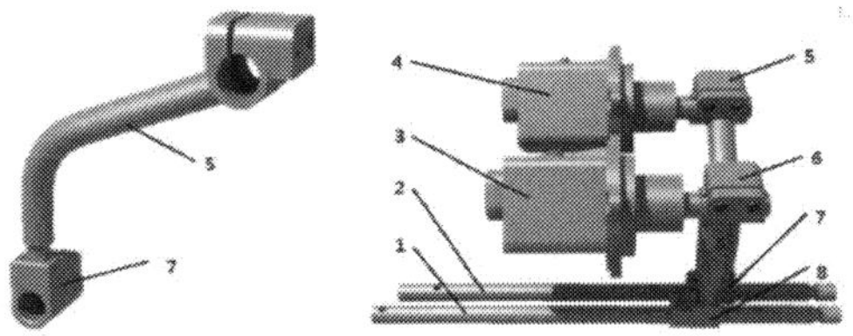

1-3/4 shift fork shaft; 2-1/2 shift fork shaft; 3-3/4 cylinder; 4-1/2 cylinder; 5-1/2 connecting part; 6-3/4 connecting part; 7-1/2 blocks; 8-3/4 blocks

Figure 8. 3D Model and assembly model of the shifting mechanism.

4.2 Strength analysis

The established 3D model is imported into ANSYS. The material for the connecting parts is 45 steel. The connection is restrained according to the actual condition. In the process of shifting, the force of the 1/2 gear shift is more complicated than that of the 3/4, so we analysis the force of the 1/2 gear shift. The direction of force is determined by the weak force region of the block. And the shift force is 1000N. The connecting parts is tetrahedral mesh, the block is hexahedral mesh. The mesh density will be increased if the force of the region of the part is weak. The constraints and loads, mesh and the force of the connecting parts are shown in figure 9.

We can see that the maximum compressive stress is 338Mpa, which is far less than the strength limit 598Mpa of the selected material. The designed connecting parts can meet the reliability requirements.

https://doi.org/10.1051/matecconf/201815304004

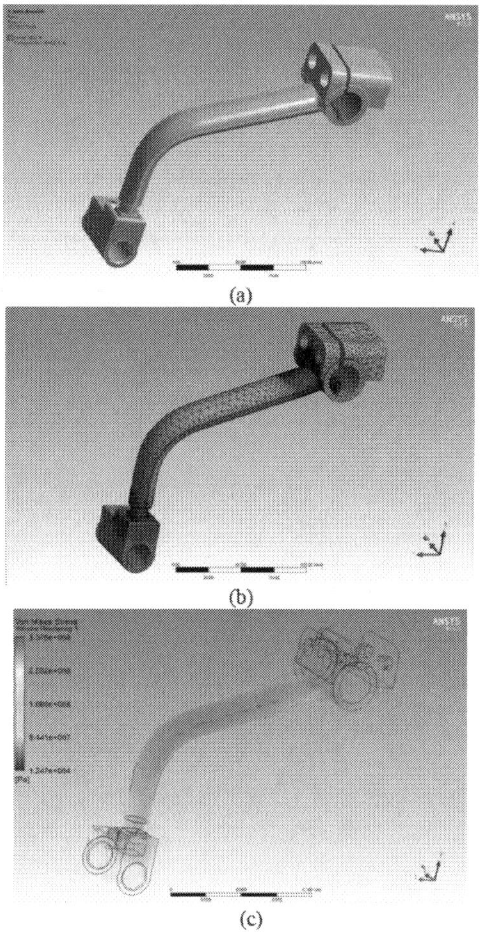

Figure 9. Strength analysis of the connecting parts.

4.3 Bench test

The new designed parts are shown in figure 10.

Figure 10. New contecting parts.

Figure 11. Bench test.

Install the newly designed connecting parts on the transmission and do reliability experiment. The bench test is shown in figure 11. The test condition is the same with old connecting parts include the pressure, engine speed. The times of the shift experiment is 6000, and the number of data acquisition is 400.

There is no fault happens after 6000 shift operation. The new designed shifting mechanism is reliable. Process the data and obtain the shifting time comparison after optimization. The comparison of the shifting fork shaft moving time after optimization is shown in figure 12.

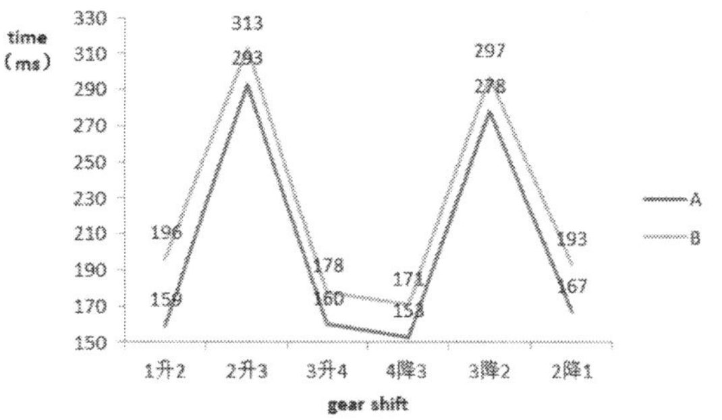

A means the shifting fork shaft moving time after optimization
B means the shifting fork shaft moving time before optimization

Figure 12. Comparison after optimization.

4.4 Experimental result analysis

When upshift from second gear to third gear or downshift from third gear to second gear, the shifting process is longer than other gear shift. This phenomenon accords with the actual situation. Because one shifting fork shaft has to arrive at a neutral position, then another shifting fork shaft began to move. And in this process, the number of the working solenoid valve is double than other gear shifting. So the shift process is long. The change trend of the switching time of other gears is same as that of theoretical synchronization. The larger the angular velocity difference of gear shifting is, the longer the movement time of the shifting fork shaft in the shift process.

Under the same shift condition, through the comparison of the shift time and reliability between the new and old parts, it can be found that after optimization, the shift time is shorter, and the shift is faster.

5 Conclusions

In this paper, we designed an automatic shift system and applied it to the TG1804 tractor successfully. We designed a new shifting mechanism because the shift failure caused by the original shifting mechanism. We did the bench test for the shift system and got enough experiment data. The newly designed shifting mechanism meets the strength requirement after simulated analysis. The test results show that the newly designed shifting mechanism has better reliability and a shorter shift time compared with the original shifting mechanism.

References

1. Kulkarni M, Shim T, Zhang Yi. Shift dynamics and control of dual-clutch transmissions. Mechanism and Machine Theory, 2007, 48: 168-182
2. Tang Xuemei. The script simulation of shifting process in transmission system. Mechanic Automation and Control Engineering. Wuhan, 2010: 3226-3228.
3. Lovas L, Play D, Marialigeti J, et al. Mechanical behaviour simulation for synchronous mechanism improvements. Automobile Engineering: 2006, 22: 17-19.
4. Hoshino H. Simulation on synchronization mechanism of transmission gearbox. International ADAMS User Conference, 1998.
5. Häggström D. Development of a program for calculating gearbox synchronization. Sweden: Luleå University of Technology, 2011.
6. T. Nitulescu, S. Talu, Applications of descriptive geometry and computer aided design in engineering graphics, Cluj-Napoca, Risoprint Publishing house, Romania, 2001

Design of an Autonomous Forklift Using Kinect

Mohamed Abdellatif[1], Metwali Shoeir[2], Omar Talaat[2], Mahmoud Gabalah[2], Mohamed Elbably[2] and Saleh Saleh[2]

[1]*Future University in Egypt, Mechanical Eng. Department, Cairo, Egypt*
[2]*Arab Academy for Science, Technology and Maritime Transport, Cairo, Egypt*

Abstract. Material handling is a necessary, but expensive activity in factories. Autonomous robot technology can help reduce the cost and relax humans from the exhaustive job of driving forklifts. In this paper, we describe the mechatronics design and implementation of an autonomous forklift. The robot can perceive the 3D dynamic world and can plan its motion autonomously to lift materials from a source to target locations. Dynamic map of the world is built using data from a Microsoft Kinect head and readings from wheel encoders, thus enabling the robot to avoid obstacles and reach target locations safely. Experiments showed success of the robot to move and load the cargo to target locations.

1 Introduction

The process of loading, unloading and transport of materials is one of the key issues for every production site and has a great impact on product cost. The material flow within the product manufacturing site is expensive; therefore, the efforts to find favorable and flexible systems continue to be of great importance. The demand toward a higher level of autonomy is strong and will change the operation of warehouses significantly [1, 2].

Forklifts are general purpose material handling machine and are generally operated manually with a driver. Operators are trained on forklifts so that they understand the limitations of the device, and the rules of safe and efficient operation. The job of driving a forklift requires long training for the drivers to ensure safe operation. Besides, the job is exhaustive and not suitable for humans and hence it is an ideal candidate for automation.

With recent decrease in prices of sensors and computing devices, several researchers realized that the mobile robot technology is matured enough to run a forklift in warehouses. Taking only high level orders from a human supervisor, an autonomous forklift can safely execute tasks that require interaction with people and objects in an unknown environment. These include the ability to detect and manipulate loaded pallets, both on the ground and on a truck bed, as well as navigating among obstacles [3, 4].

Although some product exists in the market, yet the cost is high and there is still a room to develop simple systems that are user friendly and with affordable prices [5].

In this paper, our objective is to design and develop a forklift able to move autonomously, lift the cargo and move it to target location. It should be able to map the working environment and to move autonomously or allow and ask for a human intervention when needed. The use of Kinect, though had been used in other applications [6, 7], is new to this forklift application and will be used to map the robot environment.

This paper is arranged as follows: the next section outlines the related work about building autonomous forklift technology. Then, the mechatronics design of the system is introduced in section 3. The control system logic and experiments with the system are presented in section 4 and finally conclusions are summarized in section 5.

2 Background

The Autonomous Guided Vehicle, AGV, is probably the first industrial application for mobile robots in factories. Generally, the path is predefined and can be either visible, marked by a surface with high contrast or it can be invisible like a magnetic tape. The AGV works as a line follower robot and usually has sensors for obstacle avoidance and transfer material among fixed spots in the factory. This system have limited autonomy and inconvenient when the path is changed especially in flexible manufacturing lines.

AGV concepts were combined with forklifts to build systems that enjoy both powerful material handling and simple autonomy by AGV. This was achieved in several products with even increased autonomy by replacing the line following principle by 3D scanners [8, 9]. 3D scanners enable the autonomous forklift to perceive the environment around it, build a map and plan its missions and paths in an autonomous self-controlled manner. Several manufacturers produce autonomous forklifts, and we will discuss three major products, shown in Fig.1, namely: Hitachi, Clark and CSAIL autonomous forklifts.

https://doi.org/10.1051/matecconf/201815304005

Figure 1. a) Example for AGV(By Muller), Autonomous forklifts, b) Hitachi and c) Clark

HITACHI has successfully delivered an AGV system to South Korean industrial machinery manufacturer, Hanwha. The system can handle special purpose trolleys which are placed at two different heights. Besides, it can travel without following any travel guides. The traveling route of AVGs needs to be easily modified in case of changes to the factory layout.

Clarks developed an autonomous forklift whose body is mounted on four wheels includes two castor wheels at the front side and one small castor wheel at the right rear side. To achieve autonomy, the forklift is modified so that all the necessary sensors and a PC for steering and driving motors are attached. The information about the forklift environment is obtained by using vision, laser range finder and sonar sensors.

CSAIL is an autonomous robotic forklift developed for the military. It can move pallet loads from place to place under voice command, and eventually to pack up an entire outdoor warehouse of pallets and transport it to a new location. The forklift is designed to operate in unstructured environments such as outdoor packed earth. It enables dual operation whether autonomous or manual [10-12]. Work in design of autonomous forklift interface had been made in [13, 14] using voice or other graphical interfaces. Coordination of several forklifts had been described in [15].

3 Mechatronics system design

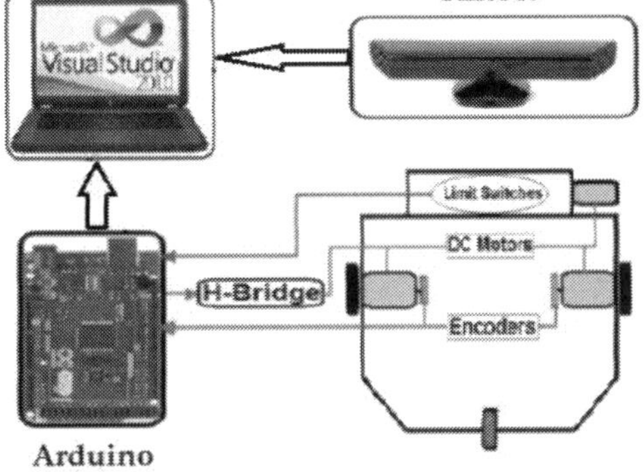

Figure 2. System Architecture for autonomous forklift equipped with Microsoft Kinect.

The system architecture is shown schematically in Fig. 2. The robot communicate its data through an arduino board which is connected to the host computer. Kinect sensor communicates directly with the host computer to send its visual measurements.

3.1. Mechanical design

The system consists of the mobile base of the forklift and the vertical moving mechanism that contains metal forks as shown in Fig. 3.

The base has three wheels, two front wheels, and one rear wheel. The front wheels are driven by DC motors with reduction gear box. The differential drive is controlled to drive and steer the base. The concept of the forklift is to have a lifting mechanism attached with two metal forks to leave the cargo. The dimensions of the forklift were made to work in indoor environment and hence the dimensions were selected to pass through normal doors. The basic dimensions are shown in Fig.3. The height is made to be 120 cm.

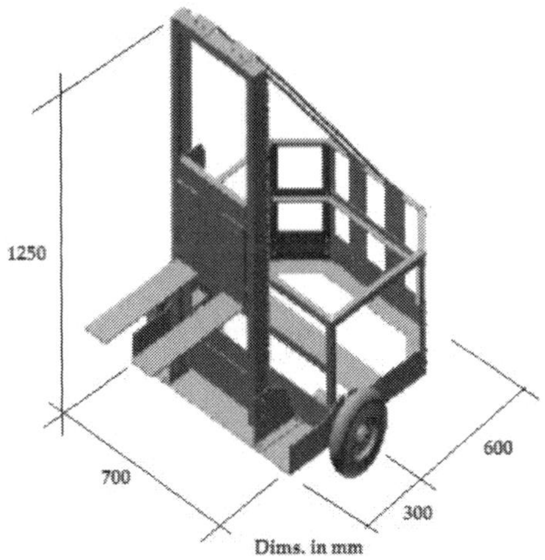

Figure 3. The mechanical design of new forklift .

Table 1. Comparison between three ranging systems

	Microsoft Kinect	Hokyu -URG-04LX	SICK LMS200
Cost, USD	150	1000	5000
Max range/Min range [mm]	3-6/0.8-0.4	4/0.06	8-80/0.07
Resolution [mm]	2.5-48	1	1-10
Accuracy [mm]	+-6(1m)/ +-130(4m)	+-30(1m) +-120(4m)	+-10(4m)
Weight, kg	0.55	0.16	4.5
Horizontal Angle, degree	57	240	100-180

3.2 Sensors and actuators

For the purpose of 3-D map building, a Microsoft X-box Kinect is attached to the forklift to detect obstacles in the near sides of the forklift. The choice of the Kinect was made in comparison to popular laser range finders used in other works having full scale forklift such as hokuyo urg-04lx, sick lms200.

The laser range finders are relatively expensive and its processing and map maintenance requires complex computations. This will also require heavy and expensive computing device. Several researchers pointed out for the use of Kinect for 3D scanning of the world at an affordable cost and satisfactory resolution [6, 7].

While to the Kinect as giving accurate maps and easy to use and require simple computing power. The Kinect give good resolution for near objects which is just enough for this application (6mm) as shown in Table 1. Therefore we decided to use the Kinect.

Optical encoders are attached to DC motors to measure the robot position and orientation.

As for the lifting mechanism, we thought of using either electric of hydraulic actuation. Since our prototype is designed to lift 10 kg maximum, therefore we exclude the hydraulic option due to light payload and the high cost and complexity of the hydraulic circuit. The electrical actuator selected is a DC motor coupled to a shaft connected to rack and pinion. Finally, we decided to choose the electrical actuation system. Because the lifting mechanism we are using depends on the rack and pinion technique, it was required to have a rotary actuator. So, we decided to use a DC motor to be attached to the mechanism to lift the cargo.

The weight of the forklift is 70 kg taking in consideration the weight to be lifted and the counter balance. So, each motor should drag 35 kg assuming that the motor size is selected to be 25kgcm.

The selected motors have a gear box with gear ratio 1:200 to increase torque and decrease the rotational speed and hence robot velocity.

3.3 Computing and interface

There are several types of controllers that can be used to compute the control logic. In selecting the robot controller, we had to select a platform compatible with the Kinect and its software. So, we used a laptop with Intel core i5 processor and windows operating system in order to connect to the Kinect using a USB interface and to the robot's microcontroller using a USB serial interface. The interface window for making commands to the forklift and monitoring its status was made through a GUI program coded in C++ language. We also used an arduino microcontroller board to control and interface with limit switches, ultrasonic sensors, encoders and DC motors.

4 Experiments

The autonomous forklift is shown in Fig. 4 with its final implementation. The Kinect is attached on to and the laptop is fitted on the back side.

The motion command for the autonomous forklift is generated based on the information of its position provided by the Kinect and motor encoders. The flow chart for robot control algorithm is shown in Fig.5. The motion control command is generated by the path planner to the goal position ensuring safety of its motion. At first, we can prepare a scenario- based behavior to accomplish a predetermined task and the other task was to go to its target and avoid obstacles using the map captured from the Kinect and create its own path using the path planner algorithm. The controller will read the data from the encoder and use these data to calculate the orientation of the robot and the distance traveled by the robot.

The robot is aware of the working environment in order to avoid obstacles while reaching its destination. The lifting mechanism has sensors to feed the controller back with the forks position.

Depth images from kinect are converted to obtain the top view image of robot front. The obtained image shows clearly the kind of obstacles and free area confronting the robot. Fig. 6 shows the depth image from Kinect and the converted to view image showing this case where a box is inserted in front of robot. Fig.7 shows the robot test path for obstacle avoidance. The Kinect images were used as input to a visual SLAM algorithms running on Extended Kalman Filter. The program can learn novel environment and recover the self- motion of the forklift. Results are not shown for space limitation.

Figure 4. Photograph of the autonomous forklift

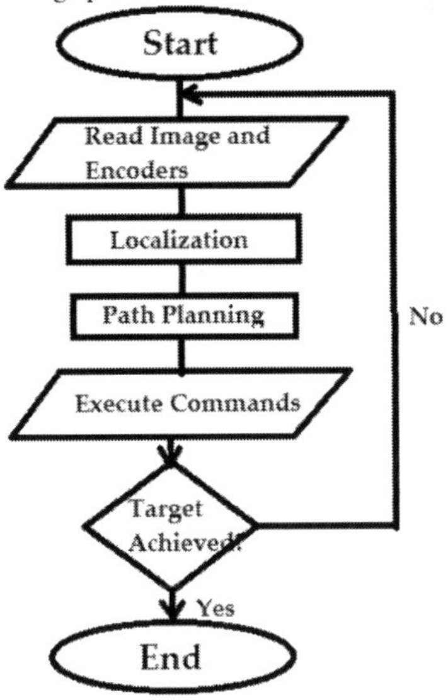

Figure 5. The flow chart for map based navigation of the autonomous forklift.

Figure 6. a. The depth map from Kinect, b. The computed top view.

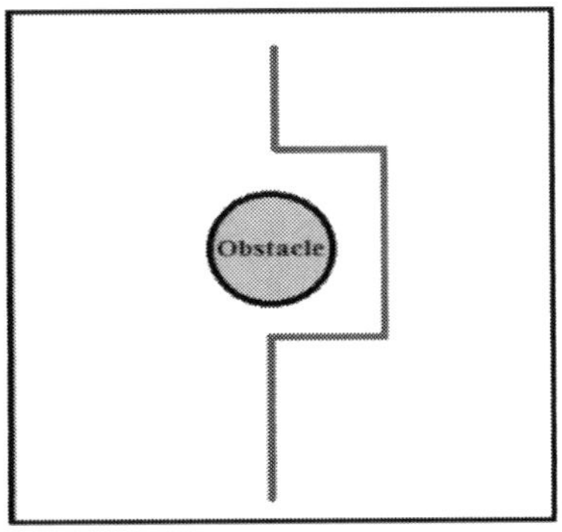

Figure 7. The robot path for testing obstacle avoidance.

5 Conclusions

We have designed and developed an autonomous forklift with a simple design, and high degree of autonomy, rigidity and stability to lift payloads and move autonomously in stores and warehouses. We exploited the Microsoft Kinect to obtain depth maps of the environment and convert these data into a top view in order to navigate the working environment for the robot. The robot self-localization was determined based on data from encoders and Kinect to mark robot position in the map. The robot self- localization was made using wheel encoders with differential drive kinematics. The robot accomplished a specific task and lifts the cargo from its origin place to target destination while avoiding static and dynamic obstacles. The actual robot path was monitored during its motion together with the required given path.

References

1. S. Berman, E. Schechtman and Y. Edan, Evaluation of automatic guided vehicle systems, Robotics and Computer-Integrated Manufacturing, 25(3), pp.522-528, 2009.
2. R. Bischoff, and G. Volker, Vision-guided intelligent robots for automating manufacturing, materials handling and services. WESIC'98 Workshop on European Scientific and Industrial Collaboration on Promoting Advanced Technologies in Manufacturing, 1998.
3. S. Datta, R. Ray, and D. Banerji, Development of autonomous mobile robot with manipulator for manufacturing environment, The International Journal of Advanced Manufacturing Technology, 38(5), pp.536-542, 2008.
4. M. Gamberi, R. Manzini and A. Regattieri, A new approach for the automatic analysis and control of material handling systems: Integrated layout flow analysis, The International Journal of Advanced Manufacturing Technology, 41(1), pp.156-167, 2009.
5. A. Kelly, B. Nagy, D. Stager and R. Unnikrishnan, Field and service applications, an infrastructure-free automated guided vehicle based on computer vision. Robotics and Automation Magazine, IEEE, 14(3), pp.24-34, 2007.

6. M. Kefer and W. Kubinger, Evaluation of Kinect depth sensor for use in mobile robotics, Proceedings of International DAAAM symposium 22(1), pp.147-148, Austria, 2011.
7. M. R. Andersen, T. Jensen, P. Lisouski, A. K. Mortensen, M. K. Hansen, T. Gregersen and P. Ahrendt, Kinect depth sensor evaluation for computer vision applications, Technical report, AARHUS University, ECE-TR-6, 2012
8. T. A. Tamba, B. Hong and K. S. Hong, A Path Following Control of an Unmanned Autonomous Forklift, International Journal of Control, Automation and Systems, 7(1), pp.113-122, 2009.
9. T. C. Pradnya and R. Ganesh, An Autonomous Industrial Load Carrying Vehicle, Advances in Electronic and Electric Engineering, 4(2), pp.169-178, 2014.
10. A. Correa, M. R. Walter, L. Fletcher, J. Glass, S. Teller and R. Davis, Multimodal Interaction with an Autonomous Forklift, Proceedings of the ACM/IEEE international conference on Human-robot interaction pp.243-250, 2010.
11. J. Pages, X. Armangues, J. Salvi, J. Freixenet and J. Marti, A computer vision system for autonomous forklift vehicles in industrial environments, MEDS 2001.
12. T. Fong, C. Thorpe and C. Baur, Multi-robot remote driving with collaborative control, IEEE Transactions on Industrial Electronics, 50(4), pp.699–704, 2003.
13. C. W. Nielsen, M. A. Goodrich and R. W. Ricks, Ecological interfaces for improving mobile robot tele-operation. IEEE Transactions on Robotics, 23(5), pp.927–941, 2007.
14. S. Teller, et al., A voice-commandable robotic forklift working alongside humans in minimally-prepared outdoor environments. In Proc. IEEE Int'l Conf. on Robotics and Automation, ICRA, May 2010.
15. P. R. Wurman, R. D'Andrea and M. Mountz, Coordinating hundreds of cooperative, autonomous vehicles in warehouses. AI Magazine, 29(1), pp.9–19, 2008.

https://doi.org/10.1051/matecconf/201815304006

Statistical Linearization of Nonlinear Stiffness in Hydropneumatic Suspension

Zhongli Cheng[1], Zonghua Li [1] and Fanqing Kong [1]

School of Mechanical Engineering, Beijing Institute of Technology Beijing, 100081, China

Abstract. Hydropneumatic springs are the elastic components of a vehicle's suspension. As the nonlinear characteristic of the spring is difficult to express accurately, the statistical linearization method is introduced to analyze the dynamic response of the hydropneumatic spring. The nonlinear stiffness of a hydropneumatic spring is approximated by a quadratic polynomial at the static equilibrium position. Parameters of the hydropneumatic spring, road roughness and vehicle velocity are provided and analytical functions for equivalent stiffness and the dynamic equilibrium position are worked out in this paper. The analytical functions are validated through numerical simulation and are shown to be more accurate than those validated by existing methods. The method proposed here could be used in the design and analysis of hydropneumatic suspensions in future.

1 Introduction

Hydropneumatic suspensions use pressed gas as the elastic medium, which exhibits the characteristic of nonlinear gradient stiffness. This provides better comfort and cross country ability so hydropneumatic suspensions are widely used in high performance cross country vehicles, military vehicles and mine trucks [1-3]. As its nonlinear characteristics restrict the application of linear system analysis, nonlinear modeling and simulation are adopted to analyse this suspension by MA, et al [4], FENG, et al [5] and useful conclusions are reached. However, explicit analytical formulae are still needed for complex hydropneumatic suspensions in engineering applications, especially at the design stage.

Vehicle suspensions can be modelled by a MDOF (multiple degree of freedom) nonlinear vibration system with wide-band random excitation, suitable for application of exact feedback linearization or statistical linearization [6-8]. Exact feedback linearization is adopted to analyze hydropneumatic suspension by LI, et al [9] and SHI, et al [10] and satisfactory results are achieved in simulation. But, exact feedback linearization has many drawbacks. It requires that all states be measurable and robustness cannot be guaranteed, so the method of statistical linearization of the nonlinear characteristic is still required. Statistical linearization is proposed by CAUGHEY, et al [11] and KAZAKOV, et al [12] independently and extended to MDOF systems by IWAN, et al [13]. ATALIK, et al [14-15] uses deformation of elastic elements to establish system differential equations so that statistical linearization of the system can be reduced to linearization of the elastic elements to simplify calculations. GAI, et al [16] concludes that the displacement of the sprung mass is a non-zero mean process in response to a zero mean unit white noise excitation because of the asymmetry of the stiffness. YOUSEFI, et al [17] and LOHMANN [18] apply a low-order approximation statistical linearization method to the nonlinear model of hydropneumatic suspension. CAO, et al [19] uses a finite Taylor series instead of the exponential function, but since only the first two terms of the series are considered, the asymmetry of the stiffness is typically neglected and this will cause considerable errors under conditions of large suspension deformation.

Based on the references mentioned above, a new method is proposed in this paper, which considers random road roughness as the excitation and a quadratic polynomial is introduced to take the asymmetry of the stiffness of a hydropneumatic suspension into consideration. The coordinates of the differential equations are composed of the deformations of the elastic element and the absolute displacement of the sprung mass. The explicit forms of the equivalent stiffness and the dynamic mean-value offset from the static equilibrium position are determined using the statistical linearization method. Lastly, the proposed model is validated through numerical simulation.

2 Nonlinear Suspension Model Development

2.1 Nonlinear elasticity of hydropneumatic suspension

The mono chamber hydropneumatic spring is shown in Figure 1. Using the ideal gas equation, the gas pressure at any position can be expressed as:

$$p_0(A_1h_0)^\gamma = p(A_1h_0 + A_2z_1)^\gamma \quad (1)$$

where γ is the gas polytropic exponent, A_1 is the cross-sectional area of the air chamber, A_2 is the ring chamber area of the hydropneumatic spring and h_0 and p_0 are the gas column length and the pressure of the hydropneumatic spring at static equilibrium position respectively.

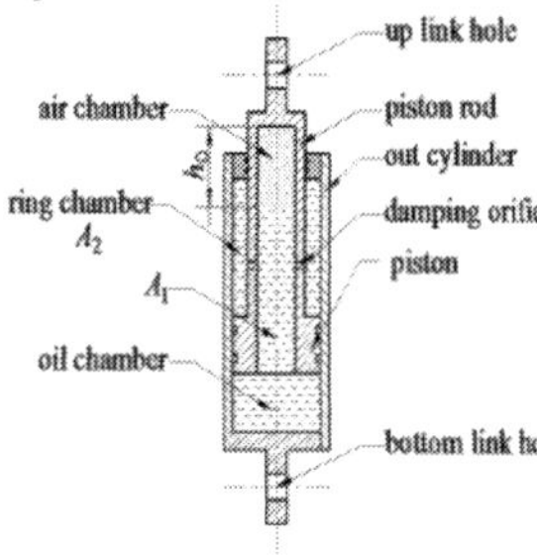

Figure 1. Cross-section of hydro-pneumatic spring

The rebound force of the hydropneumatic spring can be expressed as:

$$F_s = (p - p_0)A_1 = A_1p_0\left[\left(\frac{h_0}{h_0+\alpha z_1}\right)^\gamma - 1\right] = W_s\left[\left(\frac{h_0}{h_0+\alpha z_1}\right)^\gamma - 1\right] \quad (2)$$

where W_s is the weight of the sprung mass and $\alpha = A_2/A_1$ is the area ratio.

2.2 Two-DOF hydropneumatic suspension model

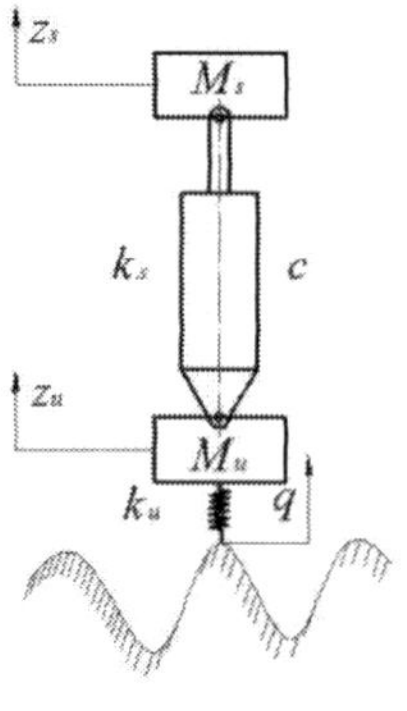

Figure 2. Model of 2-DOF hydropneumatic suspension

Ignoring the suspension links' effect and the damping of the tire, the differential equations of the system shown in Figure 2 are given in Eq. (3), assuming the damping of the hydropneumatic spring is linear.

$$\begin{cases} M_s\ddot{z}_s - F_s + c(\dot{z}_s - \dot{z}_u) = 0 \\ M_u\ddot{z}_u + F_s - c(\dot{z}_s - \dot{z}_u) = k_u(q - z_u) \end{cases} \quad (3)$$

where

M_s and M_u: sprung and unsprung masses respectively;

k_s: nonlinear stiffness of the hydropneumatic suspension;

k_u: tire stiffness;

c: linear damping coefficient of the suspension;

q: road roughness time-domain input;

z_s and z_u: the absolute displacements of M_s and M_u.

According to references [14,15], using the deformation of the spring $z_1 = z_s - z_u$ and the absolute displacement of the sprung mass z_s as new general coordinates, Eq. (3) can be rewritten as:

$$\begin{cases} M_s\ddot{z}_s = F_s - c\dot{z}_1 \\ M_u(\ddot{z}_s - \ddot{z}_1) - c\dot{z}_1 = -F_s + c\dot{z}_1 + k_u(q - z_s + z_1) \end{cases} \quad (4)$$

In Eq. (4), the nonlinear spring force is only related to the coordinate z_1, so statistical linearization of the system becomes the linearization of the nonlinear spring force. Substituting Eq. (2) into Eq. (4) gives:

$$\begin{cases} M_s\ddot{z}_s + W_s\left[\left(\frac{h_0}{h_0+\alpha z_1}\right)^\gamma - 1\right] + c\dot{z}_1 = 0 \\ M_u(\ddot{z}_s - \ddot{z}_1) - W_s\left[\left(\frac{h_0}{h_0+\alpha z_1}\right)^\gamma - 1\right] - c\dot{z}_1 = k_u(q - z_s + z_1) \end{cases} \quad (5)$$

3 Statistical Linearization of the 2-DOF Suspension System

3.1 Equivalent stiffness of the system

Define a zero-mean variable z_{10} as:

$$z_{10} = z_1 - E(z_1) = z_1 - m_{z_1} \quad (6)$$

The linearized spring force can be written as:

$$F_s^*(z_1) = -k_{eq}z_{10} - F_0 \quad (7)$$

The difference between the nonlinear spring force and the linear-equivalent spring force is:

$$\Delta F = F_s(z_1) - F_s^*(z_1) \quad (8)$$

Using the minimization of $E(\Delta F^2)$ as the statistical linearization criteria gives:

$$\frac{\partial E(\Delta F^2)}{\partial k_{eq}} = \frac{\partial E(\Delta F^2)}{\partial F_0} = 0 \quad (9)$$

Reference [6] proves that the arrest point is the minimum value point. From Eq. (9), Eq. (10) and Eq. (11) can be determined:

$$F_0 = E(F_s) = \int_{-\infty}^{+\infty} F_s(z_1)p(z_1)\mathrm{d}z_1 \quad (10)$$

$$k_{eq} = \frac{E\{[z_1 - E(z_1)]\cdot F_s(z_1)\}}{\sigma_{z_1}^2} = \frac{E[z_{10}\cdot F_s(z_{10} + m_{z1})]}{E(z_{10}^2)} \quad (11)$$

Determining the mathematical expectation of the first equation in Eq. (3) gives:

$$m_s \cdot E(\ddot{z}_s) - E(F_s) + c\cdot E(\dot{z}_1) = 0 \quad (12)$$

Based on the assumption that $\dot{z}_s$ and $\dot{z}_1$ are both Gaussian processes, reference [2] gives:

$$E(\dot{z}_s)=E(\ddot{z}_s)=E(\dot{z}_1)=E(\ddot{z}_1)=0 \qquad (13)$$

Substituting Eq. (13) into Eq. (10) gives:

$$F_0=\int_{-\infty}^{+\infty}F_s(z_1)p(z_1)\mathrm{d}z_1=0 \qquad (14)$$

Since z_1 is a stationary Gaussian random process, the probability density function of z_1 can be expressed as [20]:

$$p(z_1)=\frac{1}{\sigma_{z1}\sqrt{2\pi}}\exp\left[-\frac{(z_1-m_{z_1})^2}{2\sigma_{z_1}^2}\right] \qquad (15)$$

From Eq. (11), k_{eq} can be determined :

$$k_{eq}=\frac{\int_{-\infty}^{+\infty}[z_{10}\cdot F_s(z_1)\cdot p(z_1)]\mathrm{d}z_1}{\sigma_{z_1}^2} \qquad (16)$$

3.2 Quadratic polynomial of nonlinear spring force

Since Eq. (16) can't be expressed in an explicit form, a quadratic polynomial of non-linear spring force $F_s(z_1)$ at the static equilibrium position is used to replace the exponential function.

$$F_s(z_1)=\frac{1.478W_s\alpha z_1}{h_0}-\frac{1.697W_s\alpha^2 z_1^2}{h_0^2} \qquad (17)$$

As shown in Eq. (17), the function of nonlinear spring force is composed of a polynomial, where the variable is z_1/h_0, (i.e. the ratio of the deflection of the hydropneumatic spring to the gas height of hydropneumatic spring at the static equilibrium position). In practical situations, it is required that the maximum absolute value of z_1 be smaller than one third of h_0:

$$\max|z_1|\le\frac{1}{3}h_0 \qquad (18)$$

Setting $z_1=h_0/3$, gas polytropic exponent $\gamma=1.3$, it is noted that the nonlinear characteristic of the hydropneumatic spring cannot be approximated by the first term of a Taylor series as this will yield a maximum relative error of $\Delta F_1=20.74\%$. An approximation using the first and third terms of a Taylor series will yield a larger error of $\Delta F_{1,3}=76.47\%$ and it ignores the asymmetric characteristics of the hydropneumatic spring stiffness. An approximation using the first and second terms of a Taylor series, the even order terms of the expanding expression are certainly negative, and it exists error. An approximation using the first, second and third terms of a Taylor series, however, will enhance the accuracy of the calculation results with an increment in calculation complexity. Thus, a quadratic polynomial makes the calculations easier and accurate results can be obtained.

z_{10} is a normal random process with zero mean, and the derivative of high level power exists. Reduced order rules are used with the high level power of normal distribution [14].

$$k_{eq}=W_s\left[\frac{1.478\alpha}{h_0}-\frac{1.697\alpha^2(m_{z1}^3+m_{z1}\sigma_{z1}^2)}{\sigma_{z1}^2h_0^2}\right] \qquad (19)$$

$$F_0=\int_{-\infty}^{\infty}\left(\frac{1.478W_s\alpha z_1}{h_0}-\frac{1.697W_s\alpha^2 z_1^2}{h_0^2}\right)p(z_1)dz_1=\frac{1.478W_s\alpha}{h_0}E(z_1)-\frac{1.697W_s\alpha^2}{h_0^2}E(z_1^2)=0 \qquad (20)$$

3.3 Response of linear-equivalent suspension system

In order to determine k_{eq}, the mean value and the variance of z_1 are needed. Thus, the linear-equivalent system is introduced to determine the mean value and variance of z_1.

Define the following:

$$z_{s0}=z_s-E(z_s)=z_s-m_{z_s} \qquad (21)$$

Since z_s is a Gaussian process,

$$\dot{z}_s=\dot{z}_{s0},\quad \ddot{z}_s=\ddot{z}_{s0} \qquad (22)$$

Substituting Eq. (7), Eq. (21) and Eq. (22) into Eq. (4) gives:

$$\begin{cases}M_s\ddot{z}_{s0}+c\dot{z}_{10}+k_{eq}z_{10}=0\\ M_u(\ddot{z}_{s0}-\ddot{z}_{10})-c\dot{z}_{10}-(k_{eq}+k_u)z_{10}+k_uz_{s0}=k_u(q-m_{z_s}+m_{z_1})\end{cases} \qquad (23)$$

Noting that the terms on the left-hand side and q are all zero-mean processes, the following can be obtained by determining the mathematical expectation of the second equation in Eq. (23):

$$-m_{z_s}+m_{z_1}=0 \qquad (24)$$

So, Eq. (23) can be simplified:

$$\begin{cases}M_s\ddot{z}_{s0}+c\dot{z}_{10}+k_{eq}z_{10}=0\\ M_u(\ddot{z}_{s0}-\ddot{z}_{10})-c\dot{z}_{10}+k_uz_{s0}-(k_{eq}+k_u)z_{10}=k_uq\end{cases} \qquad (25)$$

The matrix form of Eq. (25) is as follows:

$$\begin{bmatrix}M_s & 0\\ M_u & -M_u\end{bmatrix}\begin{Bmatrix}\ddot{z}_{s0}\\ \ddot{z}_{10}\end{Bmatrix}+\begin{bmatrix}0 & c\\ 0 & -c\end{bmatrix}\begin{Bmatrix}\dot{z}_{s0}\\ \dot{z}_{10}\end{Bmatrix}+\begin{bmatrix}0 & k_{eq}\\ k_u & -(k_{eq}+k_u)\end{bmatrix}\begin{Bmatrix}z_{s0}\\ z_{10}\end{Bmatrix}=k_u\begin{Bmatrix}0\\ q\end{Bmatrix} \qquad (26)$$

The transfer function can then be worked out:

$$\begin{Bmatrix}H_{z_{s0}-q}\\ H_{z_{10}-q}\end{Bmatrix}=\frac{1}{\Delta}\begin{Bmatrix}k_u(k_{eq}-cj\omega)\\ -\omega^2M_sk_u\end{Bmatrix} \qquad (27)$$

Where

$$\Delta=\omega^2M_s(k_u-k_{eq}+jc\omega-\omega^2M_u)+(k_u-\omega^2M_u)(k_{eq}-jc\omega) \qquad (28)$$

The excitation spectrum of a vehicle travel can be expressed as:

$$G_{q0}(\omega)=(2\pi)^2G_q(n_0)n_0^2\frac{v}{\omega^2+\omega_0^2} \qquad (29)$$

where $G_q(n_0)$ is the road roughness coefficient and ω_0 is the angular frequency ($\omega_0 = 0.011\times2\pi v$) .

The variance of z_{10} can thus be determined using Eq. (27):

$$\sigma_{z_{10}}^2 = 2\pi G_q(n_0)n_0^2 v\cdot\int_{2\pi v f_1}^{2\pi v f_2}\left|\frac{H_{z10-q}}{\omega_0+\mathrm{j}\omega}\right|^2 \mathrm{d}\omega \quad (30)$$

where f_1 and f_2 are the lower and upper spatial frequency limits of the road respectively.

Since the suspension damping ratio is small:

$$\sigma_{z_{10}}^2 \approx 2\pi G_q(n_0)n_0^2 v\cdot\int_0^{+\infty}\left|\frac{H_{z_{10}-q_0}}{\omega_0+\mathrm{j}\omega}\right|^2 \mathrm{d}\omega \quad (31)$$

Noting that:

$$H_{\dot z_{10}-q_0} = \mathrm{j}\omega H_{z_{10}-q_0} \quad (32)$$

Using James Formula [21,22], the following can be obtained:

$$\sigma_{\dot z_{10}}^2 = \frac{\pi^2 G_q(n_0)n_0^2 v\cdot(M_u+M_s)}{c} \quad (33)$$

$$\sigma_{\dot z_1}^2 \approx \frac{\pi^2 G_q(n_0)n_0^2 v\cdot k_u}{c} \quad (34)$$

3.4 Explicit expression for equivalent stiffness

Substituting Eq. (33) into Eq. (14) gives:

$$\frac{1.478\alpha m_{z1}}{h_0} - \frac{1.697\alpha^2\left(m_{z_1}^2+\sigma_{z10}^2\right)}{h_0^2}=0 \quad (35)$$

Substituting Eq. (35) and Eq. (25) into Eq. (19),k_{eq}, m_{z1}, σ_{z10}^2 can be determined.

4 Validation of Model using Numerical Simulation

Using Eq. (29), filtered white noise is used to construct the road roughness time-domain excitation:

$$\dot q(t) = -2\pi n_0 v\cdot q(t) + 2\pi n_0\sqrt{G_q(n_0)v}\cdot w(t) \quad (36)$$

The suspension model parameters used in the simulation are listed in Table 1.

Table 1. Suspension and hydropneumatic spring arameters

Parameters	Value & Unit
Gas polytropic exponent	1.3
Sprung mass M_s	7000 /kg
Unsprung mass M_u	650 /kg
Stiffness of tire k_u	1×106 /(N/m)
Gas chamber length at static equilibrium position h_0	96 /mm
Area of air chamber A_1	2.5×104 /mm2
Damping coefficient c	6×104 /(Ns/m)
Area ratio α	1.2
Road roughness coefficient $G_q(n_0)$	1024×10-6/m3
Vehicle velocity v	15/(m/s)

Figure 3 shows the frequency domain input of the road roughness, It reveals that the time and frequency domain and distribution of the input is quite similar to what we want.

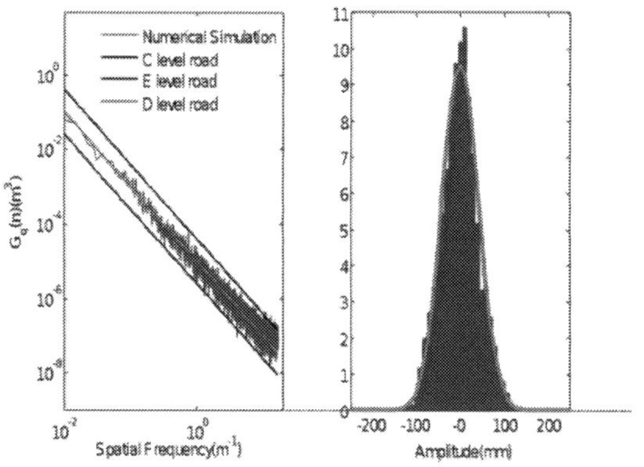

Figure 3. Validation of frequency and amplitude probability distribution

Figure 4 compares the hydropneumatic spring deformation determined through statistical linearization against that determined by the simulation and it shows that the distribution of the spring deformation closely approximates that of a Gaussian process.

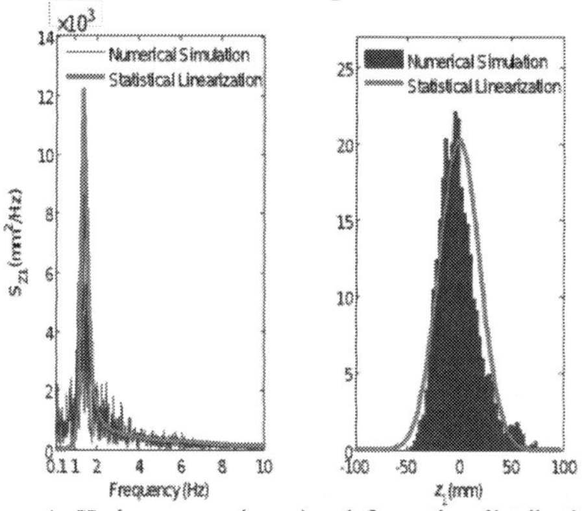

Figure 4. Hydropneumatic spring deformation distributions

Figure 5 compares the sprung mass acceleration determined through statistical linearization against that determined by the simulation and it shows that the results are of high accuracy in the frequency domain and the distribution.

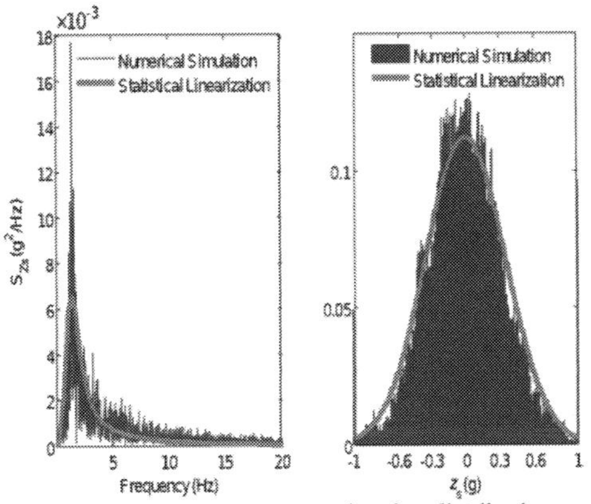

Figure 5. Sprung mass acceleration distributions

To sum up, the magnitude responses of the hydropneumatic spring deformation and the sprung mass acceleration determined through statistical linearization accurately approximate that determined by the simulation. The distributions of the hydropneumatic spring deformation and the sprung mass acceleration closely approximate that of a Gaussian process. Thus, the method of statistical linearization is validated.

To verify the response of the hydropneumatic suspension to different levels of road roughness and varying vehicle speed, two dimensionless parameters are introduced:

$$y = \frac{\sigma_{z1}}{h_0} \text{ and } t = \frac{m_{z1}}{h_0} \tag{37}$$

where y is the ratio of the hydropneumatic suspension deformation to the gas chamber length at the static equilibrium position and t is the ratio of the dynamic mean-value offset to the gas chamber length at the static equilibrium position. The relation between the relative deformation and the vehicle speed and road roughness is revealed in Figure 6 .

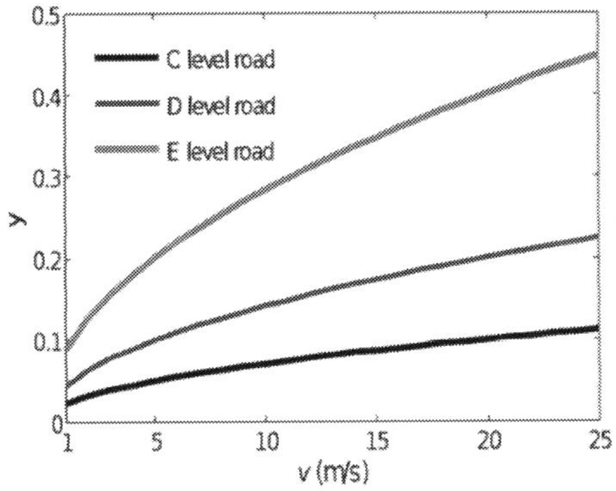

Figure 6. Relative deformation variation with vehicle velocity and road roughness

Figure 7 shows that, as the hydropneumatic suspension deformation increases, the dynamic mean-value offset increases too.

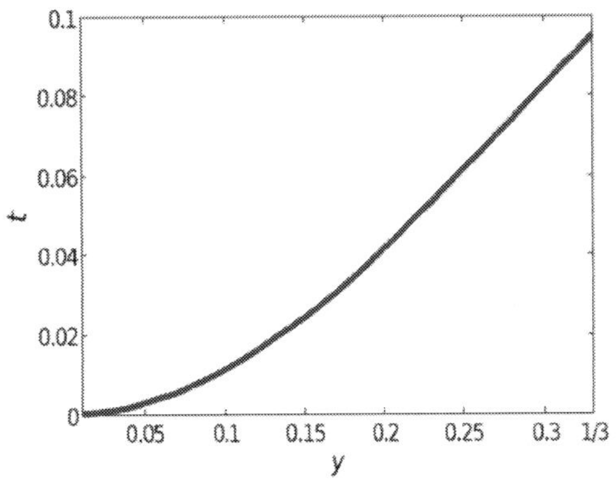

Figure 7. Relation of dynamic mean-value offset to hydropneumatic spring deformation

Figure 8 shows that, as vehicle speed and road roughness increase, the equivalent stiffness will increase as well. Thus, on a road in good condition, the small equivalent stiffness will provide a comfortable experience, but if the road is in a worse condition, the large equivalent stiffness will help the hydropneumatic suspension to absorb shock effectively.

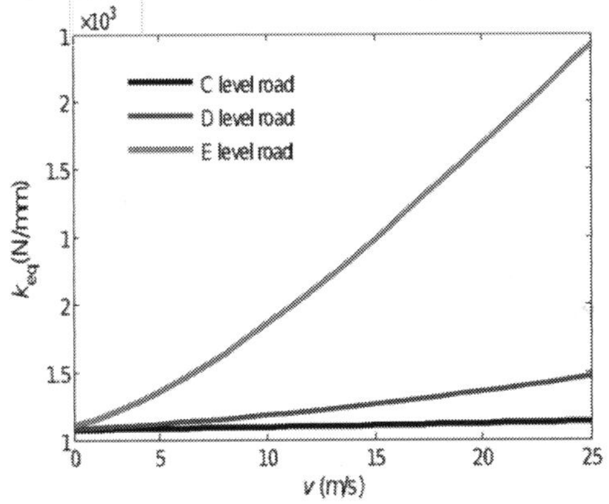

Figure 8. Equivalent stiffness with vehicle velocity and road roughness

This paper presents a novel method to depict nonlinear stiffness of hydropneumatic spring by approximating it with a quadratic polynomial. The results given in this paper can be further applied in advanced suspension system, i.e. active and semi-active suspension systems. Since the nonlinearities are affected by both road level and vehicle speed, it's believed that road adaptive advanced suspension system and adaptive observer will provide better performance with obtained results. For related topics, readers can refer to reference .

5 Conclusions

An accurate solution for the equivalent stiffness of a hydropneumatic suspension is obtained using the method of statistical linearization.A quadratic polynomial is used in computing and the results of the numerical simulation validate the accuracy of the statistical linearization method.Road roughness level and vehicle speed are considered in the analysis of the hydropneumatic suspension; as the vehicle speed and road roughness increase, the equivalent stiffness will increase too.

References

1. Solomon, U., and Padmanabhan, C., "Hydro-gas suspenseon system for a tracked vehicle: Modeling and analysis," Journal of Terramechanics, 2011, 48:125-137.
2. Konieczny, L., Burdzik, R., and Wegrzyn, T., "Analysis of structural and material aspects of selected elements of a hydro-pneumatic suspension system in a passenger car," Archives of metallurgy and materials, 2016, 61(1):79-83.
3. Daou, R.A.Z., Moreau, X., and Francis, C., "Effect of hydro-pneumatic components nonlinearities on the crone suspension," IEEE transactions on vehicular technology, 2012, 61(2): 466-474.
4. Ma, G., and Tan, R., "Nonlinear mathematic model of hydropenumatic suspension crane vehicles and its simulation," Chinese Journal of Mechanical Engineering, 2002, 38(5): 95-99.
5. Feng, S., "Research on nonlinear mathematical model and performance in hydro-pneumatic suspension," China Journal of Highway and Transport, 2002, 15(3): 122-126.
6. Zhu, W., "Random vibration," Beijing: Science Press, 1992:330-331. (in Chinese)
7. Emami, M. D., Mostafavi, S. A., and Asadollahzadeh, P., "Modeling and simulation of active hydro-pneumatic suspension system through bond graph," Mechanics, 2011, 17(3): 312-317.
8. Dong, M., and Luo, Z., "Statistical linearization on 2 DOFs hydro-pneumatic suspension with asymmetric nonlinear stiffness," Chinese Journal of Machanical Engineering, 2015, 28(3): 504-510.
9. Li, X., Shi, J., and Zhang, J., "Feedback linearization and PID control for active hydro-pneumatic suspension," Journal of ShangHai JiaoTong university, 2009, 43(10): 1521-1525.
10. Shi, J., Li, Xiaowei., and Zhang, J., "Feedback linearization control for missile launch vehicle active hydro-pneumatic suspension," Journal of System Simulation, 2009, 21(23): 7617-7622
11. Caughey, T. K., "Nonlinear theory of random vibration," Advances in Appl.Mech, 1977, 11: 209-253.
12. Kazakov, I. E., "Generalization of the method of statistical linearization to multidimensional system," Automation and Remote Control, 1965a, 26: 1201-1206.
13. Iwan, W. D., "A Generalization of the method of equivalent linearization," International J. of Non-linear Mech., 1973, 8: 279-283.
14. Atalik, T. S., "Stochastic linearization of multi-degree-of-freedom non-linear systems," Earthquake Engineering and Structure Dynamics, 1976, 4: 411-420.
15. Peng, J., Chen, S., and Chen H., "Equivalent linearization technique to a class of nonlinear multi-degree of freedom system ," Journal of Vibration and Shock, 1995, (1): 30-35. (in Chinese)
16. Gai, Z., "Non-zero response for the hydro-pneumatic suspension in vehicles," Journal of Beijing Institute of Technology, 1990, 10(1): 8-13. (in Chinese)
17. Amirhossein, Y., Boris, L., "Order reduction of nonlinear hydropneumatic vehicle suspension," IEEE Inteinational Conference on Control Applications, 2006, 08(4): 1404-1408.
18. Boris, L., "Application of model order reduction to a hydropneumatic vehicle suspension," IEEE Transactions on Control System Technology, 1995, 3(1): 102-109.
19. Cao S., Yi, M., and Luo, X., "Statistic linearization analysis of the nonlinear stiffness model for hydro-pneumatic suspensions," Huazhong Univ. Of Sci. & Tech. (Nature Science Edition), 2002, 30(6): 10-12. (in Chinese)
20. Roberts, J. B., "Response of nonlinear mechanical systems to random excitation: Part II; equivalent linearization and other methods," Shock and Vibration Digest, 1981, 13: 15-29.
21. James, H. M., Nichols, N. B., and Phillips, R. S., "Theory of Servomechanisms," New York: McGraw Hill, 1947.
22. Crandall, S. H., and Mark, W. D., "Random vibration in mechanical system," New York: Academic Press, 1963.
23. Wang, Z., Dong, M., Qin, Y., et al., "Suspension system state estimation using adaptive Kalman Filtering based on road classification," Vehicle System Dynamics, 2017, 55(3):371-398.
24. Qin, Y., Xiang, C., Wang, Z., and Dong, M., "Road Excitation Classification for Semi-Active Suspension System Based on System Response," Journal of Vibration and Control, (DOI: 10.1177/1077546317693432)
25. Qin, Y., Dong, M., Zhao, F., et al., "Road Profile Classification for Vehicle Semi-active Suspension System Based on Adaptive Neuro-Fuzzy Inference System," IEEE Control Decision Conference (CDC). Osaka, Japan. 2015, pp: 1533-1538.

Research on the compensation method of Indirect Tire Pressure Monitoring under Sinusoidal Driving Condition

Wang Liqiang, Qi Lin, Zhang Zhe, HAN Zongqi

College of Vehicles and Energy, Yanshan University, Hebei, China

Abstract. Indirect tire pressure monitoring method (ITPMS) is based on the wheel speed signal differences indirectly infer that the change of the tire pressure. In the steering mode, there is a large yaw rate, and there is a big deviation between the four wheels' speed. The traditional indirect tire pressure monitoring method based on wheel speed comparison method is invalid. That the automobiles which is matched electronic stability control system estimate the four-wheel speed according to the yaw rate and steering wheel angle become an important method to improve the accuracy of Indirect tire pressure monitoring. Four-wheel vehicle kinematics model for turning condition is established to compensate for the wheel speed in the steering mode, avoid the misjudgment of indirect tire pressure monitoring system. The seven degree of freedom vehicle model and the four wheel speed model are established firstly, and analyse the four wheel speed change of the vehicle under the condition of a certain corner input in the simulation. Then conduct a test on the vehicle road, calculated using the four wheel speed distribution module model of vehicle wheel speed on the theoretical value of the actual value of the wheel speed compensation, so it can monitor the tire pressure in turning vehicle. When the error of the wheel speed is beyond the threshold, the wheel can be judged to be in the state of lack of gas.

1 Introduction

Tire pressure monitoring system is one of the most important vehicle safety systems, which can reduce vehicle safety problems and reduce traffic accidents. At present, there are two solutions to solve the vehicle tire pressure monitoring: direct tire pressure monitoring program and indirect tire pressure monitoring program. Direct tire pressure monitoring scheme installs the tire pressure sensor in the tire directly and it measures the tire pressure directly. It has the advantages of high precision and small error, and the disadvantages are also obvious. The reliability of the wireless signal transmission is poor and it has short life and it is expensive. Indirect tire pressure monitoring scheme is to estimate the value of the four tire pressure using the sensor of automobile. Indirect tire pressure monitoring system does not require additional sensor. Low cost is suitable for large-scale application, and the disadvantage is the monitoring value has large error.

At present, there are three methods for indirect tire pressure monitoring: The first method is the rotation radius method. By calculating the effective tire rotation radius to monitor tire pressure, this method uses wheel speed signal of vehicle wheel speed sensor directly. Using the Calman filter method, observe the change of wheel rotation radius, and obtain tire pressure alarm threshold through the preliminary calibration. The second method is spectrum method. By analyzing the wheel speed resonance frequency, the purpose of tire pressure monitoring is achieved. The main method of monitoring tire pressure indirectly is to estimate tire torsional stiffness indirectly by using wheel speed. At present, the most mature indirect tire pressure monitoring solution is the rotation radius method.

Indirect tire pressure monitoring system (ITPMS) is a tire pressure monitoring device based on the wheel speed sensor (ABS) wheel speed sensor [1-2]. The advantages of the indirect tire pressure monitoring system are its simple structure, low cost and easy installation. At present, the indirect tire pressure monitoring system which is installed in a car assumes that the distance of the four rolling wheels is the same as they travel. Therefore, the indirect tire pressure monitoring system can only monitor the tire pressure when the vehicle runs in a straight line. If the vehicle is turning, the vehicle yaw speed will change greatly, which will result in the wheel speed difference. At the same time, the lateral force causes the tire side deviation and affects the speed signal error. The tire pressure monitoring method based on this principle is invalid. With the application of vehicle electronic stability program system (ESP) in automobile, the inertial unit which can measure the lateral acceleration, longitudinal acceleration and yaw angular velocity of the vehicle has been popularized. The yaw speed and steering angle of car body are important for wheel speed correction of the four wheel of vehicle [3-4]. Therefore, in this paper, the rotation radius method is applied to study the failure of the tire pressure monitoring method under the steering condition. A method of tire pressure monitoring and

https://doi.org/10.1051/matecconf/201815304007

compensation using steering signal and yaw angular speed signal is proposed. It can improve the traditional indirect tire pressure monitoring function, and enrich the monitoring range of the tire pressure monitoring method with rotary radius method.

Firstly, a theoretical model of four wheel speed under cornering condition is established. The main method is to obtain the steering angle, yaw angle and other signals of the vehicle through the seven degree of freedom vehicle model. Then, the four wheel speed model is established to calculate the change of the four wheel speed trend under the ideal turning condition, and the simulation results are obtained. Then, the steering wheel angle and yaw rate signal in the actual road test are input to the four wheel speed model, and the actual wheel speed is corrected. Finally, the tire pressure monitoring method of traditional pulse comparison method is used to correct the rear wheel speed and realize the tire pressure monitoring under cornering conditions.

2 Establish a theoretical simulation model

2.1 The establishment of seven degree equation

The seven degree of freedom model is used to describe the dynamic state of the vehicle in the control process and on-line state observation. The model takes into account both the difference in dynamics of the four wheels and the effects of load transfer [1]. This paper is based on the kinematic model of vehicle with seven degrees of freedom. A four wheel vehicle motion model is set up, which includes the input of vehicle steering wheel and the yaw rate of vehicle. The four wheeled vehicle motion model modeling method is shown in Figure 1.

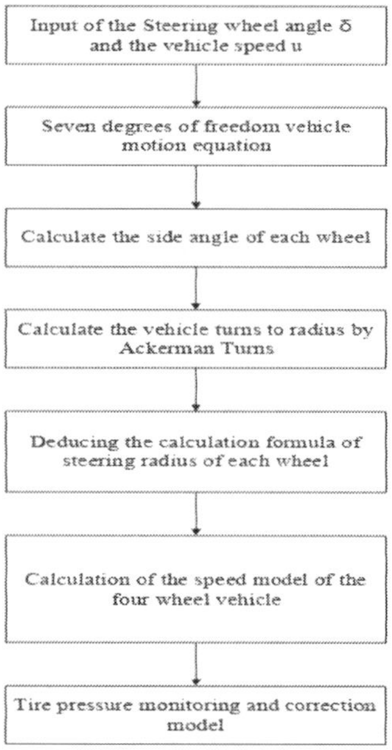

Figure 1. Modeling method for Four-wheel vehicle motion model

It is considered that the slip rate and side angle of each wheel may be inconsistent in the four wheel independent control. In the steering process and lateral load transfer. The wheel load is transferred during the braking or acceleration process, so the vertical load of the wheel is a time-varying parameter during the steering process. In addition, in the actual movement of the car, the adhesion between the tires and the pavement may be inconsistent. These factors are difficult to be quantitatively evaluated in two wheel models. Therefore, the four wheel model is needed to further describe the dynamic state of vehicle when running on a smooth road. And because of the brake intervention, the longitudinal speed of the car is time-varying, so the degree of freedom of the longitudinal motion of the car needs to be considered.

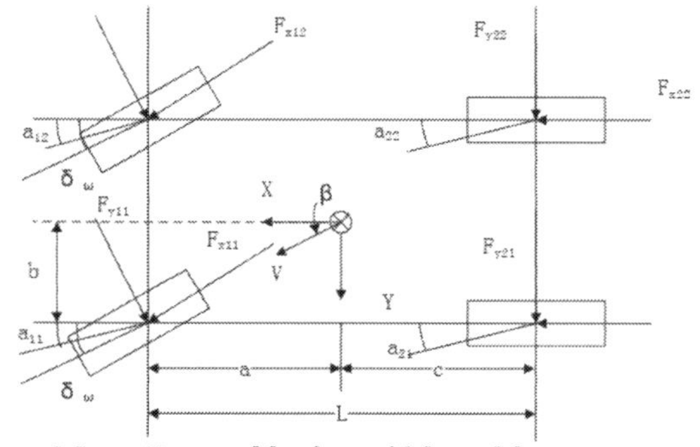

Figure.2 Seven Degree of freedom vehicle model

The significance represented by the parameters described in this paper:

a represents the distance from the center of mass to the front axle;

b represents the distance from the center of mass to the rear axle;

L represents wheelbase;

c represents half track;

I_z represents the inertia of the vehicle around the center of mass;

θ represents the yaw angle of the vehicle;

ω_r represents the yaw rate of the vehicle;

δ_f represents front wheel steering angle;

u represents a component of the speed of the vehicle in the direction of the vehicle;

v represents the speed of the vehicle in the direction perpendicular to the direction of the vehicle.

The four wheel motion equation of seven degrees of freedom is as follows:

$$m(\dot{V}_x - V_y\dot{\varphi}) = (F_{x11} + F_{x12})\cos\delta - (F_{y11} + F_{y12})\sin\delta + F_{x21} + F_{x22} \quad (1)$$

$$m(\dot{V}_x\dot{\varphi} + V_y) = (F_{y11} + F_{y12})\cos\delta + (F_{x11} + F_{x12})\sin\delta + F_{y21} + F_{y22} \quad (2)$$

$$I_z\ddot{\varphi} = -[(F_{y11}+F_{y12})a\cos\delta-(F_{y11}-F_{y12})c\sin\delta] +(F_{y21}+F_{y22})b-(F_{x11}+F_{x12})a\sin\delta +(F_{x11}-F_{x12})c\cos\delta+(F_{x21}-F_{x22})c \quad (3)$$

2.2 The establishment of four wheel equation

When the wheel is turning, assume that the following conditions are satisfied: ① The cornering stiffness of the tire is fixed, and the lateral force is linearly related to the side slip angle; ② Vehicle runs at constant speed; ③ The steering angle is characterized by low frequency and small angle.

A mathematical model of yaw rate is established according to the seven degree of freedom vehicle motion model:

$$-[(F_{y11}+F_{y12})a\cos\delta-(F_{y11}-F_{y12})c\sin\delta] +(F_{y21}+F_{y22})b-(F_{x11}+F_{x12})a\sin\delta +(F_{x11}-F_{x12})a\cos\delta+(F_{x21}-F_{x22})c = M_z \quad (4)$$

$$M_z / I_z = \ddot{\theta} \quad (5)$$

$$\int \ddot{\theta} dt = \omega_r \quad (6)$$

According to the seven degree of freedom vehicle model, the mathematical model of the sideslip angle of the center of mass is established[5]:

$$\frac{F_y}{m} = a_y = \frac{dv}{dt} - u\frac{d\theta}{dt} = v - u\omega_r \quad (7)$$

$$\frac{v}{u} = \omega_r - \frac{a_y}{u} \quad (8)$$

$$\beta = \frac{v}{u} \quad (9)$$

According to the model, the angle of sailing is calculated ξ:

$$\xi = \frac{v + a\omega_r}{u - c\omega_r} \quad (10)$$

The deflection angle of the front axle β_f and the deflection angle of the rear axle β_r can be obtained by the longitudinal and vertical velocity components of the front and rear shafts [6]:

$$\beta_f = \frac{v + a\omega_r}{u - c\omega_r} - \delta \quad (11)$$

$$\beta_r = \frac{v - b\omega_r}{u - c\omega_r} \quad (12)$$

The deflection angle of the four side tires of the vehicle can be calculated by the equation of motion of the four wheel vehicle:

$$\beta_1 = \frac{v + a\omega_r}{u - c\omega_r} - \delta \quad (13)$$

$$\beta_2 = \frac{v + a\omega_r}{u + c\omega_r} - \delta \quad (14)$$

$$\beta_3 = \frac{v - b\omega_r}{u - c\omega_r} \quad (15)$$

$$\beta_4 = \frac{v - b\omega_r}{u + c\omega_r} \quad (16)$$

According to the Ackerman steering relation, the turning radius of vehicle center of mass is obtained [7]:

$$R = \frac{L}{\delta - \beta_f + \beta_r} \quad (17)$$

Each wheel's steering radius R1~R4 has strict geometric relationship with centroid steering radius R. And each wheel's velocity can be calculated [8].

3 Theoretical model simulation analysis

The seven degree of freedom model is built in Simulink to simulate. In the simulation, we assume that the speed of the car is constant. It is 5.4m/s. In the simulation, the sinusoidal input is 15 degrees when the steering wheel angle is amplitude. The wheel speed of the four wheels is also around the speed, showing a sine change. The simulation results are shown in Figure 3.

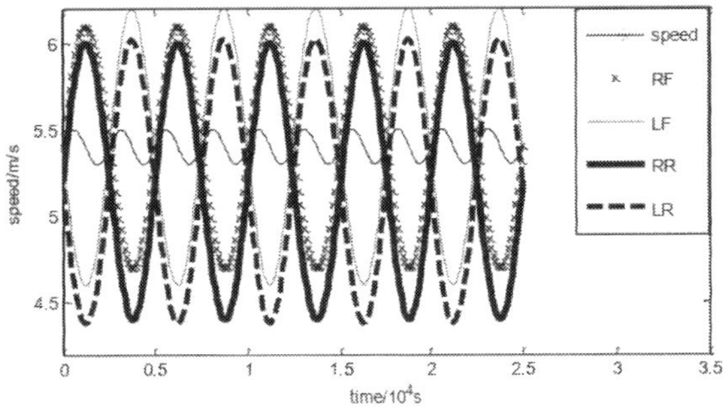

Figure 3. Wheel speed simulation curve.

As can be seen from the diagram, when the vehicle turns left, the left front wheel and left rear wheel speed are less than 5.4m/s, and the right front wheel and right rear wheel speed are greater than 5.4m/s. When the vehicle turns to the right, the result is the reverse. It can be seen that the algorithm can also calculate the wheel speed under the condition of continuous steering.

4 Revise under sinusoidal driving condition

Due to differences in tire wear and road problems, the wheel speeds measured by the four wheels are not equal.

Therefore, the four wheels should be modified to get the same speed at normal tire pressure.

In the early stage of the task group, the pulse correction under cornering condition has been studied experimentally [9]. Supposing the inside two wheels have equal turning radius, and the radius of the two wheels on the outside is approximately equal. Set the inside wheel radius as $r1$ and $r2$. The mean of the two is $r3$.

Set

$$k_i = \frac{r3}{R_i} \quad i = 1,\ 2,\ 3 \tag{26}$$

k_i is the correction factor under turn conditions in the formula [4].

$$v_{wi} = k_i * v_i \quad i = 1,\ 2,\ 3 \tag{27}$$

v_{wi} is the corrected wheel speed under turn conditions. It can call standard turn wheel speed.

After correction, the four wheel speeds eliminate the effects of road wear, wheel wear, and wheel distance between the wheels. Using the three mean comparison method to determine the tire air shortage.

5 Obtain standard wheel speed under sinusoidal driving condition

The wheel speed is judged by the three mean comparison method. The three mean comparison method compares the standard wheel speed of a wheel turning standard wheel speed with the average wheel speed of the other three wheels[10].

$$\bar{v}_{wi} = \frac{1}{3}\sum_{j=1}^{4} v_j \quad i = 1,\ 2,\ 3 \quad j \neq i \tag{28}$$

$$\Delta v_{wi}^{'} = \left| \frac{v_i - \bar{v}_{wi}}{\bar{v}_{wi}} \right| \quad i = 1,2,3,4 \tag{29}$$

$\bar{v}_{wi}$ is the standard wheel speed of the other three wheels except the i st wheel, $\Delta v_{wi}^{'}$ is the relative error between the standard wheel speed of the i st wheels and the average wheel speed average of the other three wheels. Set ε, is the threshold of the three mean comparison method. And when the relative error $\Delta v_{wi}^{'}$ exceeds the threshold, it can determine the lack of gas in the wheel[8].

6 Road test

In order to verify the correctness of the theory that the tire pressure anomaly is judged by the seven degree of freedom vehicle model, we have done the pile driving test. Test vehicle selects Havar H6. The tyre model is 225 / 65R17. By the three mean comparison method, the relative errors of the inside and outside sides are different when turning. Due to the wheel speed varies sine, there are two thresholds. Set the threshold value of the inner wheel while turning as 0.06%. And the threshold of the outer wheel is 0.09%. The reference speed is 30km/h. The test results are filtered as shown in Figure 4.

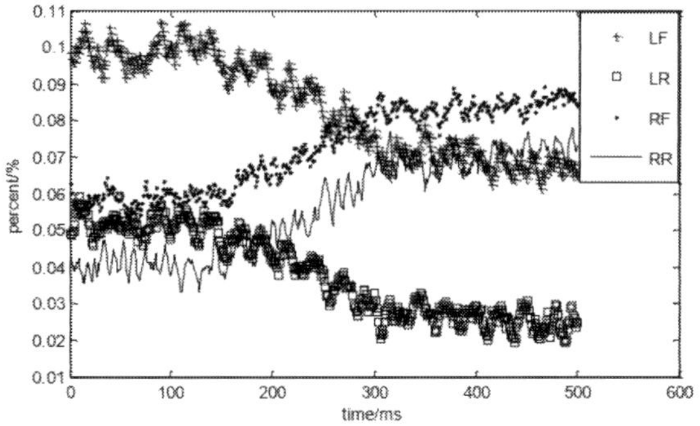

Figure 4. Relative error of wheel speed.

We can see that the left front wheel turning speed relative error exceeds the threshold from the figure, we can judge the left front gas shortage.

7 Conclusions

It can be concluded from the experiment that the wheel speed of each wheel varies greatly during cornering, so the wheel speed is judged to be abnormal when cornering, which results in the failure of the tire pressure monitoring system.

The four wheel model of the vehicle takes full advantage of the vehicle inertial unit and steering wheel angle sensor to monitor tire pressure, and when the vehicle is under steering conditions, it can successfully estimate the theoretical speed of each wheel. The seven degree of freedom model takes into account the influence of more parameters, which makes the wheel speed more accurate.

In this paper, wheel speed correction is made by formula (23) and (24), and then the three mean method is used to compare whether the tire pressure is normal or not.

8 Acknowledge

This work was support by The national natural science foundation of China (Grant No. 51175450) and Young teachers independent research project topics of Yanshan University, Institute of technology, Class A (Grant No. 14LGA019)

References

1. National Highway Traffic Safety Administration of USA. Tire pressure maintenance: A statistical investigation [R]. Tire technology International, 2009.
2. Li Liang. Research on state observation and controlling method of vehicle dynamics stability control system [D].Qin Huangdao: Huangdao: Yanshan University, 2008:**20-23**.

3. Ren Guo-xin. Study in HIL-simulation technology for ABS/TCS integrative control system of Commercial Vehecle[D]. Jilin University.2008
4. Hu Jian-jun, Zhao Yu-sheng,Qin Da-tong. Hardware-In-Loop Simulation of HEV System Based on CAN.[J]. China Mechanical Engineering.,2008,19(3): **300-304**.
5. Yu Zhuo-ping, Gao Xiao-jie. An overview of the state estimation of the vehicle running [J]. Chinese Journal of Mechanical Engineering, 2009; 45(5), **20-33**.
6. Toshihiro Hiraoka, Osamu Nishihara, Hiromitsu Kumamoto. Automatic path-tracking controller of a four-wheel steering vehicle [J]. Vehicle System Dynamics, 2009, 47(10): **1205-1227**.
7. Yu Zhi-sheng. The theory of the automobile [M]. Beijing: Machinery Industry Press, 2009:**144-146**.
8. Zhang Zhe. Monitoring Method of Indirect TPMS under Steering Situation[J]. International Conference on Mechanical, Industrial and Manufacturing (MIME 2016).2016.1,**187-190**
9. Chen Hui, Gao Bo-Lin, Xu Fan. The summary of the estimation of The vehicle sideslip angle[J]. Chinese Journal of Mechanical Engineering., 2013;49(24):**76-94**.
10. Han Zong-qi, Liu Quan-you, Wang Li-qiang, Zhang Zhong-xiao, Ju Xue-kun. Study on Tire Pressure Monitoring and Alarming System of Automobile Based on the Comparison among Standard Pulse Numbers[J]. Qin Huangdao: Yanshan University,2010，21(2)

Research on Vibration of Automobile Suspension Design

Jun Yang, Mingming Dong

School of mechanical and Vehicular Engineering, Beijing Institute of Technology, Beijing, 100081, China

Abstract. In this paper, a seven-degree-of-freedom three-dimensional vibration model of a vehicle is established based on the parameters of a vehicle. Considering the hysteresis effect of pavement incentives of rear wheel, the random vibration of the vehicle is simulated by MATLAB / SIMULINK. After changing the stiffness, the acceleration of the vehicle body and the seat, suspension dynamic deflection and relative dynamic load are compared to select the suitable stiffness to achieve optimal ride performance. It can provide a reference for the one who is designing vehicle suspension.

1 Introduction

When the automobile is running, the vibration of the car is excited by the uneven surface and the rotating parts such as the transmission system and the wheels. Typically, uneven pavement is the basic input of automotive vibration. Solving the vibration problem of vehicle and parts under the excitation of random uneven road is an important foundation to improve vehicle ride comfort, safety and reliability of parts [1]. In this paper, the 7-DOF dynamics model of the vehicle's vehicle vibration system will be established, and the parameters of suspension will be optimized by using the vehicle body acceleration root mean square value, suspension dynamic deflection and wheel displacement as the ride comfort evaluation index.

2 Vehicle vibration mechanics model

In order to more realistic reflection of the vertical vibration of the vehicle, it is necessary to establish a three-dimensional vehicle vibration model. The vibration input of the whole vehicle is the excitation of the road surface, the difference of road excitation between the front and rear wheels, and the influence of the roll of the vehicle on the vibration of the vehicle body are taken into account, which is a more realistic reflection of the actual situation of vehicle vibration[2]. Seven-DOF three-dimensional vehicle vibration model is shown in Figure 1.

m_b, J_b, I_b are respectively the body mass, pitching inertia and yawing inertia. m_i, c_i, k_i (i=1,2,3,4) are respectively the mass, damping and stiffness of the suspension. c_{ti}, k_{ti} (i=1,2,3,4) are respectively damping and stiffness of the wheel; b_f is half of front wheel track. b_r is half of front wheel track. a is the distance between body mass center and front suspension. b is the distance between body mass center and rear suspension. Z_b, θ, φ are respectively vertical displacement of car body, rate of pitch and yaw velocity. z_{bi}、z_i are respectively vertical displacement of four supporting points of the car body and vertical displacement of the tire. q_i are four wheel input of pavement excitation.

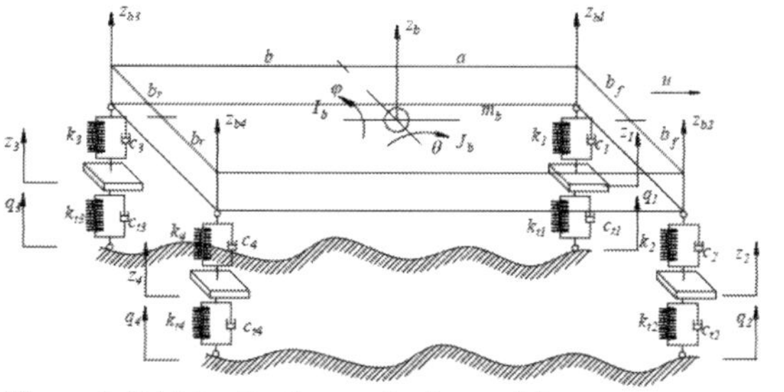

Figure 1. Vehicle vibration mechanics model.

According to the previous symbol definition,

$$\begin{cases} z_{b1} = z_b - a\theta + b_f\varphi \\ z_{b2} = z_b - a\theta - b_f\varphi \\ z_{b3} = z_b + b\theta + b_r\varphi \\ z_{b4} = z_b + b\theta - b_r\varphi \end{cases} . \tag{1}$$

The total kinetic energy of this system is:

$$T = \frac{1}{2}\left[m_b\dot{z}_b^2 + J_b\dot{\theta}^2 + m_1\dot{z}_1^2 + m_2\dot{z}_2^2 + m_3\dot{z}_3^2 + m_4\dot{z}_4^2\right] . \tag{2}$$

The potential energy of this system is:

$$U = \frac{1}{2}\left[\sum_{i=1}^{4} k_i\left(z_{bi} - z_i\right)^2 + \sum_{i=1}^{4} k_{ti}\left(z_i - q_i\right)^2\right] . \tag{3}$$

The total dissipation function of the system is:

https://doi.org/10.1051/matecconf/201815304008

$$R=\frac{1}{2}\left[\sum_{i=1}^{4}c_i\left(\dot{z}_{bi}-\dot{z}_i\right)^2+\sum_{i=1}^{4}c_{ti}\left(\dot{z}_i-\dot{q}_i\right)^2\right]. \quad (4)$$

After the above equations are arranged, the matrix form is obtained according to Lagrange's law:

$$M\ddot{Z}+C\dot{Z}+KZ=K_tQ+C_t\dot{Q}. \quad (5)$$

where k_1=k_2, k_3=k_4, c_1=c_2, c_3=c_4,

$$M=diag\{m_b \quad J_b \quad I_b \quad m_1 \quad m_2 \quad m_3 \quad m_4\}, \quad (6)$$

$$K=\begin{bmatrix} 2(k_1+k_3) & 2(bk_3-ak_1) & 0 & -k_1 & -k_1 & -k_3 & -k_3 \\ 2(bk_3-ak_1) & 2(a^2k_1+b^2k_3) & 0 & ak_1 & ak_1 & -bk_3 & -bk_3 \\ 0 & 0 & 2(b_f{}^2k_1+b_r{}^2k_3) & -b_fk_1 & b_fk_1 & -b_rk_3 & b_rk_3 \\ -k_1 & ak_1 & -b_fk_1 & k_1+k_{t1} & 0 & 0 & 0 \\ -k_1 & ak_1 & b_fk_1 & 0 & k_1+k_{t1} & 0 & 0 \\ -k_3 & -bk_3 & -b_rk_3 & 0 & 0 & k_3+k_{t3} & 0 \\ -k_3 & -bk_3 & b_rk_3 & 0 & 0 & 0 & k_3+k_{t3} \end{bmatrix}, \quad (7)$$

$$C=\begin{bmatrix} 2(c_1+c_3) & 2(bc_3-ac_1) & 0 & -c_1 & -c_1 & -c_3 & -c_3 \\ 2(bc_3-ac_1) & 2(a^2c_1+b^2c_3) & 0 & ac_1 & ac_1 & -bc_3 & -bc_3 \\ 0 & 0 & 2(b_f{}^2c_1+b_r{}^2c_3) & -b_fc_1 & b_fc_1 & -b_fc_3 & b_rc_3 \\ -c_1 & ac_1 & -b_fc_1 & c_1+c_{t1} & 0 & 0 & 0 \\ -c_1 & ac_1 & b_fc_1 & 0 & c_1+c_{t1} & 0 & 0 \\ -c_3 & -bc_3 & -b_rc_3 & 0 & 0 & c_3+c_{t3} & 0 \\ -c_3 & -bc_3 & b_rc_3 & 0 & 0 & 0 & c_3+c_{t3} \end{bmatrix}, \quad (8)$$

$$C_t=\begin{bmatrix} 0 & 0 & 0 & 0 \\ 0 & 0 & 0 & 0 \\ 0 & 0 & 0 & 0 \\ c_{t1} & 0 & 0 & 0 \\ 0 & c_{t2} & 0 & 0 \\ 0 & 0 & c_{t3} & 0 \\ 0 & 0 & 0 & c_{t4} \end{bmatrix}, \quad K_t=\begin{bmatrix} 0 & 0 & 0 & 0 \\ 0 & 0 & 0 & 0 \\ 0 & 0 & 0 & 0 \\ k_{t1} & 0 & 0 & 0 \\ 0 & k_{t2} & 0 & 0 \\ 0 & 0 & k_{t3} & 0 \\ 0 & 0 & 0 & k_{t4} \end{bmatrix}, \quad Z=\begin{bmatrix} z_b \\ \theta \\ \varphi \\ z_1 \\ z_2 \\ z_3 \\ z_4 \end{bmatrix}, \quad Q_t=\begin{bmatrix} q_1 \\ q_2 \\ q_3 \\ q_4 \end{bmatrix}. \quad (9)$$

3 Simulation optimization

3.1. The establishment of pavement simulation model

In the process of vehicle running, the roughness of the road surface varies with different grades of road and different driving speed. The pavement model is a mathematical description of the change of the road roughness with the class and speed of the road. The Seven Free 3D model of vehicle is built, which considers the rear wheel lag road roughness input, without considering the left and right coherent road input.

The pavement excitation is generated by the white noise method[3-6] and the mathematical model of pavement excitation is.

$$\dot{q}(t)=-2\pi n_l u q(t)+2\pi n_0\sqrt{G_q(n_0)u}\,w(t), \quad (10)$$

where $q(t)$ is four wheel input of pavement excitation; $w(t)$ is white noise signal which the mean is zero; n_l is lower cut-off frequency, the value of which is 0.011m^{-1}; u is speed of vehicle, the value of which is 110 km/h; $G_q(n_0)$ is the road power spectral density value under the reference road frequency, which is called the unevenness parameters of road surface. The road grade is usually divided according to the unevenness parameters of road surface.

According to Equation (10), the MATLAB / SIMULINK model is as follows:

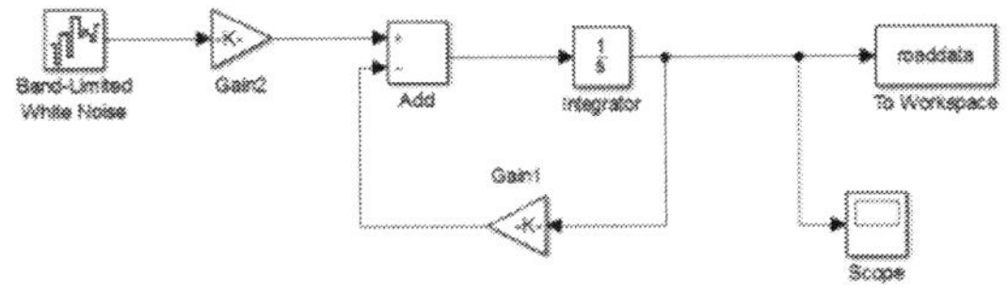

Figure 2. Pavement simulation model.

where $\text{Gain1}=-2\pi n_l u$,$\text{Gain2}=2\pi n_0\sqrt{G_q(n_0)u}$, n_0= 0.1m^{-1},$G_q(n_0)$=16×10^{-5} m^2/m^{-1}, u=100 km/h.

The rear wheel road irregularity excitation delay for a period of time compared to the front [7]. The delay time τ is:

$$\tau=\frac{a+b}{u}. \quad (11)$$

3.2 The establishment of vehicle simulation model

In the second chapter, the three-dimensional vibration model of seven-degree-of-freedom vehicle was established and the differential equation of vehicle vibration was obtained. Ignoring tire damping, it is:

$$M\ddot{Z}+C\dot{Z}+KZ=K_tQ. \quad (12)$$

In order to facilitate the calculation and simulation, it is necessary to convert equation (12) into the form of state equation by multiplied M^{-1}. After transformation, the equation is:

$$\ddot{Z}=-M^{-1}C\dot{Z}-M^{-1}KZ+M^{-1}K_tQ. \quad (13)$$

The seven - freedom vehicle vibration simulation model is built by using the MATLAB/SIMULIK module, as shown in Figure 3.

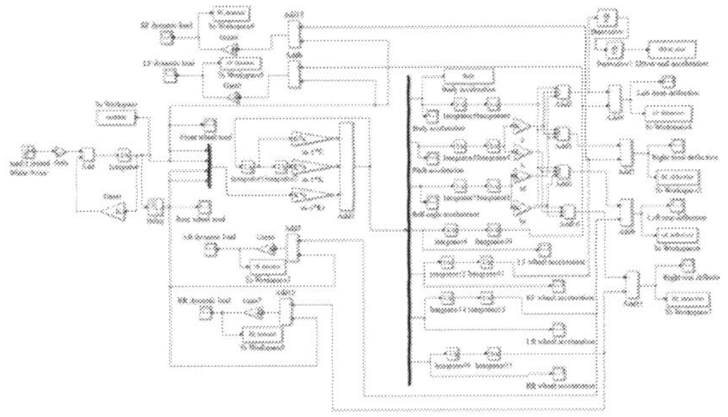

Figure 3. Vehicle simulation model

3.3 Optimization

The objective function is $\min(\ddot{z})$, which is to meet the requirements of ride comfort. Design variable is :

$$X=[k_1,k_3]^{\mathrm{T}}. \quad (14)$$

Static deflection of front and rear suspension f_e is bigger than 0.10 m and lower than 0.30 m . So the stiffness limit of the front and rear suspension is

$$m_1 g / 0.30 \le k_1 \le m_1 g / 0.10, m_3 g / 0.30 \le k_3 \le m_3 g / 0.10, \quad (15)$$

where m_1, m_3 are respectively left front and left rear suspension sprung mass. In order to improve the comfort of the driver, the frequency of the front suspension is often lower than that of the rear suspension. There is :

$$k_1 / k_3 < m_1 / m_3, \quad (16)$$

The parameters of a car are shown in Table 1.

Table 1. Parameters of a car

m_b / kg	m_1 / kg	m_3 / kg	J_b / (kg·m^2)
1380	40.5	45.4	2440
I_b / (kg·m^2)	k_1 / (kN/ m)	k_3 / (kN/m)	c_1 / (kN/ s·m^{-1})
80	17	22	1.5
c_3 / (kN/ s·m^{-1})	k_{t1} / (kN/ m)	k_{t2} / (kN/ m)	a / m
15	40.5	45.4	1.51
b / m	b_r / m	b_f / m	
1.51	0.74	0.74	

According to the above requirements, the vehicle model is simulated and the simulation results by using improved and unimproved data are compared[7-8]. The simulation time is 30s. To see the details, let's take 5~10s as a contrast.

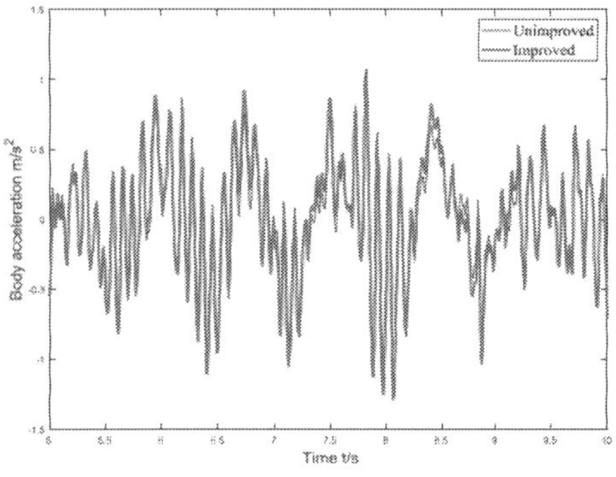

Figure 4. Comparison of body acceleration.

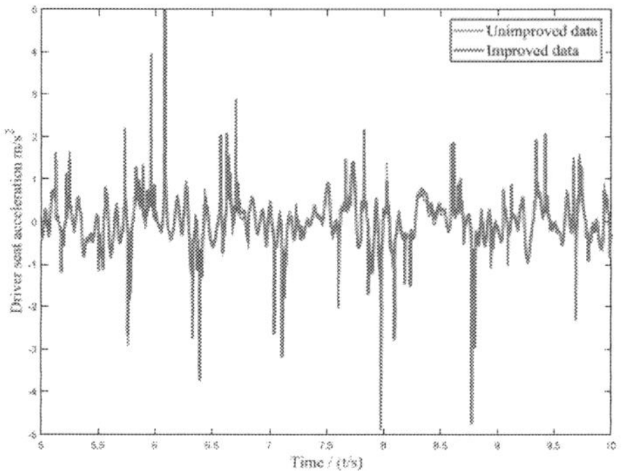

Figure 5. Comparison of driver seat acceleration.

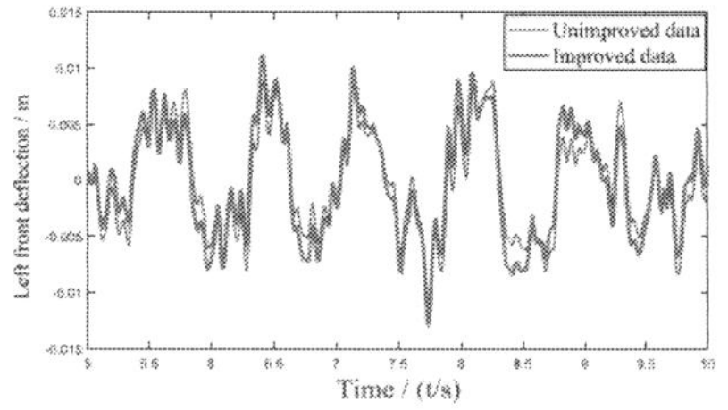

Figure 6. Comparison of left front dynamic deflection.

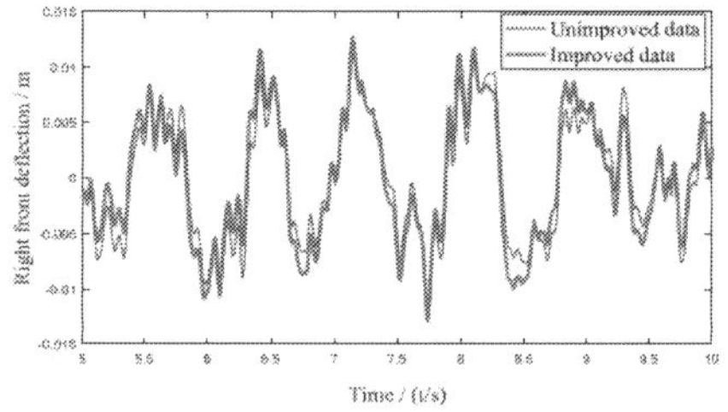

Figure 7. Comparison of right front dynamic deflection.

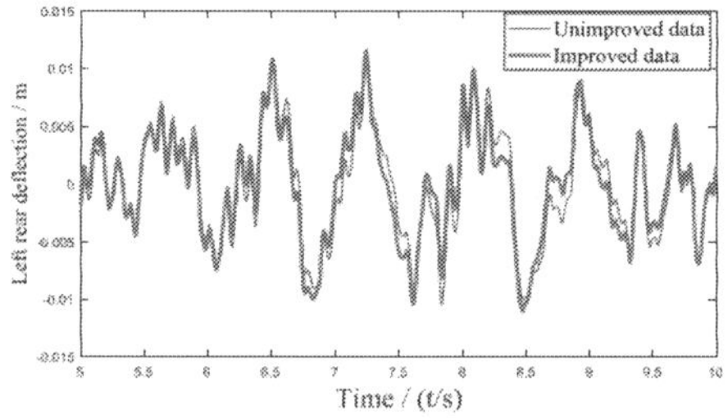

Figure 8. Comparison of left rear dynamic deflection.

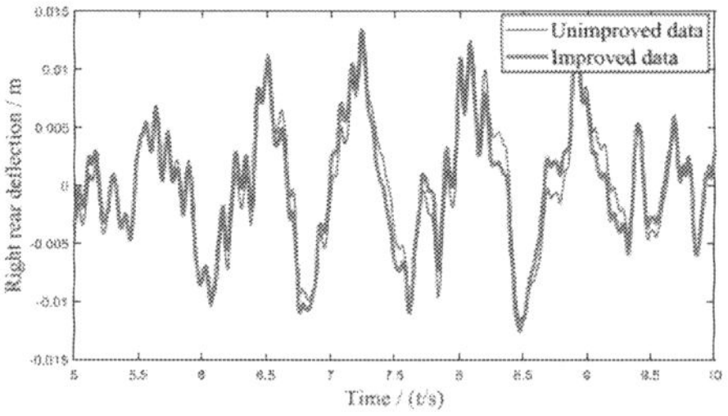

Figure 9. Comparison of right rear dynamic deflection.

https://doi.org/10.1051/matecconf/201815304008

In order to perform more clearly, we calculate the root mean square values of the body accelerations, the RMS values of the driver seat accelerations and the RMS values of the dynamic deflections, as shown in the Table 2.

Table 2. Comparison of performance indicators

Stiffness N/m	**Body acceleration $m \cdot s^{-2}$**	**Driver seat acceleration $m \cdot s^{-2}$**	**Left front deflection mm**
$k_1 = k_2 = 17$ $k_3 = k_4 = 22$	0.4150	0.6364	4.4
$k_1 = k_2 = 13$ $k_3 = k_4 = 18$	0.3899	0.5944	4.3

Stiffness N/m	**Left rear deflection mm**	**right front deflection mm**	**Right rear deflection mm**
$k_1 = k_2 = 17$ $k_3 = k_4 = 22$	4.3	5.3	4.9
$k_1 = k_2 = 13$ $k_3 = k_4 = 18$	5.3	4.4	5.0

From Table 2, it can be seen that, as far as possible without affecting the dynamic deflection of the suspension, the root mean square values of the body accelerations and the driver seat accelerations have been reduced and ride comfort of the vehicle have been improved.

4 Conclusion

The simulation design of the vehicle suspension system is carried out by using MATALAB / SIMULINK. The simulation curve of the vehicle suspension system is obtained, so that the vehicle parameter designer can predict the ride comfort of the vehicle at the design stage and adjust parameters so that the suspension parameters are optimized.

The seven-degree-of-freedom three-dimensional vibration model selected in this paper only considers the hysteresis effect of the input of the front and rear wheel surfaces, and does not take the difference in the input of the left and right wheel into account. Compared with the eleven degrees of freedom model, can not reflect the effect of the seat. The vehicle's mass, center of mass, moment of inertia, and vertical vibration of the seat system, as well as the rotation and body force, are changed in the case of manned or loaded vehicles.

References

1. J.C. Dixon, *Suspension Geometry and Computation*, (2009)
2. M. Mitschke, H. Wallentowitz, *Dynamik der Kraftfahrzeuge*, 378(2004)
3. W. Zhao, D. Zhao, Y.W. Qu, Mod. Manuf. Tech. Eqpt. (2015).
4. J. Peng, J. He, X.H. Li, Y.K. Chen, Y. Cong, J. Pla. Univ. Sci. Tech., **10**, 77(2009)
5. Y. Zhang, J. Zhang, Mech. Syst. Signal Pr., **20**, 363(2006)
6. J. Chen, W. Chen, H. Zhu, M. Zhu, Trans. Chin. Soc. Agr. Mach., **41**, 11(2010)
7. X.P. Wang, W.B. Liu, J.J. Huang, Z.Y. LI, J. Shandong Jiaotong Univ., (2010).
8. J. Feng, Y. H. Chen, Adv. Mater. Res., **510**, 250 (2012)

Development Identification Method of Inertia Properties for Heavy Truck Engine Based on MIMS Test Rig

Tianjun Zhu[1,3], Fudong Zhang[1], Jianying Li[1], Fei Li[2], Changfu Zong[1]

[1] *Department of Mechanical and Automotive Engineering, Zhaoqing Univeristy, Zhaoqing, China*
[2] *China Automotive Technology and Research Center, Tianjin 300300, China*
[3] *Department of Mechanical and Equipment Engineering, Hebei University of Engineering, Handan, 056021, China*

Abstract. A new development for the accurate measurement of the inertia parameters for heavy truck engine is presented. It is specifically intended for measuring the inertia properties of vehicle bodies, and it has the potential to be applied to the measurement of the properties of vehicle bodies, such as the vehicle powertrain, engine, and gearbox. This paper, based on CATARC MIMS test rig, develops an accurate measuring method to identify inertia parameters of heavy truck engine. Firstly corresponding tests are carried out and the lever principle and moments of inertia parallel theorem are employed to calculate and analyze the test results, which leads to the accurate value of inertia parameters. Secondly the accuracy of proposed method is verified through the calibration system. As a result the method shows high accuracy, which provides an experimental basis for accurate inertia parameters measurement of heavy truck engine.

1 Introduction

Modern automobile technology used to isolate the vibration caused by vehicle engine is usually elastic suspension system, which effectively reduces the vehicle vibration and noises leading to the improvements in ride comfort and vehicle quality. Consequently, the design and optimization analysis of vehicle engine mounting and suspension system has been a tough issue and received constant concern in the modern vehicle development and research area. While the design of heavy truck engine mounting and suspension system includes more than mounting position, static and dynamic stiffness and damping parameters design, it is also closely related to the inertia parameters of heavy truck engine such as the position of the center gravity, moment of inertia. The application of the inertia parameters test rig enables the accurate measurement of inertia parameters, then it provides the basic parameters required in the establishment of heavy truck model and the design of heavy truck engine mounting and suspension system, so as to make a valid way for the heavy truck engine dynamics analysis and suspension system optimization design.

In recent studies, researchers and automotive manufactures pay close attention to the inertia parameters measurement technique, which plays an important role in the dynamic performance of vehicle comfort and drivability[1]-[3]. The methodologies for measuring the C.G position and inertia tensor of vehicle bodies are usually divided into three categories: calculation method based on 3D solid model [4], multi-cable pendulum[5], trifilar torsional pendulum[6]-[7]. It is a huge and difficult job to accomplish the 3D solid modelling of vehicle bodies, such as powertrain, engine and gearbox. Hence the calculation method based on the CAD model is not popular used. The most reliable way to get accurate estimation of the inertia properties is via experimental tests. Among the inertia properties only the mass is indirectly measurement with very high accuracy[8]-[9]. The center of gravity can be identified by complex test rigs. For exmaple, M. Gobbi[5] uses the non-linear motion of multi-cable pendulum to measure the inertia properties of rigid bodies. A validation of the measurement procedure is performed by means of experimental tests. Raquel A.B. Almeida[10] provides the group of modal methods, where the rigid body properties are estimated from a modal analysis identification procedure. Renato Brancati[11] proposes an equipment and a procedure for the determination of large object inertia parameters. Measurement is based on the application of a least-square optimization technique on data from force and displacement sensors on the test bench. The ABD company [12] develops the Moment of Inertia (MOI) test system. The MOI test system is capable of finding the position of the center of gravity of the body, the mass and principal inertias, and the orientation of the principal axes. In the latter case, better accuracy would be expected, however the identification of the center of gravity location is a direct operation only for the two coordinates in the horizontal plane. For the third coordinate, if the rigid bodies cannot be rotated at will, quite complicated approaches are needed and poor measurement accuracy is to be expected [13].

The paper is organized as follows. At first the hardware of the test rig is introduced, and then the lever principle and moments of inertia parallel theorem are employed to calculate and analyze the test results. A validation of the measurement procedure is performed by means of experimental tests. Finally, Error analysis and the technical specifications of the test rig are given.

2 Test hardware system

2.1 The MIMS test system

The Moment of Inertia Measurement System (MIMS) is an enhancement to ABD's Suspension Parameter Measurement Machine (SPMM), which extends the machine's capability beyond its original Kinematics and Compliance testing role [13]. As illustrated in Figure 1, MIMS inertia parameters test system based on K&C test rig in CARTAR, with four six axis force calipers (see Figure 2), and can attach the vehicle bodies to the central table of the test rig where carries out six axis motions (X、Y、Z , roll, pitch and yaw).A series of test conditions are exerted by six electrical cylinders and the relevant parameters can be figured out through calculating the data generated in the calipers. The excitation is pseudo-static for measurement of the center of gravity height, and is applied at 1 Hz (typically) for inertia measurements.

Figure 1. The moment of inertia measurement system test rig

2.2 The inertia parameters test system of truck engine

As the heavy truck engine is composed of irregular parts, it is difficult to calculate the inertia parameters, which makes it obtainable through the inertia parameters testing system. The proposed test system takes the heavy truck engine of specific vehicle as example, based on the inertia parameters MIMS test rig, through adjusting the joint angle and length of the test base and heavy truck engine (see Figure 2), and the heavy truck engine is mounted to the test base as in a real vehicle. As illustrated in Figure 3, the first step is to measure the moment of inertia parameters of heavy truck engine and the base; second step is to remove the heavy truck engine, however, as the connecting device is not part of heavy truck engine, it is left together with the base, and to measure the inertia parameters of only the base as illustrated in Figure 4. The test data collected above are put to analysis, then the center of gravity position is identified and the moment of inertia of heavy truck engine can be figured out.

Figure 2. Heavy truck engine installation on the base

Figure 3. The inertia parameters test system (heavy truck engine + the base)

Figure 4. The inertia parameters test system (only the base)

3 Identification of the inertia properties

Coordinate system is established according to the right hand rule, choosing the forward direction along the platform as the positive direction of X axis, the direction to the left as the positive direction of Y axis, the vertical upward direction as the positive direction as Z axis, and the origin coordinate point as the center symmetric point. The identification procedure of MOI parameters is as the following:

(1) Step 1: According to the inertia parameters test data (heavy truck engine and the base together) and measured data of only the base, the center of gravity position of heavy truck engine can be identified. The calculating lever principle is shown as the follow.

$$mX_{CG} = m_1 X_{CG1} + m_2 X_{CG2} \quad (1)$$

$$mY_{CG} = m_1 Y_{CG1} + m_2 Y_{CG2} \quad (2)$$

$$mZ_{CG} = m_1 Z_{CG1} + m_2 Z_{CG2} \quad (3)$$

Where, m is the mass of body (heavy truck engine and the base together); m_1 is the mass of body (heavy truck engine) and m_2 is the mass of body (the base) respectively;

(X_{CG}, Y_{CG}, Z_{CG}) are the distances from the C.G point of the body (heavy truck engine and the base together) to the origin coordinate of the test rig along three directions; $(X_{CG1}, Y_{CG1}, Z_{CG1})$ are the distances from the C.G point of body (heavy truck engine) to the origin coordinate of test rig along three directions; $(X_{CG2}, Y_{CG2}, Z_{CG2})$ are the distances from the C.G point of body (the base) to the origin coordinate of test rig along three directions. From the above equations, the C.G point position (X_{CG}, Y_{CG}, Z_{CG}) relative to the origin coordinate of the test rig can be identified.

(2)Step 2: According to the parallel axis theorem, the body of mass is made to rotate about an axis passing through the body's center of mass. The body (heavy truck engine and the base together) has a moment of inertia J_0 with respect to this axis. The parallel axis theorem states that if the body is made to rotate instead about a new axis which is parallel to the first axis and displaced from it by a distance d , then the moment of inertia J with the new axis is related to J_0 by

$$J = J_0 + md^2 \tag{4}$$

The product of inertia J_{xy}、J_{xz}、J_{yz} have the similar calculation as the follow

$$J_{xy} = J_{xy0} + mx_1y_1 \tag{5}$$

$$J_{xz} = J_{xz0} + mx_1z_1 \tag{6}$$

$$J_{yz} = J_{yz0} + my_1z_1 \tag{7}$$

Where, J_{xy}、J_{xz}、J_{yz} are the products of inertia of the body, and J_{xy0}、J_{xz0}、J_{yz0} are the products of inertia of the body which origin coordinate point moves to reference point R (x_1, y_1, z_1).

(3) Step 3: Working on the same principle, the parallel axis theorem is used to figure out the moment of inertia of only the base body (products of inertia is included) in the test rig coordinate system.

(4) Step 4: The moment of inertia (products of inertia is included) obtained from step 2 is used to minus the moment of inertia (products of inertia is included) obtained from step 3, and the result is the moment of inertia of heavy truck engine (products of inertia is included) in the test rig coordinate system.

(5) The CG/MOI testing process takes less than 10 minutes.

4 Test results

4.1 The test results of mass and C.G position

Based on the Inertia parameters measuring test rig, the experimental case study is illustrated in Figure 5, with the corresponding location of the reference point. In the test rig measuring system, the multi-axis load cells used are piezoelectric devices manufactured by Kistler. Piezoelectric are used because of their unique ability to measure small forces accurately in the presence of the large static load generated by the weight of the vehicle bodies, such as heavy truck engine. The test excitation is pseudo-static for measurement of the center of gravity height, and is applied at 1 Hz (typically) for inertia parameters measurements. The measurement result of mass and C.G. position is shown in Table 1.

Figure 5. CATARC test rig during the experimental case

Table 1. The experimental result of mass and C.G. position

	heavy Truck engine+the base	the base
Mass	1674.4(kg)	459.8(kg)
C.G. position (X_{CG})	-76.5(mm)	-5.6(mm)
C.G. position (Y_{CG})	5.4(mm)	-2.7(mm)
C.G. position (Z_{CG})	1060.6(mm)	489.7(mm)

From Table 1, it is illustrated that the mass of heavy truck engine and the base is 1674.4kg, and the center of gravity is 76.5mm backward from the test rig origin coordinate point, 5.4 mm to the left and 1060.6 mm upward. Simultaneously the mass of only the base body is 459.8 kg and the center of gravity is 5.6 mm backward from the test rig origin coordinate point, 2.7 mm to the right and 489.7mm upward.

Based on the lever principle according the test data of mass and C.G position, the mass and center of gravity of heavy truck engine are listed in Table 2 choosing the reference point on the heavy truck engine to identify the C.G. position of heavy truck engine.

Table 2. The experimental result of mass and C.G. position (Heavy Truck Engine)

	heavy Truck engine+the base	the reference point
Mass	1214.6 (kg)	
C.G. position (X_{CG})	-103.34 (mm)	-650.34 (mm)
C.G. position (Y_{CG})	8.47 (mm)	15.47 (mm)
C.G. position (Z_{CG})	1276.72 (mm)	165.72 (mm)

From Table 2, the mass of heavy truck engine and the base minus the mass of the base equals the mass of heavy truck engine which is 1214.6kg, and the center of gravity is 103.34 mm backward from the test rig origin coordinate point, 8.47 mm to the left and 1276.72 mm upward.

Notes: The center of the intersection between the center line of engine crankshaft and flywheel housing end plane is chosen as reference point as illustrated in Figure 6 which is marked by a red circle. The position of the reference point in the coordinate system of test rig is (547,-7, 1111). As the reference point is the origin coordinate point in the 3D model coordinate system, so the center of gravity of heavy truck engine in the 3D model system is (-650.34, 15.47, 165.72).

4.2 The identification results of inertia parameters

The experimental data by the Inertia parameters measuring test rig are used in the identification of inertia parameters for heavy truck engine based on the parallel axis theorem. The identification results are shown in Table 3.

Table 3. The identification results of inertia parameters

	heavy Truck engine+the base body	heavy truck engine	the base body
J_x(kgm^2)	477.1	81.37	189.1
J_y(kgm^2)	730.3	244.8	275.7
J_z(kgm^2)	657.6	207.3	447.1
J_{xy}(kgm^2)	-0.8	-2.14	1.7
J_{xz}(kgm^2)	-34.1	-8.94	0.5
J_{yz}(kgm^2)	6.4	4.57	-1.1

From Table 3, it is shown that heavy truck engine moment of inertia is 81.37 kgm^2 relative to X axis, 244.8 kgm^2 relative to Y axis, and 207.3 kgm^2 relative to Z axis. The three products of inertia of J_{xy}、J_{xz}、J_{yz} are -2.14 kgm^2, -8.94 kgm^2 and 4.57 kgm^2 respectively.

5 Conclusions

A new approach for the accurate measurement of the inertia parameters for heavy truck engine is presented. Compared to the conventional inertia parameters measurement method, this new approach has several advantages:

(1) By using the MIMS measuring instrument, the measurement accuracy is significantly enhanced.

(2) The calibration procedure with standard specimen showed little deviation from conventionally true values, the developed method showed good repeatability.

(3) The MOI measurement system enables the center of gravity coordinates and inertia tensor components to be measured for kinds of vehicle bodies such as powertrain, truck cabs, engine, gearbox, etc.

(4) The CG/MOI testing process takes less than 10 minutes, so the test method has good working efficiency.

Acknowledgments

The authors would like to acknowledge the financial support of Collaborative Innovation and Platform Environment Construction Project of Guangdong Province (2015A050502053), Hebei science and technology plan project(17394501D), Science and technology research project of Hebei higher education (ZD2017213) Hebei provincial High Level Talents Foundation of Hebei Province (A2016002025), Characteristic innovation project of Guangdong Province (2016KTSCX154), The authors also would like to acknowledge the financial support of Natural Science Foundation of Guangdong Province (2014A030311045) and project of Education Department of Guangdong Province.

References

1. Wen-Bin, Shangguan, Y. Zhao, "Dynamic analysis and design calculation methods for powertrain mounting systems," International Journal of Automotive Technology, Vol. 8 , No. 6 , pp. 731 – 744, 2007.
2. Rozyn, Matthew, Zhang, Nong, "A method for estimation of vehicle inertial parameters," International Journal of Vehicle System Dynamics, Vol. 48, No. 5, pp. 547-565, 2010.
3. Wegener, Daniel, "Vehicle inertia measurement machine (VIMM)," Society of Allied Weight Engineers - 71st International Conference on Mass Properties, pp. 153-162, 2012.
4. J.P. Pegram, W.A. Anemaat, "Preliminary estimation of airplane moments of inertia using CAD solid modeling," SAE paper 2000-01-1700, 2000.
5. M. Gobbi, G. Mastinu, G. Previati, "A method for measuring the inertia properties of rigid bodies, "Mechanical Systems and Signal Processing ,Vol. 25 ,No.1, pp. 305 – 318,2011.
6. S. Carsten, L. Michhael, "A survey of current inertia parameter identification methods, "Mechanical Systems and Signal Processing, Vol. 15, No. 1, pp. 189 – 211, 2001.
7. Zhi-Chao. Hou, Yi-ning. Lv, Yao-xin. Lao, Dan. Liu, "A new trifilar pendulum approach to identify all inertia parameters of a rigid body or assembly," Mechanism and Machine Theory,Vol.44, No.3 pp.1270 – 1280, 2009.
8. R.A.B. Almeida, A.P.V. Urgueira, N.M.M. Maia, "Evaluation of the performance of three different methods used in the identification of rigid body properties,"Shock and Vibration Journal (Special Issue ICED 2007), pp. 467–479, 2008.
9. H. Lee, Y.B. Lee, Y.S. Park, "Response and excitation points selection for accurate rigid body inertia properties identification," Mechanical

Systems and Signal Processing,Vol.13,No.4 , pp.571–592 ,1999.

10. Raquel A.B.Almeida ,Antonio P.V.Urgueira, NunoM.M.Maia, "Further developments on the estimation of rigid body properties from experimental data," Mechanical Systems and Signal Processing,No.4,pp.1391–1408, 2010.
11. Renato Brancati,Riccardo Russo, SergioSavino, "Method and equipment for inertia parameter identification," Mechanical Systems and Signal Processing,No.24,pp. 29–40, 2010.
12. Anthony Best, "Addendum 1 to SPMM User Guide (MIMS) v8.12,"Anthony Best Dynamics Company, pp.63-68, 2012.
13. Xiao Shuchen, Qing Yuxun, Han Jiqing, " A test analysis method based on Grubbs in experiments ". Journal of Projectiles; Rockets; Missiles and Guidance, Vol.27, No.1， pp.275-277, 2007.

Design, Fabrication and Testing of Carbon Fiber Reinforced Epoxy Drive Shaft for All Terrain Vehicle using Filament Winding

Suhas Yeshwant Nayak, Nishank Minil Amin, Srinivas Shenoy Heckadka, Vishal Shenoy P, Ch. Sravan Prakash and Ruthvik Mabbu

Department of Mechanical and Manufacturing Engineering, Manipal Institute of Technology, Manipal Academy of Higher Education, Manipal, Karnataka, India - 576104

Abstract Filament winding is a composite material fabrication technique that is used to manufacture concentric hollow components. In this study Carbon/Epoxy composite drive shafts were fabricated using filament winding process with a fiber orientation of $[85_2/\pm45_2/25_2]_s$. Carbon in the form of multifilament fibers of Tairyfil TC-33 having 3000 filaments/strand was used as reinforcement with low viscosity epoxy resin as the matrix material. The driveshaft is designed to be used in SAE Baja All Terrain Vehicle (ATV) that makes use of a fully floating axle in its rear wheel drive system. The torsional strength of the shaft was tested and compared to that of an OEM steel shaft that was previously used in the ATV. Results show that the composite shaft had 8.5% higher torsional strength in comparison to the OEM steel shaft and was also lighter by 60%. Scanning electron microscopy (SEM) micrographs were studied to investigate the probable failure mechanism. Delamination, matrix agglomeration, fiber pull-out and matrix cracking were the prominent failure mechanisms identified.

1 Introduction

Drive shafts serve the purpose of connecting the differential to the wheel hubs in automotive vehicles [1]. SAE Baja vehicle has a competition regulated standard 10HP engine [2] and makes use of a centrifugal continuously variable transmission (CVT) coupled to a custom designed two stage reduction single speed gearbox in a spool configuration to drive its rear wheels. A conventional OEM drive shaft has three major components – inboard CV joint, shaft element, outboard CV joint [1]. In order to minimise transmission losses and unsprung mass of the car which ultimately improves ATV performance it is essential to lower the weight of every component possible due to the limited power available. The shaft element of the OEM drive shaft being used weighs about 2.5kg, providing ample room for weight reduction by the use of composite materials which have a lower density in comparison to steel. Carbon fiber is a reinforcing material that is widely used with a suitable resin due to its high strength and modulus [3]. Carbon fiber – epoxy resin combination provides the best strength to weight ratio in comparison to glass fiber-epoxy or aramid – epoxy systems [4, 5]. Since the shaft is to be used as a fully floating axle it has to take up only torsional loading caused by acceleration and braking of the vehicle. Hence ideal stacking sequences for a shaft undergoing torsional loading along with negligible axial and bending loads is $[90_2/\pm45_2/0_2]_S$ [6, 7, 8]. Calculations for all material parameters are based on the type of fiber and epoxy. In addition to this the design parameters of the shaft like length, diameter and thickness are also considered [5,9].

This study aims to evaluate and compare the torsional strength of steel drive shaft to that of a carbon/epoxy composite shaft. The composite shaft is fabricated by filament winding process with a winding pattern of $[85_2/\pm45_2/25_2]_s$. against an ideal winding pattern of $[90_2/\pm45_2/0_2]_S$ due to machine constraints. The composite shafts were cured in a hot air oven. SEM micrographs are used to study the potential failure mechanisms, to verify if the composite shaft can be a potential replacement of the OEM steel shaft.

2 Materials and Methods

2.1 Materials

Tairyfil TC33 carbon fibers were chosen as the reinforcement. Fibers were procured from Hindoostan Composites, Mumbai. It has 3000 filaments in each strand with a yield Tex of 800g/1000m. The tensile strength of the fiber is 3450 MPa, tensile modulus is 230 GPa, density is 1.8g/cm^3, elongation of 1.5% and each filament has a diameter of 7μm [10]. A low viscosity resin is preferred while performing filament winding [11] and hence the L12- K6 epoxy-hardener system was chosen and procured from Atul Ltd. India. The neat polymer has a tensile strength of 55-70 MPa, tensile modulus of 2.5-4.0 GPa and tensile elongation of 1.5-3.0% [12].

2.2 Calculations

All nine material parameters E_{11}, E_{22}, E_{33}, G_{12}, G_{13}, G_{23}, v_{12}, v_{13} and v_{23} use the following nomenclature: the first subscript represents the loading direction and the second subscript represents the direction in which the particular property is measured as seen in Figure 1.

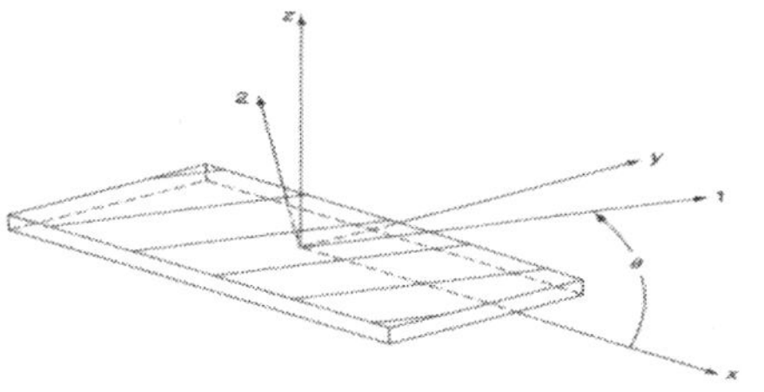

Figure 1: Definition of principle axes (1 and 2) and loading axes(x and y) for a lamina [9]

The following parameters were taken as inputs and base reference values:
Length of shaft – 300mm
Maximum external diameter – 28±0.1mm
Thickness of shaft – 4±0.05mm
Laminate thickness (t) ~ 0.33mm
Maximum torque applied by gearbox to shaft – 650Nm
Failure of torque of OEM steel shaft – 1630Nm
Factor of safety for design torque – 2.5
Volume fraction of fiber – 25%
Poisson's Ratio of reinforcement – 0.2 [13]
Poisson's Ratio of matrix – 0.35 [13]

$$E_{11}=E_fV_f+E_mV_m \quad (1)$$

E_{11}: Young's modulus of the composite in the direction of fiber
E_f: Young's modulus of the fiber
V_f: volume fraction of the fiber
E_m: Young's modulus of the matrix
V_m: volume fraction of the matrix

$$E_{22}= E_fE_m/(E_fV_m+E_mV_f) \quad (2)$$

E_{22}: Young's modulus of the composite perpendicular to the fiber

$$v_{12}= v_f*V_f+ v_m*V_m \quad (3)$$

v_{12}: the ratio of strain in direction 2 to the applied strain in direction 1
v_f: Poisson's ration of the fiber
v_m: Poisson's ratio of the matrix

$$v_{21}= (E_{22}/E_{11})* v_{12} \quad (4)$$

$$G=E/2(1+ v) \quad (5)$$

$$G_{12}=G_f*G_m/(G_f*V_m+G_m*V_f) \quad (6)$$

G = rigidity modulus
For special orthotropic lamina (θ= 0 or 90) the equation can be written as:

$$\begin{bmatrix}\sigma_{xx}\\ \sigma_{yy}\\ \tau_{xy}\end{bmatrix}=\begin{bmatrix}Q_{11} & Q_{12} & 0\\ Q_{21}(=Q_{12}) & Q_{22} & 0\\ 0 & 0 & Q_{66}\end{bmatrix}\begin{bmatrix}\xi_{xx}\\ \xi_{yy}\\ \gamma_{xy}\end{bmatrix} \quad (7)$$

Stiffness matrix
'σ' represents the stress in a specific direction
'ξ' represents the strain in the specific direction

$$E_{11}/(1-v_{12}*v_{21}) =Q_{11} \quad (8)$$

$$v_{12}*E_{22}/(1-v_{12}*v_{21}) = Q_{12} =Q_{21} \quad (9)$$

$$E_{22}/(1-v_{12}*v_{21}) =Q_{22} \quad (10)$$

$$G_{12}=Q_{66} (8)$$

For general orthotropic lamina (θ ≠ 0 or 90) the values of Q_{11}, Q_{22}, Q_{12}, Q_{22} *change to* q_{11}, q_{22}, q_{12}, q_{21}.

$$q_{11}=Q_{11}\cos^4\theta+2(Q_{12+}2Q_{66}\sin^2\theta\cos^2\theta+ Q_{22}\sin^4\theta \quad (11)$$

$$q_{22} =Q_{22}\cos4\theta+2(Q12+2Q66)\sin2\theta\cos2\theta+Q_{11}\sin4\theta \quad (12)$$

$$q_{12}=Q_{12}(\sin^4\theta+\cos^4\theta) + (Q_{11}+Q_{22}-4*Q_{66})\sin^2\theta\cos^2\theta \quad (13)$$

$$\begin{bmatrix}Nx\\ Ny\\ Nxy\\ Mx\\ My\\ Mxy\end{bmatrix}=\begin{bmatrix}A_{11} & A_{12} & A_{31} & B_{11} & B_{12} & B_{13}\\ A_{21} & A_{22} & A_{23} & B_{21} & B_{22} & B_{23}\\ A_{31} & A_{32} & A_{33} & B_{31} & B_{32} & B_{33}\\ B_{11} & B_{12} & B_{13} & D_{11} & D_{12} & D_{13}\\ B_{21} & B_{22} & B_{23} & D_{21} & D_{22} & D_{23}\\ B_{31} & B_{32} & B_{33} & D_{31} & D_{32} & D_{33}\end{bmatrix}\begin{bmatrix}\xi_x\\ \xi_y\\ \gamma_{xy}\\ K_x\\ K_y\\ K_{xy}\end{bmatrix} \quad (14)$$

N_{xx} =normal force resultant in the x direction (per unit width)
N_{yy} =normal force resultant in the y direction (per unit width)
N_{xy} =shear force resultant (per unit width)
M_{xx}=bending moment resultant in the yz plane (per unit width)
M_{yy}=bending moment resultant in the xz plane (per unit width
M_{xy} =twisting moment (torsion) resultant (per unit width)
K_x, k_y = bending curvatures of the laminate
K_{xy} = twisting curvature of the laminate as shown in Figure 2.

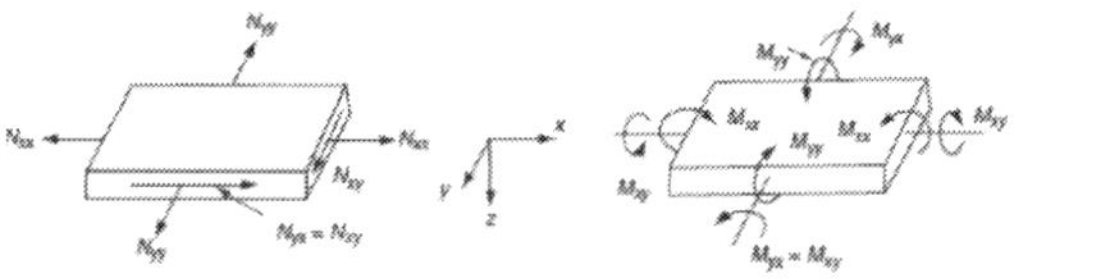

Figure 2: In-plane, bending, twisting loads applied on a laminate [9]

[A]= extensional stiffness matrix for the laminate [B] =coupling stiffness matrix for the laminate
[D]=bending stiffness matrix for the laminate

$$A_{mn}=\Sigma^N_{j=1}(qmn)_j(h_j-h_{j-1}) \quad (15) [9]$$

h_j: distance of the top end of the jth laminate from the centre plane
h_{j-1}: distance of the bottom end of the jth laminate from the centre plane as seen in Figure 3.

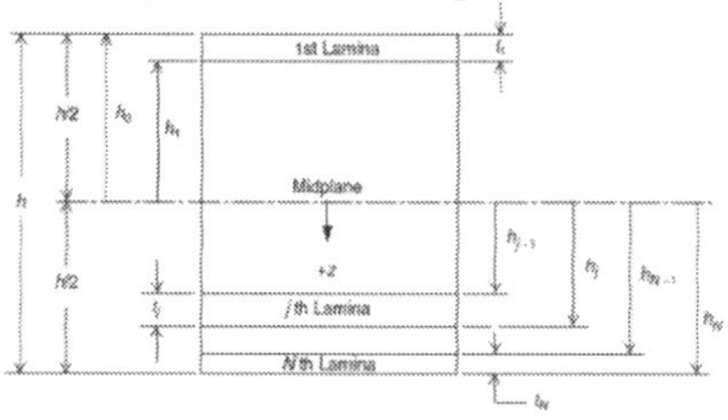

Figure 3: Thickness of laminate [9]

$E_{xx}= 1/t\ [A_{11}-A_{12}^2/A_{22}]$ (Young's modulus of laminate along loading (x) direction) (16)

$E_{yy}= 1/t\ [A_{22}-A_{12}^2/A11]$ (Young's modulus of laminate perpendicular to loading (y) direction) (17)

For inner layers the critical buckling torque:

$T_{cri}=2.28/\sqrt{L} * (E_{xx})^{0.375} * (E_{yy})^{0.625} * (t)^{2.25} * (D)^{1.25}$ (18)

For outer layers the critical buckling torque

$T_{cro}=2\pi r^2 t * 0.272 * [E_{xx}*E_{yy}]^{0.25} * [t/r]^{1.5}$ (19)

[14]

E_{xx}=15.910GPa
E_{yy}=23.4GPa
T_{cri}=3900Nm
T_{cro}=4152.17Nm
Average buckling torque, Tav=4026.09Nm

2.3 Fabrication of Composites

An aluminium rod of diameter 19.5mm and length of 450mm was used as the mandrel. The length was chosen so as to clamp the mandrel at either of the chucks of the filament winding machine. It was first wrapped in cling film to prevent contact between epoxy and mandrel. Next a layer of peel ply is applied and coated evenly with Polyvinyl alcohol (PVA) which acts as a release agent. The fibers are passed through directing combs and the resin pot. The resin pot was then filled with the epoxy resin-hardener mixture taken in the ratio 10:1 by weight as specified by the manufacturer. The tensioner is then fixed into position and the fibers are tied to the mandrel.

The motor is switched on causing the spindle to rotate and a small amount of fibers are wound onto the mandrel until the epoxy coated fibers reach the mandrel. The carriage movement is then started and fibers are wound along the mandrel. A breaker switch is connected to a pushrod that is actuated when the carriage reaches the designated end point for winding, causing the direction of winding to be reversed. After the first two layers of 85° are wound the carriage speed is adjusted to change the angle of winding to 45° and later to 25°. After all 12 layers have been wound, fibers are disconnected from the mandrel and the mandrel is rotated in the machine for about 5 minutes to ensure uniform distribution of matrix material and to start the precuring phase. The mandrel is then placed in a hot air oven that had been preheated to 120°C for about 15 minutes to cure. Post cure the mandrel is cooled rapidly using water at a temperature of about 1°C causing the mandrel to shrink slightly and enabling its extraction from the shaft.

Next the shaft is cleaned on the inner surface using sandpaper of 80, 120 and 240 grit followed by cleaning of the of surface using acetone as a degreasing agent. Metal inserts are prepared according to the torsion testing machine specifications (as shown in Figure 4) and are bonded to the shaft using Araldite AV 130M and hardener HV 998 which is a bonding agent. It has low out gassing and cures at room temperature to form strong adhesive bonds between the shafts and inserts. The reason for using this type of bonded inserts lies in the application of the shaft being tested. The final use of the shaft will involve bonding to aluminium inserts which have splines that mate with the CV joint for power transmission [15].

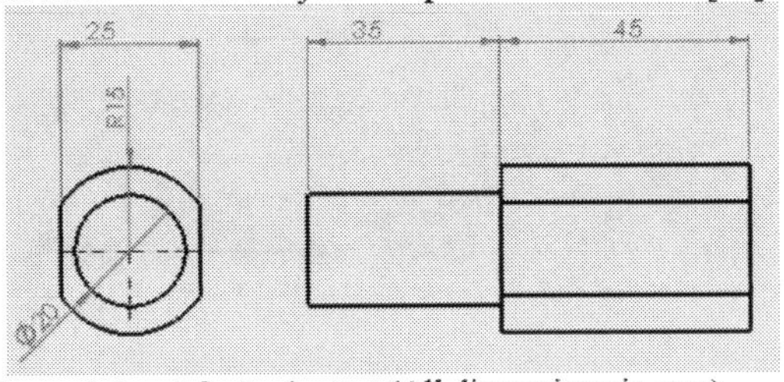

Figure 4: Insert for torsion test (All dimensions in mm)

2.4 Torsion Test

Torsional strength was determined using a torsion testing machine. The shaft was fixed into the machine and a torsional load was applied to the tailstock using a hydraulic motor at a rate of 0.5°/sec. The shaft is said to have failed when a crack starts to form on the surface of the shaft and prevents the shaft from carrying any further loads. Tests were conducted on conditioned samples at a temperature of 25°C. Five samples were considered for testing. Visual inspection for defects was done prior to testing.

2.5 Surface morphology of fractured surfaces

Specimens for morphological studies were prepared from the fractured area of the shaft. The specimens were sized accordingly so as to be mounted in the SEM to analyse and understand the probable failure mechanism. The specimen surfaces were coated with a thin layer of silver by ion sputtering and a 15kV accelerating voltage was used. The SEM was used in the variable pressure mode.

3 Results and discussion

3.1 Torsion Test

Figure 5 shows a failed specimen after torsional loading. The torsional strength of the carbon fiber (CF) shafts compared to the OEM steel shaft is presented in Figure 6. The average torsional strength of the carbon fiber shafts is 1770Nm, which is 8.5% more than the OEM steel shaft (1630Nm). The carbon fiber shaft fabricated weighs just 0.8kg which is 60% lighter than the OEM steel shaft which weighs 2.5kg. This can be attributed to the fact that the carbon fiber shaft has a lower density and it is specifically designed to take up the complex loading conditions of a drive shaft. The experimental values are 56% lower than calculated values. The most likely cause being a lower fiber volume fraction caused by a lower fiber tension provided during the filament winding process. Also there appears to be a collection (agglomeration) of matrix on the surface of the shaft during the curing process which could be caused by the following reasons: (i) excess resin take up during filament winding or (ii) lack of rotation of the shaft during curing process. This leads to an uneven transfer of

https://doi.org/10.1051/matecconf/201815304010

loads between layers of fibers. The strength to weight ratio of composite shaft and OEM steel shaft are 2212 Nm/Kg and 652 Nm/Kg respectively.

Figure 5: Failed specimen after torsional loading

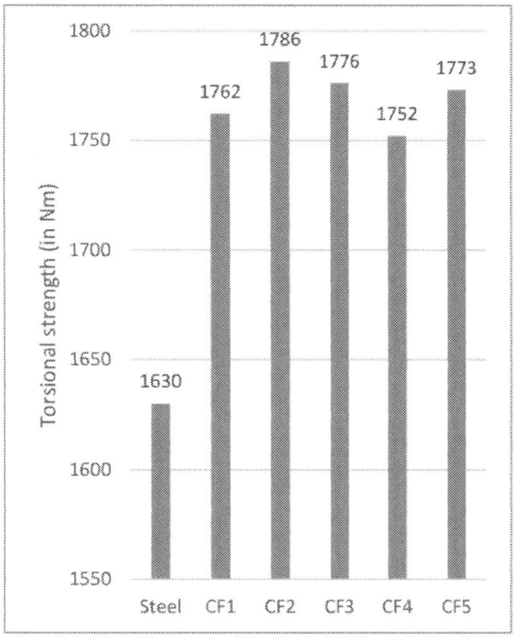

Figure 6 : Torsional strength of shafts

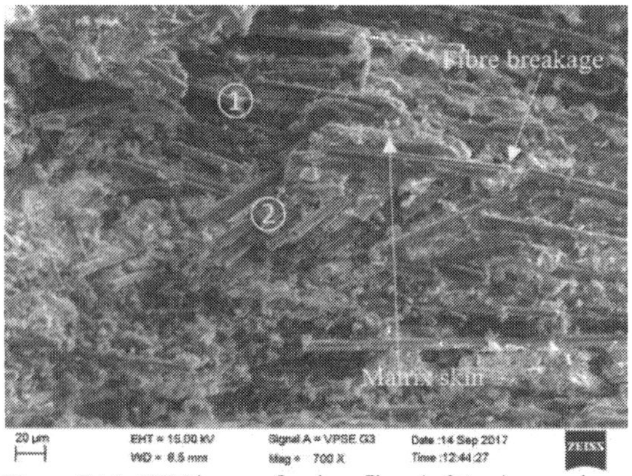

Figure 7 (a): SEM image of carbon fiber shaft torsion specimen

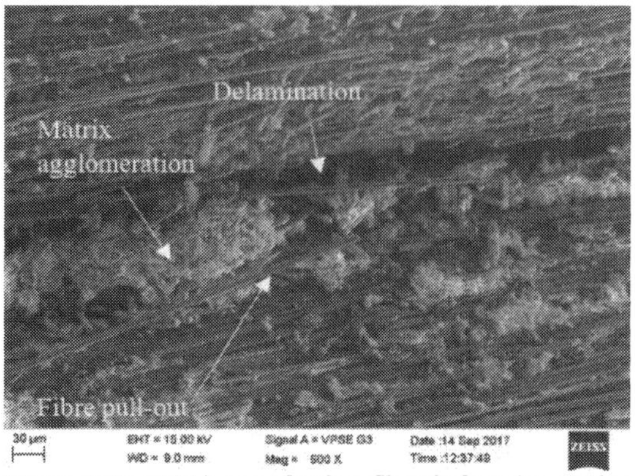

Figure 7 (b): SEM image of carbon fiber shaft torsion specimen

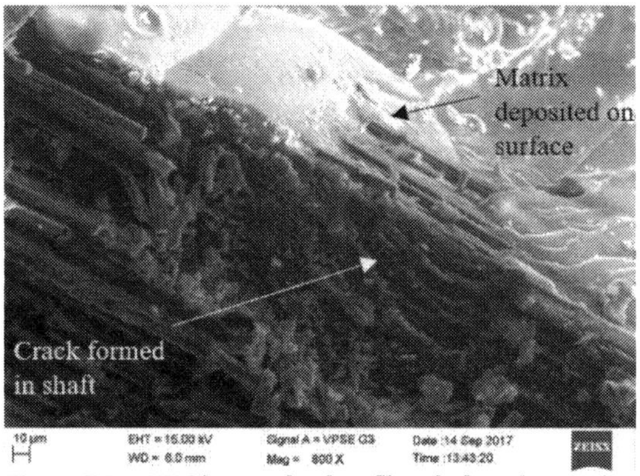

Figure 7 (c): SEM image of carbon fiber shaft torsion specimen

Figure 7 (d): SEM image of carbon fiber shaft torsion specimen

3.2 Morphology of fractured specimens

Figures 7 shows micrographs of fractured surfaces. Figure 7(a) depicts the multiple layers of the shaft oriented at the different angles in which each layer of fibers is covered with matrix skin showing good adhesion between the matrix and fibers, zone① represents 85° and zone② represents 45°. Fiber breakage due to lateral shift is also seen in Figure 7 (a). Delamination between the layers of the shaft leading to its failure during torsional loading is seen in Figure 7(b). Matrix agglomeration is visible at multiple locations which would lead to a lower torsional strength due to uneven transfer of loads between layers. Delamination has further lead to the occurrence of fiber pull-out.

Figure 7(c) represents a section closer to the outer edge of the shaft. Collection of matrix on the surface can be seen which is undesirable and a crack has been formed in the shaft which then propagates towards the lower layers. Figure 7(d) shows the fractured fibers along the loading direction with fiber breakage clearly visible.

4 Conclusion

An increase in torsional strength by 8.5% along with a weight reduction of 60% leads to an increase of strength to weight ratio by 339% which indicates that the carbon/epoxy shaft can successfully be used to replace the OEM steel shaft. A deviation from calculated torsional strength of the shaft could be attributed to lower fiber volume fraction obtained during fabrication caused by lower than ideal winding tension. The accumulation of matrix on surface needs to be avoided by making use of an oven with the capability to continuously rotate the shaft while curing. Also excessive resin pickup needs to be prevented during the winding process. Use of shrink tape may help to remove excess resin from the surface of the shaft and further improves dimensional accuracy. Delamination between layers of different orientations of fibers has been identified as the primary mode of failure. Fiber breakage, fiber pull-out and matrix agglomeration further contribute to weakening of the shaft. Increasing the number of filaments in the fibers will help in reduction of delamination as there will be greater surface area for adhesion between matrix and fibers. This is supported by the SEM images which clearly show that there is a matrix skin forming on the fibers indicating good interfacial bonding.

Acknowledgement

The authors are thankful to Dr. K Jagannath, Head of the Department, Mechanical and Manufacturing Engineering for permitting us to make use of the Advanced Material Testing and Research Laboratory. The authors are also indebted to Dr. Satish Shenoy B, Head of the Department and Mr. Padmaraj N. H., Assistant Professor, Department of Aeronautical and Automobile Engineering for allowing us to use their Advanced Composite and Material Testing Laboratory. The authors thank Dr. M. Vijaya Kini, Associate Professor, Department of Mechanical and Manufacturing Engineering. The authors would also like to thank Dr. Mohandas Chadaga, Head of the Department, Civil Engineering, for permitting us to use the torsion testing machine. The authors thank Mr. Shivprasad Kamath and Mr. Mahesha for assisting us in fabrication of the drive shafts.

References

1. Carroll Smith "*Tune to Win*" Aero Publishers, INC. (1978) ISBN:978-0-87938-071-7
2. Baja, S. A. E. "*2015 Collegiate Design Series.*"(2014)
3. Daniel, Isaac M., Ori Ishai, Issac M. Daniel, and Ishai Daniel. *Engineering mechanics of composite materials*. **3**. New York: Oxford university press, (1994)
4. Moorthy, R. Srinivasa, Yonas Mitiku, and K. Sridhar. "*Design of automobile driveshaft using carbon/epoxy and kevlar/epoxy composites.*" . AJER **2,** 10 (2013): 173-179.
5. Hatwar, Pankaj K., and R. S. Dalu. "*Design and Analysis of Composite Drive Shaft.*". IJSR **4,** 4 ISSN (Online): 2319-7064(2013)
6. Rajmane, S. M., and M. A. Chopade. "*Investigation of Effect of Stacking Sequence and Fiber Orientation on Maximum Torque Transmission Capacity of a Composite Automobile Driveshaft by Analytical & FEA Methods.*" IJAIEM **3,** 11 (2014): 31-37.
7. Reddy, P. Satheesh Kumar, Ch Nagaraju, and T. Hari Krishna. "*Optimum design and analysis of filament wound composite tubes in pure and combined loading.*" IJERT **1,** 8 (2012).
8. Bankar, Harshal, Viraj Shinde, and P. Baskar. "*Material optimization and weight reduction of drive shaft using composite material.*" IOSR-JMCE **10,** 1 (2013): 39-46.
9. Mallick, Pankar K. *Fiber-reinforced composites: materials, manufacturing, and design*. CRC Press 3rd Edition ISBN:1420005987 (2007)
10. Formosa Plastics Corporation,(2014) "*Tairyfil Data Sheet*".
11. Cai, Z. H. O. N. G., T. Gutowski, and S. Allen. "*Winding and consolidation analysis for cylindrical composite structures.*" *J. Compos. Mater.* **26,** 9 (1992): 1374-1399.
12. Atul Limited "*LAPOX L12/K6*". **2** (2012)
13. Patil, Sagar D. ,DS Chavan, MV Kavade,"Investigation of Composite Torsion Shaft for Torsional Buckling Analysis using Finite Element Analysis"." *IOSR-JMCE.***4,** 3 (2012): 26-31.
14. Badie, M. A., E. Mahdi, and A. M. S. Hamouda. "*An investigation into hybrid carbon/glass fiber reinforced epoxy composite automotive drive shaft.*"Mater.Des. **32,** 3 (2011): 1485-1500.
15. Sharma, Ambuj, James M. Nelson, Douglas P. Goetz, Amit Patel, Paul Sedgwick, Wendy Thompson, Travis Gorsuch, and Dave Knauff. "*The Impact of Nanosilica-Modified Epoxy Resins on the Design and Performance of Filament-Wound Drive Shafts.*" CAMX Conference Proceedings, pp. 26-29. (2015).

MATEC Web of Conferences 153, 04011 (2018) https://doi.org/10.1051/matecconf/201815304011
ICMME 2017

Aerodynamic research of a racing car based on wind tunnel test and computational fluid dynamics

Jianfeng Wang[1,2], Hao Li[1], Yiqun Liu[1,2], Tao Liu[1], and Haibo Gao[2]

[1] *School of Automotive Engineering, Harbin Institute of Technology, Weihai, 264209, China*
[2] *State Key Laboratory of Robotics and System, Harbin Institute of Technology, Harbin, 150080, China*

Abstract. Wind tunnel test and computational fluid dynamics (CFD) simulation are two main methods for the study of automotive aerodynamics. CFD simulation software solves the results in calculation by using the basic theory of aerodynamic. Calculation will inevitably lead to bias, and the wind tunnel test can effectively simulate the real driving condition, which is the most effective aerodynamics research method. This paper researches the aerodynamic characteristics of the wing of a racing car. Aerodynamic model of a racing car is established. Wind tunnel test is carried out and compared with the simulation results of computational fluid dynamics. The deviation of the two methods is small, and the accuracy of computational fluid dynamics simulation is verified. By means of CFD software simulation, the coefficients of six aerodynamic forces are fitted and the aerodynamic equations are obtained. Finally, the aerodynamic forces and torques of the racing car travel in bend are calculated.

1 Introduction

Aerodynamics is one of the most important factors affecting dynamic performance of a racing car. Especially in some open-cab and open-wheel car, air flows through car body should be properly channelled to provide reasonable downward pressure. Wing through car body can be divided into three parts, top wing, both sides' wing and bottom wing. Upper wing moves along the surface of the car, reaching the tail-wing area at the end of the car. The wing on both sides of the car is also divided into two parts, the flow front of the wheel and the flow to the radiator box through the wheel and the body. Using the diversion of the front wing to guide the wing inward or outward in front of the wheel, make the wing bypass the tire to reduce the air drag force caused by the tire. The bottom of the air is the need for reasonable guidance to allow the air to flow smoothly through the car chassis.

For research of aerodynamics characteristics of racing car, mainly use wind tunnel experiment and simulated by computational fluid dynamics (CFD) software. CFD is a system analysis method and tool for simulating fluid flow, heat transfer and related transfer phenomena by computer. The basic idea of CFD is the original in time domain and space domain continuous physical quantity field, replaced by a series of discrete points on the value of the variable, and through the principles and methods to establish certain algebraic equations reflect the discrete points on the relationship between variables, and then solving algebraic equations to obtain the approximation of the field variables solution. The relationship between discrete points conforms to fluid mass conservation, momentum conservation and energy conservation, which are the fluid control equations. In order to verify the accuracy of the calculation of fluid analysis software, using the wind tunnel experiment on wing was tested. Mohrfeld-Halterman described a high fidelity vehicle aerodynamic model and presented a comparison between the effects of this proposed model and a conventional quasi steady-state aerodynamic model on race vehicle simulation results [1]. J. D. Kee, et al [2] researched driving stability of a passenger car at high-speed and under crosswind conditions by wind tunnel test. Roberts, et al. [3] established the fundamental aerodynamic flow features of a racing car's front wing operating at yaw. A. Buljac, et al. [4] developed a computational model to analyse aerodynamic loads and flow characteristics of the rear wing. M. D. Soso and P. A. Wilson [5] researched aerodynamic factors which influence the creation of overtaking opportunities in the racing car by wind tunnel test. A. Altinisik, et al. [6] studied the drag forces and the pressure distributions of the model FIAT Linea at increasing yaw angle and two-vehicle platoon. Experiments were performed in the wind tunnel for the yaw angles of 0 deg, 5 deg, and 10 deg due to the test section dimensional restriction. He obtained the drag coefficients for three different test configurations, and the comparison of the drag coefficients for all three coastdown configurations were in good agreement with the CFD results and FIAT wind tunnel results [7]. Zhang Yingchao, et al. [8] researched on the aerodynamic drag both in wind tunnel test and CFD, and obtained the reasons for the drag coefficients increased.

2 Six aerodynamic forces

The influence of aerodynamics on racing car is realized mainly through aerodynamic and aerodynamic moment. The driving speed, body shape and sideslip angle of the car are important factors to influence the aerodynamic force and aerodynamic moment. Aerodynamic forces and moments are composed by three vertical forces and torques around X axis and Y axis and Z axis, also known as six aerodynamic component forces, mainly includes drag force F_D, lift force F_L, side force F_S, rolling moment T_{RM}, pitching moment T_{PM} and yawing moment T_{YM}, as shown in Figure 1.

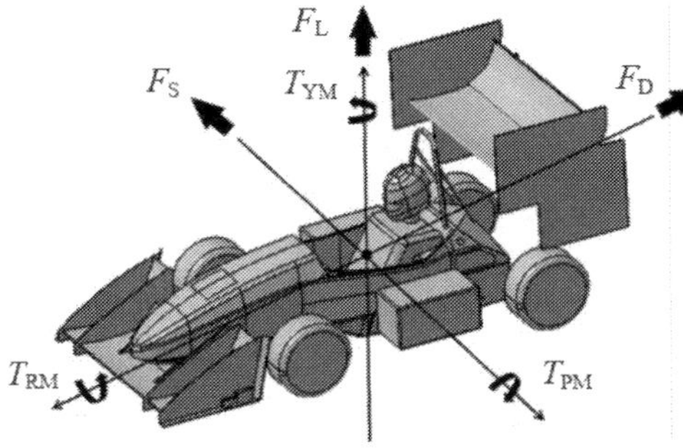

Figure 1. Diagram of six aerodynamic component forces.

According to the ISO standard, the origin of aerodynamic coordinate is located in the center of front and rear axles. Six aerodynamic component forces can be defined in Equation (1).

$$\begin{cases} F_D = \frac{1}{2} C_D A \rho u^2 \quad F_L = \frac{1}{2} C_L A \rho u^2 \quad F_S = \frac{1}{2} C_S A \rho u^2 \\ T_{RM} = \frac{1}{2} C_{RM} A \rho L u^2 \quad T_{PM} = \frac{1}{2} C_{PM} A \rho L u^2 \\ T_{YM} = \frac{1}{2} C_{YM} A \rho L u^2 \end{cases} \tag{1}$$

where F_D (N) is drag force, F_L (N) is lift force, F_S (N) is side force, C_L is lift force coefficient, C_D is drag force coefficient, C_S is side force coefficient, T_{RM} (Nm) is rolling moment, T_{PM} (Nm) is pitching moment, T_{YM} (Nm) is yawing moment, C_{RM} is rolling moment coefficient, C_{PM} is pitching moment coefficient, C_{YM} is yawing moment coefficient, A is car's projection area (taken 1.15041m^2), ρ is air density (taken 1.18415 kg/m^3), L is wheelbase (taken 1.53m), u (km/h) is velocity.

3 Wind tunnel test

3.1 Experimental installation

The wind tunnel test is the most direct method to research aerodynamics of automobile and automobile parts. In order to study aerodynamic characteristics of the wing and the correctness of the comparison of CFD software, the wing experiment using wind tunnel. The wind tunnel in School of Automotive Engineering of Harbin Institute of Technology at Weihai is used in experiment, as shown in Figure 2. The wind tunnel is open type wind tunnel, wind tunnel test section size: 2.5m×1.2m×1m, wind speed can be adjusted continuously at the range of 2-30 m/s. The test system uses Donghua DH5902 test data acquisition system.

Figure 2. Photograph of wind tunnel.

In order to test the aerodynamic characteristics of wing, three aerodynamic component forces bar-shaped strain gauge balance was designed. A. L. Smith, et al. [9] designed a six component force balance using a combination of modelling and analysis techniques. The balance was designed for the measurement of impulsive aerodynamic forces and moments characteristic. The wing sample is well connected on the upper support by mechanical connection. The tray can be installed at the bottom of the tunnel. The rotation of the tray can realize the adjustment of wing's yaw angle. The lower support connects tray and strain gauge balance, and ensures the tested wing lift height can get rid of the influence of surface boundary layer.

The middle three-component bar-shaped strain gauge balance is the main body of the balance assembly. The balance is composed of three pieces of beam element and the vertical beam, the middle beam of three beams is used to measure normal force, under the symmetric set of two beams are used to measure the pitching moment. Four vertical beams are used to measure axial force element were used to measure the axial force. The strain sensors are arranged in the corresponding position to measure the wing's force and torque.

3.2 Comparison between wind tunnel test and simulation results

The pressure and drag force of the wing numbered NACA7048 is tested in wind tunnel experiment. At the same time, in accordance with the wind tunnel test conditions using computational fluid analysis STAR-CCM to establish flow field simulation model shown in Figure 3.

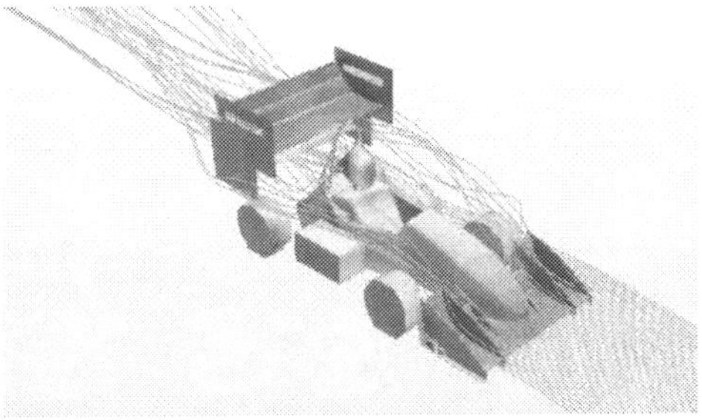

Figure 3. Simulation model in STAR-CCM.

Wind tunnel experiment results and simulation results are compared as shown in Figure 4. The average relative error in less than 4%. Therefore, simulation results are reliable and can be used for detailed analysis of the car in different conditions. Therefore, the following analysis of aerodynamic characteristics of fluid simulation can use STAR-CCM software.

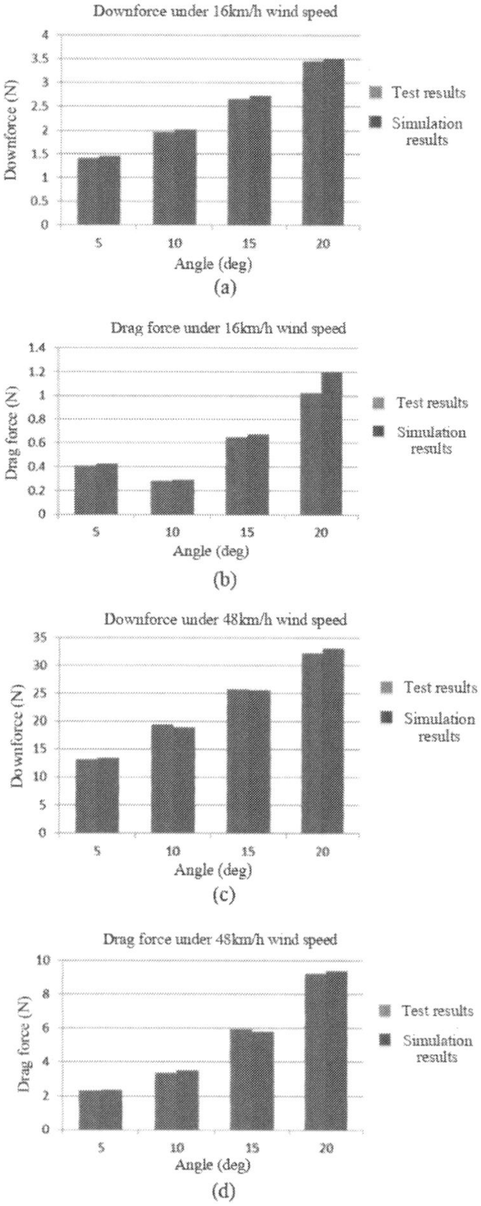

Figure 4. Comparison of wind tunnel test with simulation.

4 Aerodynamic parameter fitting and performance analysis

4.1 Aerodynamic coefficients calculation

The moving car's speed and direction is often inconsistent. In the analysis, discussing the crosswind for the car, the car in the steering process can be simplified to the wind blowing the car. The angle between the wind and the car is vehicle sideslip angle β. The body is fixed when solving the problem changing the direction of wind, and meanwhile simulates the cornering conditions. The vehicle coordinate system is same with flow field coordinate. X axis of flow field and X axis of vehicle has angle β as the same. Entrance velocity V of flow field is decomposed into windward speed V_a along the X axis and crosswind velocity V_w along the Y axis as shown in Figure 5.

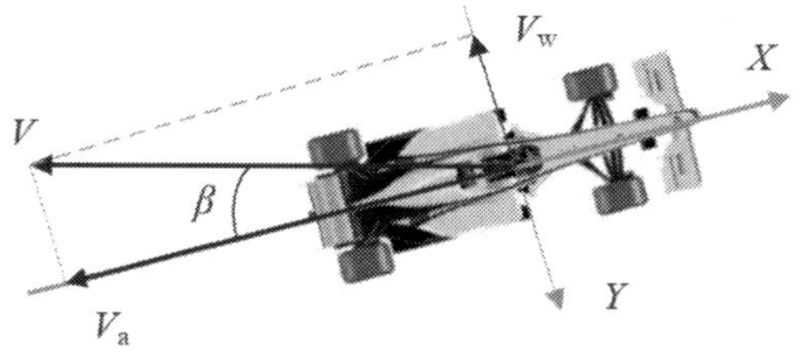

Figure 5. Diagram of racing car moving in bend.

This paper use CFD software STAR-CCM to compute the influence of aerodynamic coefficients by car's speed and sideslip angle quantitatively. Set gradient as 5km/h, speed increased from 25km/h to 60km/h. To set gradient as 2.5°, the sideslip angle increases from 0° to 20°. The corresponding set of crosswind velocity V_w and windward velocity V_a are in Table 1 and Table 2.

Table 1. Crosswind velocity values of different sideslip angles.

Crosswind velocity (km/h)		Sideslip angles (°)			
		2.5	5	7.5	10
Inlet velocity (km/h)	25	1.090	2.179	3.263	4.341
	30	1.309	2.615	3.916	5.209
	35	1.527	3.050	4.568	6.078
	40	1.745	3.486	5.221	6.946
	45	1.963	3.922	5.874	7.814
	50	2.181	4.358	6.526	8.682
	55	2.399	4.794	7.179	9.551
	60	2.617	5.229	7.832	10.419
Crosswind velocity (km/h)		Sideslip angles (°)			
		12.5	15	17.5	20
Inlet velocity (km/h)	25	5.411	6.470	7.518	8.551
	30	6.493	7.765	9.021	10.261
	35	7.575	9.059	10.525	11.971
	40	8.658	10.353	12.028	13.681
	45	9.740	11.647	13.532	15.391
	50	10.822	12.941	15.035	17.101
	55	11.904	14.235	16.539	18.811
	60	12.986	15.529	18.042	20.521

Table 2. Windward velocity values of different sideslip angles.

Windward velocity (km/h)		**Sideslip angles (°)**			
		2.5	5	7.5	10
Inlet velocity (km/h)	25	24.976	24.905	24.786	24.620
	30	29.971	29.886	29.743	29.544
	35	34.967	34.867	34.701	34.468
	40	39.962	39.848	39.658	39.392
	45	44.957	44.829	44.615	44.316
	50	49.952	49.810	49.572	49.240
	55	54.948	54.791	54.529	54.164
	60	59.943	59.772	59.487	59.088
Windward velocity (km/h)		**Sideslip angles (°)**			
		12.5	15	17.5	20
Inlet velocity (km/h)	25	24.407	24.148	23.843	23.492
	30	29.289	28.978	28.612	28.191
	35	34.170	33.807	33.380	32.889
	40	39.052	38.637	38.149	37.588
	45	43.933	43.467	42.917	42.286
	50	48.815	48.296	47.686	46.985
	55	53.696	53.126	52.454	51.683
	60	58.578	57.956	57.223	56.382

The relationship among aerodynamic force coefficient, speed and sideslip angle are shown in Figure 6. The relationship among aerodynamic moment coefficient, speed and sideslip angle are shown in Figure 7. When the sideslip angle is fixed, there is no correlation among aerodynamic force and moment coefficient and speed. When the speed is fixed, with the increase of sideslip angle, aerodynamic force and torque coefficients showed a certain change regulation. Therefore, aerodynamic coefficients were fitted in this paper. The relationship between aerodynamic coefficient and sideslip angle is represented by semi-empirical equations. Select the average speed for calculating aerodynamic coefficients. The aerodynamic coefficients as shown in Equation (2), six aerodynamic component forces fitting coefficients are shown in Table 3.

$$\begin{cases} C_L = C_{L1}\cdot\beta^3 + C_{L2}\cdot\beta^2 + C_{L3}\cdot\beta + C_{L4} \\ C_D = C_{D1}\cdot\beta^3 + C_{D2}\cdot\beta^2 + C_{D3}\cdot\beta + C_{D4} \\ C_S = C_{S1}\cdot\beta + C_{S2} \\ C_{YM} = C_{YM1}\cdot\beta^2 + C_{YM2}\cdot\beta + C_{YM3} \\ C_{PM} = C_{PM1}\cdot\beta^2 + C_{PM2}\cdot\beta + C_{PM3} \\ C_{RM} = C_{RM1}\cdot\beta + C_{RM2} \end{cases} \tag{2}$$

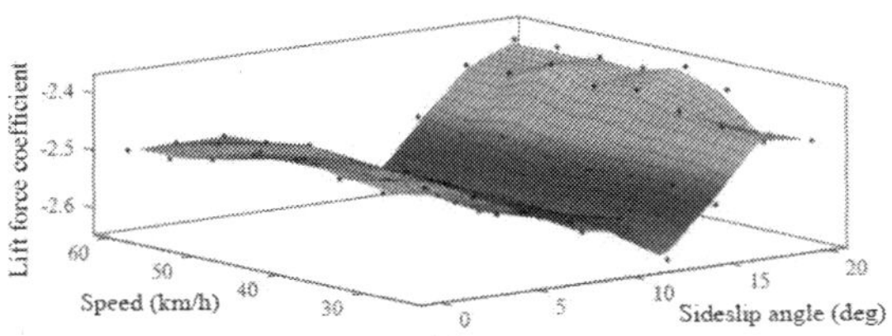

(a)

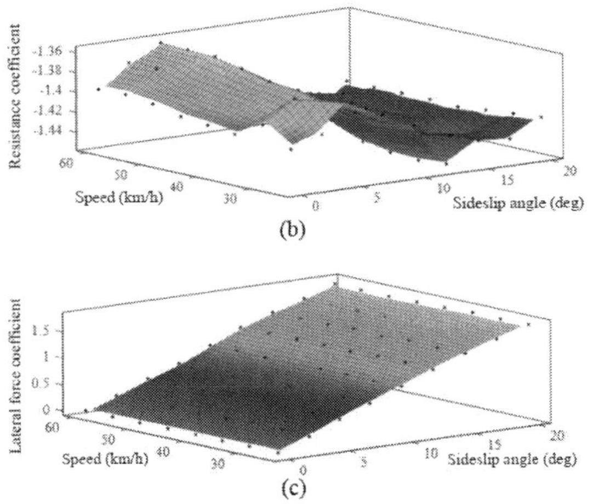

(c)

Figure 6. Relationship among aerodynamic force coefficient, speed and sideslip angle.

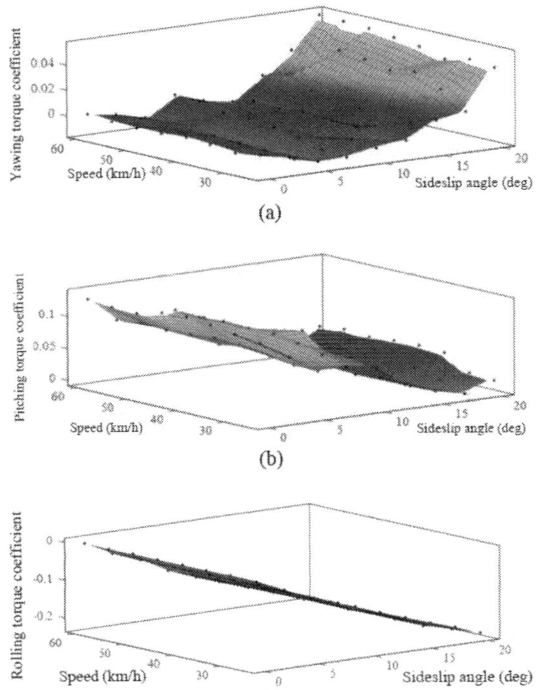

(c)

Figure 7. Relationship among aerodynamic torque coefficient, speed and sideslip angle.

Table 3. Windward velocity values of different sideslip angles.

C_{L1}	C_{L2}	C_{L3}	C_{L4}	C_{D1}	C_{D2}
15.04	-3.4	-0.418	-2.464	13.37	-7.064
C_{D3}	C_{D4}	C_{S1}	C_{S2}	C_{YM1}	C_{YM2}
0.7923	-1.404	5.103	0.02057	0.8002	-0.1573
C_{YM3}	C_{PM1}	C_{PM2}	C_{PM3}	C_{RM1}	C_{RM2}
0.002	0.7141	-0.5481	0.1252	-0.6358	-0.011

4.2 Aerodynamic forces calculation in bend

Data of this section are from the simulation results in previous section. Considering the driving path under different speeds, solve the change regulation of six aerodynamic component forces under the corresponding driving path. The car speed and the windward speed are the same, it can be seen that windward speed decreases slower from Table 3. It can be assumed that vehicle move at a constant speed in whole process. The vehicle route is selected in accordance with the windward speed and set the maximum allowable speed as windward speed. Road adhesion coefficient is selected 0.9. Corresponding curve radius R can be given by:

$$R = V_a^2 / \mu g \tag{3}$$

Corresponding driving time of side slip angle change from 0° to 20° can be calculated by:

$$T = \frac{\pi R}{6V_a} = \frac{\pi V_a}{6\mu g} \tag{4}$$

Sideslip angle β can be expressed as:

$$\beta = \frac{\pi}{6T} \cdot t \tag{5}$$

where t is driving time at a certain angle in cornering. Substituting β into Equation (2), coefficients variation with time can be obtained. So, aerodynamic forces and torques under 45km/h windward speed can be written as:

$$\begin{cases} F_D = 499.869t^3 - 374.298t^2 + 59.497t - 149.423 \\ F_L = 562.306t^3 - 180.155t^2 - 31.389t - 262.235 \\ F_S = 383.207t + 2.189 \\ T_{RM} = -73.050t - 1.791 \\ T_{PM} = 57.892t^2 - 62.947t + 20.387 \\ T_{YM} = 64.872t^2 - 18.073t + 0.126 \end{cases} \tag{6}$$

Aerodynamic forces and torques under windward speed of 45km/h are shown in Figure 8 and Figure 9.

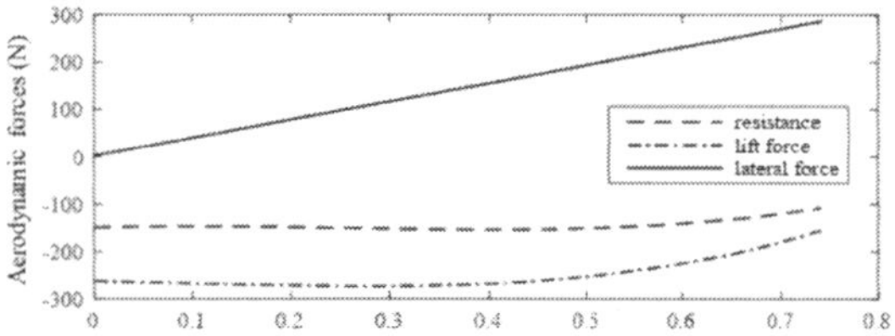

Figure 8. Aerodynamic forces under windward 45km/h.

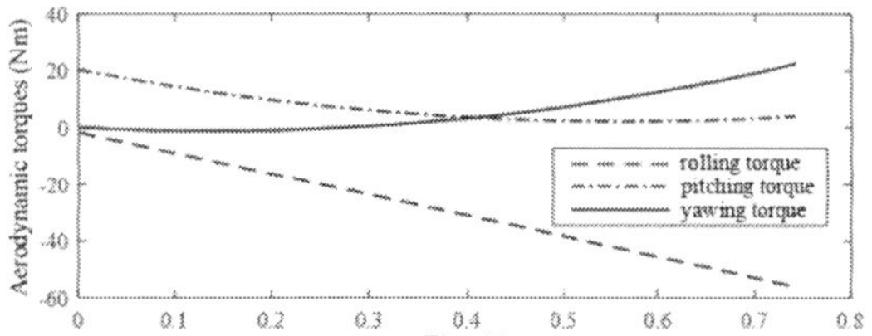

Figure 9. Aerodynamic torques under windward 45km/h.

5 Conclusion

This paper presents an effective method to study the aerodynamic characteristics of a racing car based on wind tunnel test and computational fluid dynamics simulation. Aerodynamic model of a racing car is established, and six aerodynamic component forces are defined. Wind tunnel experiment is completed, and accuracy of computational fluid dynamics simulation is verified based on the comparison between experimental and simulation results. Flow field model of racing car is established and the coefficients of six aerodynamic forces under transient conditions are fitted, and the aerodynamic equations are obtained. Finally, the aerodynamic forces and torques of the racing car travel in bend under 45km/h windward are calculated and analysed.

Reference

1. J. A. Mohrfeld-Halterman, M. Uddin, High fidelity quasi steady-state aerodynamic model effects on race vehicle performance predictions using multi-body simulation. Vehicle System Dynamics. **54**(7): 963-981, (2016).
2. J. D. Kee, J. H. Rho, et al., High speed driving stability of passenger car under crosswind effects. International Journal of Automotive Technology. **15**(5): 741-747, (2014).
3. Luke S Roberts1, Joao Correia, et al., Aerodynamic characteristics of a wing-and-flap configuration in ground effect and yaw. Journal of Automobile Engineering. **230**(6): 841-854, (2015)
4. A. Buljac, I. Džijan, et al., Automobile aerodynamics influenced by airfoil-shaped rear wing. International Journal of Automotive Technology. **17**(3): 377-385, (2016).
5. M. D. Soso, P. A. Wilson, Aerodynamics of a wing in ground effect in generic racing car wake flows. Proceedings of the Institution of Mechanical Engineers Part D-Journal of Automobile Engineering. **220**(D1): 1-13, (2006).
6. A. Altinisik, O. Yemenici, H. Umur, Aerodynamic Analysis of a Passenger Car at Yaw Angle and Two-Vehicle Platoon. Journal of Fluids Engineering-Transactions of the ASME. **137**(12): FE-15-1219, (2015).
7. A. Altinisik, Aerodynamic coastdown analysis of a passenger car for various configurations. International Journal of Automotive Technology, **18**(2): 245-254, (2007).
8. Ying-Chao Zhang, Jing Zhao, et al., Wind tunnel tests and aerodynamic numerical simulations of car opening windows. Nonlinear Dynamics, **58**(1): 62-78, (2012).
9. A. L. Smith, D. J. Mee, et al., Design, modelling and analysis of a six component force balance for hypervelocity wind tunnel testing. Computers & Structures, **79**(11): 1077-1088, (2001).

Numerical Investigation on Vortex-Induced Vibration Energy Extraction Efficiency of Double Circular Cylinders In Tandem Arrangement at Low Reynolds Number

Wu Junwu, Yin Zhongjun

School of Mechanical Engineering, University of Science and Technology Beijing, 100083, China

Abstract. Vortex shedding from a bluff body results in fluctuating forces acting on the bluff body, which may induce vibration of the bluff body when the bluff body is elastically mounted or deformable. Researchers put forward an idea that we can ex-tract energy from the water flow based on VIV at low flow velocity. Although plenty of researches on parameters of VIV are already presented, however, the improvement of energy extraction efficiency still needs further study. According to the previous research, this essay has simulated flow-induced vibration of tandem double circular cylinders when Reynolds number is 100. Working condition has been considered as the fixed upstream cylinder and the free vibration of the downstream cylinder. The influence of the mass coefficient and the two cylinders spacing ratio on the downstream cylinder's energy obtained from the fluid is studied. Analysis results show that, the maximum value of the energy extraction efficiency is before the frequency locked range. In the case of large spacing ratio (L/D=7~9), the phenomenon of "beat vibration" appears on the downstream cylinder. The results of this work could provide reference for the improvement of energy extraction efficiency and the design of VIV converter.

1 Introduction

Vortex shedding from a bluff body results in fluctuating forces acting on the bluff body, which may induce vibrations of the bluff body when the bluff body is elastically mounted or deformable [1]. Vibrations that induced by vortex shedding and wake instability widely exist in nature and engineering field, such as cable bridges, marine risers and ocean platforms [2], and they may destroy the integrity and reliability of structures. However, based on vortex-induced vibration (VIV), renewable energy can be harvested from the flow water at low velocity, which has caused abroad attention in recent years. Bernitsas proposed a single cylinder converter which harvesting energy from water flow based on VIV in the Marine Renewable Energy Laboratory (MRELab) at the University of Michigan [3, 4]. The effects of high damping and high Reynolds number were researched, and the highest efficiency of energy con-version and the corresponding damping value were found (the converter was at a speed between 0.4m/s and 1.1m/s and the spring stiffness between 400N/m and 1800N/m). A.Barrero-Gil [5] measured the forces acting on the cylinder and the vibration response of single cylinder in crossflow, and the effects of mass ratio and damping ratio on the energy extraction efficiency were dis-cussed. It was found that there is a maximum efficiency when $m^*\xi$ is 0.25. Moreover, the effects of parameters like m^*, ξ, $m^*\xi$ and nature frequency were studied by 2D simulation [6]. It was found that the vibration amplitude could increase when double circular cylinders in tandem arrangement ac-cording to experiment and numerical simulation [7, 8].

Now there are many researches on the influence parameters of VIV energy extraction of single cylinder based on experimental investigation and numerical simulation. However, it is hard to obtain expression of vibration response of double circular VIV systems because the vibration response is not a single sinusoidal variation and the improvement of energy extraction efficiency still needs further study. In this paper, it focuses on two circular cylinders oscillating at low Reynold numbers by numerical simulation method. The influences of mass ratio and spacing ratio on the energy extraction efficiency were discussed, and the working condition has been considered as the fixed upstream cylinder and the free vibration of the downstream cylinder [9].

2 Mathematical modeling and numerical simulations

The numerical simulation of the governing equations of fluid-structure interaction with a freely oscillating cylinder is hard to obtain, because the boundary conditions change as the cylinder moves. The fluid flow is governed by 2-D incompressible Navier-Stokes equations in terms of primitive variables, which can be expressed as follows:

$$\frac{\partial u}{\partial t} = -(\boldsymbol{u}\cdot\nabla)\boldsymbol{u} - \frac{1}{\rho}\nabla p + \nu\nabla^2\boldsymbol{u} + f \qquad (1)$$

$$\nabla \cdot \boldsymbol{u} = 0 \qquad (2)$$

Where $\boldsymbol{u}$ is the velocity vector, t is the time, p is the pressure, ρ is density for fluid, ν is the kinematic viscosity, ∇ is the gradient operator, and f is the additional volume force vector, which represents the fluid-structure coupled boundary conditions. Laminar model and SIMPLEC methods was adopted.

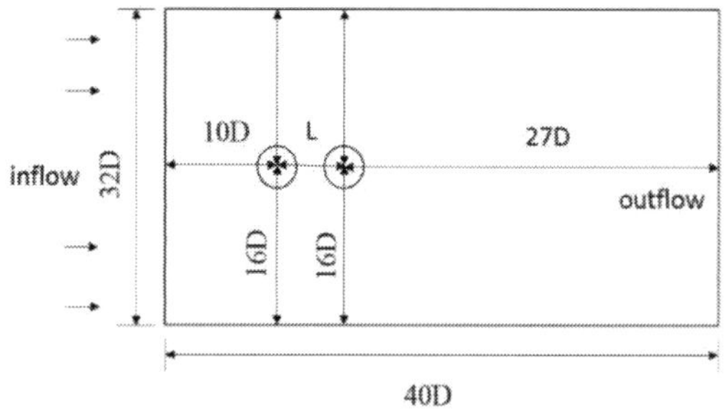

Figure 1. Computational domain in present simulation.

As shown in Figure 1, the sizes of the computational domain are $40D \times 32D$, and double circular cylinders are placed in tandem arrangement (the upstream cylinder is fixed and the downstream cylinder single is free in cross flow). Where D is the diameter of cylinder, L is the distance between two cylinders.

For the free vibration of the downstream cylinder, the equation of motion for an elastically mounted cylinder of mass m can be defined as

$$m\ddot{y} + c\dot{y} + ky = F_y(t) \qquad (3)$$

Where F_y is provided by the fluid pressure and viscous forces, y is the direction normal to the flow, $\dot{y}$ and $\ddot{y}$ are the velocity and acceleration of the cylinder. To obtain the vibration response of the downstream cylinder, Newmark-β was used to solve the Equation 3.

The ratio of harnessed power to the power available in the fluid is

$$\eta = \frac{P_{harness}}{P_{fluid}} = \frac{\frac{1}{2}U_\infty^3}{\frac{1}{n\Delta t}\sum_{t}^{t+n\Delta t} F_y \dot{y} \Delta t} \qquad (4)$$

Where η is the energy conversion efficiency, $P_{harness}$ is the harnessed power, P_{fluid} is the total power in fluid, Δt is the time step.

3 Results and discussions

In this study, the effects of mass ratio and spacing ratio were investigated, and the Reynolds number is 100. The values of amplitude and the energy conversion of cylinders that placed in tandem arrangement were examined. The working conditions were based on references, in order to verify the numerical simulations results. As it can be seen from Figure 2, the amplitude of the downstream cylinder increases at first and then decreases with increasing Ur (reduced velocity), the trend is in agreement with the references [7, 10].

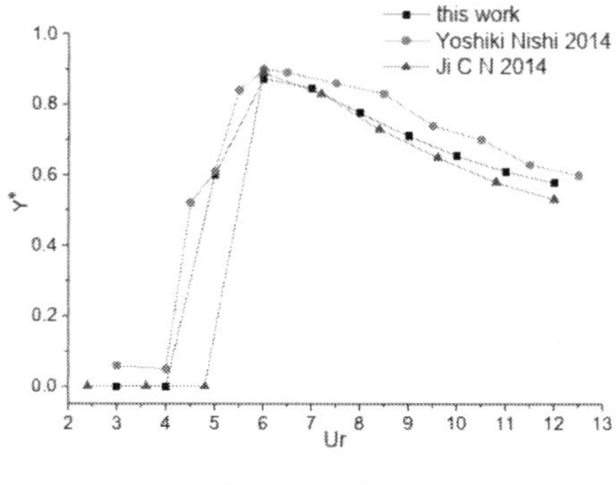

Figure 2. Variation of the amplitude of the downstream cylinder with the reduced velocity When the cylinder spacing ratio is L/ D=3.

3.1 Mass coefficient

The conditions are listed in Table 1, and the vibration response and forces are shown in Figure 3 and Figure 4. The efficiency is related with F_y and y . Figure 3 shows the Y* versus Ur for different m*, when m* is fixed, the amplitude of the downstream cylinder increases at first and then decreases with increasing Ur. Y* decreases with the increasing of mass coefficient, when Ur is fixed. When Ur is 6~8, Y* peaks at approximately 0.85. The vibration was strong in this condition, but Lift coefficient dropped to a low value. The variation of lift coefficient is presented in Figure 4. The Lift coefficient increases to the peak with the increasing of Ur, and then drop fast, finally increases slowly, and the Lift coefficient decreases with the increasing of mass coefficient with fixed Ur. Those cause peak values of the efficiency when Ur are 5and 8, and the minimum value when Ur =6.

Table 1. Numerical simulation scheme

Case	Fixed parameter	Variable parameter
1	L/D=3, ζ=0	m*=1, m*=2, m*=4
2	m*=2, Ur=6, ζ=0	L/D=6~9

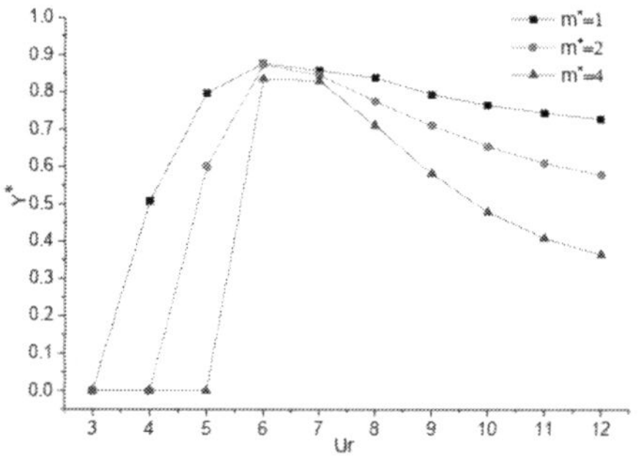

Figure 3. Variation of dimensionless vibration amplitude with the reduced velocity in case 1

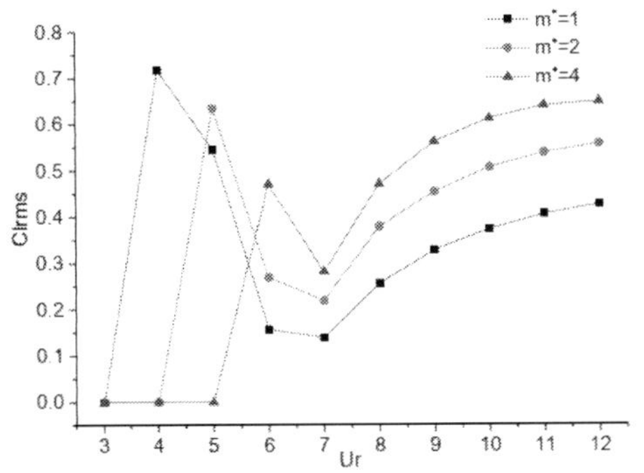

Figure 4. Variation of Lift coefficient root mean square with the reduced velocity in case 1.

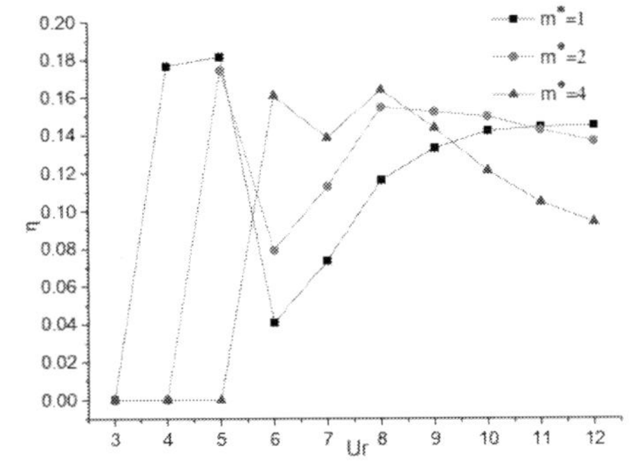

Figure 5. Variation of energy acquisition efficiency with the reduced velocity in case 1.

3.2 Space ratio

The influence of the space ratio is investigated at this part. The conditions are listed in Table 2. Figure 6 shows that the value of efficiency reaches to the maximum when space ratio L/D=4. When L/ D=2~4, the Y*and η increase with the increasing of L/D, and when L/D=4~6, though the value of Y* is still high, the efficiency decreases as a result of the sharply decrease of lift coefficient. While L/D is greater than 6, there is another peak of efficiency, then the efficiency deceases. In all, a higher Y* can be obtained when the cylinders are placed in tandem arrangement, and the values of the energy extraction efficiency have two maximums when L/D=4 or 7.

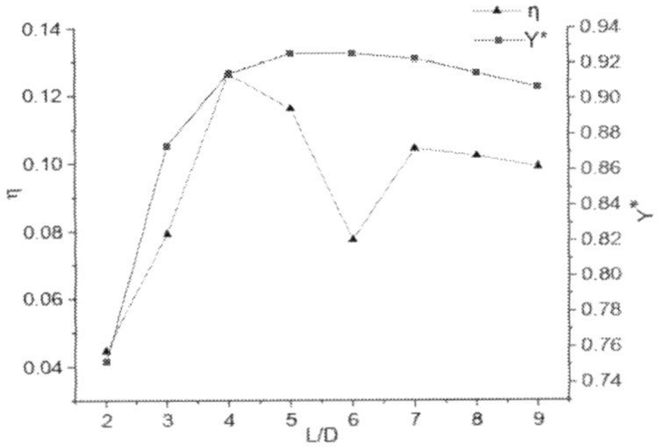

Figure 6. Variation of energy acquisition efficiency and dimensionless vibration amplitude with the space ratio in case 2

In addition, the harnessed power curve of the downstream cylinder when the double circular cylinders that placed in tandem arrangement is not sinusoidal. The phenomenon of "beat vibration" appears on the downstream cylinder when the spacing ratio is large (L/D=7~9).

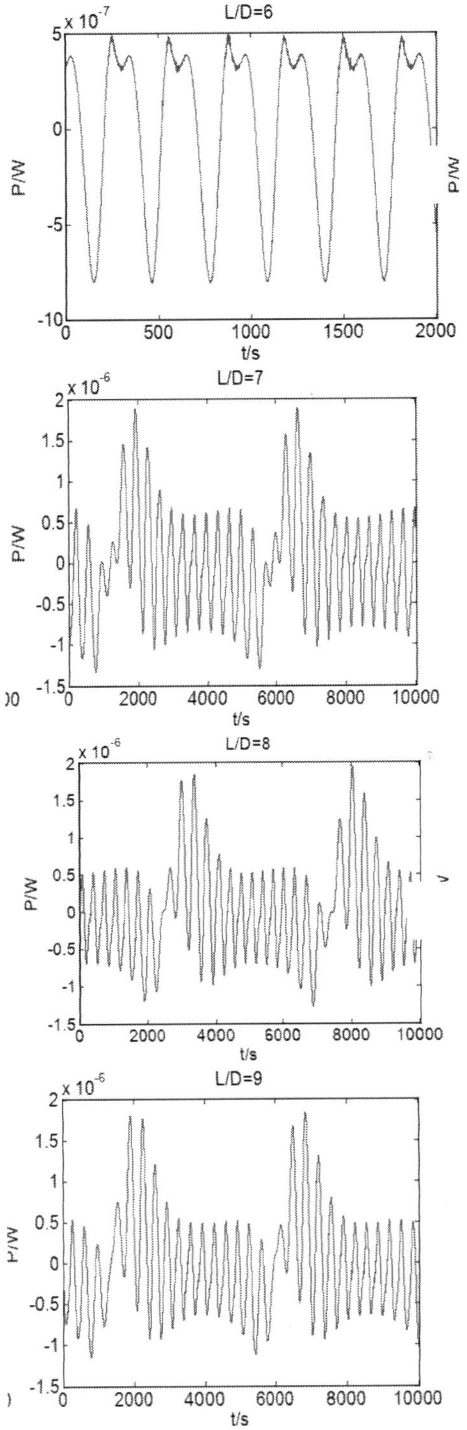

Figure 7. The harnessed power of case 2.

4 Conclusion

In this work, the influence of the mass coefficient and the spacing ratio of two cylinders on the energy extraction efficiency were discussed, and the following conclusions are obtained:

• The value of the energy extraction efficiency reached maximum when the Ur is 6 or 8, and the optimum spacing ratio is 4 in this work condition.

• The instantaneous power curve of the downstream cylinder when the double circular cylinders that placed in tandem arrangement is not sinusoidal. The phenomenon of "beat vibration" appears on the downstream cylinder when the spacing ratio is large.

References

1. P. Bearman, J FLUID STRUCT **27** (2011).
2. Z. Han, D. Zhou, T. He, J. Tu, C. Li, K.C. Kwok, C. Fang, OCEAN ENG **96** (2015).
3. M.M. Bernitsas, K. Raghavan, Y. Ben-Simon, E. Garcia, J OFFSHORE MECH ARCT **130** (2008).
4. J. Lee, M. Bernitsas, OCEAN ENG **38** (2011).
5. A. Barrero-Gil, G. Alonso, A. Sanz-Andres, J SOUND VIB 329 (2010).
6. Z.M. Luo, L.X. Zhang, Journal of Vibration & Shock **33** (2014).
7. Y. Nishi, Y. Ueno, M. Nishio, L.A.R. Quadrante, K. Kokubun, J SOUND VIB **333**(2014).
8. B.S. Carmo, S.J. Sherwin, P.W. Bearman, R.H.J. Willden, J FLUID STRUCT **27** (2011).
9. G.R.S. Assi, J FLUID STRUCT **50** (2014).
10. C. Ji, W. Chen, J. Huang, W. Xu, Chinese Journal of Theoretical & Applied Mechanics **46** (2014)

Experiment and simulation study on relationship between diameter of high-pressure water abrasive jet nozzle and cutting capacity

Zhang Yu Xian[1], Li Nan[1], Wang Hong[1], Liu Bin Bin[1]

[1]*Guilin University of Electronic Technology, Guilin. Guangxi, China*

Abstract: High pressure water abrasive jet nozzle is key part of the particles acceleration. The diameter of nozzle has important effect for jet fluid. Through experiment and simulation explore the relationship between nozzle diameter and jet fluid cutting capacity. The result indicate: ①the cut depth and broad reduced in linearly relationship with the diameter of the jet nozzle . ②In same pressure ,decrease the nozzle diameter will reduce the cutting ratio energy ;③The acceleration and attenuation of the nozzle axial flow in different diameters are basically consistent, The smaller nozzle diameter, The flow acceleration will slower ,the attenuation will faster, the is velocity core segment will shorter and the cutting ability will lower;Consider from energy consumption, cutting efficiency and other factors, for common material cutting the nozzle preferred diameter is: 0.6~1.0mm, it's unfavourable to select the 0.2mm diameter

1 Introduction

High pressure water jet technology is a new technique in latest half century, which broad used in cutting, broken and washing. The technique separate with pure water jet and abrasive jet .The nozzle is generation part of AWJ, it's important in abrasive particle acceleration. The nozzle's diameter has deep effect on Jet power and jet machining quality. By experiment and simulation, the article researches the relationship between nozzle diameter and cutting capacity.

2 The experiment facility and condition

2.1 experiment facility

The experiment use tiny nozzle pro-mix high pressure jet cutting system.(*shown in picture 1*). The jet float system consist of high pressure pump, abrasive kettle, abrasive jet nozzle ,overflow valve .Throttle and high pressure flexible tube. High pressure water from pump 1separates two aside, one flows into relief valve7, 8 and another one flows to part 5. The abrasives through the bottom throttle 9 then flow into mixing cavity10 and mixture with high pressure water that come from throttle 8.The mixed liquor jet through abrasive nozzle 12.By control the nozzle or work plant can realize cutting workpiece.

When the abrasive jet is required could open the throttle 9, when only use the pure water cut is used to close it. The combined action of 5, 7 and 8 can control the abrasive concentration in the jet stream. When the abradant is needed reduce, close 5, 7 part.

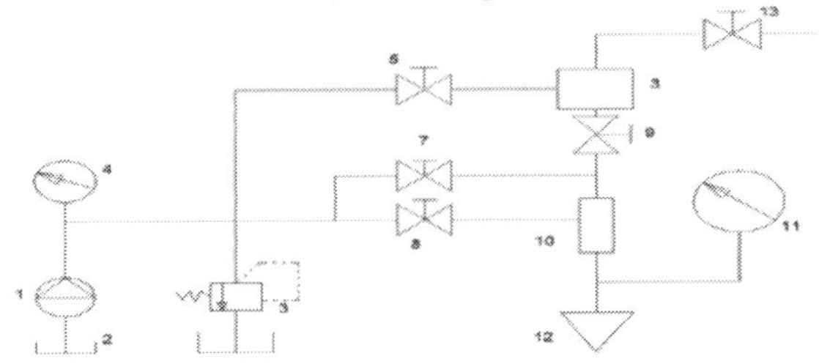

1. bump;
2. water-tank;
3. overflow valve;
4, 11. Piezometer;
5, 7, 8, 9, 13. Throttle;
6. abrasive-tank;
10. .mix-cavity;
12. Abrasive jet nozzle

Figure 1. The front mixing abrasive jet system of small nozzles

Under the static water pressure and self-weight effect the abradant arrive mix-cavity 10.If throttle 7 opening enlarged and the throttle 8 opening reduced, the abrasive concentration will increase accordingly. Throttle 5 is used to adjust the abrasive concentration of the jet flow to meet the cutting requirements. The throttle 13 has exhaust effects. Overflow valve 3 is used to control the working pressure of the system.

2.2 The experimental materials

In this experiment, a variety of brittle and plastic materials were cut, including marble, granite, aluminum and copper plate.

2.3 The experimental nozzle

A variety of taper nozzle were used (chart 2), the nozzle entrance is transitioned by arc. Conic section shrinkage angle for 15~13, nozzle material is hard alloy, the inner hole surface has the coating, Internal surface quality of nozzle Ra is 0.8 μm. The nozzle diameter has 0.2, 0.6, 1.0, 1.4, 2.0 mm, etc.

Figure 2. Various experimental nozzles

3 Experiment and analysis of jet cutting

3.1 The influence of nozzle diameter on cutting depth

Figure 3 (a) is the relationship about five different diameter nozzles and correspond cutting depth. The experiment cutting pressure is 15 MPa.

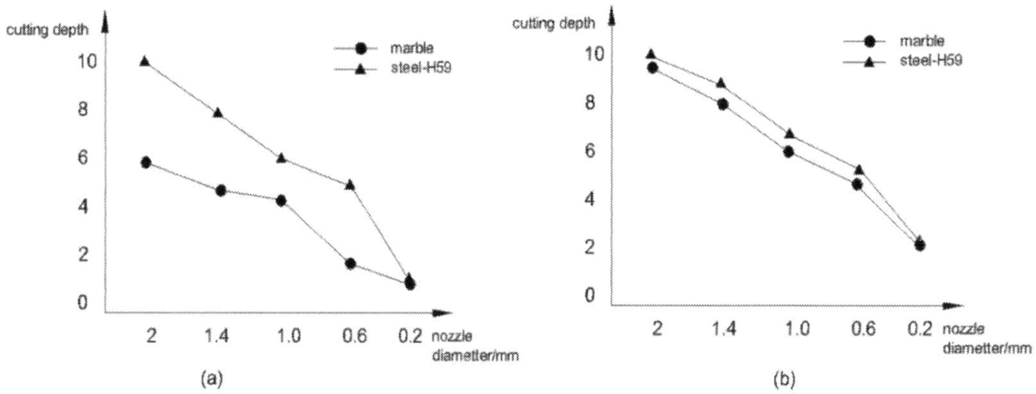

Figure 3. The incision depth and width of the specimen changed by the nozzle diameter variation

It can be seen that the diameter of nozzle has a significant influence on cutting ability. The cutting depth of marble and brass follow the nozzle diameter changed present similar linearly decreases.

The nozzle Ø0.6mm compared with Ø2mm, its cutting ability has decreased observably. When cutting marble, the cutting depth of the former is about 1/3 of the latter, when cutting the copper plate, it is only 1/6.When the nozzle diameter is less than 0.2mm, the cutting ability of the jet is very low. Therefore, to achieve effective cutting of the Ø2mm nozzle, the driving pressure of the system must be further improved.

3.2 Influence of nozzle diameter on cutting width

A single factor correlation between nozzle diameter and incision width of specimen is shown in FIG.3b, the width of specimen incision decreases linearly with the decrease of nozzle diameter basically. Through test, The incision width of specimen is larger than jet diameter. When nozzle diameter d≥0.6mm.The incision width of the specimen is 1.08~ 1.27 multiple the jet nozzle diameter; When nozzle diameter d≤0.6mm, the incision width of the specimen is 1.33~ 2.5 multiple the jet nozzle. When d=0.2mm the multiple get maximum.

The abrasive water jet cutting material mainly depends on the grinding effect of abrasive particles on the material. On the one hand, the abrasive particle is discontinuous in the water jet. High speed particles composed of abrasive particles also have high frequency impact on materials. The workpiece material is gradually destroyed by the high frequency shock and grinding of high energy abrasive particles. When the nozzle diameter is reduced to 0.2mm, the boundary layer of jet flow increases the influence of particles. As the complexity of the boundary flow, the increase of the width of the specimen incision is made, another hand, due to the continuity of the specimen material, the particles needed to produce enough energy can tear apart the connections between the specimens. Therefore, when d≤0.2mm, through decrease the nozzle diameter to reduce the incision width and improve the stock utilization is meaningless.

3.3 Cutting ratio energy consumption

Cutting ratio energy consumption is the energy consumed by cutting or crushing unit volume workpiece material, It is the comprehensive quality index of water jet technology, and reflects the rationality and technical level of the system. Its expression is:

$$E = \frac{\pi \sqrt{\frac{2}{\rho}} p^{\frac{3}{2}} d^2}{4bHv} \tag{1}$$

b is the cutting depth, H is the incision depth, v is the value.

The cutting ratio energy consumption of different diameter in simple pressure:

$$E\frac{E_2}{E_1} = \frac{H_1}{H_2}\frac{b_1}{b_2}[\frac{d_2}{d_1}]^2 \tag{2}$$

The table 1 shows two different diameter nozzle's cutting ratio energy consumption (cutting material: marble)

Table 1. The cutting ratio energy about Ø0.6mm and Ø1mm diameter nozzle E2/E1

Pressure(MPa)	5	10	15	20	25
E2/E1	0.668	0.611	0.827	0.887	0.666

As the pressure increases, the jet cutting ration energy consumption shows decrease tendency. From table 1, in the same pressure, as this nozzle diameter decrease, the ratio of cutting energy consumption is also reducing. In the experiment, the work stress is 5~20MPa.The cutting ratio energy consumption of Ø0.6mm jet nozzle compared with Ø0.1mm is lower than 11.3%~38.9%.

4 Simulation and analysis of jet nozzle

4.1 The simulation conditions

4.1.1 The geometric model

The most critical parts of the efflux are the pipes, nozzles, and the cut workpiece, and the key is nozzle. The Internal surface shape and roughness for the pipe and nozzle has great effect on jet-flow. The geometry parameter in the simulated of this paper have: pipe diameter 8mm,nozzle diameter 2mm, 1mm, 0.6mm, 0.2mm etc., the target distance 5mm.

4.1.2 Fluid condition and numerical simulation environment

Water: incompressible flow, except boundary layer, flow state is turbulent. Grinding material: Non-crushing solid, density 2600 kg/m^3. Air: room temperature gas. The CFD package is fluent.

4.1.3 Jet physics model

The abrasive jet is actually a three-phase jet couple constitutes with water, air and abrasive. Due to the complexity of the problem, this paper only considers the flow of water and air. The abrasive particles are added to the flow field in discrete phase.

Continuity equation:

$$\frac{\partial a_q}{\partial_t} + \vec{v} \bullet \nabla a_a = 0 \tag{3}$$

V is speed, a_q is the volume ratio of q phase fluid. The equation satisfies the constraint $\sum_{q=1}^{2} a_q = 1$

The momentum equation of the two phases of water and air is:

$$\frac{\partial}{\partial_t}(\vec{\rho}_v) + \nabla□(\vec{\rho}_v + \vec{v}) = -\nabla p + \nabla□[\mu(\nabla\vec{v} + \nabla\vec{v}T)] + \vec{\rho}g + \vec{F} \tag{4}$$

P, μ, F respectively are average density, viscosity and unit mass of the two-phase fluid.

The turbulent kinetic energy equation (5) of RNG k-ε model and the diffusion equation (6):

$$\frac{\partial}{\partial_t}(\rho k)+\frac{\partial}{\partial_t}(\rho k u_i)=\frac{\partial}{\partial_{xj}}(\alpha_k \mu_{eff}\frac{\partial_k}{\partial x})+G_k+G_b-\rho\varepsilon-Y_M+S_k \quad (5)$$

$$\frac{\partial}{\partial_t}(\rho k)+\frac{\partial}{\partial_{x_i}}(\rho\varepsilon u_i)=\frac{\partial}{\partial_{xj}}(\alpha_k \mu_{eff}\frac{\partial_\varepsilon}{\partial x_j})+G_\varepsilon\frac{\varepsilon}{K}(G_k+C_{3\varepsilon}G_b)-C_{2\varepsilon}\rho\frac{\varepsilon^2}{k}-R_\varepsilon+S_\varepsilon \quad (6)$$

In the equation, G_k represents the turbulent energy generated by laminar velocity gradient, Gb is turbulent energy produced by buoyancy, Y_M generated by the fluctuations in the transition of compressible turbulence,C1,C2,C3 Is the model constant, α_k and α_ε is the turbulent Prandtl number of the equation k and equation ε, μ_{eff} is effective viscosity, Sk and Sε is the effective parameter. The simulation condition is basically consistent with the above experimental conditions.

4.2 Simulation results and analysis

Start simulation after modeling, A model simulation is about spend 6~ 32 h, FIG. 4 is a CDF simulation diagram shows the axial velocity variation of four different nozzle jet diameters .Through figure known The acceleration and attenuation of the flow axis of different nozzle diameters are basically consistent. The smaller of the nozzle diameter, the jet acceleration is slower, the highest speed is also smaller, and the attenuation is faster, this is o-velocity core is shorter .When the nozzle diameter is reduced to 0.2mm.The isokinetic segment has completely disappeared.

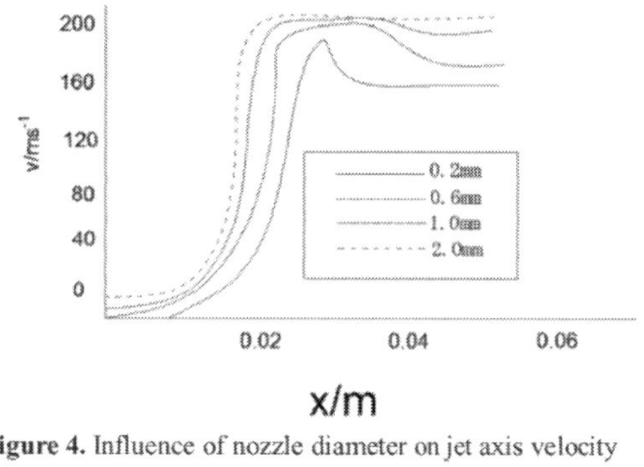

Figure 4. Influence of nozzle diameter on jet axis velocity

To better understand the relationship between jet cutting ability and nozzle diameter.FIG.5 shows the dimensionless dynamic pressure diagram of different diameters with different distance. The x-coordinate is the inverse coordinate, using 2 mm nozzle diameter for length scale, the vertical coordinate adopts 2mm nozzle maximum outlet dynamic pressure as ratio ruler. The target distance is L = 0mm, 5mm, 10mm, 20mm. The analysis as follows:

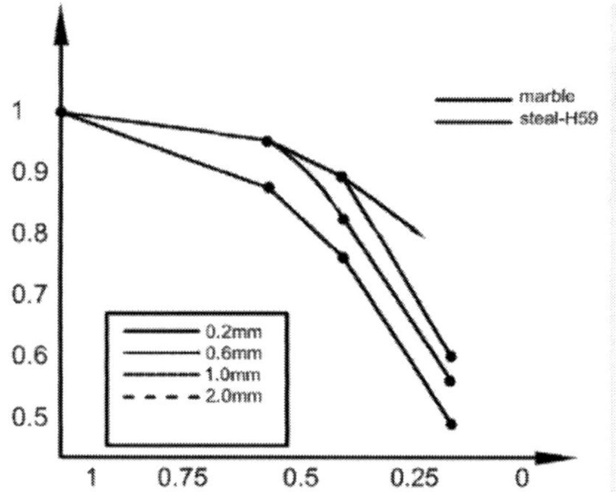

Figure 5. The influence of nozzle diameter on outlet dynamic pressure

If the target distance not changed, the steeper of the curve shows the greater influence of the dynamic pressure on the nozzle diameter. When the nozzle scale among1~ 0. 5mm (the nozzle diameter is 2~ 1 mm),the curve flattens out, when the target is increased to 10 mm, The dynamic pressure ratio of efflux still maintains 0.98.When the nozzle diameter among 0. 6~ 0.2 mm, the curve section more steeper, the dynamic pressure drops quickly more. At 5 mm target distance, 0.2mm nozzle dynamic pressure ratio is just 0.65.Cutting ability significantly decreased. At the nozzle exit, the dynamic pressure ratio of the 2 mm nozzle is 0.88; therefore, the best target distance is within 5 mm. When the target distance is more than 10 mm, the dynamic pressure ratio of various nozzles is obviously reduced.

When we want the dynamic pressure ratio reaches 0.90 in 5 mm distance, the nozzle diameter should above 0.28 *0.2 mm=0.56mm(using linear interpolation).If the ratio of dynamic pressure reached 0.80,the nozzle diameter is above 0.18 *0.2mm=36mm approximately.Experiments indicate, using 0.6mm nozzle. In the condition of 25M Pa, the plastic and brittle materials such as copper, aluminum and marble can be effectively cut. Therefore, to achieve effective cutting of above material by 0.5mm nozzle, the driving pressure of the system must be increased to more than 25x1/0.86≈30MPa.

5 Conclusions

The following conclusions can be drawn from the study of cutting experiment and simulation in different nozzle diameters:

(1) The influence of nozzle diameter on cutting ability: The depth and width of the incision are reduced linearly

by the nozzle diameter .The incision wide of the specimen is larger than the diameter of the efflux, and smaller the diameter is, the more obvious the trend is. When the nozzle diameter is less than 0.2 mm, the jet cutting ability is very low.

(2) The acceleration and attenuation of the different nozzle diameters axial flow are basically consistent. The smaller the nozzle diameter, the jet acceleration will slower, the highest speed is also smaller, but the faster the attenuation, the isovelocity core section get shorter, and the cutting ability get reduced. The best target distance is within 5 mm ranges. To achieve the effective cutting of the small nozzle (0. 2~ 0.5mm), the driving pressure of the system must be above 30~ 38 MPa.

(3) Consider the energy consumption, the processed materials saved, cutting efficiency and equipment investment and other factors. The better choice of cutting ordinary material is 0.6~1.0mm nozzle diameter. The 0.2mm nozzle is not suitable; under the same pressure, reducing nozzle diameter will decrease the cutting ratio of the jet-flow.

Reference

1. WANG J X, ZHANG S X. Experimental study on abrasive jet cutting of small nozzles [J]. Fluid machinery St. 2007, 35 (9) : 9- 11.
2. ZHANG B Z, YIN J A, ZHAGN H J. Numerical method of fluid mechanics[M].BEI JING: Mechanical industry press, 2003
3. Li Yang. Numerical simulation of fluid jet in pingement on a wall [A]. Pro jet Report for M E608 Mechanical Engineering [C]. Indiana USA: Purdue University, 2004.
4. LIU S D, HU D, TANG C L, etc. The numerical analysis of the pre-submerged abrasive jet [J].Journal of Hunan university of technology, 2008, 22 (4): 102- 104, 108.
5. ZHANG S X, ZHANG Y X, WANG J X. Experimental study of abrasive water jet particle motion simulation and cutting experiment[J] Mining machinery, 2008, 36(484): 32-
6. Wang J S, ZHANG S X. Experimental studies on cutting performance of a fine abrasive water jet [A]. Advanced Design and Manufacturing for Sustainable Development [C]. Sydney Australia: ICFDM, 2006.

MATEC Web of Conferences **153**, 05003 (2018) https://doi.org/10.1051/matecconf/201815305003
ICMME 2017

Study on Side-entering Agitator Flow Field Simulation in Large Scale Biogas Digester

Li Xinxin[1], Cao Yadong[1], Huang Zhenfeng[1], Liang Jingfu[2], Cheng Gang[2], and Pan Rui[2]

[1]*School of Mechanical Engineering, Guangxi University, China 530004*
[2]*Guangxi Bossco Environmental Protection Technology Co.,Ltd, China 530004*

Abstract. Based on the CFD software Fluent, the multi-reference system method, the RNG κ-ε turbulence model and the pressure-velocity coupled SIMPLEC algorithm are applied to simulate the flow field of a single side-entering agitator in a large biogas digester. And the impact of different installation angle, agitator-to-bottom height, agitating speed and agitator diameter on the agitating power, the effective area percentage and the effective power ratio are analyzed. The results show that: 1) when the agitating speed is more than 450r/min and the agitating diameter is more than 750 mm, the effective area percentage reaches the maximum value; 2) when the horizontal angle is 30 °, the agitating effect is the best; 3) when the vertical angle is inclined to the bottom and the agitator-to-bottom height is less than 8 m, it is helpful to prevent the sediment from appearing. The results of numerical simulation provide a theoretical basis for the design to optimize the agitator.

1 Introduction

With the rapid development of computational fluid dynamics (CFD) in recent years, some scholars at home and abroad have applied computational fluid dynamics technology to the optimization design of mixing tank through numerical simulation. At present, the numerical simulation of vertical agitation is common, but not the research on the side-entering agitator [1-3]. Wesselingh[4] studied the mixing time of different size single-side agitator, and analysed flow field with multiple factors such as paddle type, propeller angle, Reynolds coefficiency, and so on. The simulation results are in agreement with the experimental results. Asghar[5] studied the effect of the different factors on mixing time of two kinds of crude oil with one side-entering agitator through RNG turbulence model. Fang Jian[6-7] carried out the numerical simulation of the fluid flow and mixing process in the tank with 4 side-entering agitators, and the power curve was obtained. Also, the influence of different tracer feeding point, monitoring point position and agitator shaft deflection angle on the mixing time were studied. Zhang Linjin[8] used computational fluid dynamics (CFD) technology to simulate the side-entering agitator flow filed in the bottom of the flue gas desulfurization absorption tower, the influence of the agitating speed and agitating paddle installation angle and other factors on the three-dimensional flow field were studied. Zheng Xiaodong[9] studied the effect of the installation position, the ventilation rate, the solid particle concentration and the liquid level on the particle suspension performance in the side-entering mixing tank. Chen Jia[10] uses the computational fluid dynamics technique to calculate the mesoscopic macroscopic flow field in the large-scale side-entering agitated tank with the diameter and height of 13 m, and the effects of different agitating speed, agitating angle and number on the distribution of low dead zone in the kettle were investigated at the same time. In summary, there are some research reported on the side entering agitator, but a thorough study on single agitator is still needed, which should analysis the parameters of agitating system and provide the base for optimization design.

In this paper, a simplified model of the biogas digester with a single side-entering agitator is built and applied to simulate flow field. With the help of CFD software Fluent, the impact of horizontal and vertical installation angles, height from tank bottom, agitator speed and diameter on power consumption and effective area are studies, which provided reference for the optimization design of side entering agitator and its practical application.

2 Model and Simulations

2.1 Model

The calculation model of this paper is based on a simplified large-scale biogas mixing tank designed by one of the authors' cooperative company. The cylinder tank is with 16m diameter and 16m high. For the single side entry agitator, the angle of the agitator between the XZ plane and the negative direction of the X axis is set to horizontal angle α, and the positive direction of the Y axis is defined as positive. The angle of the agitator between the XY plane and the negative direction of the X axis is set to the vertical angle β, and the negative direction of the Z axis is defined as negative. The distance between the agitator's agitating shaft and the bottom of mixing tank is defined as the agitator-to-bottom height h, as shown in Fig. 1. The agitator type uses the industry standard three-leaf right-handed propeller(HG/T 3796.8—2005), and the simplified three-dimensional model of the agitator is shown in Fig. 2. The shaft length of the agitator inserted into the tank is fixed at 2.8 m as required.

In order to simplify the calculation, it is assumed that the mixing tank is a closed cell body filled with liquid, and no feed port and waste discharge port are provided. The three-dimensional physical model of numerical simulation is established by UG, and then, imported into Fluent, preparing for simulations.

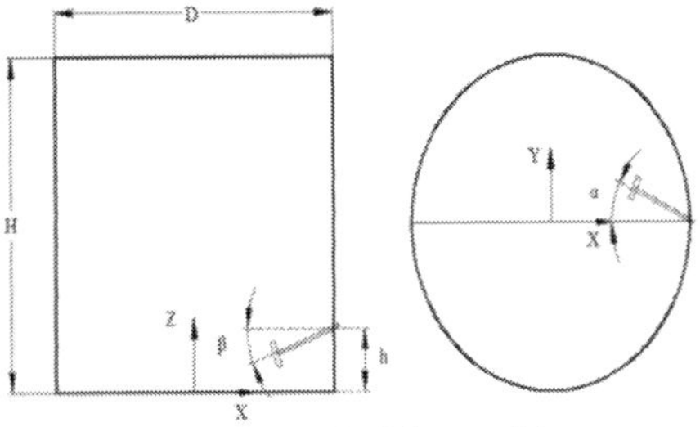

Figure 1. Schematic diagram of mixing model

Figure 2. Three dimensional model of propeller agitator

2.2 Simulations

2.2.1 Design experiment

There are several main factors which are most important for the agitating power and the agitating effect of the side-entering agitator, such as viscosity of the liquid in the mixing tank, agitating diameter, agitating speed, agitator-to-bottom height, and angle between agitating shaft and tank. In this paper, the solid content in in the anaerobic tank is about 10%, and the temperature in the tank is about 60 ℃. According to Liu Yi[11]'s results, it can be seen that the viscosity of the anaerobic tank is about 0.05 Pa*s, which belongs to the low viscosity liquid. Hence, compared with other factors, the viscosity impact on agitating flow field and agitating power would be negligible. As a result, five simulation parameters, horizontal angle α, vertical angle β, agitator-to-bottom height h, agitating speed n, and agitating diameter d, are selected to analyse, and their impacts on flow field and agitating power are simulated separately. Table 1 lists the simulation values for each single factor.

Table 1. Design of test parameters

Experimental group	Test parameters	
1	Horizontal angle α / °	0、5、10、15、20、25、30、35、40
	Fixed value	β=0°、h=2.5 m、n=360 r·min^{-1}、d=650 mm
2	Vertical angle β / °	-40、-30、-20、-10、0、10、20、30、40
	Fixed value	α=0°、h=2.5 m、n=360 r·min^{-1}、d=650 mm
3	Agitator-to-bottom height h / m	2、4、6、8、10、12、14
	Fixed value	α=20°、β=20°、n=360 r·min^{-1}、d=650 mm
4	Stirring speed n/ r·min^{-1}	200、250、300、400、450、500
	Fixed value	α=30°、β=0°、h=2.5 m、d=650 mm
5	Agitating diameter d / mm	400、450、500、550、600、650、700、750、800
	Fixed value	α=30°、β=0°、h=2.5 m、n=360 r·min^{-1}

2.2.2 Solution

The software Fluent provides three modelling methods to solve the problem with coexistence of stationary and moving regions, namely: multiple reference system (MRF) model, mixed plane model and sliding mesh model. This paper adopts the multi-reference method for steady-state calculation: firstly, the mixing device is divided into two non-overlapping regions, representing for agitating region and the external region, respectively; secondly, calculations are carried out in these two regions with two different reference systems-the paddle blade and its adjacent fluid areas with rotated coordinate system, and the other fluid regions with the stationary coordinate system. It is noted that the interface of the two regions is set to the inner surface (interface), the blade surface, the bottom of the tank and the outer surface of the pool set to walls (wall), and the upper surface of the tank set to free exit (outflow).

The RNG κ-ε turbulence model is applied to predict the effects of buoyancy, strong rotation, high shear and low Reynolds number. The anaerobic fermentation broth is replaced with incompressible water as the working medium, and its viscosity is set as 0.05Pa·s. Because of self-weight of the liquid, the Z-axis negative acceleration is $9.8m/s^2$. The SIMPLEC algorithm and the second-order windward difference scheme are used to improve convergence speed and accuracy of the unstructured grid. When the residual becomes stable or less than 1×10^{-3}, it is regarded as stable statue or calculation completion.

3 Results and discussion

3.1. Power

The size of the agitating power is an important indicator of the degree of mixing and flow of the material in the mixing tank. It is also the key basis for selecting the motor power.The agitator is subjected to pressure and viscous forces in the flow field so that the power of the agitator can be calculated by the torque acting on the blade[12]. The torque on the agitator and the velocity of the blades in the rotational tangential direction are obtained, and the power is calculated through Eq.(1).

$$P = M\omega = M \times 2\pi n / 60 \qquad (1)$$

Where, M represents torque, ω represents angular velocity, and n represents speed.

When the calculation is stable, the torque on the agitator can be obtained by post-treatment in Fluent, then is applied to calculated the agitating power. The simulation power consumptions of the test groups, are shown as (a) to (e) in Fig. 3, corresponding to No. 1 to 5 group.

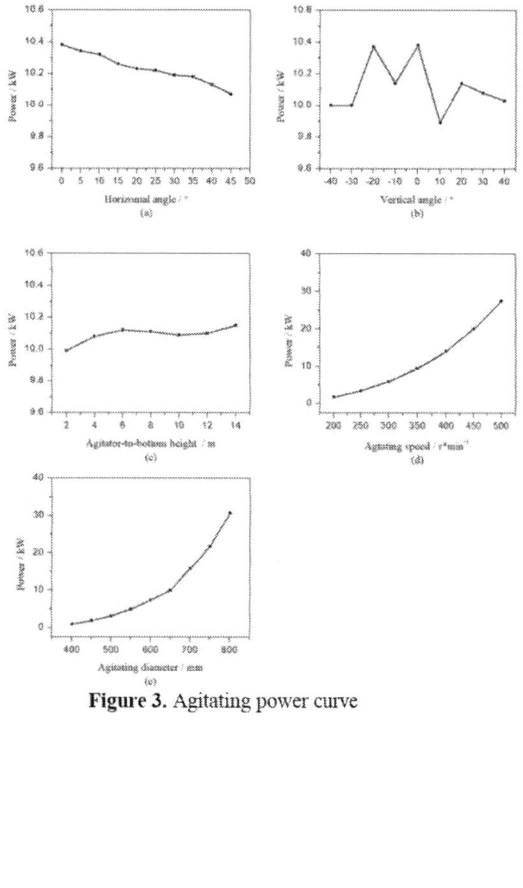

Figure 3. Agitating power curve

According to Fig. 3(a), the agitating power is gradually reduced as the horizontal angle increases from 0° to 40°. With the vertical angle increasing or decreasing, the agitating power does not change with the same direction, but just fluctuated around 10 kW in a small range, as seen in Fig. 3(b). From Fig. 3(c), the influence of the agitator-to-bottom height on the agitating power is very small. However, from Fig. 3(d) - (e), with the increase of the agitating speed and the agitating diameter, the agitating power increases rapidly in the form of parabola.

In summary, simulation results show that:

1) A certain horizontal angle is conducive to reducing the agitating power.

2) The vertical angle and the agitator-to-bottom height have little effect on the agitating power.

3) The change of agitating speed and agitating diameter has great influence on the agitating power, the agitating power increases rapidly with the increase of agitating speed and agitating diameter; hence, reducing the agitating speed and agitating the diameter can effectively reduce the agitating power.

3.2 The effective area percentage

According to the People's Republic of China urban construction industry standards (CJ/T 109—2007): the area where the water flows greater than 0.3 m/s is known as the effective working area of the agitation. To meet the requirement, in this paper, the area where the fluid flow rate is no less than 0.3 m/s is set as the effective agitation area, i.e., in which the mixing can be effectively agitated and the precipitation is prevented. The percentage of effective agitation zone in whole agitated area is referred as the effective area percentage, that is, the greater the percentage, the better the agitating effect.

After the numerical simulation, the effective area percentage of each test group can be worked out by post-processing. The simulation results are plotted as shown in Figure 4. As can be seen from Fig. 4(a), with the increase of the horizontal angle, the effective area percentage increases from 3% to 16%, and then reduces to 5%, the minimum value at 5°and the maximum at 30°. Seen from Fig. 4(b), as the vertical angle changes from -40°to 40°, the effective area percentage increases first and then reduces, showing the maximum value 7% at -20°; From Fig. 4(c), with the increase of the agitator-to-bottom height, the effective area percentage increases first and then decreases, and the effective area percentage is 8% at the bottom height of 6 m and the agitating effect is the best; And from Fig. 4(d), with the increase of the agitating speed, the effective area percentage increases slowly with the speed between 200 r/min and 300 r/min and then increases rapidly between 300 r/min and 400 r/min, finally reaches its maximum value at 450 r/min and tends to be stable. From Fig. 4(e), as the agitating diameter increasing, the effective area percentage increases as well, and reaches its maximum value at 750 mm and tends to be stable.

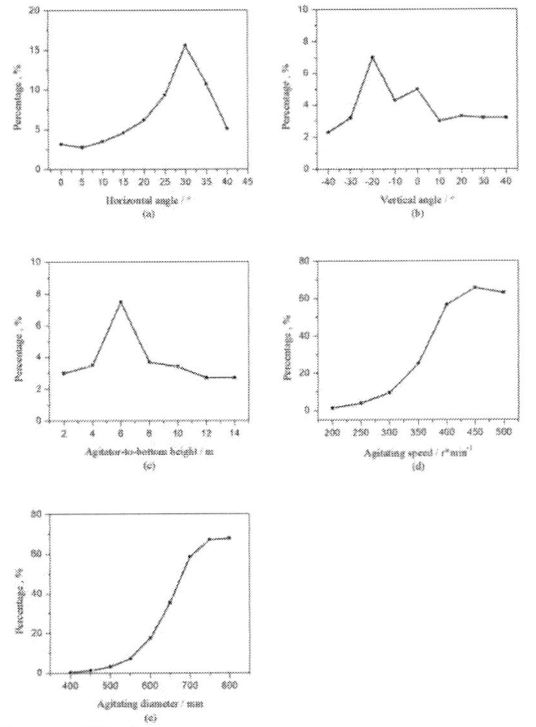

Figure 4. Effective area percentage curve

Simulation results show that:

1) When the horizontal angle, the vertical angle and the bottom height respectively are 10° ~ 40°, -30° ~ 0° and 2 m ~ 8 m, the effective area percentage is relatively large and the agitating effect is better.

2) When the agitating speed is between 300 r/min and 450 r/min or the agitating diameter is between 550 mm and 700 mm, increasing the agitating speed or agitating diameter is very beneficial to improve the effective area percentage.

3) When the agitating speed is greater than 450 r/min or the agitating diameter is greater than 750 mm, increasing the agitating speed or agitating diameter will not only reduce the effective area percentage, but also increase the agitating power.

3.3 The effective power ratio

Combining the mixing power and the effective area percentage, the ratio of the effective area percentage to the agitating power can be used to characterize the efficiency of unit agitating power, referred to as the effective power ratio. The larger the ratio is, the better the agitating efficiency. From Fig. 5(a), when the vertical angle between -30° and 0°, the effective power ratio is larger, but for the other angles, the ratio is smaller. From Fig. 5(b), when the agitating speed is between 300r/min and 500r/min, the effective power ratio is relatively large and the ratio for other rotational speed is small. As can be seen from Fig. 5(c) when the agitating diameter is between 600 mm and 750 mm, the effective power ratio is relatively large and the ratio for the other mixing diameter is small.

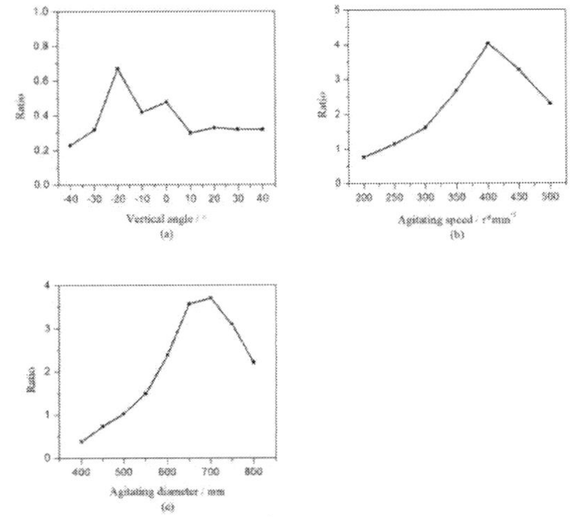

Figure 5. Effective power ratio curve

It can be known from above simulation results that:

When the vertical angle lies between -30° and 0°, the agitating efficiency would be better if the agitating speed is in the range of 300 r/min and 500 r/min or the agitating diameter in the range of 600 mm and 750 mm.

4 Conclusion

In this paper, the agitating process of the single side-entering agitator in the large anaerobic agitated tank was simulated by Fluent software. By analyzing the influence of each factor on the agitating power, the effective area percentage and the effective power ratio is worked out, and the conclusions are drawn:

1) The horizontal angle has little effect on the agitating power, but its effect on the effective area percentage is relatively significant. When the horizontal angle is 30°, the effective area percentage is the largest and the agitating flow field is the best.

2) The effect of vertical angle on the agitating power is very small. When the vertical angle between -30° and 0°, the agitating efficiency is relatively high.

3) The effect of agitator-to-bottom height on the agitating power is very small. When the agitator-to-bottom height is less than 8m, it would be helpful to improve the effective area percentage and prevent the appearance of the precipitation.

4) The agitating speed and the agitating diameter have great impacts on the agitating power and the effective area percentage. Increasing these two factors can effectively increase the effective area percentage. When the agitating speed lies between 300 r/min and 500 r/min and the agitating diameter between 600 mm and 750 mm, the agitating efficiency is better.

References

1. G. Montante, K. C. Lee, A. Brucato, et al. Numerical simulations of the dependency of flow pattern on

impeller clearance in stirred vesseds[J]. Chem Eng Sci, 2001, 56(12):3751-3770.

2. A. Kukekova, M. Mostek, M. Jahoda, V. Machon. CFD prediction of flow and homogenization in a stirred vessel: Part I vessed with one and two inpellers[J]. Chemical Engineering and Technoloyg, 2005, 28(10):1125-1133.
3. Mingzhong L. Gracme W. Derek Wilkinson, et al. Scale up study of retreat curve impeller stirred tanks using LDK measurements and CFD simulation[J]. Chem Eng Sci. 2005, 108(1-2):81-90.
4. Wesselingh J A. Mixing of Liquids in Cylindrical Storage Tanks with Side-entering Propellers[J]. Chem Eng Sci, 1975, 30(8): 973-981.
5. Asghar A D. Masoud R. CFD simulation of homogenization in large-scale crude oil storage tanks[J]. J Pet Sci Eng, 2004, 43(3-4): 151-169.
6. Fang Jian, Sang Zhifu, Yang Qubao, Numerical simulation of three-dimensional flow field of the side entering agitator[J]. Petroleum Machinery, 2009, 37(1):30-34.
7. Fang Jian, Ling Xiang, Sang Zhifu, et al. CFD simulation of the mixing process of side entry mixer [C].Proceedings of the 2010 annual meeting of the Chinese Society for Environmental Science (volume fourth). 2010.
8. Zhang Linjin, Chen Gongguo, Ye Xuchu.Numerical Simulation of Flow Field in Side-Entering Stirred Reactor[J]. Chemical Reaction Engineering and Technology, 2011, 27(2):97-102.
9. Zheng Xiaodong, Huang Xiongbin, Du Rongli. Suspension of Solid Particles by Side-entering Agitators[J]. The Chinese Journal of Process Engineering, 2009, 9(3):417-423.
10. Chen Jia, Xiao Wende. Numerical simulation of turbulent flow field in industrial-scale side-entering stirred tank[J].Chemical Engineering, 2013, 41(8):38-42.
11. Liuyi. High concentration of biogas fermentation material mixture stirring and influencing factors of research[D]. Chinese Academy of Agricultural Sciences, 2009.
12. Chen Nianjin. The Power Calculation of the Stirrer Based on CFX[J]. Chemical equipment technology 2014, 35(3):27-29.
13. CJ/T 109-2007, Diving mixer[S].

Optimization and Numerical Simulation of Outlet of Twin Screw Extruder

Yuan Zhang , *Xiaohan Jiang* , Huaping Fan and Xihan Li

School of Mechanical and Power Engineering,Harbin University of Science and Technology,150080Harbin ,China

Abstract. In view of the unreasonable design of non-intermeshing counter-rotating twin screw extruder die, the problem of productivity reduction was discussed. Firstly, the mathematical model of extruder productivity was established. The extruder die model was improved. Secondly, the force analysis of twin screw extruder physical model was carried out. Meanwhile, A combination of mechanical analysis and numerical simulation was adopted. The velocity field, pressure field and viscosity field were calculated by Mini-Element interpolation method, linear interpolation method and Picard iterative convergence method respectively. The influence of die model on the quantity of each field before and after improvement was analyzed. The results show that the improved model had increased the rheological parameters of the flow field, the leakage and reverse flow decreased. Through post-processing calculation, the productivity of the third dies extruder was 10% higher than before. The research results provide a theoretical basis for the design and optimization of die model of non intermeshing counter-rotating twin screw extruder.

1 Introduction

At present, the extrusion processing as an important means of transport to work[1], screw extruder is widely used in polymer, food and medicine[2-6], while in the traditional Chinese medicine pill making machine, its sliver quality and efficiency directly affects the quality and output rate of the pill.If the model design or the selection of the screw extruder die is not reasonable, not only will the reverse flow and leakage flow increase, but also the efficiency will be reduced. The method of numerical simulation was used to simulate the internal flow field of double screw by Many scholars[7, 8].However, the influence of different die models on the flow field in twin screw extruder is not considered in these study.

In order to improve the efficiency of extrusion of the extruder, the die model was optimized with the ANASYS[9] POLYFLOW software as a platform to polymer rheology as the foundation, by means of computer simulation analysis, the influence of the model before and after optimization on rheological parameters of extruder flow field was studied, and the flow characteristics of non Newtonian viscous polymer in screw rod under isothermal condition were analyzed. The simulation results are verified by experiments. It is of great significance to optimize the die model of non intermeshing twin-screw extruder.

2 Numerical model of extruder

2.1 Mathematical model

In the process of extrusion, the water content and the temperature of the material change constantly, and the viscosity changes. The material belongs to the rheological body. In order to simplify the calculation, the following assumptions are made:

(1) The interior of the runner is filled with material;

(2) The fluid is a steady flow with time independent;

(3)The temperature in the flow field remains constant;

(4) flow is turbulent flow, and Re is between 8000~12000;

(5) it is assumed that the fluid is incompressible;

(6) the wall of runner is free of sliding;

(7) because of the large viscosity of the fluid, the bulk forces such as gravity and inertia are neglected.

Based on the above hypothesis, in the Cartesian coordinate system, the continuity equation is:

$$\frac{\partial v_x}{\partial x}+\frac{\partial v_y}{\partial y}+\frac{\partial v_z}{\partial z}=0 \tag{1}$$

Equation of motion is:

$$-\frac{\partial P}{\partial x}+\left(\frac{\partial \tau_{xx}}{\partial x}+\frac{\partial \tau_{yx}}{\partial y}+\frac{\partial \tau_{zx}}{\partial z}\right)=\rho\frac{\partial v_x}{\partial t} \tag{2}$$

$$-\frac{\partial P}{\partial y}+\left(\frac{\partial \tau_{xy}}{\partial x}+\frac{\partial \tau_{yy}}{\partial y}+\frac{\partial \tau_{zy}}{\partial z}\right)=\rho\frac{\partial v_y}{\partial t} \quad (3)$$

$$-\frac{\partial P}{\partial z}+\left(\frac{\partial \tau_{xz}}{\partial x}+\frac{\partial \tau_{yz}}{\partial y}+\frac{\partial \tau_{zz}}{\partial z}\right)=\rho\frac{\partial v_z}{\partial t} \quad (4)$$

Considering that the material belongs to food processing, the shear rate is not too high, and the constitutive equation is simplified power law model:

$$\eta=\zeta\left(\frac{1}{2}\dot{I}_2\right)^{\frac{n-1}{2}} \quad (5)$$

The volumetric flow rate formula of twin screw extruder is:

$$Q_{总}=2Q=i\pi^2d^2nh\cos\theta\sin\theta-\frac{\Delta P}{6\eta L_1}(ih^3d\pi\sin^2\theta+\pi^2qd^2\delta^3\tan\theta/e) \quad (6)$$

i - screw head number d - screw outer diameter

n - screw speed h - groove depth θ - helix angle
η - viscosity e - screw width L - pitch

d_1 - screw diameter L_1 - measuring section length

ΔP –Pressure drop at both ends of metering section

q - meshing coefficient, which indicates the influence of the meshing area on the material flow due to the engagement of two screws

2.2 Physical model

Through the experiment of dual-mode extrusion machine efficiency is not very high, the extrusion speed is slow, so the die was optimized, the optimized die model eliminates the elbow, will shorten the length of extrusion, and on the basis of the original increase in a die model, as shown in figure 1.

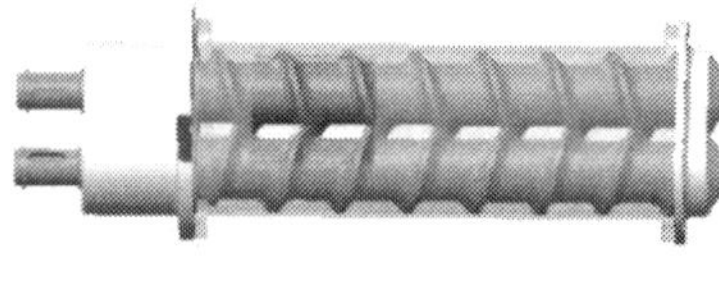

(a) two dies

(b) three dies

Figure 1. A model for the twin screw extruder

The main performance parameters of twin-screw extruder include screw diameter, center distance, pitch and so on, as shown in figure 2:

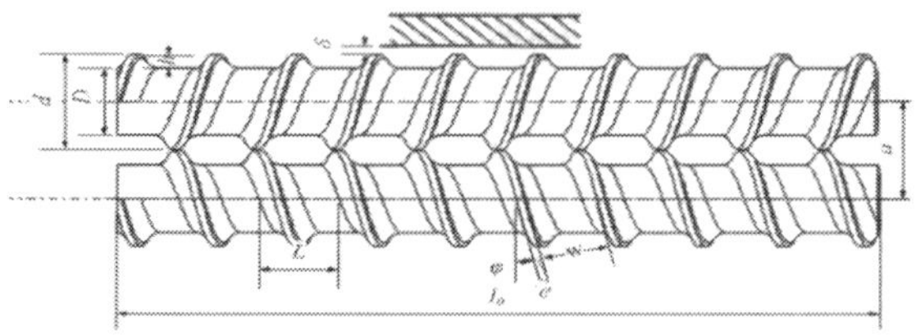

D- screw internal diameter; δ -clearance; *a*- center distance; φ -thread lift angle; *w*- spiral groove width; l_0 - screw length

Figure 2. Twin-screw extruder structure parameters

3 Simulation analysis of flow field in extrusion process

3.1 Material parameter

Since the ingredients of the powder are complex, flour is used as a substitute. By referring to the relevant literature[10], the parameters of the flour melt material were obtained. The consistency coefficient was 25680Pa. Sn, the flow index was 1.42 and the density was 850kg/m^3.

3.2 boundary condition

In the flow field simulation of the extruder runner, the inlet volume flow rate and outlet pressure are given, and the material has no slippage with the inner wall of the cylinder and the surface of the screw:

(1) Define the entrance boundary of computational domain by volume flow rate, $V=6.42\times10^{-5}\mathrm{m}^3/\mathrm{min}$;

(2)The exit boundary is defined by the pressure value,; $P=6.25\times10^6\mathrm{Pa}$

(3) Set the speed boundary at the head of the twin screw. In the direction of extrusion, the speed of the left screw is -20r/min, the direction is clockwise, the speed of the right screw is 20r/min, and the direction of the screw is counterclockwise.

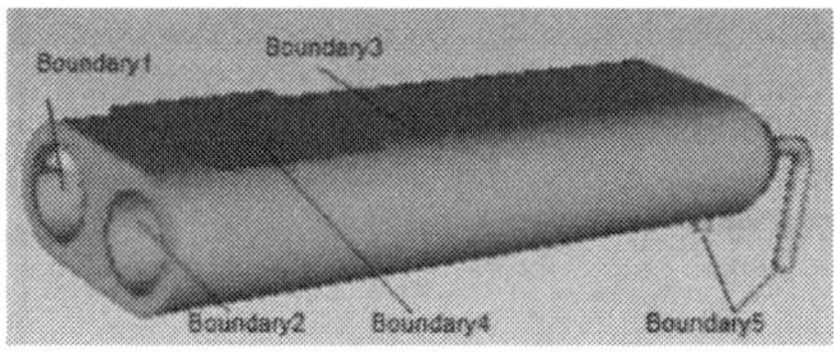

(a) two dies

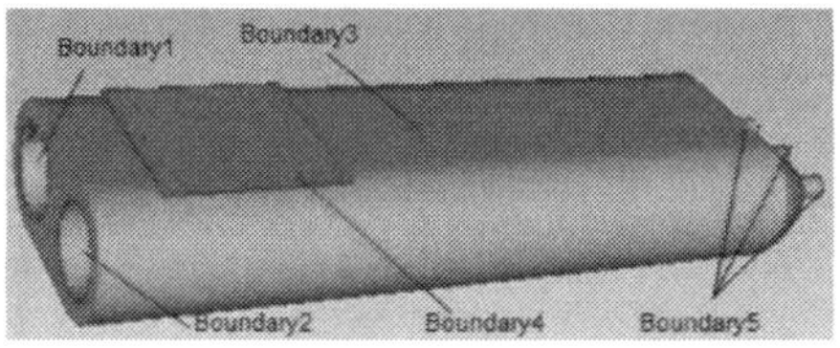

(b) three dies

Figure 3. boundary names

Figure 3 is the runner part boundary name, Boundary4 is runner inlet, Boundary5 is runner outlet,

Boundary3 is runner and cylinder contact surface, Boundary1 and Boundary2 are runner and left and right screw contact surface. Table 1 is the channel flow boundary condition.

Table 1. Flow channel boundary condition setting

Boundary position	Flow boundary condition
Boundary1	Descartes speed input boundary
Boundary2	Descartes speed input boundary
Boundary3	no slip at wall
Boundary4	Volume flow rate
Boundary5	Natural flow like fixed outlet pressure

3.3 Analysis of rheological parameters of materials during steady compression and extension

FEM method was used to model nonlinear control coupling equation of power-law solution in the mesh module of the screw and the runner model grid and the grid overlap, set the parameters in polydata, in order to improve the accuracy of the calculation, calculation of single element velocity field using Mini-Element interpolation method to calculate the pressure field by using the linear interpolation method, calculation of viscosity using Picard iterative method, using algebraic equation group implicit Euler method for solving the discrete after. Through the post-processing module CFD-POST, the results of each rheological parameter were analyzed, and the flow velocity, pressure, viscosity parameter nephogram and the change curve of parameters with the axial distance were obtained.

3.3.1 Velocity field distribution

After using CFD-POST to visualize the results, the velocity vector distribution of the flow channel was shown in Figure 4 and the contour map of velocity distribution at the die mouth are obtained.

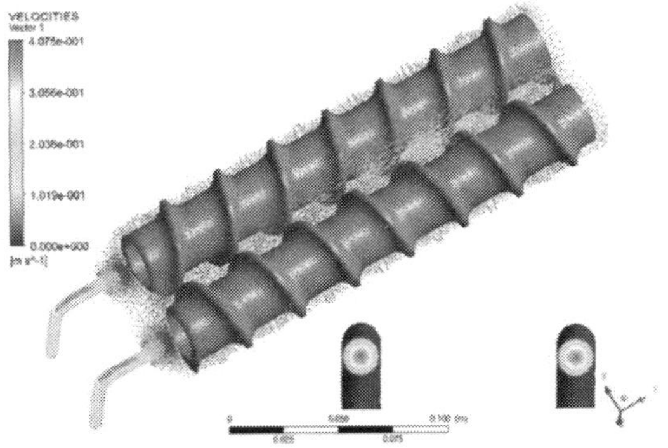

(a) two dies

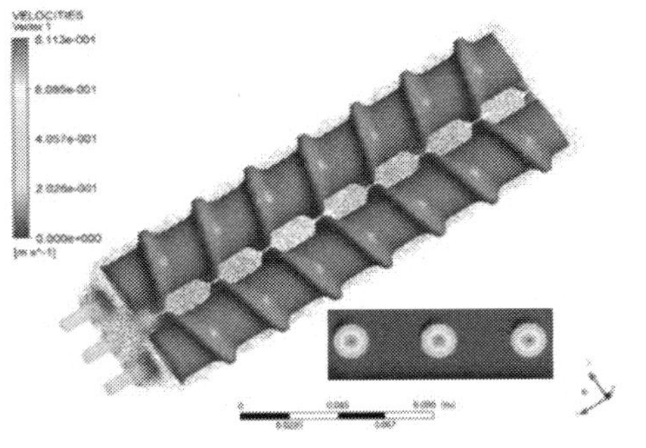

(b) three dies

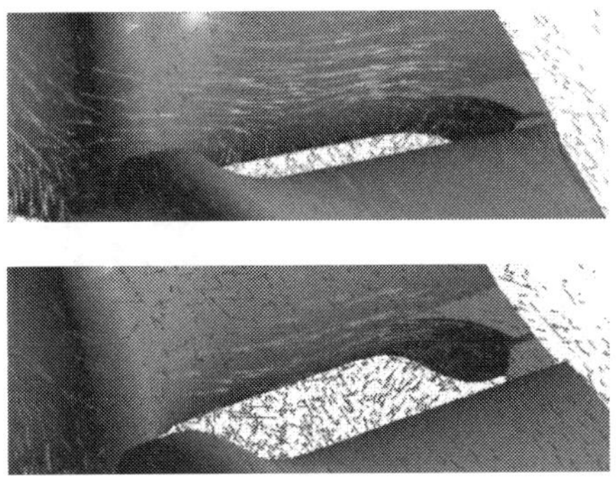

(c) Local amplification of the screw tip near the entrance

Figure 4. Flow velocity vector diagram and die speed contour map

The change trend of the drawing speed showed that maximum speed of the die in the axis, and the trend was from the wall to the axial velocity increasing gradually, a small amount of leakage at the entrance and measurement section. After the measuring section, the materials were gathered to the die, got the material in the metering section average velocity, the velocity of double dies was 0.02579m/s, and three dies was 0.06372m/s. Showed that the velocity vector images, material appeared counter current phenomenon at the entrance, and two port inverse flow and speed was obviously higher than that of the three dies in the model, as shown in figure 7 (c) shows.

3.3.2 Pressure field distribution

Pressure contours showed that both the two and the three dies, the maximum pressure appeared in the die, and from the entrance to the die pressure decreased first and then increased, figure 5 showed the screw top near the entrance area darker, larger pressure. From the color, it can be seen that the pressure drop between the measuring section and the die mouth was different.

 https://doi.org/10.1051/matecconf/201815305004

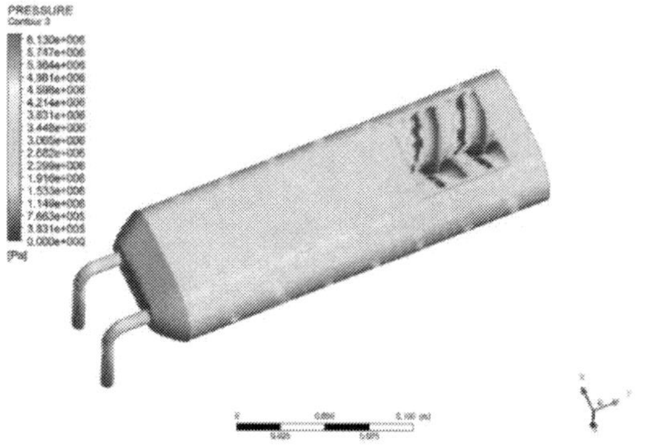

(a) two dies

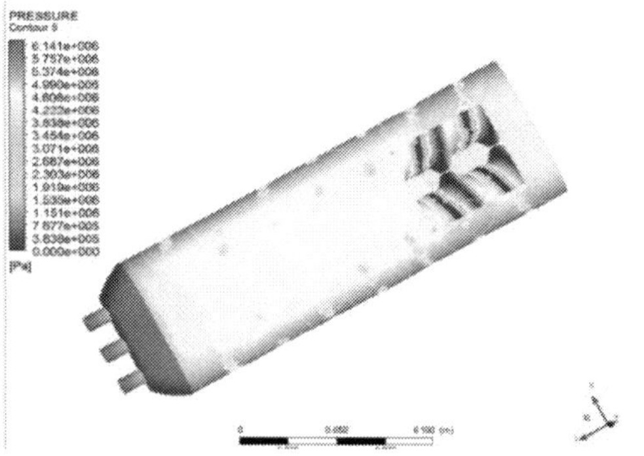

(b) three dies

Figure 5. Pressure distribution diagram of flow channel

Figure 6 was a two dies and the three dies extrusion machine from the metering section to the die pressure trend comparison, the graph showed that the three dies export metering section of the pressure was higher than the double export, but export dual extruder die pressure drop was bigger, its value was 3.175MPa, the three dies to 1MPa, because the double die model had elbow, and the elbow was too long to the exit, the long elbow caused a sharp change in the die pressure.

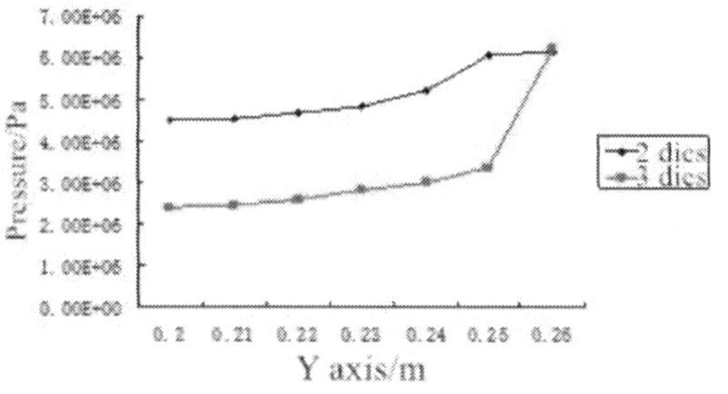

Figure 6. Pressure distribution diagram of flow channel

3.3.3 Viscosity distribution

The material viscosity had some influence on the extrusion efficiency. The material chosen for the simulation was an expanding plastic fluid, which had the characteristics of shear thickening, that was, the apparent viscosity of fluid increases with the increase of shear rate. Figure 7 showed the distribution of channel viscosity.

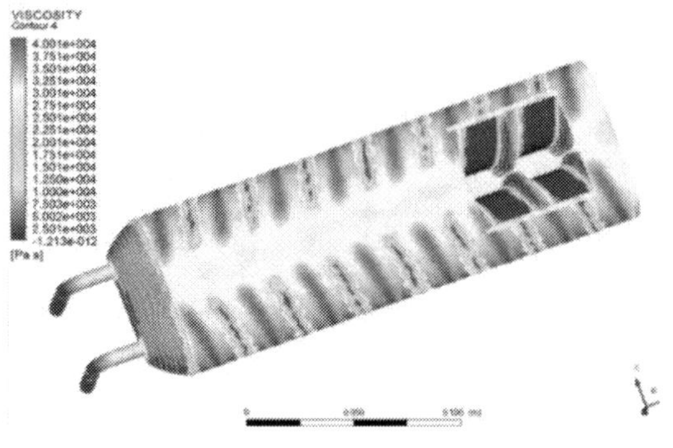

(a) two dies

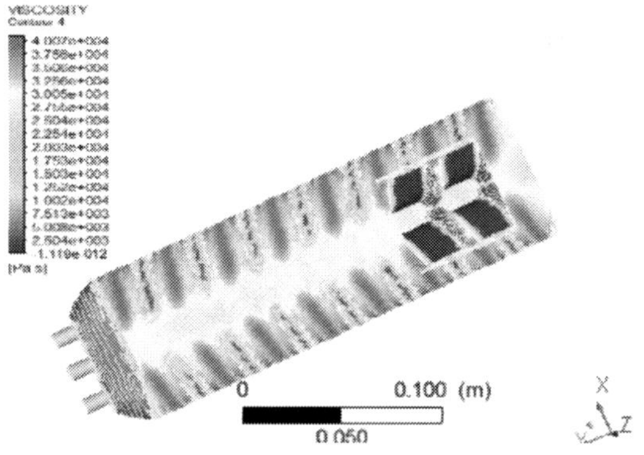

(b) three dies

Figure 7. Viscosity distribution

The viscosity of cloud display, two dies and three dies extrusion runner viscosity reached the maximum at the edge, the maximum values were 40010Pa·s and 40070Pa·s, when the die size was constant, the viscosity increased with the increase of the pressure, so the viscosity change flow rate has little effect, but the die diameter is small and it is not directly affected by the screw, the greater the viscosity, the retarding effect of the material will be stronger, as the die length, roughness, elbow caused by the loss of the process and local losses will increase. After optimization, the elbow part of the double dies orifice was removed and the straight part length of the dies was shortened, the blocking effect of the three dies was far lower than the export by two.

3.3.4 Comparison of flow field calculation results with actual volumetric flow rate

The above analysis results and the numerical formula (6) was calculated, respectively, and the positive flow, reverse flow, leakage current and volume flow rate of twin-screw extruder with different die numbers were obtained, as shown in table 2.

After calculation, the volume flow rate before and after optimization were 15.912kg/h and 17.442kg/h, the improved model of the reverse flow in the three dies was reduced by 28%, the leakage flow is reduced by 28%, volume flow rate increased by 10%.

Table2. Productivity comparison of twin screw extruder with different dies

Die number	Positive flow (m^3/s)	Against the Current (m^3/s)	Leakage current (m^3/s)	Volume flow rate (m^3/s)
2	7.0×10^{-6}	1.8×10^{-6}	5.7×10^{-9}	5.2×10^{-6}
3	7.0×10^{-6}	1.29×10^{-6}	4.1×10^{-9}	5.7×10^{-6}

In order to verify the accuracy of the calculated productivity values, the actual flow rates of different die extruders were measured several times at different time, as shown in table 3, from the measurement results can be seen, the volume flow rate calculation is slightly higher than the actual volume flow rate, two dies and export the three dies error is 5.35% and 5.53% respectively.The numerical calculation model compared to the actual measured by the existence of local differences in the structure of extruder, extruder and material surface roughness, the material cannot fill machine and other factors will also affect the results, therefore, under ideal conditions, the simulation value will be higher than the measured value, the error was small, the simulation results can reflect the real situation.

Table 3. Measurement of volume flow rate at different time

Time(min)	10	20	30	40	50	60
2dies ($10^{-6}m^3/s$)	4.72	4.85	4.90	5.13	5.07	4.86
3dies ($10^{-6}m^3/s$)	5.30	5.31	5.48	5.52	5.37	5.33

4 Conclusion

(1) By comparing the two dies and the three dies export extruder numerical simulation results, two dies and export the three dies in the extruder the material in the average speed of metering section were 0.02579m/s and 0.06372m/s, and was significantly higher than that of the three dies mode, it showed that the flow velocity at the measuring section is affected by the die resistance when the viscosity of the measuring section is about the same, double mouth length, roughness and bend along the path of energy loss and increase the local energy loss, resulting in slow flow rate. Therefore, the optimized die model can increase material flow rate and speed up extrusion.

(2)Analyzed the impact on the pressure field of different die model, it was found that the pressure of flow channel of three dies was higher than that of double dies,the pressure distribution showed the trend of decreasing first and then increasing. Through actual measurement the error is within 6%, the calculation results showed that the flow field calculation results have reference value. Display the pressure screw top pressure near the entrance area, showed that the resistance caused by the die resistance moves to the top of the screw, so in the design of extruder entrance should consider this factor, it is necessary to prevent the vacuum of the tip of the screw from rotating when it continues to inhale the material at the entrance, resulting in an increase in counter flow.The entrance can be designed at the top of the screw.

(3) Post processing calculation results showed that the three dies after optimization of the extrusion machine to reduce leakage current and counter current, the volume flow rate increased, indicating the die pressure drop impact on the metering section of the pressure drop, the improved die model makes the metering section of the pressure drop decrease, and to reduce reflux and leakage flow. The three dies after optimization of the extrusion machine can not only improve the output, save time, but also can improve the output efficiency of traditional Chinese medicine pill making machine, which laid a theoretical foundation for the production practice.

References

1. G.D. Zhang, J.P. Lin , Y.H. Ge , X.Y. Hao. *Mold-flow equilibrium coefficient correction of polymer extrusion*, Journal of Mechanical Engineering,**75-80**,48(2012)
2. M.Y. Yue, P. Xue, X.D. Song, J. Zhang, M.Y. Jia, *Research progress in screw extrusion technology of explosives,* China Plastics, **1-7**(2015)
3. Z.B. Chen, Q. Xie, J.J. Du, Y. Li, T. Liu, *Melting process of PVCwood-plastic single-screw extruder ' s melting section,* Plastics, **84-89**(2015)
4. A.D. Zhao, X.L. Li, G.Y. Zheng, L.Q. Zhang, *Preparation of liquid desulfurization rubber by co-rotating twin screw extruder*,China Synthetic Rubber Industry, **78-82**, 39(2016):
5. T. Yang, J.M. Xin, Q. Xu, H.Y Luo, *The research application status of in food industry*, Journal of Food Science and Biotechnology, **733-740** (2009)
6. H.W. Wang, *Application and research progress of twin screw extruder*, Cereals and Oils Processing , **131-134**(2009)
7. J.F. Shi, Z.G. Huang, W.X. Jiang, M.L. Li, H.J. Zhang, *Study on flow field of polylactic acid in the nose transition body of co-rotating twin-screw extruder*, Food & Machinery, **72-75**(2017)
8. S.A. Salahudeen, A.R. Rahmat, *Design Parameter to Develop Secondary Flow in Twin Screw Extruder*, Applied Mechanics and Materials, (2015)
9. G. Wang , X.Z. Zhu , Y. S. Chun, *Numerical Simulation of Mixing Performance of Intermeshing Co-Rotating Tri-Screw and Twin-Screw Extruders*, Advanced Materials Research, (2012)
10. S.H. Chen, *Theoretical and experimental study of NI-MPE in intermeshing co-rotating twin-screw extrusion*, Beijing University of Chemical Technology, (2003)

Analysed statistically Modelling and Optimization of Laser Machining by Response Surface Methodology

Hani Mizhir[1], Kamil Jawad[1], Zuhair H Obaid[1]

[1]*Al-Furat Al-Awsat Technical University, Iraq*

Abstract. One of the important goals of this research is to predict a relationship between the process input parameters and resultants from surface roughness features through developing a laser cutting model. In most engineering applications, natural sciences and computing; statistical methods, which are one of mathematical branch are widely used for investigating the results. Laser cutting process of stainless steel (2205) is a machining process selected for this study. The technique which adopted here is a response surface methodology (RSM). The main portion for this study is the influence of cutting speed on surface quality. To study the model response, and for statistical approach with further prediction; a mathematical based model has been developed through regression analysis. It's found that as one of the important results in this research, that cutting speed and surface roughness has a significant rule on the model response. To produce a good surface roughness, it's approved that the high cutting speed connected with high power regardless of high pressure has a high influence on surface quality.

1 Introduction

The emergence of advanced engineering material, complex design, sophisticated shape and unusual size of work material make more restriction for using the conventional machine in industry. Advance machining process (AMPs) is become urgent in manufacturing process as unconventional method (Choudhury and Shirley, 2010). One of the important (AMPs) is a laser machining which used to cut different types of material economically. Comparing with other types of conventional method; laser beam machining gives good properties to the final product like a fine surface (Choudhury and Shirley, 2010). In many industrial applications, and especially for purpose of cutting the hard materials, laser machining is one of the commonly method which used due to its advantages comparing with other methods (Eltawahni et al., 2012). Laser cutting process involves highly localized heat input to the work peace, no tool wear, remove less material, and minimize distortion.

It's well suited for many manufacturing process because it's highly automated, thermal and non- contact process. This method is always used to produce high surface finish products, especially in mass production with high dimensional accuracy (Madic et al., 2012). During any manufacturing process, like laser cutting, the majority of metals have resistance to the maximum operation temperature due to their structure and boundary layers (Samant and Dahotre.2009). Cutting process can be defined as a method to produce a specific geometry with some specifications by removing the material using different types of tools. The waste or access material and the finished part is the net results of any cutting process. There are different types of manufacturing process, some of these can called machining and others can be classified as cutting process, and others are coating or burning. Oxy-fuel cutting torch can be called as burning. Others process like use of chemical masking or etching to remove the excess material is called chemical process (Tulasiramarao et al., 2013).

Advance machining process (AMPs) like electron beam machine, laser beam machine, ion beam machining, plasma beam machining, ultra sonic machining (USM), chemical blanking, photo chemical machining and others are wildly used nowadays in manufacturing process (Dubey and Yadava 2008; Samant and Dahotra, 2009). Some metals like stainless steel cannot or very difficult to cut or machine by classical methods due to high melting point and low viscosity of the formed oxide; so laser cutting method is the most suitable method in this case (Ghany and Newish, 2005).

2 Initial Designs

RSM is a collection of Mathematical and statistical techniques that are useful for modelling and analysis in application will be developed to investigate the selected parameters effect on the responses represented by surface roughness, and will be used to predict the surface roughness.. The response will be predicted by statistical means that refer to an empirical method to describe the relationship between the parameters and their effect on the response. RSM To establish the surface roughness prediction model, MINITAB 16 statistical software package will be used. Based on these runs with the particular parametric values, the experiment will be performed and the surface roughness values taken were re-entered into the matrix. After entering the surface roughness values in the modelling matrix, which will be developed by the central composite design CCD method, a complete analysis can be executed using RSM to define the importance of each parameter of the response.

3 Practical Laser Cutting Type Materials

The duplex stainless steel (DSS), is a mixed microstructure of about equal proportions of austenite and ferrite. The early grades were alloys of chromium, nickel and molybdenum. Therefore, Duplex stainless steel the product description (AK) steel 2205 is a ferritic-austenitic. It provides a valuable combination of high strength, excellent corrosion resistance and stress-corrosion cracking resistance as well as very good pitting resistance. In addition, it is highly resistant to stress and provides a low level of thermal expansion. Stainless steel duplex 2205 will be used in the research experiments. The initial dimensions of the work piece where; length of 2000 mm, width of 500 mm, thickness of 8.0 mm and Non-Linear incline cutting of 32° for cut nozzle

4 Cutting Parameters

Cutting speed, laser power and gas pressure are the main parameters are studied . in addition; to find a statistical representation of the parameter to the responses, the effect of these parameters on the surface roughness are the major aspect of this study.. the high and low values of the input parameters in experiment should be identified. the high and low values are depends on the material used. in general, the range of the parameter's values are shown in table 1. the other parameters will kept at a constant values through the experiment as shown in table 2.

Table 1. Parameters Range.

Paramètre	Range (Low- High)
Laser Power (watt)	1250 - 2500
Cutting Speed (mm/min)	500 - 1000
Gas Pressure bar	100 - 175

Table 2. Parameters values during the central composite design.

Central Composite Design			
Factors:	3	Replicates:	1
Base runs:	20	Total runs:	20
Base blocks:	1	Total blocks:	1
Two-level factorial: Full factorial			
Cube points:	8		
Centre points in cube:	6		
Axial points:	6		
Centre points in axial:	0		
Alpha: 1.68179			

Std Order	Run Order	Pt Type	Blocks	Power	Speed	Pressure
1	1	1	1	1503	601	115
14	2	-1	1	1875	750	175
2	3	1	1	2246.	601	115
5	4	1	1	1503.	601	159
11	5	-1	1	1875	500	137
10	6	-1	1	2500	750	137
15	7	0	1	1875	750	137
6	8	1	1	2246	601	159
3	9	1	1	1503	898	115
17	10	0	1	1875	750	137
12	11	-1	1	1875	1000	137
4	12	1	1	2246	898	115
9	13	-1	1	1250	750	137
20	14	0	1	1875	750	137

The various values within the range of specified parameters, are shown in Table 1, above and Table 2, down. After generation of the parameter values as in Table 2, it will be the data input which used into the Minitab 16 software, then a DOE model will be automatically generated with a particular run number coupled with specific parametric settings. In this research, 20 runs were generated as shown in Table 2. DOE have been applied by many authors to define the factors effect of a specified process with many outputs and to develop the mathematical relationship in order to define the optimal conditions (Eltawahni et al., 2011). Design Table 3 (randomized):

Table 3. Experimental Layout for Response.

Run	Blk	A	B	C
1	1	-1.00000	-1.00000	-1.00000
2	1	0.00000	0.00000	1.68179
3	1	1.00000	-1.00000	-1.00000
4	1	-1.00000	-1.00000	1.00000
5	1	0.00000	-1.68179	0.00000
6	1	1.68179	0.00000	0.00000
7	1	0.00000	0.00000	0.00000
8	1	1.00000	-1.00000	1.00000
9	1	-1.00000	1.00000	-1.00000
10	1	0.00000	0.00000	0.00000
11	1	0.00000	1.68179	0.00000
12	1	1.00000	1.00000	-1.00000
13	1	-1.68179	0.00000	0.00000
14	1	0.00000	0.00000	0.00000
15	1	1.00000	1.00000	1.00000

The experiment is achieved based on the runs given with the specific parametric values, then re- entering all the surface roughness values into the matrix. A complete analysis can be performed after entering the surface roughness values by using RSM to define each parameter effect on the response. . A laser cutting machine LVD Model 2513 with a maximum output power of 4000 Watt, 10.6 μm wavelength and100 μs-us pulse duration will be used to conduct the CO2 laser cutting experiments with the nitrogen gas assistance instead of acetylene oxygen. The nitrogen can ensure the metal surface safety and maintain the metal physical characteristics. CO2 laser cutting will be used in cutting the stainless steel. Figure

(1), Figure (2) and Figure (3); are shown the laser cutting machine, design material and control operating system..

Figure 1. Cutting speed machine CNC laser.

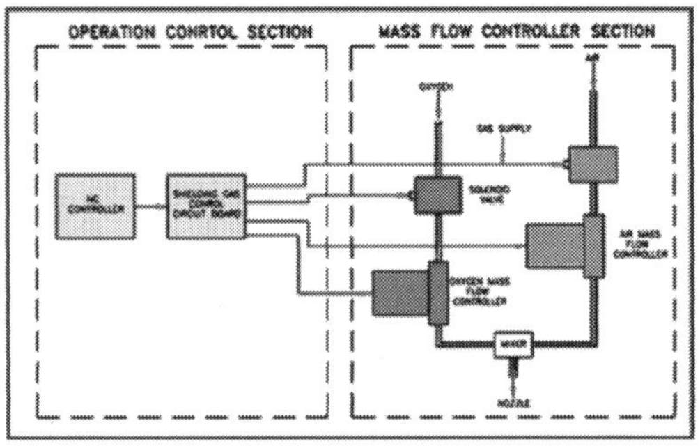

Figure 2. Operation control section and mass flow controller.

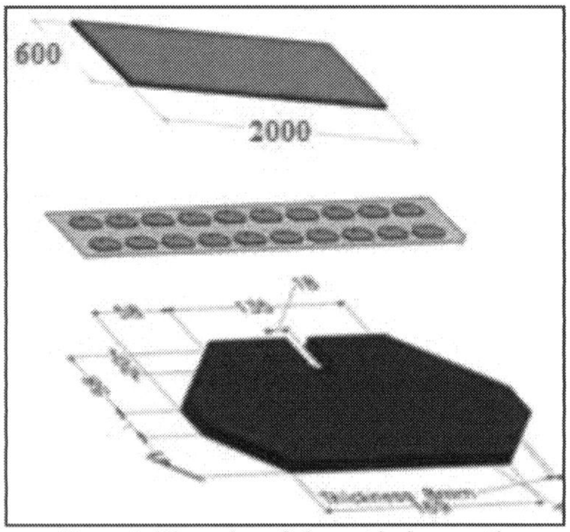

Figure 3. Dimension of plate cut by laser cutting.

5: Result & Discussion

5.1 Response Surface Regression

Twenty experimental runs were carried out as listed in Table 4, Table 5, and Table 6, are show the relation between Roughness versus Pressure, Power, and Speed. The surface roughness of the developed stainless steel cutting in each experimental run was analysed by using instrument inspection and the surface roughness data were obtained. Three surface roughness measurements were collected per each sample and the average roughness were calculated. The output response of the process as tabulated in Table 4, Table 5, and Table 6. ; Experimental Results for Laser Process surface roughness Non-Linear Incline Cutting and Reading Measured Specified in Three cutting Area x1, x2, and x3 at Bevel Angle 32° are plotted in figure 4.

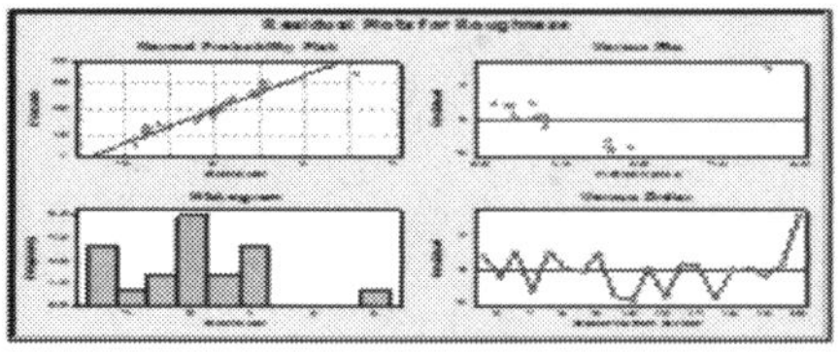

Figure.4. Residual Plots for Roughness.

Table 4. Experimental results for the area x1 x2, x3

Run	X_1 mm	X_2 mm	X_3 mm	Average mm
1	0.93	1.31	1.75	1.34
2	0.88	0.75	0.76	0.79
3	2.19	1.92	1.85	1.98
4	1.22	1.17	1.16	1.18
5	1.18	1.21	1.33	1.24
6	0.80	0.75	0.65	0.75
7	1.00	1.18	1.39	1.19
8	1.35	1.40	1.62	1.46
9	1.54	1.13	1.11	1.26
10	0.71	0.64	0.69	0.69
11	1.25	1.27	1.31	1.28
12	0.94	0.73	0.84	0.84
13	1.42	0.80	0.59	0.94
14	1.84	1.15	0.92	1.30
15	1.54	1.25	0.81	1.20

Table 5. Experimental run and results of surface Roughness

Run	Std Ord er	Run Ord er	Pt Ty pe	Bl oc ks	power	spee d	press ure	Rough ness
2	7	2	1	1	1503	898.	5.1	0.79
3	19	3	0	1	1875.	750.	4.	1.98
4	14	4	-1	1	1875	750	6	1.18
5	3	5	1	1	1503.	898	2.	1.24
6	4	6	1	1	2246.	898.	2.	0.75
7	17	7	0	1	1875	750	4.	1.19
8	8	8	1	1	2246	898	5.	1.46
9	15	9	0	1	1875.	750.	4.	1.26
10	20	10	0	1	1875	750.	4	0.69
11	16	11	0	1	1875	750.	4.	1.28
12	11	12	-1	1	1875	500.	4	0.84
13	18	13	0	1	1875	750.	4.	0.94
14	5	14	1	1	1503	601.	5.	1.30
15	6	15	1	1	2246	601.	5.	1.20

Table 6. Estimated Regression Coefficients for Roughness.

Term	Coef.	SE Coef.	T	P
Constant	1.18692	0.6572	1.806	0.101
Pressure	0.07895	0.4360	0.181	0.860
Power	-0.10724	0.4360	-0.246	0.811
Speed	-0.90223	0.4360	-2.069	0.065
Pressure*Pressure	-0.30510	0.4244	-0.719	0.489
Power*Power	-0.15308	0.4244	-0.361	0.726
Speed*Speed	0.96062	0.4244	2.263	0.047
Pressure*Power	0.02375	0.5697	0.042	0.968

 https://doi.org/10.1051/matecconf/201815305005

Pressure*Speed	-0.12125	0.5697	-0.213	0.836
Power*Speed	-0.09875	0.5697	-0.173	0.866

S = 1.61129 PRESS = 194.375

R-Sq = 51.76% R-Sq (pred) = 0.00% R-Sq (adj) = 8.34%

5.2 Interpreting the Results and Discussion

Since the non-linear model is consider as a higher-order model, it will be needed to adequately model the response surface as well, and full the quadratic model. For the full quadratic model, p-value for lack of fit is 0.000 will be suggested for this model to adequately fits the data shown in Table 6 and Table 7.

The results shown in Table 6 are the estimated regression coefficients for roughness to all the terms in the model. By using an orthogonal design, each effect is estimated independently. Therefore, the coefficients of linear terms are the same as in the linear model. The error term, S = 1.61129, is smaller because of reducing the variability accounted for by the error. For versus fits, the residuals do not appear to be randomly scattered about zero.

The analysis of variance for surface roughness in Table 7 summarizes the linear terms, the squared terms, and the interactions. The large p-values for the interactions P = 0.994 and the squared terms P = 0.165 suggests there is curvature in the surface roughness. Large p-values of the pressure by power interaction P = 0.968, pressure by speed interaction P= 0.836 and power by speed interaction P= 0.866. Squared P = 0.489, and power squared P = 0.726 and speed squared smallest P= 0.047 indicate that these effects are statistically significant. The p-value of 0.000 indicates that the factor of green colour is significant.

Table 7. Analysis of Variance for Roughness.

Source	DF	Seq SS	Adj SS	Adj MS	F	P
Regression	9	27.8	27.8	3.09	1.19	0.39
Linear	3	11.3	11.3	3.78	1.46	0.28
Pressure	1	0.08	0.08	0.08	0.03	0.86
Power	1	0.15	0.15	0.15	0.06	0.81
Speed	1	11.1	11.1	11.1	4.28	0.06
Square	3	16.2	16.2	5.43	2.09	0.16
Pressure* Pressure	1	2.09	1.34	1.34	0.52	0.48
Power* Power	1	0.89	0.33	0.33	0.13	0.726
Speed* Speed	1	13.2	13.2	13.2	5.12	0.04
Interaction	3	0.20	0.20	0.06	0.03	0.99
Pressure* Power	1	0.00	0.00	0.00	0.00	0.96
Pressure* Speed	1	0.11	0.11	0.11	0.05	0.83
Power* Speed	1	0.07	0.07	0.07	0.03	0.86
Residual Error	10	25.9	25.9	2.59		
Lack -of-Fit	5	25.5	25.5	5.11	63.8	0.00
Pure Error	5	0.40	0.40	0.08		
Total	19	53.8				

Figure (5) is the contour plots which indicate that the highest roughness is obtained when pressure levels are low and power levels are high. This area appears at the upper left corner of the plot. The analysis for surface roughness (Ra) was done by using the response surface regression as shown in Table 5 and Table 6.

Whereby; P – value is denoted the importance of the parameter having less than 0.05 is deemed significant the important factor if focused in this note here. It can be observed from the estimated regression coefficients for Ra that the (cutting speed * cutting speed) has the lowest p – value of 0.047 only. This indicates that this parameter has considerable effect on the surface roughness when the p – value is less than 0.05. That's mean the model is fit and the defect in this model is not too significant. from the analysis of variance section, the linear regression has a smallest p – value with 0.047, and that's means the empirical model is going to be a square model.

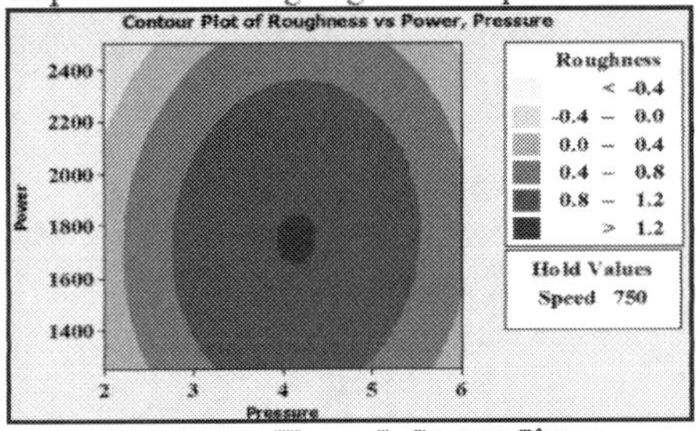

Figure 5. Contour Plot.

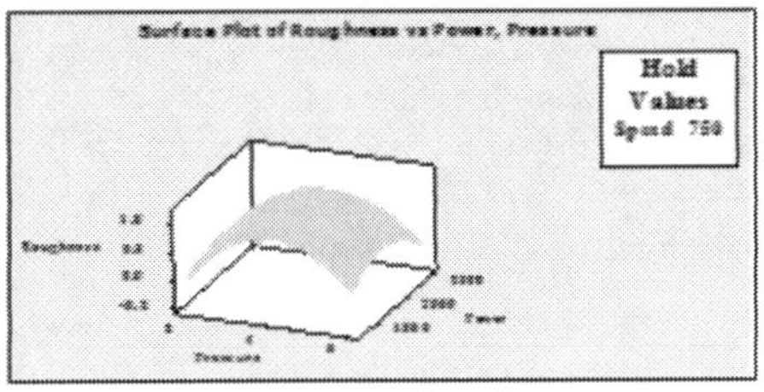

Figure 6. Surface Plot.

Contour plots in Figure (6) indicate that the highest yield is obtained when pressure levels are low and power levels are high. This area appears at the upper left corner of the plot. Surface plots in Figure (7) indicate that the highest yield is obtained when pressure levels are low and power levels are high. This area appears at the upper left corner of the plot. In addition, It's important to keep in mind that these plots should be based on a model equation, and the model is adequate before interpreting the plots

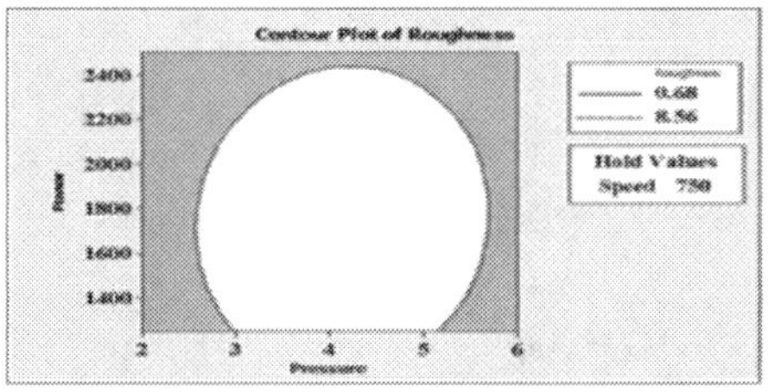

Figure 7. Contour of Roughness.

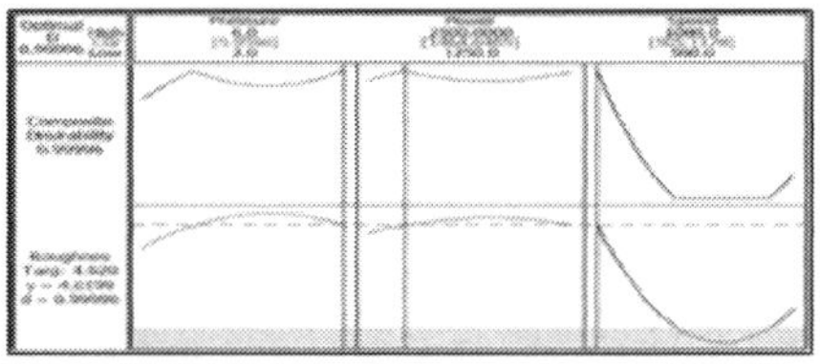

Figure 8. Surface Roughness Optimization.

In Figure (7), the white area in the mid square of the plot shows the range of pressure and power where the criteria for both response variables are satisfied. You may increase or decrease the holding value to see the range change. To understand the feasible region formed by the three factors, it's important to repeat the process to obtain plots for all pairs of factors.

6 Response Optimization

Table 8 shows the results surface roughness optimization

Table 8. Surface Roughness Optimization.

Roughness	Goal	Lower	Target	Upper	Weight	Import
	Target	0.68	4.62	8.56	1	1

As show in Figure (8), the individual desirability of both surface roughness and the variance in surface roughness is 0.99996; therefore the combined or composite desirability of these two variables is 0.99996. To obtain this desirability, you would set the factor levels at the values shown under Global Solution. That is, pressure would be set at 5.9596, power at 1464.6465, and speed at 505.1178. Move the vertical bars to change the factory settings and see how the individual desirability of the responses and the composite desirability change. In Figure (8) you may want to see if you can reduce the speed which would save costs and still meet the product specifications.

7 Conclusions

The research findings suggest that RSM can be used to model the laser cutting process for stainless steel duplex 2205 materials. Three separate models in the form of polynomial equations were successfully developed to relate the relationships between machine, laser input process parameters (Co_2 pressure, cutting speed, and power supply) and output responses (surface roughness).
The models were also qualitatively validated by justifying the behaviour of the output responses (roughness) with respect to variation of input variables, based on published work by researchers and practitioners in this field. This study also identified the significant parameters that influenced the cutting roughness through ANOVA analysis during the development of the models. Surface roughness was influenced by Co2 pressure and power supply, but cutting performance was influenced by speed. The ANOVA analysis also suggested that there was a significant interaction between the Co2 pressure and power supply which cannot be ignored and has not been reported in published journals. The ANOVA analysis indicated that the interaction between the pressure and power significantly influenced the resultant cutting roughness. The correlation study between cutting characteristics and the non-linear inclined suggested that the cutting performance correlated most significantly to the surface roughness with stainless steel type value. The study also suggested some correlation between cutting performance with percentage ratio of roughness measurement in stainless steel.

References

1. Choudhury, I. A., & Shirley, SLaser cutting of polymeric materials: An experimental investigation. Optics & Laser Technology, 42(3), 503-508. . (2010).
2. Eltawahni, H. A., Hagino, M., Benyounis, K. Y., Inoue, T., & Olabi, A. G. Effect of CO2 laser cutting process parameters on edge quality and operating cost of AISI316L. 44(4), 1068-1082. (2012).
3. Madić, M., Radovanović, M., & Nedić, B. Correlation between Surface Roughness Characteristics in CO2 Laser Cutting of Mild Steel. Tribology in Industry vol. 34, No. 4. (2012).
4. Samant, A. N., & Dahotre, N. B. Laser machining of structural ceramics: a review. Journal of the European Ceramic Society, 29(6), 969-993. (2009).
5. Tulasiramarao, B., Srinivas, K., Effect of Processing Parameters on Surface Finish of the Components Processed By CNC Turning Machine. IJMEAR. Vol 04, Issue 01. (2013),
6. Dubey, A. K., & Yadava, V. Laser beam machining: a review. International Journal of Machine Tools and Manufacture, 48(6), 609-628. (2008).
7. Samant, A. N., & Dahotre, N. B. (2009). Laser machining of structural ceramics: a review. Journal of the European Ceramic Society, 29(6), 969-993
8. Ghany KA, Newish M. Cutting of1.2 mm thick austenitic stainless steel sheet using pulsed and CWNd:YAG laser. Journal of Materials Processing Technology; 168:438–47. (2005).
9. Gok, A., Gologlu, C., Demirci, H. I., & Kurt, M. Determination of Surface Qualities on Inclined Surface Machining with Acoustic Sound Pressure. Strojniški vestnik - Journal of Mechanical Engineering, 58(10), 587-597. (2012).
10. Eltawahni H. A. , A. G. Olabi, and K. Y. Benyounis, 2011. "Investigating the CO2 laser cutting parameters of MDF wood composite material," Optics & Laser Technology, vol. 43, no. 3, pp. 648–659, Apr. (2011).

Axial and Pressure Thrust Stiffness of Metal Bellows for Vibration Isolators

Macháček Ondřej, Kubík Michal, Strecker Zbyněk, Roupec Jakub, Novák Petr and Mazůrek Ivan

Brno University of Technology, Institute of Machine and Industrial Design, Technická 2896/2 Brno, Czech Republic

Abstract. Metal bellows are used as a spring element and simultaneously as a container (shell) for a damping medium in vibration isolators, especially in applications where any leakage is inadmissible. Dynamic behaviour of these isolators is affected by axial stiffness of bellows and by a resistance against axial deformation of bellows filled with fluid, which is in this article called pressure thrust stiffness. A method of the pressure thrust stiffness determination is discussed in this study. The method uses FEM model, which has been verified by stiffness measurement of a chosen bellows. Consequently, the sensitivity analysis of bellows dimensions to axial and pressure thrust stiffness was performed to find a dimension parameter of bellows which allows to adjust the ratio between axial and pressure thrust stiffness. Sensitivity analysis shows that the stiffness ratio of metal bellows can be adjusted by two dimensional parameters - mean diameter of bellows and corrugation width.

1 Introduction

Bellows is a flexible shell component which is usually made of stainless steel by welding or forming. Shape, dimensions and number of corrugations (waves) affect mechanical properties of the bellows [1]. Except well know applications such as pipeline systems, flexible couplings etc. [2], the metal bellows have been used in vibration isolators, especially for space applications [3, 4].

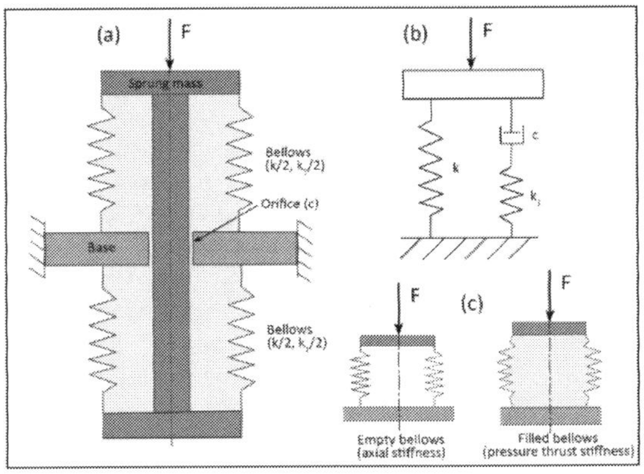

Figure 1. Scheme of vibration isolator with bellows.

The isolators with bellows, see Fig. 1 (a), have several benefits compared to conventional vibration isolators with a piston and a piston rod which must be sealed by sliding seals. The isolators with bellows are sealed by static seals, thus they are frictionless [5]. Therefore, the risk of leakage is minimal. Last but not least, their isolation ability at high frequencies is much better than the isolators with piston [3].

Sum of axial stiffness of both bellows result in primary stiffness of the isolator k. The bellows are filled with a damping medium (fluid), see Fig. 1 (b). Flow of the fluid through the orifice causes a damping c. The load force F increases the pressure of fluid in upper bellows. The pressure increase causes the fluid flow and a pressure thrust (force in axial direction due to pressure), which forces the bellows to extend. Ratio between damping force caused by flow and spring force caused by pressure thrust depends on deformation and velocity of the isolator. Considering that both bellows are connected rigidly due to the red part, the bellows can extend or shorten only because of change their internal volume. Resistance of bellows filled with fluid against compression or tension is in this study called a pressure thrust stiffness, see Fig. 1 (c).

The damper c is elastically connected. The secondary stiffness k_1 of the isolator involves the compressibility of the fluid and the pressure thrust stiffness of both bellows. In case of oil as damping medium the compressibility of oil is much higher than the pressure thrust stiffness of metal bellows. Therefore, the bellows pressure thrust stiffness is almost equal to secondary stiffness of the isolator k_1.

Ratio between the secondary and the primary stiffness (1) specifies dynamic behaviour of the isolator [7].

$$N = \frac{k_1}{k} \qquad (1)$$

This study was created during a design of vibration isolator for space application. Bellows proved to be the advantageous solution for securing desired stiffness and damping in the vibration isolation system. However, pressure thrust stiffness has not been declared by the bellows manufacturers. Therefore, it was necessary to create FEM model of bellows which allows to determine the pressure thrust stiffness from the bellows dimensions, verify the model by measurement and perform a sensitivity analysis to find dimension parameters which allow to adjust the stiffness ratio. After these steps, it is possible to find the most suitable bellows for the proposed vibration isolator.

2 Methods

The FEM model was compared with stiffness measurement of two bellows with serial number 324125 by Wizenmann Company. Dimensions of both bellows were measured almost identical, but it was slightly different than the values listed in the product list [8]. Measured values of the bellows parameters are summarized in Table 1 and shown in Figure 2. The bellows was made by forming, therefore the wall thickness is lower than the initial thickness of the sheet blank [9]. The value *t* in Table 1 represents average value of the thickness along the bellows corrugation, which is experimentally determined by the manufacturer.

Table 1. Measured parameter of the bellows.

Parameter	Symbol	Value	Unit
Outer diameter	D	36.5	[mm]
Inner diameter	D	23.6	[mm]
Mean diameter	D_m=(d+D)/2	30.05	[mm]
Corrugation width	w=(D-d)/2	6.45	[mm]
Corrugation length	L	3.2	[mm]
Wall thickness	T	0.22	[mm]
Number of corrugations	I	13	[-]

Figure 2. Metal bellows no. 324125 and its dimensions.

Material of bellows is Stainless steel:

Stainless steel	1.4571
Modulus of Elasticity	E = 200 000 MPa
Poisson ratio	μ = 0.28
Yield stress after forming	Re = 500 MPa

Bellows was filled by shock absorber oil:

Shock absorber oil	Paramo TL 15
Volume modulus of Elasticity	K = 2 200 MPa
Density	ρ = 885 kg/m3

Result of axial stiffness simulation was compared with the experiment, calculation provided by manufacturer and also with the product list. The pressure thrust stiffness can be compared only with the measurement.

2.1 FEM Model

The model consists of two parts: wall (steel) and the fluid inside (oil). During determination of axial stiffness the fluid mesh is not considered in the model. The model has following parameters:

Dimensional:	2D
Symmetry:	Axisymmetric (y)
Element type:	SOLID 182 (wall)
	HSFLD 241 (fluid)
Solution:	Non linear

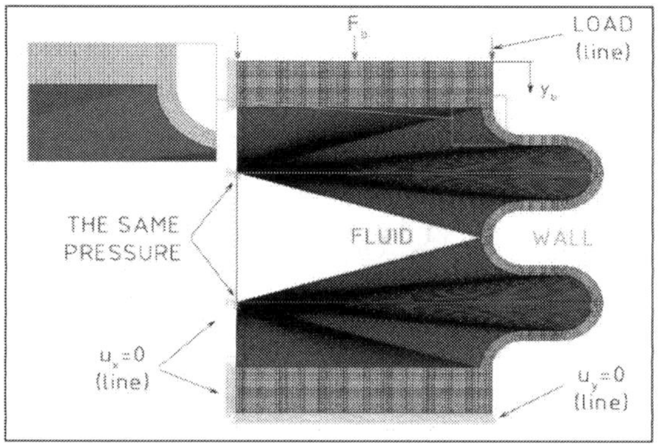

Figure 3. FEM model of metal bellows.

Load force is applied by the pressure on the top line, which corresponds to the force *Fb*. The displacement of the top line y_b is used for the stiffness calculation. The bottom line cannot move in vertical direction *Y*. Nodes located on *Y* axis cannot move in the horizontal direction *X*. There are several points (depends on number of corrugations) with the same pressure in mesh of fluid. It ensures identical pressure in the fluid, as shown in Figure 3.

Both stiffnesses are obtained by interpolation of several points calculated as ratio of Load force F_b and displacement of top line y_b. The dependence of load force and displacement of top line is assumed to be linear.
The difference between pressure thrust and axial stiffness is that the first one is evaluated when the fluid elements are activated. On the other hand, axial stiffness is evaluated when the fluid is not active.

2.2 Measurement

Measurement as well as the simulation was conducted in two configurations. The difference between them was filling and sealing of the bellows. In the case of axial stiffness measurement, the bellows was empty and the sealing was not used. But, when the pressure thrust stiffness was measured, the bellows was sealed and filled by the oil.

2.2.1 Axial stiffness

Two bellows with dimensions corresponding to Table 1 were consecutively compressed by hydraulic workshop press. The force acting on the bellows front surface was measured by the strain gauge load cell HBM - BMT mounted between the press and the front surface of bellows. Linear position transducer Penny & Gilles SLS150 was placed between the force sensor and press table as shown Figure 4. Both sensors were connected to the data acquisition analyzer Dewetron DEWE-800. Each measurement took 30 seconds. During this period, several strokes of hydraulic press were carried out for higher statistical significance. Bellows was slowly pressed in the range from 0 to 10 millimetres with corresponding force 0-230 Newton.

2.2.2 Pressure thrust stiffness

Pressure thrust stiffness measurement was conducted with similar methodology which was used for the axial stiffness, see Figure 4. Steel bellows were pressed in the range from 0 to 1 mm which corresponds to 0 and 560 N respectively.

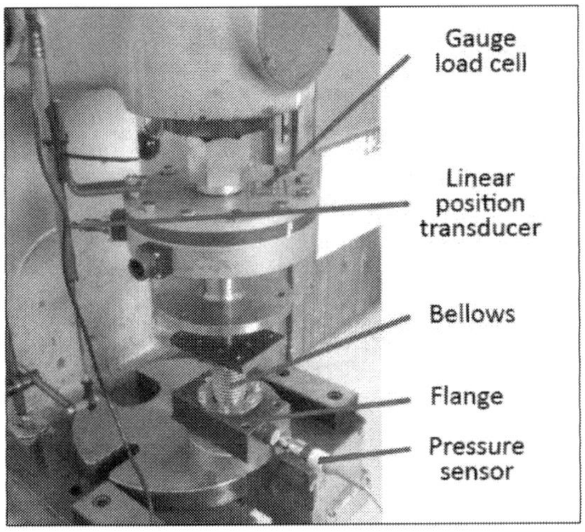

Figure 4. Pressure thrust stiffness measurement.

The measured force is caused by the sum of pressure thrust stiffness and the axial stiffness. Therefore, the pressure thrust stiffness was calculated as a subtraction of measured force and force caused by axial stiffness.
The value of pressure thrust stiffness was similarly to axial stiffness determined as the average of measurements of two bellows.The pressure of the oil inside the bellows was measured by pressure sensor *HBM* 154210274 to check that the pressure inside the bellows did not exceed the allowed value.

3 Results and discussion

Result of axial and pressure thrust stiffness was determined from measured or simulated data by linear approximation. Values of deformation higher than zero corresponds to the compressing of the bellows, deformations lower than zero means tension of bellows.

3.1 Measurement

Force-deformation dependency was measured for two similar bellows and the measured data was linearly approximated by line. The slope of line represents axial or pressure thrust stiffness respectively. Force-deformation dependency of empty bellows, shown in Figure 5, was almost linear in whole range of measurement.

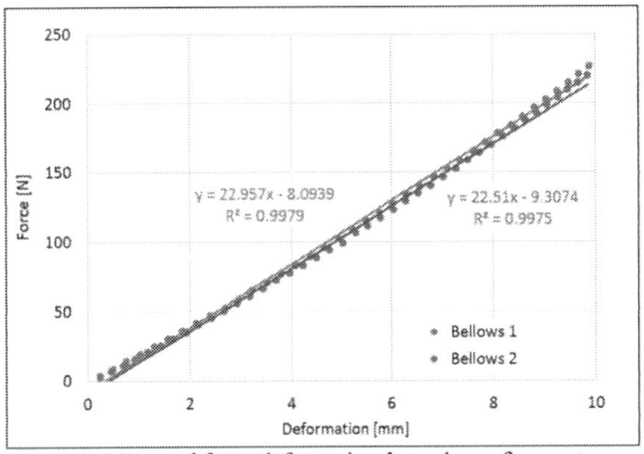

Figure 5. Measured force-deformation dependency for empty bellows (axial stiffness).

Force-deformation dependency of bellows filled by oil is nonlinear at the beginning of bellows compression. This behaviour was probably caused by non-parallel faces of flanges, which are screwed to the bellows. Therefore, the force grew gradually, until the moment when the flanges become parallel to each other. Therefore, the less saturated points in Figure 6 are not included for the approximation.

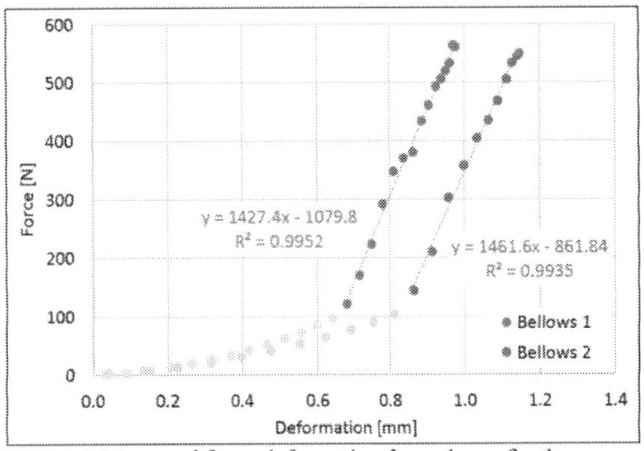

Figure 6. Measured force-deformation dependency for the bellows filled by the oil (pressure thrust stiffness).

3.2 Simulations

The same load force F_b=300 N was applied to compare bellows behaviour with or without the filling. Membrane stress in bellows wall is shown in Figure 7. The filling of bellow affects the stiffness, but also the shape of corrugation deformation.

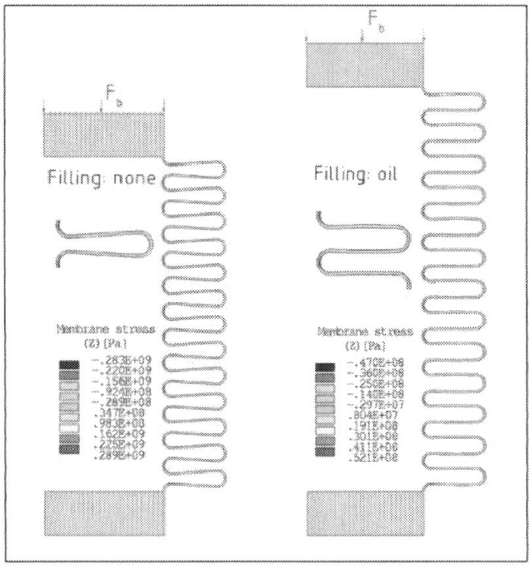

Figure 7. Simulated force-deformation dependency for empty bellows (axial stiffness).

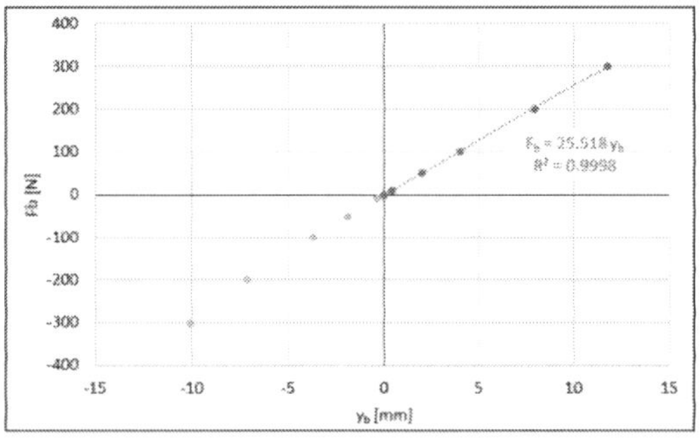

Figure 8. Simulated force-deformation dependency for empty bellows (axial stiffness).

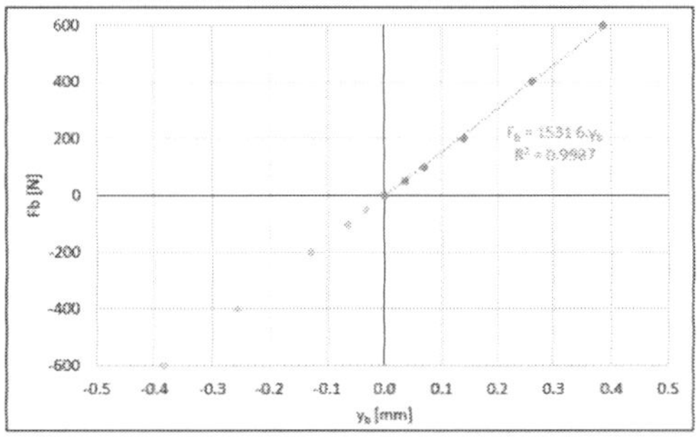

Figure 9. Simulated force-deformation dependency for the bellows filled by the oil (pressure thrust stiffness).

Besides the compression of bellows, the tensile load was simulated too. The results of force-deformation dependency of bellows load by compression and tension differs minimally. However, to achieve the same conditions of simulation and measurement, only the values of force and deformation which are higher than zero (compression of bellows) were compared.

3.3 Results comparison

The results obtained by simulations or measurement were compared. The comparison of axial stiffness was complemented by values from the product list or manufacturer calculations, listed in Table 2.

Table 2. Axial stiffness of bellows.

Procedure	Axial stiffness	Value [N/mm]	Deviation [%]
Product list [8]	k_L	22.3	12.6
Manufacturer calc.	k_C	29.5	15.6
FEM model	k_F	25.52	ref. value
Measurement	k_M	22.7	11

Axial stiffness determined by FEM model differs from all other methods less than 30 %, which is allowed tolerance of axial stiffness, which is guaranteed by manufacturer.

As already mentioned, the pressure thrust stiffness is not declared. Therefore, the pressure thrust stiffness of bellows determined by FEM model k_{IF} was compared only with the measurement k_{IM}, see Table 3.

Table 3. Pressure thrust stiffness of bellows

Procedure	Axial stiffness	Value [N/mm]	Deviation [%]
FEM model	k_{IF}	1531.6	ref. value
Measurement	k_{IM}	1444.5	5.7

3.4 Sensitivity analysis

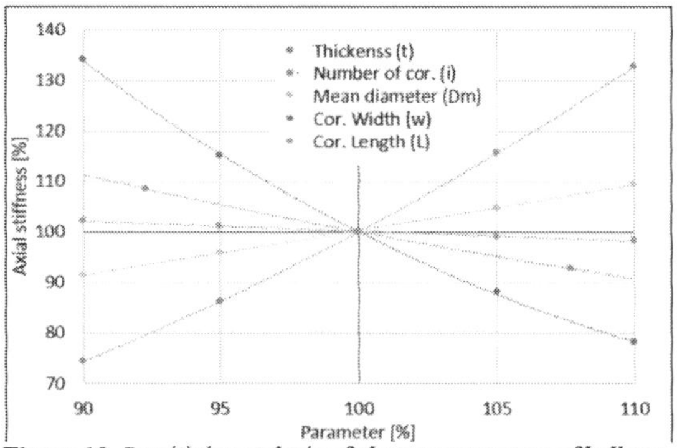

Figure 10. Sensitivity analysis of chosen parameters of bellows to axial stiffness.

The FEM model mentioned above was created for selecting suitable bellows which is going to use in vibration isolator. Therefore, the stiffness ratio N is one of important parameter. However, to adjust the stiffness ratio of isolator in Figure 1, it is necessary to find a dimension parameter of bellows which has a different influence on axial and pressure thrust stiffness of the bellows. Therefore, sensitivity analysis of bellows dimension parameters was performed separately for axial and pressure thrust stiffness. Examined parameters are listed in Table 1. Their values were chosen as 90, 95, 105 and 110% of initial value. The stiffness change is also plotted in percent of initial value (100% of each parameter) in Figure 10 - 12.

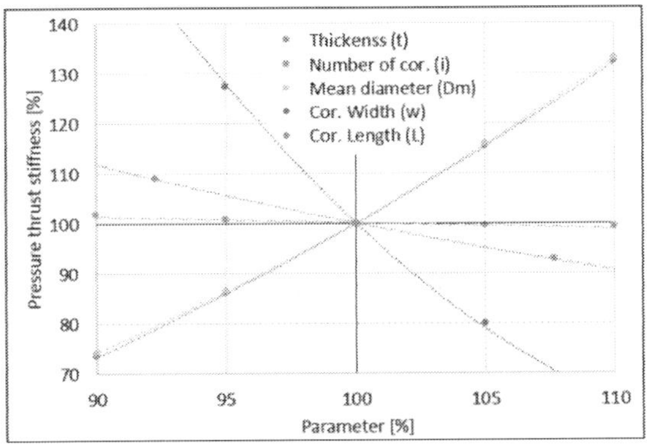

Figure 11. Sensitivity analysis of chosen parameters of bellows to pressure thrust stiffness.

The analysis confirms that the wall thickness t has significant influence on the stiffness of bellows. However, the influence is almost the same for axial and pressure thrust stiffness. Similar match was observed for two other parameters: number of corrugation i or its length L. On the contrary, the analysis revealed that change of the bellows mean diameter D_m or corrugation width w has higher influence on the pressure thrust stiffness than on the axial one.

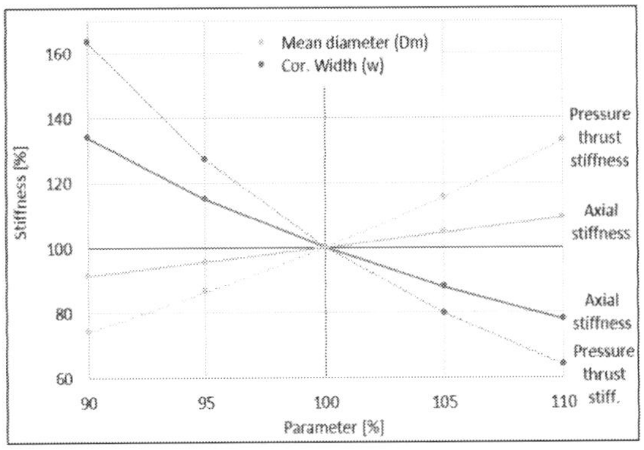

Figure 12. Parameters which allow to adjust the stiffness ratio of isolator.

4 Conclusions

The axial and pressure thrust stiffness of metal bellows have been analyzed in this study. The pressure thrust stiffness (resistance against compression of bellows filled with fluid) is not commonly referenced value. Therefore, the FEM model was created to determine this variable from the bellows dimensions. The model was verified by comparison of axial stiffness obtained by the FEM model, product list, measurement and calculation provided by the bellows manufacturer. The results of axial stiffness obtained by FEM model was 11% higher than the measured value. Simulated pressure thrust stiffness was approximately 5 % higher than value obtained by measurement. This deviation is presumably caused by the imperfect determination of average wall thickness. Moreover, the real thickness of bellows wall is not constant along corrugation.

Sensitivity analysis revealed that the parameters: wall thickness, number of corrugations and its length affect the axial and pressure thrust stiffness almost identically. On the other hand, mean diameter of bellows and corrugations width affect the pressure thrust stiffness more than axial stiffness. Therefore, the last two parameters can be used for adjustment of the ratio between axial and pressure thrust stiffness, which is important for some application of bellows, for example in vibration isolators, which uses the metal bellows as a shell for damping medium. The method mentioned in this study helps to design the most suitable dimensions of the bellows for a specific application.

Acknowledgement

This work could be provided thanks to the kind sponsorship of various grants and numerous agencies. We would like to thank explicitly to the GAČR 17-10660J, GAČR 17-26162S, FCH/FSI-J-17-4534 and FSI-S-17-4428.

References

1. M. Geni, M. Kikuchi, Advances in Fracture and Demage Mechanics VII, **385-387** (2008)
2. Y. Shi, B. Yao, L. Wang, Q. Zhang, Y. Che, Applied Mechanics and Materials, **130-134**, 2353-2357 (2011)
3. P. Davis, D. Cunningham, H. Head et al. *35th AIAA SDM Conf.* (1994)
4. DO. Lee, G. Park, JH. Han, Aerospace Science and Technology, **45**, 1-9 (2015)
5. M. Kubík, O. Macháček, Z. Strecker, J. Roupec, I. Mazůrek, Smart Materials and Structures **26**, 4 (2016)
6. E. A. Sibru, S. Horga, G. Vrabioiu, 7th Conf. on Advanced Concepts in Mechanical Engineering, **147** (2016)
7. M. J. Brennan, A. Carrella, T. P. Waters, V. Lopes, Journal of Sound and Vibration **306**, 823-837 (2008)
8. http://www.witzenmann.cz/cs/media/Kovove_vlnovce_a_kompenzatory_8307cz_2_02_10_10.pdf
9. S. Wei, L. Xu, K. He, J. Li, W. Feng, R. Du, International Journal of Advanced Manufacturing Technology, **84**, 129-133

MATEC Web of Conferences **153**, 06002 (2018) https://doi.org/10.1051/matecconf/201815306002
ICMME 2017

Transient Magnetic Model of Magnetorheological Damper and Its Experimental Verification

Kubík Michal , Macháček Ondřej, Strecker Zbyněk, Roupec Jakub, Novák Petr and Mazůrek Ivan

Brno University of Technology, Institute of Machine and Industrial Design, Technická 2896/2 Brno, Czech Republic

Abstract. The present paper deals with the transient magnetic model of the magnetorheological (MR) damper and its experimental verification. The response time of MR damper affects the quality of semi-active control of this damper. The lower the response time, the higher the system efficiency. The most important part of the response time of the MR damper is the response time of magnetic field of the MR damper which can be determined by transient magnetic model. The transient magnetic model was created by the software Ansys Electromagnetics 17.1 as 2D axisymmetric and verified by measurement of magnetic field in the gap of MR damper piston. The maximum difference between the model and the experiment was 28 %. The response time depends on the electric current in the coil of MR damper. The transient magnetic model was used for determination of influence of MR fluid type, material of cover and material of magnetic circuit on the response time of magnetic field of MR damper. The type of MR fluid has a significant influence on the response time. The lower the mass concentration of ferromagnetic particles, the lower the response time of magnetic field. A material selection of magnetic circuit is always a trade-off between the response time and the maximum magnetic flux density (dynamic force range) in the gap of the MR damper. According to the verified transient magnetic model, it is possible to find a suitable material of magnetic circuit for specific application (response time).

1 Introduction

Magnetorheological fluid (MRF) is a suspension of micro-scale, non-colloidal ferromagnetic particles in a non-magnetic carrier fluids and additives. The MR fluid exhibits a rapid change of rheological behaviour under the influence of external magnetic field. Ferromagnetic particles in MR fluid form chain-like structures in the direction of magnetic field with a rapid increase in yield stress [1]. The value of yield stress is strongly dependent on the level of applied magnetic field. The increase of yield stress in the MR fluid located in the external magnetic field is a basic feature of MR technology and is usually called a magnetorheological effect.

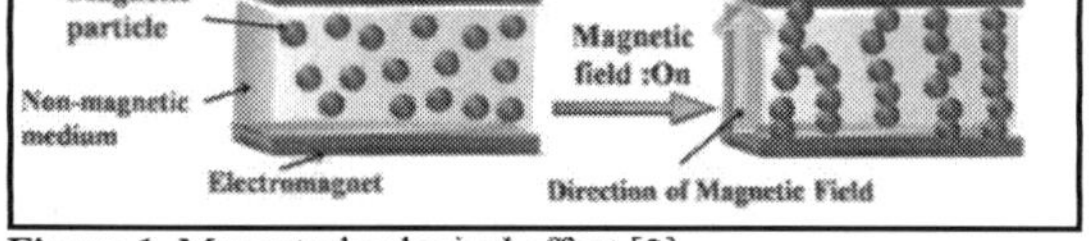

Figure 1. Magnetorheological effect [2].

An interest in MR fluid derives from its ability to provide a simple interface between the electronic and mechanical system. A magnetorheological (MR) damper is a device which profits from the unique properties of MR fluid.

This damper uses strong and rapid changes in yield stress of MR fluid in the direction of magnetic field which leads to an increase in damping force [3]. The MR damper is composed of piston, seal piston rod, floating piston, and MR fluid. The piston consists of magnetic circuit and electromagnetic coil. There is a gap in the magnetic circuit which is flooded with the MR fluid (through the annular orifice in piston). The piston divides the MR fluid volume between the compression chamber and the rebound chamber. The gap in the piston allows for the MR fluid to flow between the chambers during the piston motion. The value of yield stress (hydraulic resistance) can be controlled by the electric current in the coil during the passage of MR fluid through the gap. This causes a rapid increase in damping force. The MR damper is mechanically simpler than the traditional design of the hydraulic damper because it does not contain mechanical moving parts. For this reason, the MR damper is noise-free [4]. MR dampers are commonly used for adaptive control. MR dampers with semi-active control have been mentioned in scientific papers in the last years. The most significant factors of quality semi-active control of MR damper are as follows: response time of MR damper and dynamic force range of MR damper; the latter is well described in [3], [1]. The magnetic flux density in the gap of MR damper has a significant influence on the dynamic force range (yield stress of MR fluid).

Response time of MR damper

The response time of MR damper is defined as the time necessary for transition from the initial state to 63.3 % (in some publications 95 % [7]) of the final state [1] represented by a steady-state damping force. The response times of published designs of MR dampers are in the range of 20-100 ms [1],[3], [4]. From the existing literature, it can be assumed that the response time of MR damper has three sources [5], [6]: inductance of MR damper coil (response time of exciting current), response time of MR fluid, eddy currents in the coil core (response time of magnetic field in the active zone). The response time of MR fluid was estimated in the range of 0.45-0.6 ms by Goncalves [7]. Yang [1] published his findings that the use of a suitable current controller can significantly reduce the response time of exciting current. Strecker [5] stated that the response time of exciting current can be reduced to 1 ms. Eddy currents in the magnetic circuit create a time lag between the course of electric current and the course of MR damping force [8]. These currents produce a magnetic flux which flows in the opposite direction to the magnetic flux in the magnetic circuit. A selection of the suitable material or the geometry of magnetic circuit of MR damper can eliminate eddy currents (a short response time of magnetic field in the gap). One of the methods how to select the suitable material and the geometry of MR damper is to use a magnetic simulation. According to the available information, only Naoyuki [9] and Strecker [5] dealt with the magnetic modelling of eddy currents in the magnetic circuit of magnetoreological devices. However, in their contribution, this issue was solved only marginally.

The main goal of this paper is to create a transient magnetic model which allows to determine the response time of magnetic field in the active zone of MR damper, and its experimental verification. The influence of type of MRF, material of cover and material of magnetic circuit on the response time will be determined by a verified model.

2 Materials and method

2.1 Geometry of MR damper

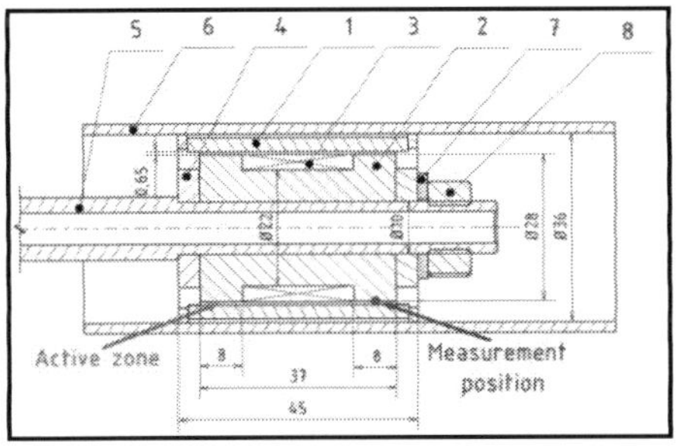

Figure 2. Geometry of MR damper used for verification.

The geometry of the MR damper was used to compare the proposed transient magnetic model and the experiment, according to fig. 2.

The MR damper is composed of outer cylinder (1), coil core (2), coil (3), covers (4), piston rod (5), hydraulic cylinder (6), washer 10 (7), and nut M10 (8). The coil (3) was wound on the coil core (2) with 120 turns of cooper wire (diameter 0.5 mm). The coil core (2) and the outer cylinder were made of steel 11SMn30 (fig. 3, red color). The B-H curve of this material was experimentally determined. The covers (4) were made of aluminum alloy. The piston rod (5) and the hydraulic cylinder (6) were made of steel 34CrMo4 (fig. 3, yellow color). The power line of coil passes through the hollow piston rod which is sealed.

2.2 Transient magnetic model

A transient magnetic model was created in software Ansys Electromagnetics 17.1. A simplified geometry of MR damper in this transient model was created (fig.3) as 2D axisymmetric.

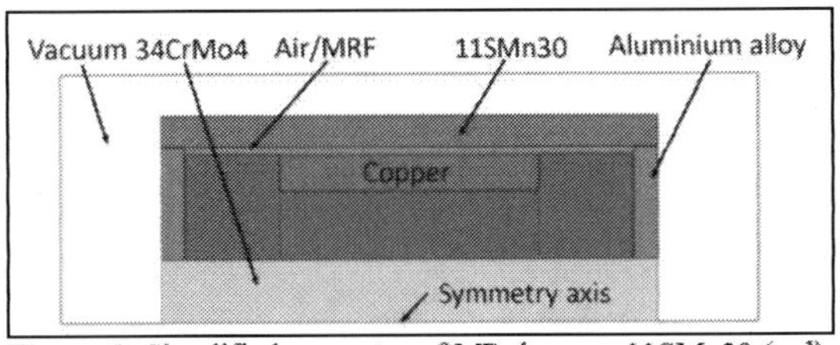

Figure 3. Simplified geometry of MR damper; 11SMn30 (red), MR fluid or air (grey), aluminium alloy (blue), copper (green), 34CrMo4 (yellow) and vacuum (white).

A boundary condition balloon was set on the outside part of the vacuum region. The excitation on the coil was a magnetomotive force I*120 A.turns. The external electric circuit was connected to the coil winding. The electric circuit is composed of resistor (resistance of winding coil) and current source. A rise time of the current source was set at 0.2 ms. This value was determined experimentally as an average value using the current clamp and the current controller. The important value for transient magnetic model is electric conductivity (resistivity) of magnetic circuit material. In our case, the electric conductivity of material 11SMn30 was set at 5.75 MS/m according to the datasheet. The eddy current effect in terms of the magnetic transient solution was applied to the entire geometry. 3584 elements were used for discretization of geometry. The time step of 0.1 ms and nonlinear residual at 0.0001 were used for the analysis.

2.3 Measurement of response time of magnetic field in the gap of MR damper

The magnetometer Bell 5180 and the ultrathin Hall probe were used for experimental verification of transient magnetic model. Magnetic flux density was measured in the middle of the active zone of the MR damper free off MR fluid.

The relative permeability of Hall probe (μ_r = 1) is significantly lower than that of MR fluid (μ_r = 3-10) which causes the magnetic flux to flow around the Hall probe (fig. 4). This will result in inaccurate measurement. For this reason, the experiment is carried out in the regime with the air in the active zone of MR damper ($\mu_r = 1$). It is assumed that the change of material in the active zone does not affect the accuracy of the transient magnetic model.

The coil of MR damper was powered by laboratory source Manson SDP 2603 (fig. 5). The current controller was made by Strecker [5] and controlled by the Arduino system, which was switched between the maximum and zero current in the coil and, vice-versa, with the frequency of 2 Hz.

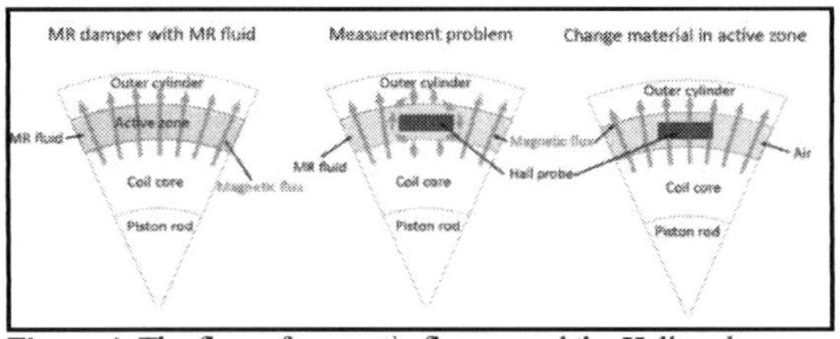

Figure 4. The flow of magnetic flux around the Hall probe.

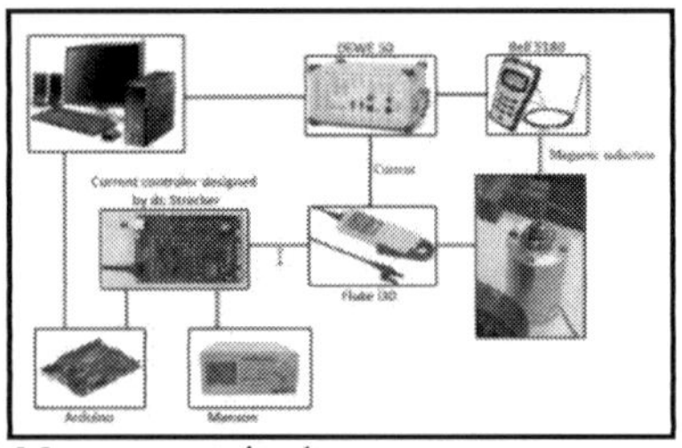

Figure 5. Measurement circuit.

The magnetic flux density in the middle of gap of MR damper was measured with the magnetometer Bell 5180 and the ultrathin Hall probe (fig. 2, fig. 6). The current clamp Fluke i30 was used to measure the electric current. Data from the magnetometer and the current clamp were acquired using Dewe 50 in the sampling frequency of 200 kHz; both channels were synchronized. The measured data were evaluated in software DeweSoft.

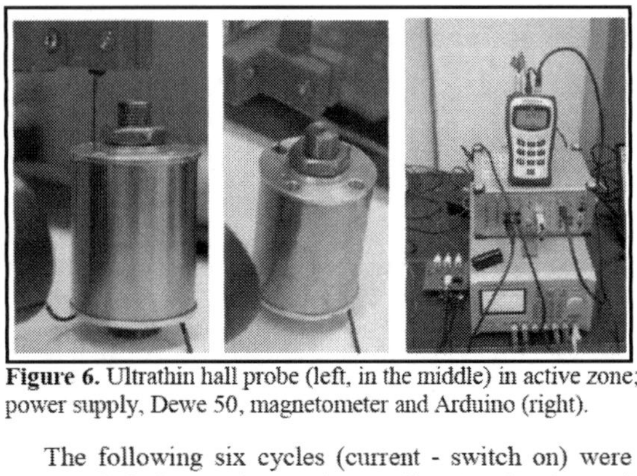

Figure 6. Ultrathin hall probe (left, in the middle) in active zone; power supply, Dewe 50, magnetometer and Arduino (right).

The following six cycles (current - switch on) were measured during each experiment (fig. 7). The rise and drop of magnetic flux density in the active zone were measured. The coil of MR damper was supplied with 0.3 A, 0.6 A, 1 A and 1.6 A. The measured data of magnetic flux density have been offset due to the residual magnetism in the magnetic circuit. For verification, the offset was neglecting. The response time of magnetic field of MR damper was determined as 63.3 % of steady-state value of magnetic flux density.

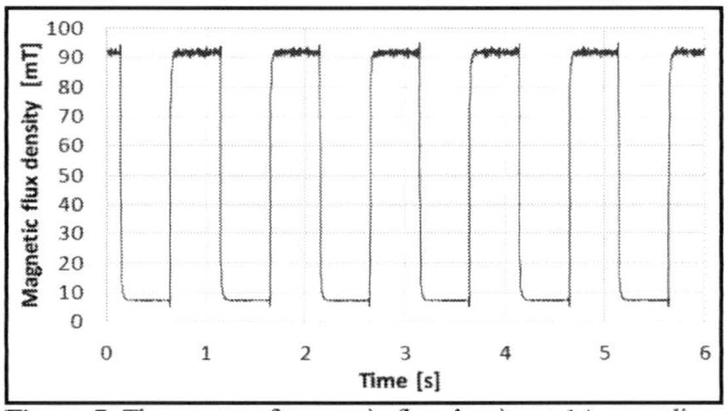

Figure 7. The course of magnetic flux density at 1A according to control.

3 Results

3.1 Experimental verification of model

Fig. 8 shows a comparison of magnetic flux density courses between the transient magnetic model (blue) and the experiment (red) for the rise and drop of the supplied current 1 A (green).

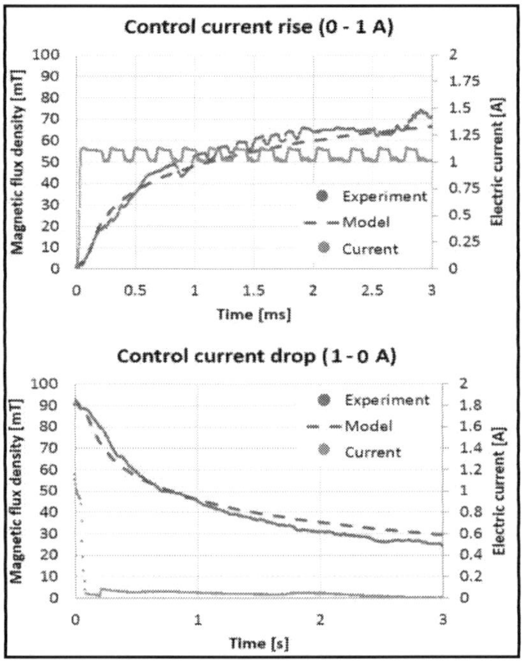

Figure 8. Rise of magnetic flux density; drop of magnetic flux density.

The response time of magnetic field in the gap was determined from the measured data and the transient magnetic model for different currents on the coil (fig. 9).

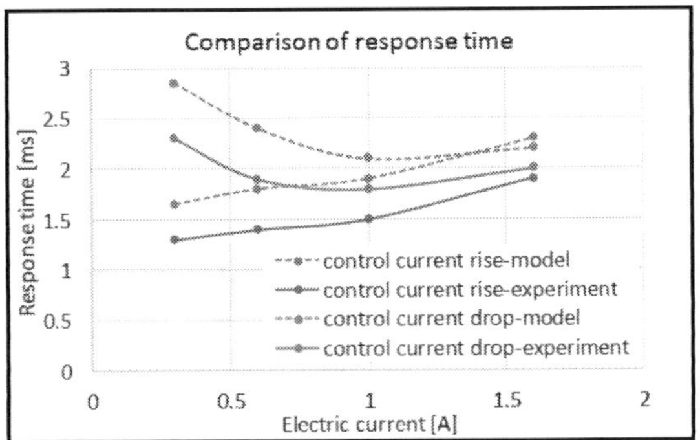

Figure 9. The response time of magnetic field from the model and the experiment for different currents.

A maximum difference between the model and experiment is 28 %. However, the slope of measured data and that of the experiment is similar. The difference between the model and the experiment can be justified by inaccurate specification of electrical conductivity, which was used according to the material product list. Interestingly, the response time of magnetic field is not constant but dependent on the electric current. This dependency on the response time of magnetic field is probably caused by non-linear B-H curve of material of magnetic circuit.

3.2 Transient model with MR fluid

The material in the active zone was changed in the verified transient magnetic model from the air to MRF. The B-H curve of MR fluid MRF-132DG, produced by Lord company, was set in the gap. The response time for different currents (0.1 A, 0.3 A, 0.6 A, 1 A, 1.6 A and 2 A) was determined by transient magnetic model (fig. 10).

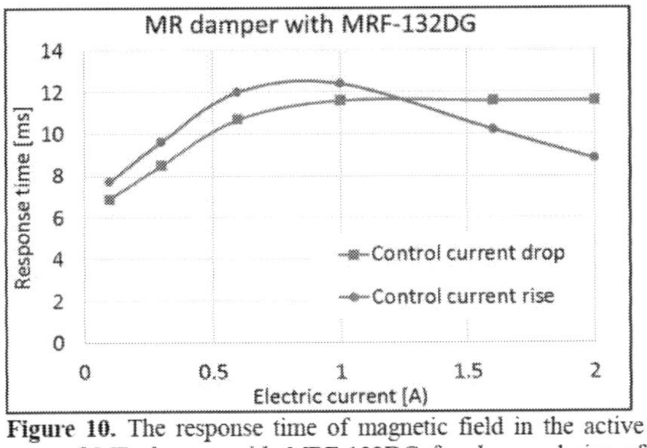

Figure 10. The response time of magnetic field in the active zone of MR damper with MRF-132DG for drop and rise of control current.

The response time of magnetic field of MR damper with MR fluid is also dependent on the electric current because of non-linear B-H curve of magnetic circuit material and MR fluid. The response time of control current rise from 0 A to 1A is increasing. After that, the response time is decreasing (fig. 10). The response time of control current drop from 0 A to 1A is increasing. After that, the response time is being stabilized (fig. 10).

The magnetic flux from the transient magnetic model at different times (0, 1, 5 and 20 ms), and the current of 1 A for control current drop was shown in fig. 11.

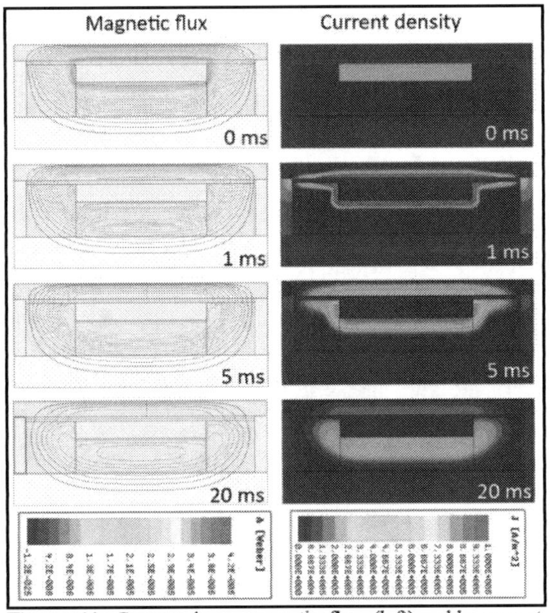

Figure 11. Current drop: magnetic flux (left), eddy current density (right).

The magnetic flux in the magnetic circuit caused by the induced eddy currents can be seen after a control current drop. The largest flow of eddy currents is near the coil surface.

3.3 Influence of MR fluid

The influence of MR fluid on the response time of magnetic field for three different MR fluids was tested. We chose the MR fluids produced by company Lord (MRF-122EG, MRF-132DG a MRF-140CG). The chosen MR fluids have different mass concentration MRF-122EG 72 %, MRF-132DG 80.98 % and MRF-140CG 85.44 %. The results are shown in tab 1.

Table 1. The influence of MR fluid on the response time.

Name	Magnetic flux density	Response time of rise 1A	Response time of drop 1A
122EG	286 mT	8.7 ms	9.8 ms
132DG	342 mT	12.4 ms	11.6 ms
140CG	387 mT	14.5 ms	13.8 ms

The lower the mass concentration in the MR fluid, the lower the response time and the magnetic flux density.

3.4 Influence of material of cover

Fig. 11 shows eddy currents in covers, which creates opposite magnetic flux in the gap. The influence of cover material was tested for this purpose. The result is that, according to the transient magnetic model, the material of cover has no influence on the response time of magnetic field.

3.5 Influence of material magnetic circuit

Five different samples of material magnetic circuit were tested: pure iron (behanit), low carbon steel (11SMn30), magnetic stainless steel (AISI 416), the alloy Fe-Co (Hyperco 50A), and the ferrite Epcos N95. These materials were compared using the previously described transient magnetic model with MRF-132DG at the current of 1 A. The results obtained from the model were the response time and magnetic flux density in the gap (tab.2). A selection of magnetic circuit material for the MR damper is always a trade-off between minimizing the response time of magnetic field and maximizing the magnetic flux density in the gap.

Table 2. The influence of material magnetic circuit on the response time and magnetic flux density in the gap.

Name	Electric conductivity [MS/m]	Magnetic flux density at 1A [mT]	Response time of drop at 1A [ms]
Pure iron	8.3	410	18.5
11SMn30	5.75	342	11.6
AISI 416	2.33	315	3.5
Hyperco 50A	2.45	466	5
Feritte N95	$0.16x10^{-6}$	112	0.2

4 Conclusion

The present paper deals with the transient magnetic model of magnetorheological (MR) damper and its experimental verification. This model was created by software Ansys Electromagnetics 17.1 as 2D axisymmetric. The verification of magnetic model was performed on the geometry of MR damper by comparing the simulated value of magnetic flux density and the value measured with the magnetometer Bell and the ultrathin Hall probe. A maximum difference between the model and the experiment was 28 %. The response time of magnetic field in the gap of MR damper with MRF 132-DG was around 12 ms, according to the transient magnetic simulation. The response time of drop and rise in the control current of MR damper has different courses, depending on the electric current. This effect is probably caused by a nonlinear B-H curve of magnetic circuit and MR fluid. The verified transient magnetic model allows to design and select the material for magnetic circuit of MR damper for a specific engineering application with respect to the response time and the dynamic force range.

The influence of type of MR fluid, material of cover and material of magnetic circuit on the response time of magnetic field was determined by verified transient magnetic model. A type of MR fluid has a significant influence on the response time of MR damper. The higher the mass concentration of ferromagnetic particles in MR fluid, the higher the time response of magnetic field of MR damper and the lower the magnetic flux density in the gap. The selection of material of cover has no influence on the response time. On the contrary, the analysis showed that the material of magnetic circuit significantly affects the response time and magnetic flux density in the gap. The ferrite material had the lowest response time of magnetic field for chosen materials. However, the magnetic flux density of this material in the active zone was almost three times lower than that of other materials. The ferrite material is suitable for the magnetic circuit of fast semi-active control of MR damper. Unfortunately, a dynamic force range is very low. The material of magnetic circuit Hyperco 50A is suitable for the adaptive regime of MR damper because of high dynamic force range. The selection of material of magnetic circuit of MR damper is always a trade-off between the response time and the dynamic force range.

Further research is focused on finding the method of how to reduce the response time while maintaining a similar dynamic force range. A promising method to use is a shape approach (grooving magnetic circuit).

Acknowledgement

This experiment could be performed thanks to a kind sponsorship of various grants and numerous agencies. We would like to thank explicitly to the GAČR 17-10660J, GAČR 17-26162S, FEKT/FSI-J-17-4483 and FSI-S-17-4428.

References

1. G. Yang, S. Jr. D. Carlson, K. Sain, Eng. Struct. **24,** 309-323 (2002)
2. B. J. Park, F. F. Fang, H. J. Choi, Soft Matter **6,** 5246-5253 (2010)
3. M. Kubík, O. Macháček, Z. Strecker, J. Roupec, I. Mazůrek, Smart Mater. Struct. **26,** 1-9 (2017)
4. J. Gołdasz, B. Sapiński, *Insight into Magnetorheological Shock Absorbers, 1-224 (2015)*
5. Z. Strecker, J. Roupec, I. Mazurek, O. Machacek, M. Kubik, M. Klapka, J. Intell. Mater. Syst. Struct. **26,** 1951-1958 (2015)
6. J. H. Koo, F. D. Goncalves, M. Ahmadian, Smart Mater. Struct. **15**, 351-358 (2006)
7. F. D. Goncalves, M. Ahmadian, J. D. Carlson, Smart Mater. Struct., **15**, 75-85 (2005)
8. J. Mass, D. Guth, *ASME 2011 Conf. smart Mater. Adapt. Struct. Intel. Syst.,*(2011)
9. N. Takesue, J. Furusho, Y. Kiyota, *IEEE International Conference on Robotics and Automation*, (2003)

Analytical Velocity Study in a Conical Diffuser with Screw Tape Inserts

Ehan Sabah Shukri

Department of electronic technology, Engineering Technical College-Baghdad, Middle Technical University (MTU) Baghdad, Iraq

Abstract. A study is made to enhance the rate of velocity distribution in a conical diffuser. In this work, a numerical analysis on screw tape inserts in a conical diffuser is presented. In the numerical simulations, the swirling flow was introduced by using rectangular screw tape placed inside the inner test wall of the conical diffuser. Screw tape with different aspect ratios (AS) 2.5, 3.5, 4.5, 6.5 and 7.5 was analysed. The simulations were carried out with constant inlet condition considering the flow turbulent and incompressible with inlet Reynolds number 3.2 x 10^5. The simulations were performed using air as a working fluid. The results obtained from the conical diffuser with screw tape inserts are compared with those without screw tape (plain conical diffuser). On the basis of the same inlet boundary conditions for the screw tape in the conical diffuser and the plain conical diffuser, it was found that the velocity distribution performance of screw tape inserts with different AS is better than plain conical diffuser. It is also observed that the screw tape with AS 3.5 offered the best velocity distribution rate.

1 Introduction

Turbulent flows in passages with gradually varying cross-sectional area are present in several applications in different engineering branches, especially where the geometric limitations are required like turbo-machinery. It has long been the subject of interest from the scientific community. Diffusers are devices of divergent area play an important role in many fluid machines to convert kinetic energy into pressure energy. Flow through diffuser has numerous industrial applications, and needs to be understood in more details. Therefore considerable amount of experimental and numerical investigations has been carried out on conical diffusers [1-4].

A swirl flow is a fluid flow contains a noticeable tangential velocity component and it can provide notable fluid mixing. In recent year, these kinds of flows have received more attention. They are attractive for many heat transfer and mixing applications. Therefore, many studies have focused on the swirl flows. It is widely used in industrial combustion devices such as gas turbine combustors, furnaces, burners to provide power generation. It has a dominant effect on the mixing process in gas turbine combustion chamber [5-8].

Eldrainy et al. [9-10] examined the effect of different swirler geometries on the flow dynamics inside a micro gas turbine combustion chamber model. Designing air swirler to produce stable and efficient combustion with low-pressure losses was a challenge. Axial flat vane swirler with different vane angles was tested. The simulated results confirmed that weak swirl with low swirl number was not sufficient to produce a strong centrifugal force. Consequently, if the swirler vane angle increased, both swirl number and flow deflection angles were increased. Moreover, extra pressure reduction was occurred, and the reversed flow was increased as well, so as a conclusion, with high deflection angles, turbulence intensity increased and the mixing process will be promoted [9]. Furthermore, Eldrainy et al. [10] confirmed that swirl number under 0.4, recirculation would not be observed. Therefore, mixing applications should be designed for swirl number more than 0.6. Thus, controlling such parameters, the mixing process can be improved. Swirling flow is one of the passive techniques that usually accompanied with high tangential velocity and turbulence intensity, which provides an additional mechanism to increase the mixing.

Most attention has been given to strong recirculating swirling flow in diffusers, for their extensively used in fluid mechanical devices. To increase mixing and combustion intensity, a study had been conducted numerically and experimentally [11]. It was carried out with flat and curved vanes. Four different vane angles (40° 50° and 60°) in the inlet of a combustion chamber were examined. Comparing the results between numerical and experimental method, it was found that the higher the vane angle, the greater the swirling flow, and the larger the recirculation zone. The results showed that the flat vane was more effective than the curved vane for numerical and experimental methods.

Numerical simulations have been successfully used by many authors to describe swirling flows. Therefore, the aim of the present numerical study is mainly on velocity distribution analysis in a conical diffuser with screw tape inserts. The effect of different aspect ratios of the screw-tape on velocity distribution are simulated and analyzed by means of CFD software. All of the simulations are achieved at the same inlet conditions with Reynolds number (Re = 3.2×10^5). Results for the velocity distributions along the conical diffuser are compared with those of the plain conical diffuser.

2 Geometrical models

Numerical calculations were carried out for the conical diffuser fitted with screw tape inserts. Model geometries presented in Figure 1 and Table 1.

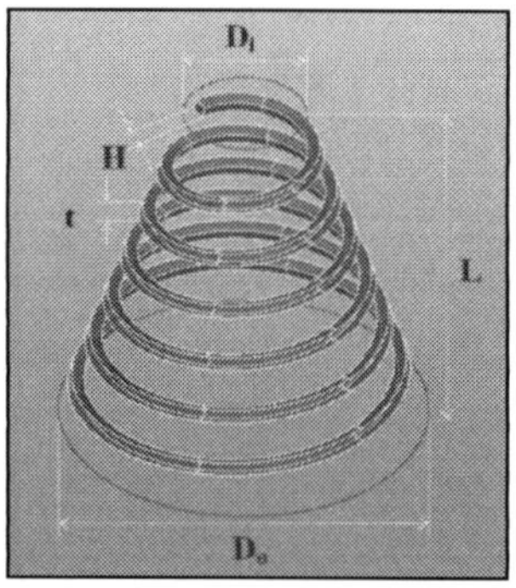

Figure 1. Diffuser and Screw tape geometries.

Table 1. Simulated model geometries.

Model geometries	Values, m
Inlet diameter (D_i)	0.048
Outlet diameter (D_o)	0.145
Diffuser length (L)	0.14
Screw tape width (t)	0.002
Screw tape height (H)	0.005, 0.007, 0.009, 0.011 and 0.013

2.1 Boundary layer

The simulations were analysed with air as the working fluid. Assumptions were made to model the velocity profile in a conical diffuser fitted with a screw tape. The main assumptions are:

1. The flow in the conical diffuser fitted with screw tape is turbulent and incompressible.
2. The flow is steady.
3. Inlet Reynolds number is (Re = 3.2×10^5) based on the hydraulic diameter of the conical diffuser (D_h) which is the inlet diameter of the diffuser (D_i).
4. The physical properties of the fluid at the conical diffuser inlet are constant.

It is set that the physical properties of the working fluid (air) have been considered as the density (1.225 kg/m^3) and the dynamic viscosity (1.7894e − 05 kg/m.s). The static pressure at the outlet is set as zero.

3 Numerical models

In the present study, numerical investigation was carried out with a screw tape inserts in a conical diffuser. The commercial software ANSYS FLUENT 16.1 was chosen as the CFD tool for this work. The standard *k*-ε model was applied as a turbulence model. The turbulence kinetic energy *k* and its rate of dissipation ε calculations followed literature reference. Standard *k*–ε turbulence model was allowed to predict the heat transfer and fluid flow characteristics. This turbulence model has been successfully applied to flows with engineering applications including internal flows [12]. The turbulence kinetic energy *k* and its rate of dissipation ε was obtained from the following transport equations [13],

$$K = 3/2\,(VI)^{2} \tag{1}$$

$$I = 0.16\,(Re)^{-1/8} \tag{2}$$

$$\varepsilon = (C\mu^{3/4}.\,K^{3/2})\,l^{-1} \tag{3}$$

$$l = 0.07L \tag{4}$$

where, V is the inlet velocity magnitude, I is the initial turbulence intensity, Re Reynolds number, $C\mu$ is a *k*-ε model parameter whose value is typically given as 0.09, l is the turbulence length scale and L is a characteristic length. For this study, L is considered as the hydraulic diameter.

In this numerical study, the finite volume method has been used to discretize the partial differential equations of the model. The unstructured tetrahedral meshes are generated as shown in Figure 2.

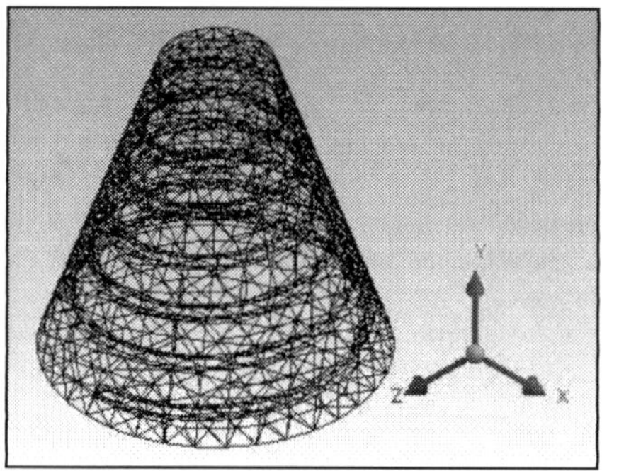

Figure 2. Model mesh generation.

4 Results and discussion

Velocity distribution is numerically investigated for a conical diffuser fitted with screw tape inserts to study flow velocity profile.

4.1 Effect of the screw tape inserts with different aspect ratios (AS)

In the present study, numerical results are analysed and discussed for different aspect ratios AS (2.5, 3.5, 4.5, 5.5, 6.5 and 7.5) in a conical diffuser fitted with screw tape inserts. They are represented by the Figures 3-8. These Figures visualized the performance of the screw tape on the axial velocity distribution. Figure 9 shows the velocity distribution in a conical diffuser without screw tape (plain conical diffuser).

https://doi.org/10.1051/matecconf/201815306003

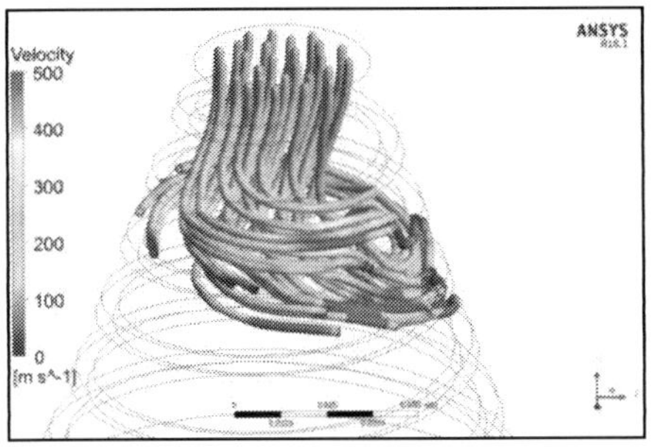

Figure 3. Velocity distribution with aspect ratio 2.5.

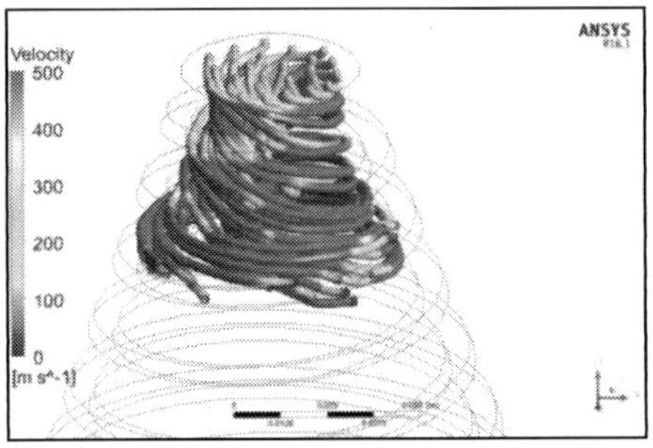

Figure 4. Velocity distribution with aspect ratio 3.5.

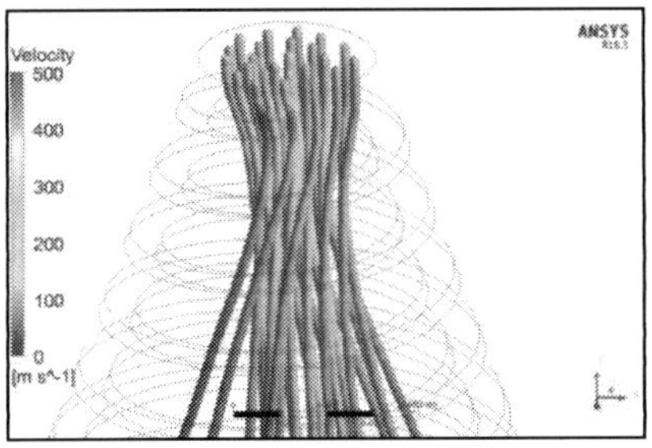

Figure 5. Velocity distribution with aspect ratio 4.5.

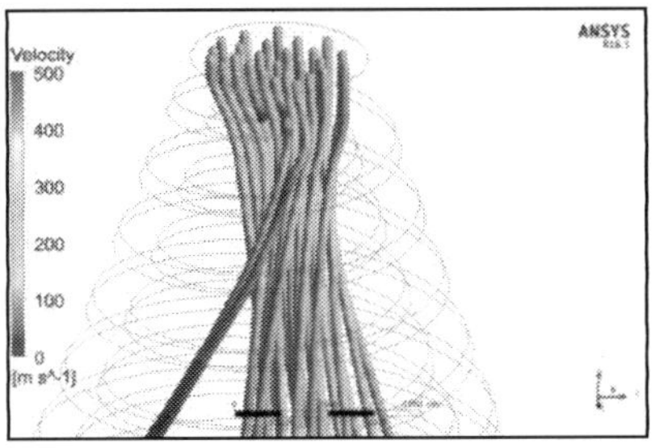

Figure 6. Velocity distribution with aspect ratio 5.5.

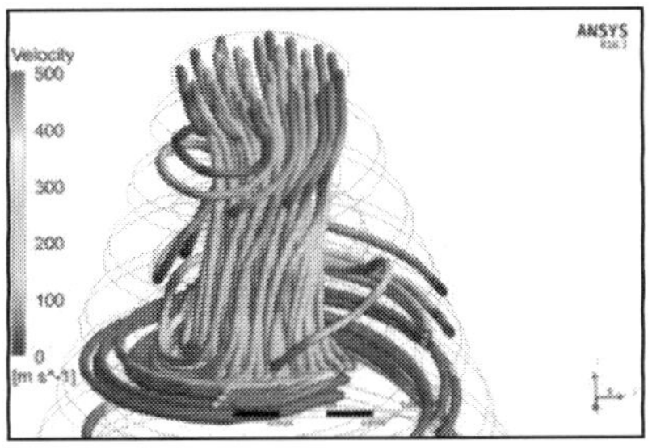

Figure 7. Velocity distribution with aspect ratio 6.5.

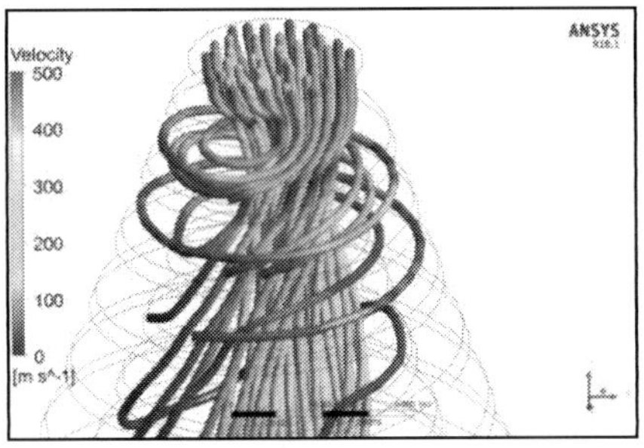

Figure 8. Velocity distribution with aspect ratio 7.5.

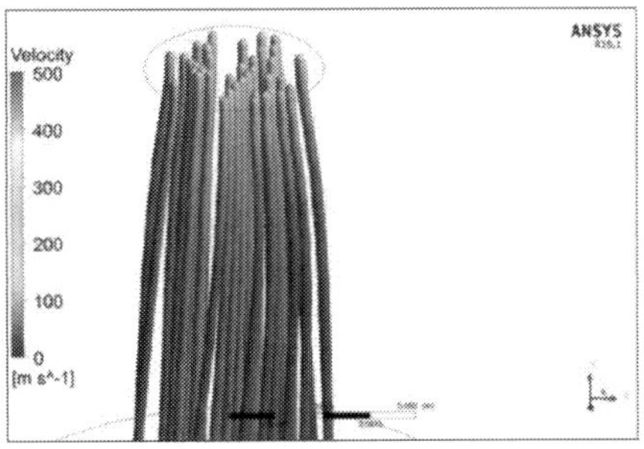

Figure 9. Velocity distribution in a plain conical diffuser.

Figure 3 shows velocity distribution for aspect ratio 2.5. It is found that with the existence of the screw tape, the flow velocity is started to distribute in a helical motion and this due to the increase of the turbulence that will be induced by the screw tape inserts.

Velocity distribution for aspect ratio 3.5 is illustrated in Figure 4. It can be seen that the phenomenon appears to be similar to the previous Figure but with a better effect.

Figures 5-6 present the effect of aspect ratios (4.5 and 5.5). On comparing these aspect ratios, it has been observed that velocity distribution performance is less than the previous aspect ratios (2.5 and 3.5).

Moreover, the velocity distribution for aspect ratios (6.5 and 7.5) is shown in Figures 7-8. In these Figures, it is clear that the velocity will distribute in the radial direction and the flow direction. For comparison purposes, the velocity distribution was simulated in a plain conical diffuser as shown in Figure 9.

It is found that the velocity distribution rate with the aspect ratio 3.5 is higher than those from other ratios (2.5, 4.5, 5.5, 6.5 and 7.5) and this means that the turbulent intensity obtained from the screw tape aspect ratio 3.5 is higher than the other tested aspect ratios.

Furthermore, the simulated results of the velocity distribution of screw tapes inserts for the tested aspect ratios are compared with simulated data for a plain conical diffuser. It is found that the screw tape inserts offered an additional velocity distribution enhancement with more helical motions and recirculation.

The variation of the velocity distribution with the flow direction for different screw tape aspect ratios AS (2.5, 3.5, 4.5, 5.5, 6.5 and 7.5) in a conical diffuser and a plain conical diffuser is displayed in Figure 10. The velocity distribution for these inserts was enhanced compared to the plain conical diffuser. Highest velocity distribution was obtained for screw tape with aspect ratio 3.5. This result is mainly due to the effects of the helical motion of the flow and turbulence generated by the screw tape inserts which leads to create better flow mixing between the fluids at the core and the diffuser inner wall.

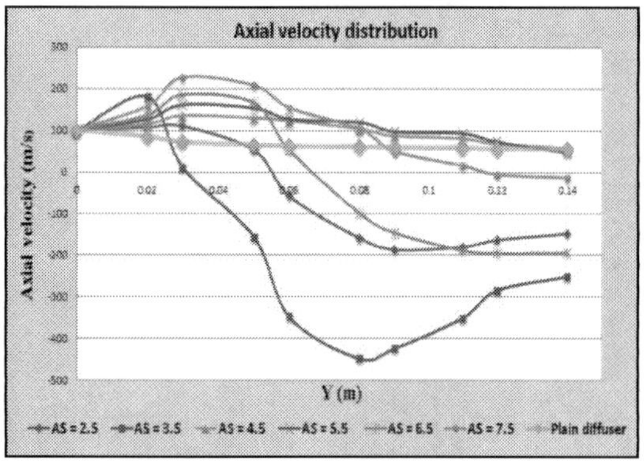

Figure 10. Simulated velocity distribution of the screw tape inserts for different aspect ratios AS.

5 Conclusions

In the present numerical study, velocity distribution in a conical diffuser was investigated. The flow was disturbed by screw tape inserts in a conical diffuser with air as working fluid. The performance of the screw tape inserts was evaluated for different aspect ratios AS (2.5, 3.5, 4.5, 5.5, 6.5 and 7.5). For comparison, numerical study with the same inlet boundary conditions for the plain conical diffuser was carried out. Based on the numerical simulation results, conclusions can be made as follows,

1. The augmentation of the velocity distribution with screw tape inserts has shown a significant improvement as compared to conical diffuser without screw tape (plain conical diffuser).
2. The values of the velocity distribution rate for the conical diffuser with screw tape aspect ratio AS 3.5 are noticeably higher than the other values of the tested aspect ratios for the conical diffuser fitted with screw tape and the plain conical diffuser as well. The best performance for the velocity distribution rate was offered by this aspect ratio.

References

1. M. Benisek, D. Ilic, D. Cantrak and I. Bozic "Investigation of the turbulent swirl flows in a conical diffuser " Thermal Science, **14**, no. suppl., pp. 141–154, (2010).
2. R. Prakash, D. Christopher and K. Kumarrathinam "CFD analysis of flow through a conical exhaust " International Journal of Research in Engineering and Technology, **3**, pp. 239– 248, 11 (2014).
3. N.V. Mahalakshmi, G. Krithiga, S. Sandhya, J. Vikraman, V. Ganesan, Experimental investigations of flow through conical diffusers with and without wake type velocity distortions at inlet, Experimental Thermal and Fluid Science, **32**, pp.133–157, (2007).
4. K. P. Skotniczny, "Numerical analysis of the impact of conical diffuser geometry change on velocity distribution in its outlet cross-section" Mechanics, **25**, 2 (2006).
5. T. Tanneberger, T. G. Reichel, O. Krüger, S. Terhaar, C. Paschereit, Numerical investigation of the flow field and mixing a swirl stabilized burner with a non-swirling axial jet, Proceedings of ASME Turbo Expo 2015: Turbine Technical Conference and Exposition GT2015, pp. 1-12, 2015.
6. R. Orl, Experimental study of passive scalar mixing in swirling jet flows, Technical reports from Royal Institute of Technology KTH Mechanics SE-100 44 Stockholm, Sweden, (2006).
7. K. G. Kulkarni, S. Y. Bhosale, Experimental Investigation of Heat Transfer Enhancement by Using Clockwise and Counter -clockwise Corrugated Twisted Tape Inserts, International journal of innovations in engineering research and technology ISSN: 2394-3696, **2**, ISSUE 5, (2015).
8. T. Parra, R. Perez, M. Rodriguez, F. Castro, R. Szasz, A. Gutkowski, Numerical Simulation of Swirling Flows -Heat Transfer Enhancement, Journal of Fluid Flow, Heat and Mass Transfer, **2**, ISSN: 2368-6111 ,.(2015).
9. Y. A. Eldrainy, J. Jeffrie and M. Ridzwan "Prediction of the flow inside a micro gas turbine combustor " Jurnal Mecchanikal, no. 25, pp. 50–63, 2008.
10. Y. A Eldrainy, H. S. Aly, K. M. Saqr and M. N. Jaafar "A multiple inlet swirler for gas turbine combustors " Proceeding of The International Conference on Fluid Mechanics, Heat Transfer and Thermodynamics, World Academy of Science, Engineering and Technology Tokyo, Japan, **3**, pp. 1260–1263, 5 (2009).
11. M. Jaafar Nazri and K. Jusoff "Combustor aerodynamic using radial swirler " *International Journal of Physical Sciences*, **6**, no. 13, pp. 3091–3098, (2011).
12. D. Erdemir, S. Gunes, V. Ozceyhan, and N. Altuntop, Numerical investigation of heat transfer enhancement and pressure drop in heat exchanger tube fitted with dual twisted tape elements, World Scientific and Engineering Academy and Society (WSEAS, ISSN: 2227-4596), pp.167-172, (2013).
13. D. Flórez-Orrego, W. Arias, D. López, and H. Velásquez, Experimental and CFD study of a single phase cone-shaped helical coiled heat exchanger: an empirical correlation, Proceedings of the 25th International Conference on Efficiency, Cost, Optimization, Simulation and Environmental Impact of Energy Systems, pp. 375–394, (2012).

Research on the Design Criterion of Packers and Slips in Pipeline Isolation Tool

Kang Zhang, Shimin Zhang, Qingxin Ding, Pu Gong and Xiaoxiao Zhu

China University of Petroleum-Beijing, No.18 Fuxue Road, Changping District Beijing, China

Abstract. With the rapid development of pipeline industry, the pipeline maintenance technology using isolation tool has got developed widely in the global scope. The packers and slips are the determining parts for isolation reliability and their design criterions are essential to investigate. This paper firstly studied the contact pressure between the packers and pipeline through theoretical and simulation methods. And then the modified model has been achieved according to the analysis results. Finally, the design criterions of packers and slips have been presented accordingly. The results of this research has great meaning and provide efficiency and convenience for tool designing in engineering.

1 Introduction

Transmission pipeline has become the main way to deliver fluids such as oil and gas due to its advantages of safety, efficiency and economy. In order to reduce the economic loss caused by the accidents like corrosions and cracks, the pipeline maintenance technology using isolation tool [1-3] has been widely developed in recent years.

The pipeline isolation tool is propelled from pig launcher via a flow of fluid and stops in the desired position inside pipeline to anchor and isolate [4, 5]. The packers and slips are the most important parts which determine the reliability of pipeline isolation. However, research on the design of packers and slips is scarcely found in the literatures, most of the available knowledge is mainly based on the experiences of designer, which made it inefficient and error prone. As a result, the research on design criterion of packers and slips is necessary.

In this paper, the analysis methods and design criterion of packers and slips have been proposed, and the results of this research will be helpful for providing some academic basis for engineering applications.

2 Analysis on contact pressure between packers and pipeline

The isolation process is shown in Fig. 1, the packers and slips expand under the axial compression load F_Z during the isolation process, the contact pressure between packers and pipeline is P_k. It is assumed that the differential pressure which the tool could isolate is equal to the contact pressure. So the contact pressure is the essential parameter and needs to be studied in advance.

2.1 Theoretical analysis

The simplified force model of packer is shown in Fig. 2. The packer will deform under the axial compression load F_Z, and the axial stress is:

$$\sigma_z = \frac{F_Z}{S} \tag{1}$$

Where, S is the section area of packer.

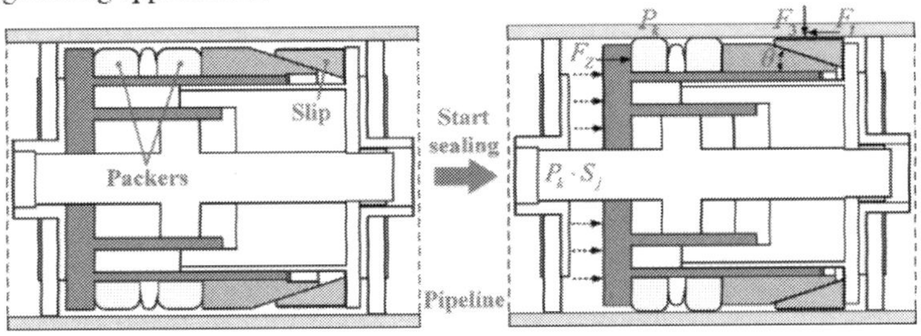

Figure 1. The schematic diagram of isolation process.

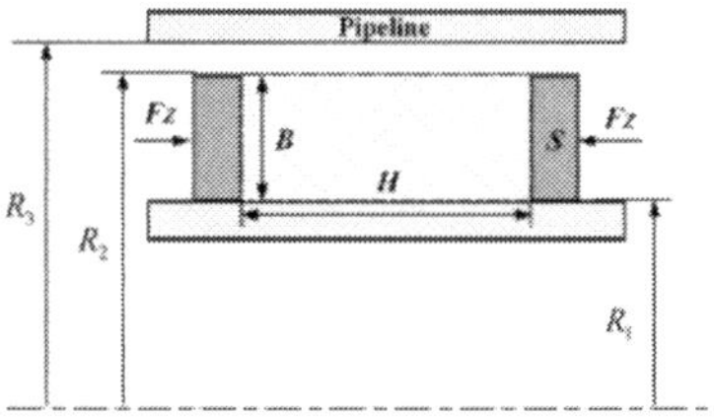

Figure 2. The simplified force model of packer.

Under the axial compression load, the deformation of packer could be divided into two phases: in phase 1, the packer will deform freely until contacting with pipeline; in phase 2, after contacting with pipeline, the contact pressure will increase with the rise of axial compression load. So the axial stress could be written as:

$$\sigma_z = \sigma_z^{\,1} + \sigma_z^{\,2} \tag{2}$$

Where, $\sigma_z^{\,1}$ is the axial stress of phase 1, and $\sigma_z^{\,1} = G\left[\frac{R_3^2 - R_1^2}{R_2^2 - R_1^2} - \left(\frac{R_2^2 - R_1^2}{R_3^2 - R_1^2}\right)^2\right]$, G is the modulus of rigidity. $\sigma_z^{\,2}$ is the axial stress of phase 2 and it could be rewritten as:

$$\sigma_z^{\,2} = \sigma_z - \sigma_z^{\,1} = \frac{F_Z}{S} - G\left[\frac{R_3^2 - R_1^2}{R_2^2 - R_1^2} - \left(\frac{R_2^2 - R_1^2}{R_3^2 - R_1^2}\right)^2\right] \tag{3}$$

According to generalized Hooke law, the radial stress σ_r and axial stress $\sigma_z^{\,2}$ could be written as:

$$\sigma_r = \frac{E}{(1-2\mu)(1+\mu)}[(1-\mu)\varepsilon_r + \mu(\varepsilon_\theta + \varepsilon_z)] \tag{4}$$

$$\sigma_z^{\,2} = \frac{E}{(1-2\mu)(1+\mu)}[(1-\mu)\varepsilon_z + \mu(\varepsilon_r + \varepsilon_\theta)] \tag{5}$$

Where, μ is the Poisson's ratio. ε_r and ε_θ are the strain of radial and circumference, respectively. And $\varepsilon_r = \varepsilon_\theta = 0$. So the radial strain could be written as:

$$\sigma_r = \frac{\mu}{1-\mu}\sigma_z^{\,2} \tag{6}$$

Therefore, the contact pressure between the packer and pipeline is:

$$P_k = \sigma_r = \frac{\mu}{1-\mu}\left[\frac{F_Z}{S} - G\frac{(R_3^2 - R_1^2)^3 - (R_2^2 - R_1^2)^3}{(R_2^2 - R_1^2)(R_3^2 - R_1^2)^2}\right] \tag{7}$$

In order to improve the accuracy, the friction of packer has been taken into consideration, and the contact pressure [6] could be rewritten as Eq. (8):

$$P_k = \begin{cases} 0 & (0 \le F_Z < \frac{2\pi E R_3 (R_3 - R_2)}{1+\mu}) \\ \frac{\mu}{1-\mu}\left[\frac{F_Z}{\pi(R_3^2 - R_1^2)} - \frac{2ER_3}{(1+\mu)}\cdot\left(\frac{R_3 - R_2}{R_3^2 - R_1^2}\right)\right] & (F_Z \ge \frac{2\pi E R_3 (R_3 - R_2)}{1+\mu}) \end{cases} \tag{8}$$

2.2 Error correction of theoretical model

In order to verify the theoretical model, the contact pressure between packer and pipeline was also analysed through simulation method. In the simulation, the Mooney-Rivlin model was used in rubber material properties. The strain energy density function is: $W = C_{10}(I_1 - 3) + C_{01}(I_2 - 3)$, W is strain potential, I_1 and I_2 are deformation tensor, C_{10} and C_{01} are parameters of Mooney-Rivlin Model. The other parameters selected in this simulation is shown in Table 1.

Table 1. The parameters selected in the simulation.

D (mm)	R_1 (mm)	R_2 (mm)	R_3 (mm)	H (mm)	B (mm)	IRHD (HS)
508	182	232	244	80	50	70

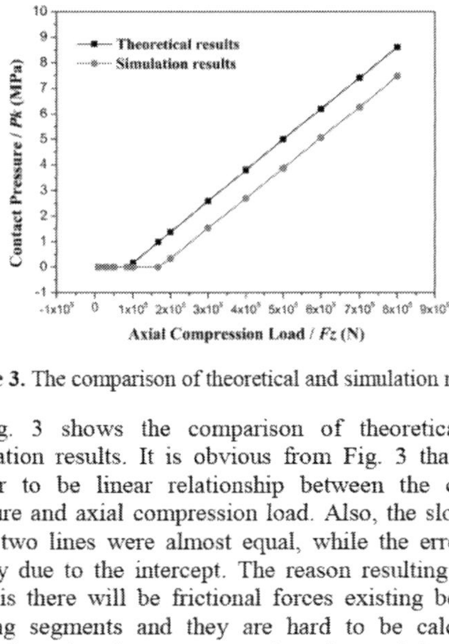

Figure 3. The comparison of theoretical and simulation results.

Fig. 3 shows the comparison of theoretical and simulation results. It is obvious from Fig. 3 that there appear to be linear relationship between the contact pressure and axial compression load. Also, the slopes of these two lines were almost equal, while the error was mainly due to the intercept. The reason resulting in the error is there will be frictional forces existing between moving segments and they are hard to be calculated precisely. Therefore, a correction factor λ has been introduced to the Eq. 9:

$$P_k = \left[\frac{\mu}{1-\mu}\cdot\frac{1}{\pi(R_3^2 - R_1^2)}\right]\cdot F_Z - \lambda\cdot\left[\frac{\mu}{1-\mu}\cdot\frac{2ER_3}{(1+\mu)}\cdot\left(\frac{R_3 - R_2}{R_3^2 - R_1^2}\right)\right] \tag{9}$$

Through comparing the results of theory and simulation, the correction factor could be determined as $\lambda = 1.898$. So the modified equation could be written as:

$$P_k = \begin{cases} 0 & (0 \le F_Z < 1.898 \times \frac{2\pi E R_3 (R_3 - R_2)}{1+\mu}) \\ \left[\frac{\mu}{1-\mu}\cdot\frac{1}{\pi(R_3^2 - R_1^2)}\right]\cdot F_Z - 1.898\left[\frac{\mu}{1-\mu}\cdot\frac{2ER_3}{(1+\mu)}\cdot\left(\frac{R_3 - R_2}{R_3^2 - R_1^2}\right)\right] & (F_Z \ge 1.898 \times \frac{2\pi E R_3 (R_3 - R_2)}{1+\mu}) \end{cases} \tag{10}$$

3 Design criterion of packers and slips

3.1 Design criterion of packers

The structural optimization of packers and slips has been widely developed [7-11]. While another important design criterion of packer is self-sealing. As shown in Fig. 4, when the axial compression load is F_Z, the contact pressure between packer and pipeline is P_k, so the axial load generated by differential pressure is $F_2 = P_k \cdot S_j$, where S_j is the sectional area of isolation tool. With the increase of F_Z and P_k, if $F_2 > F_Z$, which means that the axial compression load needed for isolation could be provided totally by differential pressure, the self-sealing achieved. Therefore, the design criterion of packer is:

$$\begin{cases} F_2 = P_k \cdot S_j \\ F_2 > F_Z \end{cases} \tag{11}$$

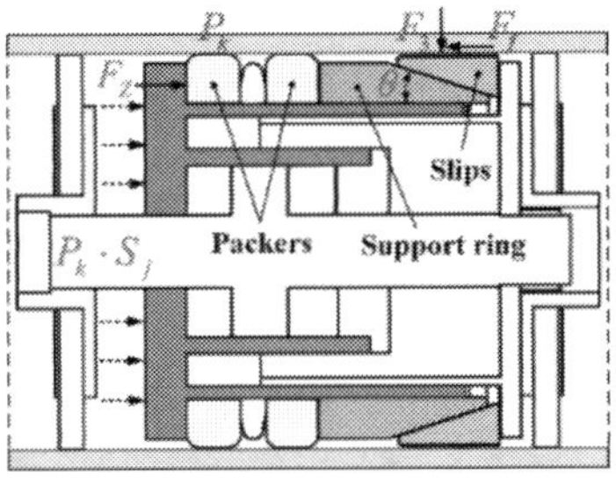

Figure 4. The schematic diagram of force analysis.

In order to make it more clear to understand easily, an example has been analyzed in this paper. The parameters selected in this analysis are shown in Table 1, and the results are shown in Fig. 5.

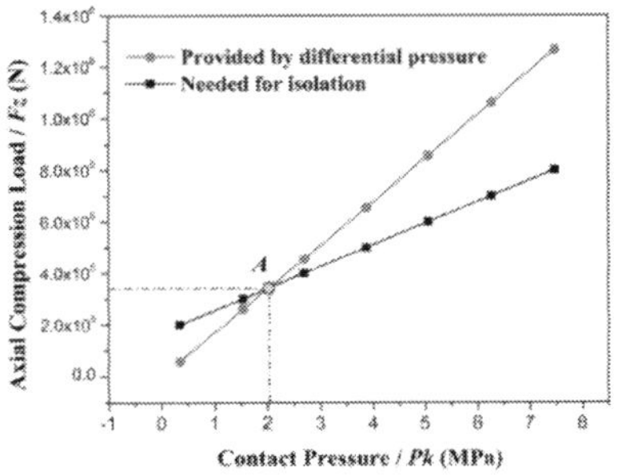

Figure 5. The analysis of self-sealing of packer.

It is clear from Fig. 5 that below the intersection point A (2.02MPa, 340kN), the axial compression load needed for isolation was smaller than that provided by differential pressure. However, with the rise of contact pressure, the gap between them narrowed gradually and when it reached beyond point A, the axial load provided by differential pressure was consistently bigger than that needed for isolation. It can be concluded that the axial compression load needed for isolation is only 340kN, generating the contact pressure 2.02MPa, the packer could achieve self-sealing.

3.2 Design criterion of slips

The force diagram of slips is also shown in Fig. 4. According to force balance, the normal force of slip is $F_3 = F_Z / \tan\theta$, and the frictional force of the slip is $F_f = \mu_f \cdot F_3 = \mu_f \cdot F_Z / \tan\theta$, where μ_f is the frictional coefficient, θ is the angle of support ring. The main function of slips is to anchor the tool in pipe, so the design criterion of slips including:

① In order to ensure the reliability of anchor, the frictional force needs to be larger than axial load provided by differential pressure.

$$F_f > F_2 \tag{12}$$

② In order to achieve self-anchoring, the axial load provided by differential pressure needs to be larger than axial compression load needed for anchoring.

$$F_2 > F_Z \tag{13}$$

③ The maximum contact stress needs to be smaller than the allowable stress of slip and pipeline.

$$\frac{F_{3\max}}{S_n} < 80\% \cdot \min\{\sigma_{S1}, \sigma_{S2}\} \tag{14}$$

Where, S_n is radial sectional area of slips, σ_{S1} is the allowable stress of slip and σ_{S2} is the allowable stress of pipeline.

④ The contact pressure of packer needs to be larger than design requirements under maximum axial compression load.

$$P_{k\max} \geq P_D \tag{15}$$

⑤ The angle of support ring needs to be larger than self-lock angle.

$$\theta > \text{self-lock angle } (18°) \tag{16}$$

Above all, the design criterion of slips can be written as Eq. (17):

$$\begin{cases} \dfrac{3.796\pi E \cdot R_3(R_3 - R_2)S_j \cdot \mu}{(1+\mu)\left[S_j \cdot \mu - \pi(1-\mu)(R_3^2 - R_1^2)\right]} < F_Z < \dfrac{3.796\pi E \cdot R_3(R_3 - R_2)S_j \cdot \mu \cdot \tan\theta}{(1+\mu)\left[S_j \cdot \mu \cdot \tan\theta - \mu_f \cdot \pi(1-\mu)(R_3^2 - R_1^2)\right]} \\ \left(\dfrac{1-\mu}{\mu} P_{k\max} + \dfrac{3.796E \cdot R_3(R_3 - R_2)}{(1+\mu)(R_3^2 - R_1^2)}\right)\dfrac{\pi(R_3^2 - R_1^2)}{\tan\theta \cdot S_n} < 80\% \cdot \min\{\sigma_{S1}, \sigma_{S2}\} \\ P_{k\max} = \left[\dfrac{\mu}{1-\mu} \cdot \dfrac{1}{\pi(R_3^2 - R_1^2)}\right] \cdot F_{Z\max} - 1.898\left[\dfrac{\mu}{1-\mu} \cdot \dfrac{2ER_3}{(1+\mu)} \cdot \left(\dfrac{R_3 - R_2}{R_3^2 - R_1^2}\right)\right] \geq P_D \\ \theta > \text{self-lock angle } (18°) \end{cases} \tag{17}$$

4 Conclusion

This paper investigated the design criterion of packers and slips. The conclusions can be drawn as follows:

1). The contact pressure between packers and pipeline has been analyzed through theoretical and simulation methods. And the error correction model has been achieved by comparing the results to improve its precision.

2). The design criterions of packers and slips have been presented in this paper, it is essential for the design of packers and slips, and make it more efficient and convenient.

The results of this research provided not only theoretical basis for design criterion, but also guidance for pipeline isolation tool in engineering application.

Acknowledgements

This work was supported by the National Natural Science Foundation of China (No. 51509259).

References

1. E. Tveit, J. Aleksandersen, *SPE Asia Pacific Oil and Gas Conference and Exhibition* (2000)
2. B.J. Zhao, C.R. Li, J.H. Zhang, *International Offshore and Polar Engineering Conference* (2010)
3. J. Aleksandersen, H. Syse, U.S. Patent 7,878,221. (2011)
4. J. Aleksandersen, E. Tveit, *International Offshore and Polar Engineering Conference* (2001)
5. M.R.G Lie, M.N. Muangsuankwan, *ASME India International Oil and Gas Pipeline Conference* (2015)
6. S.K. Tong, Oil Field Equipment. **41**, 7 (2012) (In Chinese)
7. Z.C. Lin, Applied Mechanics & Materials. **423**, 4 (2013)
8. J. Wang, X. Han, J. Wang, *Asia-Pacific Energy Equipment Engineering Research Conference* (2015)
9. D.X. Li, H.L. Zhao, S.M. Zhang, Adv. Mater. Res. **199**, 6 (2011)
10. M. Ma, W. Jia, Y. Bu, Open Access Library Journal. **01**, 8 (2014)
11. Z.P. Guo, W. Guo, Y.F. Wang, Applied Mechanics & Materials. **252**, 3 (2013)

MATEC Web of Conferences **153**, 06005 (2018)
ICMME 2017
https://doi.org/10.1051/matecconf/201815306005

Vertical Vibration Control of Cold Rolling Mill's Rolls Based on Time-delay Feedback Control Method

Dongxiao Hou

School of Control Engineering, Northeastern University at Qinhuangdao, China

Abstract. In this paper, a two degree of freedom nonlinear vertical vibration equation of the cold rolling mill with the dynamic rolling force was established, then the delay feedback control method was introduced into the equation to controlled the vertical vibration of the system. The amplitude-frequency equations of primary resonance of system was carried out by using the multi-scale method, and the resonance characteristics of different parameters of delay feedback control method were obtained by adopting the actual parameters of rolling mill. It is found that the size of the resonance amplitude value was effectively controlled and the resonance region and jumping phenomenon of the system were eliminated by selecting the appropriate time-delay parameters combination, which provides an effective theoretical reference for solving mill vibration problems.

1 Introduction

The vibration of rolling mill often occurs in rolling process. The occurrence of vibration not only affects the quality of rolling products, but also leads to breakdown of the rolling equipment. There are mainly two kinds of vibration in rolling mill, namely torsional vibration of main drive system and vertical vibration of rolls. Especially, the vertical vibration of rolls directly effects the exit thickness of strip. In order to restrain vertical vibration of rolls and improve stability of rolling process, the active control theory and technology has been used and to control the vertical vibration phenomenon. Zhang adopted the robust theory to control the strip tension of cold rolling mill and then inhibited the occurrence of vertical vibration [1]. Fuzzy neural network is used in hydraulic AGC system and the accuracy of strip is improved[2]. Time-delay phenomenon is very common in the control system. Plenty of researches have been studied on it in different research fields [3, 4]. Time-delay phenomenon also exists in the control system of rolling mill.

In this paper, in order to control resonance characteristics of rolling mill, the time-delay feedback control method is applied to the two degrees of freedom vertical vibration equation of cold rolling mill. By using the multi-scale method, the primary resonance amplitude-frequency characteristic equation is solved out, and the impact of time-delay parameters on the primary resonance are studied. The simulation results indicate that the size of resonance amplitude is effectively controlled, and the resonance region and jumping phenomenon of the system are eliminated.

2 The time-delay feedback control model of rolling mill's two degree freedom vibration system

Considering the effects of dynamic rolling force [5] and the up-down symmetry of rolling mills, a two degree of freedom nonlinear vertical vibration model of rolling mill is established in Figure 1.

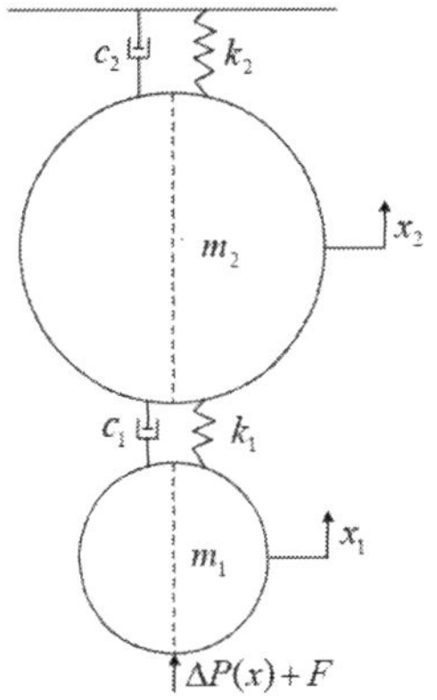

Figure 1. The mechanical vertical vibration model of rolls with nonlinear dynamic force.

The nonlinear dynamic equation of rolling mill in Figure 1 can be established as

$$\begin{cases} m_1\ddot{x}_1 + c_1(\dot{x}_1 - \dot{x}_2) + k_1(x_1 - x_2) = \Delta P(x) + F \\ m_1\ddot{x}_2 + k_2x_2 + c_2x_2 - k_1(x_1 - x_2) - c_1(\dot{x}_1 - \dot{x}_2) = 0 \end{cases} \quad (1)$$

Where, m_1 and m_2 are defined as the equivalent mass of top work roll and top back-up roll, respectively. k_1 is defined as equivalent stiffness between the top work roll and top back-up roll. k_2 and c_2 are defined as equivalent stiffness and equivalent damping between the top back-up roll and top beam of mill housing, respectively. F is defined as external excitation. $\Delta P(x)$ is dynamic variation of rolling force, $\Delta P(x)=b_1x+b_2x^2+b_3x^3$, $x=2x_1$.

The time-delay feedback control is applied in equation (1) and it can be expressed as

$$\begin{cases} m_1\ddot{x}_1 + c_1(\dot{x}_1-\dot{x}_2) + k_1(x_1-x_2) = \Delta P(x) + F \\ \quad + g_1x_1(t-\tau_1) + g_2\dot{x}_1(t-\tau_2) \\ m_1\ddot{x}_2 + k_2x_2 + c_2\dot{x}_2 - k_1(x_1-x_2) - c_1(\dot{x}_1-\dot{x}_2) = 0 \end{cases} \tag{2}$$

Where, τ_1 and τ_2 are time-delay parameters, g_1 and g_2 are gain factor.

The equation (2) can be simplified as

$$\begin{cases} \ddot{x}_1 + \omega_1^2 x_1 = \alpha_1 x_2 + \beta_1(\dot{x}_1-\dot{x}_2) + \gamma_1 x_1^2 \\ \quad + \eta_1 x_1^3 + F_0 + d_1x_1(t-\tau_1) + d_2\dot{x}_1(t-\tau_2) \\ \ddot{x}_2 + \omega_2^2 x_2 = \alpha_2 x_1 + \beta_2\dot{x}_1 + \gamma_2\dot{x}_2 \end{cases} \tag{3}$$

Where, $\omega_1^2 = (k_1-2b_1)/m_1$, $\omega_2^2 = (k_1+k_2)/m_2$, $\alpha_1 = k_1/m_1$, $\beta_1 = -c_1/m_1$, $\gamma_1 = 4b_2/m_1$, $\eta_1 = 8b_3/m_1$, $\alpha_2 = k_1/m_2$, $\beta_2 = c_1/m_2$, $\gamma_2 = -(c_1+c_2)/m_2$, $d_1 = g_1/m_1$, $d_2 = g_2/m_1$, $F_0 = F/m_1$.

3 Primary resonance analysis

Assuming equation (3) is a small nonlinear system, the right side of the equation (3) multiply by small parameter

$$\begin{cases} \ddot{x}_1 + \omega_1^2 x_1 = \varepsilon[\alpha_1 x_2 + \beta_1(\dot{x}_1-\dot{x}_2) + \gamma_1 x_1^2 \\ \quad + \eta_1 x_1^3 + F_0 + d_1x_1(t-\tau_1) + d_2\dot{x}_1(t-\tau_2)] \\ \ddot{x}_2 + \omega_2^2 x_2 = \varepsilon[\alpha_2 x_1 + \beta_2\dot{x}_1 + \gamma_2\dot{x}_2] \end{cases} \tag{4}$$

In the case of primary resonance, $F_0=F_1\cos(\omega t)$, set $\omega_1^2 = \omega^2 + \varepsilon\sigma$, and substituting it in equation (4)

$$\begin{cases} \ddot{x}_1 + \omega^2 x_1 = \varepsilon[\alpha_1 x_2 + \beta_1(\dot{x}_1-\dot{x}_2) + \gamma_1 x_1^2 + \eta_1 x_1^3 \\ \quad + F_0 + d_1x_1(t-\tau_1) + d_2\dot{x}_1(t-\tau_2) - \sigma x_1] \\ \ddot{x}_2 + \omega_2^2 x_2 = \varepsilon[\alpha_2 x_1 + \beta_2\dot{x}_1 + \gamma_2\dot{x}_2] \end{cases} \tag{5}$$

Where, σ is frequency modulation parameter,

By using multiple scales method, assuming it has the first-order approximate solution as follows

$$\begin{cases} x_1 = x_{11}(T_0,T_1) + \varepsilon x_{12}(T_0,T_1) \\ x_2 = x_{21}(T_0,T_1) + \varepsilon x_{22}(T_0,T_1) \end{cases} \tag{6}$$

Substituting equation (6) into equation (5), separating terms each order of ε, it has

$$\begin{cases} D_0^2x_{11} + \omega^2 x_{11} = 0 \\ D_0^2x_{21} + \omega_2^2 x_{21} = 0 \end{cases} \tag{7}$$

$$\begin{cases} D_0^2x_{12} + \omega^2 x_{12} = -2D_0D_1x_{11} + \alpha_1 x_{21} + \beta_1 D_0(x_{11}-x_{21}) \\ \quad + \gamma_1 x_{11}^2 + \eta_1 x_{11}^3 + F_1\cos\omega t + d_1x_{11}(t-\tau_1) \\ \quad + d_1D_0x_{11}(t-\tau_2) - \sigma x_{11} \\ D_0^2x_{22} + \omega_2^2 x_{22} = -2D_0D_1x_{21} + \alpha_2 x_{11} + \beta_2 D_0 x_{11} + \gamma_2 D_0 x_{21} \end{cases} \tag{8}$$

Set the solution of equation (7) was

$$\begin{cases} x_{11} = A_1(T_1)e^{i\omega T_0} + \varepsilon\bar{A}_1(T_1)e^{-i\omega T_0} + cc \\ x_{21} = A_2(T_1)e^{i\omega_2 T_0} + \varepsilon\bar{A}_2(T_1)e^{-i\omega_2 T_0} + cc \end{cases} \tag{9}$$

Substituting equation (8) into equation (7), the following equation can be expressed as

$$\begin{cases} D_0^2x_{12} + \omega^2 x_{12} = (-2i\omega D_1A_1 + i\omega\beta_1A_1 + 3\eta_1A_1^2\bar{A}_1 + 0.5F_1 \\ \quad + d_1A_1e^{-i\omega\tau_1} + i\omega d_2A_1e^{-i\omega\tau_2} - \sigma A_1)e^{i\omega T_0} + (\alpha_1A_2 \\ \quad - i\omega_2\beta_1A_2)e^{i\omega_2T_0} + \gamma_1A_1^2e^{i2\omega T_0} + \eta_1A_1^3e^{i3\omega T_0} + 2\gamma_1A_1\bar{A}_1 + cc \\ D_0^2x_{22} + \omega_2^2 x_{22} = (-2i\omega_2D_1A_2 + i\omega_2\gamma_2A_2)e^{i\omega_2T_0} \\ \quad + (\alpha_2A_1 + i\omega\beta_2A_1)e^{i\omega T_0} + cc \end{cases} \tag{10}$$

The secular term of equation (10) can be obtained as

$$\begin{cases} -2i\omega D_1A_1 + i\omega\beta_1A_1 + 3\eta_1A_1^2\bar{A}_1 + 0.5F_1 \\ \quad + d_1A_1e^{-i\omega\tau_1} + i\omega d_2A_1e^{-i\omega\tau_2} - \sigma A_1 = 0 \\ -2i\omega_2D_1A_2 + i\omega_2\gamma_2A_2 = 0 \end{cases} \tag{11}$$

The polar coordinate of A_1 and A_2 of the equation (11) is introduced as follows

$$\begin{aligned} A_1 &= 0.5a(T_1)e^{i\varphi_1(T_1)} \\ A_2 &= 0.5b(T_1)e^{i\varphi_2(T_1)} \end{aligned} \tag{12}$$

Substituting equation (12) into equation (11), separate the real part and the imaginary part and get

$$\begin{cases} \dot{a} = \dfrac{\varepsilon}{2\omega}[a\beta_1\omega - F_1\sin\varphi_1 - d_1a\sin(\omega\tau_1) \\ \quad + a\omega d_2\cos(\omega\tau_2)] \\ \dot{\varphi}_1 = -\dfrac{\varepsilon}{2a\omega}[\dfrac{3}{4}\eta_1a^3 + F_1\cos\varphi_1 + d_1a\cos(\omega\tau_1) \\ \quad + a\omega d_2\sin(\omega\tau_2) - a\sigma] \\ \dot{b} = \dfrac{1}{2}\varepsilon\gamma_2 b \\ \dot{\varphi}_2 = 0 \end{cases} \tag{13}$$

In the steady-state, existing $\dot{a} = \dot{b} = \dot{\varphi}_1 = \dot{\varphi}_2 = 0$, eliminate φ_1 and φ_2, then the primary resonance amplitude-frequency equation of the controlled equation of system can be obtained as

$$\begin{aligned} &[\frac{3}{4}\eta_1a^3 + d_1a\cos(\omega\tau_1) + a\omega d_2\sin(\omega\tau_2) - a\frac{\omega_1^2-\omega^2}{\varepsilon}]^2 \\ &+[a\beta_1\omega - d_1a\sin(\omega\tau_1) + a\omega d_2\cos(\omega\tau_2)]^2 = F_1^2 \end{aligned} \tag{14}$$

4 Simulation

Taking the actual structure parameters and technological parameters of the rolling mill as an example, the detail parameters are as follows:

m_1=2.182×10^{4}kg, m_2=1.225×10^{5}kg, k_1=6.0186×10^{10} N/m, k_2=3.1825×10^{10}N/m, c_1=1.01×10^6 N•s/m, c_2=1.25×10^6 N•s/m, F_0=0.5×10^6 N.

The nonlinear parameters of dynamic rolling force are as follows:

b_1=-4.478×10^{9}N/m, b_2=-2.910 ×10^{12} N/m^2, b_3=-3.256×10^{15} N/m^3, $\varepsilon = 0.01$.

Figures 2 illustrates primary resonance amplitude-frequency curves in different delay-time parameters.

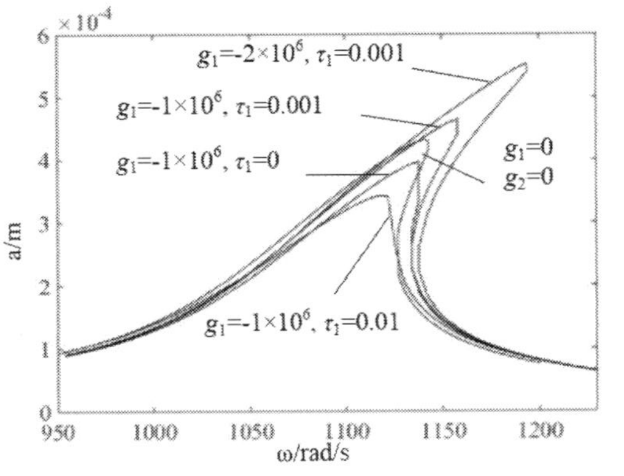

(a) g_2=-1×10^3 and τ_2 =0.001

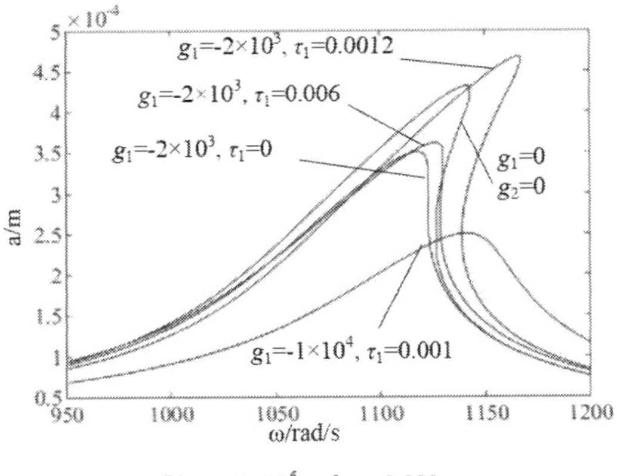

(b) g_1=-1×10^6 and τ_1 =0.001

Figure 2. Primary resonance amplitude-frequency curves.

Figure 2(a) is the primary resonance amplitude-frequency curves of system when g_1 and τ_1 take different combinations and the time delay parameters g_2=-1×10^3, τ_2=0.001. When g_1=0 and g_2=0, it is the primary resonance amplitude-frequency curve of original system. If τ_1=0(namely no delay), when g_1=-1×10^6, the amplitude and resonance zone decreases. As τ_1 increased, the amplitude and resonance zone increases firstly, then the amplitude and resonance zone decreases, it means adjusting the parameters g_1 and τ_1 can effectively control the size of the resonance amplitude values.

Figure 2(b) illustrates the primary resonance amplitude -frequency curves of system when g_2 and τ_2 take different combinations and the time delay parameters g_1=-1×10^6, τ_1=0.001. when g_2 is constant and τ_2 increases, firstly the amplitude decreases with τ_2 is no delay, then the amplitude increases with τ_2 increases. When τ_2 increase to a certain value, the amplitude decreases. when g_2 increases, the amplitude and resonance zone decreases.

Figures 3 are the time-domain curves of m_1 in the case of no control and control condition with ω is equal to 1140 rad/s.

Figure 3(a) is g_1 and g_2 are both equal to zero, namely no control. Figure3 (b) is time-domain curve in control, it can be seen that the resonance amplitude is less than no control, it means it could be improve resonance problem by adopting time-delay feedback method. So it can be seen adjusting the time-delay parameters can effectively reduce the amplitude of system, eliminate the resonance phenomenon and make the system tend to be stable more quickly.

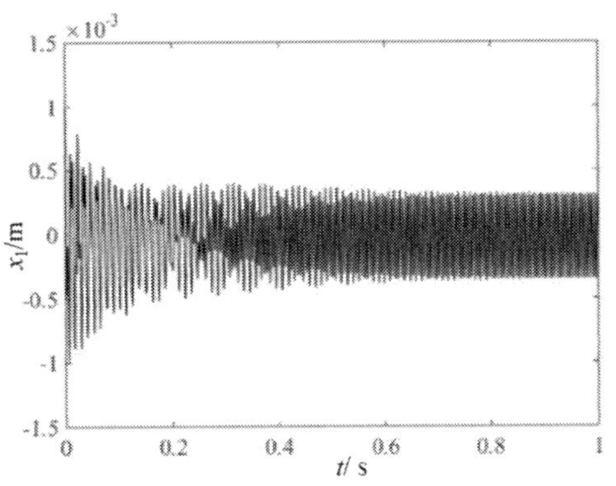

(a) g_1=0, g_2=0, τ_1 =0, τ_1 =0

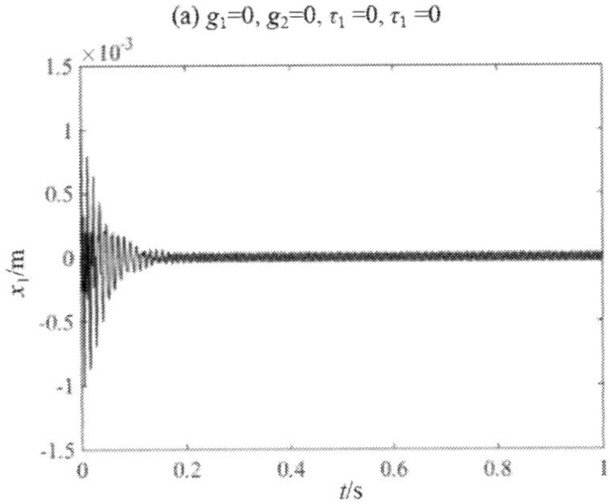

(b) g_1=-2×10^6, g_2=-1×10^3, τ_1 =0.001, τ_1 =0.001

Figure 3. The time-domain curves of m_1 with uncontrolled and controlled.

In summary, by using the time-delay feedback method to control the two degree of freedom vibration system and selecting the appropriate time-delay feedback parameters can effectively improve the primary resonance amplitude frequency characteristic of system and make the mill system in more stable state.

5 Conclusion

This paper firstly established a two degree of freedom vibration equation of the cold rolling mill with the dynamic rolling force. then the delay feedback control method is introduced into the equation to control the vertical vibration of system. In addition, the primary resonance amplitude-frequency characteristic controlled equation of system is solved out and the effect of time-delay parameters on the primary and internal resonance are analyzed.

The appropriate time-delay parameter combination is selected to effectively control the amplitude value of resonance. It eliminates the internal resonance region and jumping phenomenon of system, improved the stability of system, which provides an effective theoretical reference for the study of inhibiting the mill vibration problems.

5 Acknowledgments

This research is supported by National Natural Science Foundation of China (Grant No. 51405068), Natural Science Foundation of Hebei Province of China (Grant No.E2014501006), and Technology Research Project of Colleges and Universities of Hebei Province of China (Grant No. 2014202).

References

1. Zhang X.F Robust Control of Strip Tension for Tandem Cold Rolling Mill. *Mater Thesis*, University of Science and Technology Beijing, China (2008)
2. Yang Y and Zhang X.M Fuzzy Neural Networks Control for Hydraulic AGC System of Aluminum Cold Rolling Mill. In: *Proceedings of 2008 3rd International Conference on Intelligent System and Knowledge Engineering*(ISKE 2008), Xiamen, China, **17-19** November 2008, pp. 92-97. Piscataway: Inst. of Elec. and Elec. Eng. (2008)
3. Liu S, Liu B, Zhang Y.K, et al. Hopf bifurcation and stability of periodic solutions in a nonlinear relative rotation dynamical system with time delay. *Acta Physica Sinica* **59**(1):38-43 (2010)
4. Shi P.M, Li J.Z, Liu B, et al. Stability and time-delayed feedback control of a relative-rotation nonlinear dynamical system under quasic-periodic parametric excitation. *Acta Physica Sinica* **60**(9):094501(2011)
5. Hou D.X, Zhu Y and Liu H.R. Research on Nonlinear Vertical Vibration Characteristics of Cold Rolling Mill Based on Dynamic Rolling Force. *Chinese Journal of Mechanical Engineering* **49**(14):45-50. (2013)

Analysis of Vertical Stiffness of Air Spring Based on Finite Element Method

Jiatong Ye, Hua Huang, Chenchen He, Guangyuan Liu

School of mechanical and Vehicular Engineering, Beijing Institute of Technology,100081 China

Abstract. In this paper, a finite element model of membrane air spring in the vehicle is established , and its vertical stiffness characteristics under a certain inflation pressure are analysed. The result of finite element simulation method is compared with the result of the air spring bench test. The accuracy and reliability of the finite element simulation method in nonlinear analysis of air spring system are verified. In addition, according to the finite element method, the influence of the installation of the air spring limit sleeve on its stiffness is verified.

1 Introduction

As the air spring has great advantages in the vibration performance and energy storage and other aspects, relative to other elastic components, the air spring gains more and more popularity in consumers and developers of all ages. The stiffness of the air spring is an important performance parameter involved in the design and application of air springs. Analysing the stiffness of the air spring with finite element method can take the impact of each parameter on the stiffness into account synthetically, greatly shortening the development cycle[1]. Therefore, the finite element method has great theoretical and practical value for the research and analysis of air springs. Based on the advantages above, the stiffness of a certain type of spring is analysed by finite element method, and the correctness of the analysis is proved by comparison with the experimental results.

2 Vertical stiffness of air spring

The force diagram of air spring is shown in figure 1.

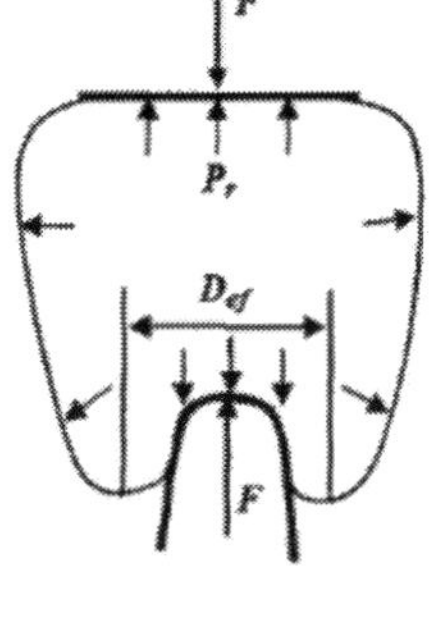

Figure 1. The force diagram of air spring.

The relationship between the vertical load F, the relative internal pressure P_r, the effective bearing area A_{ef} and the effective diameter D_{ef} on the air spring is:

$$F = \mathrm{P}_r A_{ef} = \frac{\pi}{4} \mathrm{P}_r D_{ef}^{\ 2} \quad (1)$$

If the quality of compressed gas does not change, the load change will cause the height change of the air spring, which will lead to changes in the volume and pressure of the airbag. the process of change meets the ideal gas equation:

$$(\mathrm{P}_r + \mathrm{P}_a)V^n = (\mathrm{P}_0 + \mathrm{P}_a)V_0^n \quad (2)$$

Where: P_a is the atmospheric pressure, P_0 is the initial internal pressure of the air spring, V_0 is the initial volume of the air spring; V is the volume of the airbag at any position, P_r is the internal pressure in the airbag at any position; n is the polytropic index, in the range of 1 to 1.4.

Sort out the formula above, its vertical stiffness can be obtained by deriving the stroke x:

$$k = \frac{dF}{dx} = -nA_{ef} \frac{\mathrm{P}_r + \mathrm{P}_a}{V} \frac{dV}{dx} + \mathrm{P}_r \frac{dA_{ef}}{dx} \quad (3)$$

Where $\frac{dV}{dx}$ is the volume change rate of the air spring and A is the effective area change rate of the air spring[2].

3 Finite element analysis of membrane air spring

3.1 The establishment of finite element model for membrane air spring

In this study, the effects of the emergency stop, stud and bolt holes are ignored because the emergency stop only operate under extreme conditions, and the stud and bolt holes were independent of the product performance. As a result, the key point of finite element analysis is focused on the shape of the piston, the capsule and the edge of the lid. Experience shows that the piston and the lid should be seen as rigid body structure, because the upper and lower bead ring fixed in the lid edge and the piston, so it can be ignored in finite element analysis[3].

The airbag of the membrane air spring is made of a composite material obtained by vulcanization of rubber and cord. Because of the limit deformation degree of the rubber of the membrane air spring airbag, the rubber has the same nature in the small deformation period, and the order of the macromolecular chains has not been affected. The superelastic mechanical properties can be described by Mooney-Rivlin model[4] strain energy density function U, that is

$$U = C_{10}(I_1 - 3) + C_{01}(I_2 - 3) + \frac{1}{D}(J^{el} - 1)^2 \quad (4)$$

Where C_{10} 、 C_{01} and D are temperature-dependent superelastic material constants, which can be obtained by testing the material; J^{el} is the elastic volume ratio; I_1 and I_2 are the first and second strain invariants, which are defined as:

$$I_1 = \lambda_1^2 + \lambda_2^2 + \lambda_3^2 \quad (5)$$

$$I_2 = \lambda_1^{-2} + \lambda_2^{-2} + \lambda_3^{-2} \quad (6)$$

Where λ_1 、 λ_2 and λ_3 are the tensile coefficients of the three main stretching directions.

The initial shear modulus and bulk modulus of elasticity are:

$$\mu_0 = 2(C_{10} + C_{01}) \quad (7)$$

$$K_0 = \frac{D}{2} \quad (8)$$

In this experiment, in order to control the deformation degree of the air spring balloon better, so as to obtain the air spring stiffness curve which is more in accordance with the actual situation, a more rigid limit sleeve made by aluminum alloy is set on the airbag surface. The discrete rigid body unit RAX2 can be used to simulate the limit sleeve. The piston and the upper cover are also simulated using the RAX2 discrete rigid body unit. The cord layer is a very important bearing component of the airbag. The ABAQUS software provides a rebar unit for simulating the stiffened ribbed structure. The rebar unit in the cord layer structure is usually embedded in the rubber shell model to form the airbag compound material[5]. The cord parameters of the air spring selection are shown in Table 1.

Table 1. Membrane air spring cord parameters

Cord angle / (°)	±34
Number of layers	2
Spacing /mm	0.30
Diameter /mm	0.39
Distance from the neutral surface /mm	±0.38

There's no heat transfer to the outside when the air spring is compressed or stretched in a short time period ,so the process of change is similar to the adiabatic process. A hydrostatic fluid unit may be used: a model such as F3D3 (three-dimensional three-node fluid unit) or F3D4 (three-dimensional four-node fluid unit) as this fluid.

3.2 Finite element analysis of air spring

This paper uses ABAQUS for finite element analysis. Based on the piston restraint condition and the rigid freedom level of the upper cover, the corresponding initial pressure is determined in conjunction with the gas unit reference point data to meet the system working requirements and is adjusted to the working condition by inflating the rubber air bag. According to a similar method, the vertical displacement parameter of the upper cover plate is evaluated and the expected vertical movement distance of the air spring is determined, and the ideal height is determined. Then, with reference to the constraints of upper cover and the free movement of the piston, the ideal working pressure of the air spring is determined and inflate to ideal value[6]. Here, select the same conditions with the test, that is, inflate to the static balance height H = 320mm, inflation pressure is 0.6MPa. This inflated pressurization process is shown as figure 2.

Figure 2.The stress cloud at the equilibrium position.

Based on the vertical travel of the upper cover, the load of air spring can be determined, and the confirmation method is similar to the analysis method of the rigid surface constraint condition of the upper cover. This paper analyzes the load fluctuation caused by stretching and compression with the change of displacement of the air spring, and then evaluates the vertical stiffness performance of the air spring. After the spring is inflated, the compressive stress and the tensile stress are shown in figure 3.

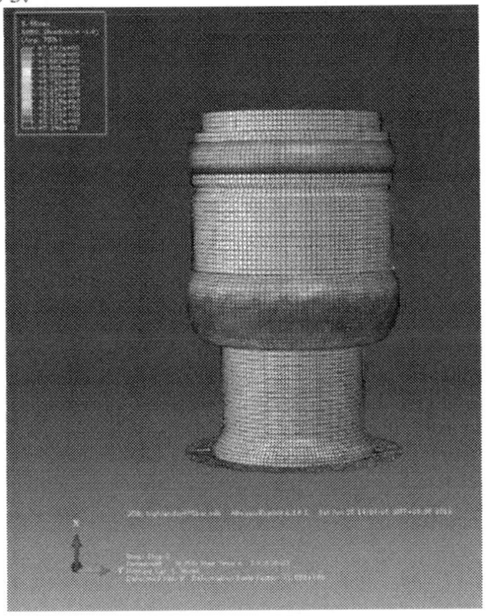

Figure 3.1. Compressive stress cloud.

Figure 3.2. Tensile stress cloud.

3.3 Influence of limit sleeve on vertical stiffness of air spring

In order to verify the effect of the air spring limit sleeve on its stiffness, the limit sleeve is removed and the finite element calculation is performed again. To ensure that the two prerequisites for calculation such as the boundary conditions, displacement levels and other parameters are kept fixed. The results of the two finite element analysis are processed, and the force-displacement data are shown in figure 4.

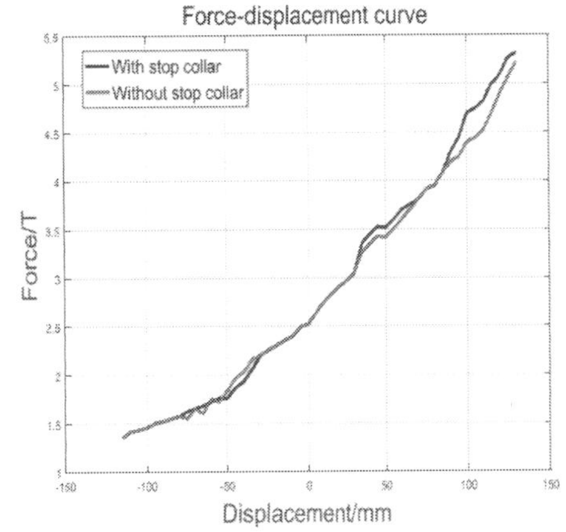

Figure 4. Force - displacement curve of air spring.

In order to facilitate the comparison with the subsequent test data, here take the data in the range of 50mm, draw the air spring static stiffness curve of the two finite element calculations ,which are shown in figure 5.

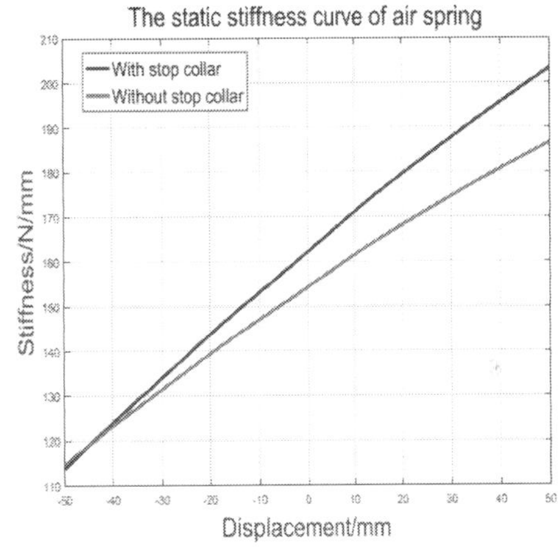

Figure 5. Comparison curve of static stiffness of air spring.

As can be shown from the curve comparison chart, the rigidity level of the air spring is improved after the installation of the limit sleeve, and its overall performance is improved synchronously. Besides, the excessive deformation of the airbag can be effectively suppressed to avoid the effect on the safety of the equipment.

4 Vertical static characteristic test of air spring

4.1 Test of air spring

The air spring is connected to the test bench through the flange connector and test bench gantry, as shown in figure 6.

Figure 6. Test piece installation diagram.

The parameters of air spring assembly shown in Table 2.

Table 2. Air spring assembly parameters

Name	Parameters
Air spring assembly maximum height	470mm
Air spring assembly minimum height	170mm
Static balance height	320mm
Stroke	±150mm
Piston diameter	180mm

Adjust the air spring to the maximum height, inflate to 6bar. The specimen is run under 0.1Hz sine excitation 10 times in the 80% of the stroke range . Adjust the H to 460mm, each time when the test bed rise 5mm, read the spring elastic force, until H reaches 180mm. And similarly drop the test bed every 5mm, read again until H reaches 460mm. In the compression and recovery process, due to airbag capsule friction and air through the orifice , the resulting damping forces are different. In addition, the existence of a variety of uncertain factors, the static force and pressure in test will also be different.

According to the original data obtained from the experiment, the static displacement - elastic curve in ± 50mm stroke is drawn and the corresponding fitting curve equation is obtained[7]. Based on the fitting equation, the derivation is obtained, and the expression of the stiffness curve is obtained and the stiffness curve can be seen in figure 7.

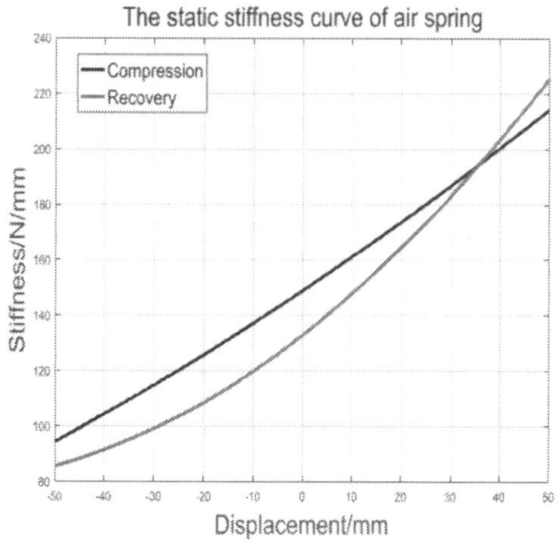

Figure 7. Static stiffness curve of air spring.

4.2 Contrastive analysis of experiment and simulation data

The static stiffness curve obtained during the compression process is compared with the static stiffness curve obtained by finite element simulation, as shown in figure 8.

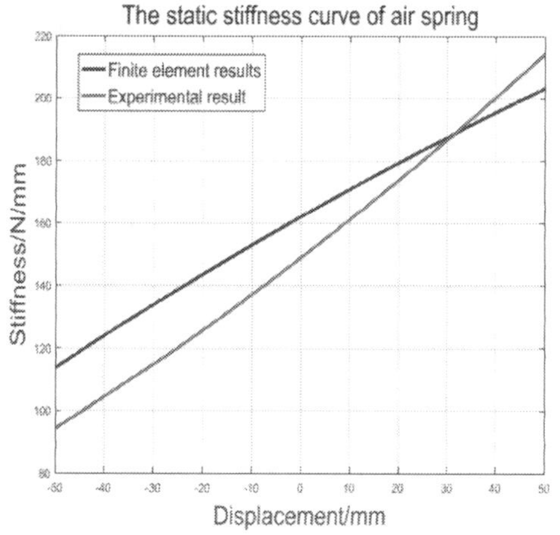

Figure 8. Comparison of static stiffness curve of air spring.

From the analysis above, it can be seen that finite element simulation analysis based on ABAQUS can effectively carry out the analysis of the model air spring. The result of the finite element simulation analysis shows good agreement with the test value obtained by the test, which fully demonstrate the scientific nature of this method.

5 Conclusion

In this paper, the finite element model of the membrane air spring is established, and the relationship between the elastic force and the displacement is analyzed with 0.6MPa as the inflation pressure. Further, the static stiffness curve of air spring is calculated and the effect of the installation of the limit sleeve on its stiffness is verified. After that, the air spring bench test is carried out at the same inflated pressure, and the static stiffness curve of the air spring in the process of compression and recovery is fitted. The accuracy and reliability of the finite element simulation method in the analysis of the nonlinear problem of the air spring system are verified by the comparison between the simulation and the experiment. In addition, the use of finite element simulation methods can not only greatly shorten the development cycle of membrane air spring products, reduce the cost of R&D but also avoid the risk of possible problems during the test.

References

1. B. Liu , J CVGI,**07**,112-114 (2007)
2. Z. S. Yu , *Automobile Theory*(2009)
3. C. Y. Yuan, K. K .Zhou, L. Q. Wu, D. F. An, G. L. Wang,CJME,**06**,262-266(2009)
4. E. Richer, Y. Hurmuzlu, J. Dyn. Syst. Meas. Control,**122**,416-425(2000)
5. J. L. Zhang, H. Y. Wang, Jou, **06**, 115-119(2013)
6. D. Chen, L. I. Fu, Y. H. Huang, EL.MTV（2010）
7. K. Wang, E. F. Chen, Y. Zhang, PVT, **12**, 1-6(2016)

Measurement of six degree bi-directional motion error of linear stage

Ryoshu Furutani

Tokyo Denki University, School of Engineering, Senju-Asahi 5, Adachi, 120-8551 Tokyo, Japan

Abstract. We proposed the measurement system of the six degree of motion errors which is based on distance measurement by the laser interferometer. The system has six parallel laser beams and six ball lenses as the retroreflectors on the linear stage, which reflect the corresponding laser beams. In the proposed system, the error of axial direction is measured with the ordinary distance measurement method by laser interferometer. The vertical errors to the axial direction and the roll errors around the optical axis are measured by tilted beams using the wedge prism. The pitch and yaw errors in the vertical plane to the optical axis are measured by the difference between distance of two ball lenses. The former system can measure the displacement and the error angle in one-direction. The propose system are expanded and bi-directional displacement and error angle can be measured. In this paper, it is shown how to expand the measurement system. As a result, the maximum displacement errors in x, y and z directions are 242nm, 179nm and 90nm. The maximum rotational errors around x, y, z axes are 1.75 arcsec, 2.35 arcsec and 1.67 arcsec.

1 Introduction

Six Degree-of-freedom (6DOF) motion errors of a linear stage of a machine tool, a coordinate measuring machine and so on should be measured simultaneously in order to improve their accuracies. As a result, the more accurate position and orientation can be obtained and higher compensation can be expected. However, it is difficult to measure 6DOF motion errors simultaneously because the stages have no enough space to setup conventional measuring instruments such as displacement sensors and autocollimators. Therefore, it is expected to develop the measuring instrument that can simultaneously measure the 6DOF motion errors. For these reasons, many researchers have developed the measuring method of the 6DOF motion errors [1-5] and the measuring system of 6DOF motion errors that employs the principle of the Michelson interferometer to detect all direction displacement and rotational angle errors is developed because of the advantage of constructing high accurate measuring system with comparative ease [6-7]. However, the system proposed in [6-7] can measure 6 DOF motion errors in only one direction. In this paper, the system can be expanded to measure the 6 DOF motion errors in bi-direction.

2 Measurement Principle

2.1 Optical System

The schematic diagram of the system is shown in Figure 1. The system is based on the length measurement by Michelson interferometer. The moving plate (MP) which consisted of six target mirrors (TM) and two plates, was mounted on a stage as a measuring object in situ situation. The six ball lenses were adapted as the target mirrors. The refractive indices of those are nearly 2. To fulfil retro reflection of a laser beam, the ball lenses were coated with silver film in a hemispheric part as shown in Figure 2. The other optical devices were fixed on an optical table. The laser beam emitted from laser source was expanded by two plano-convex lenses. The expanded laser beam was separated into the two laser beams by a beam splitter(BS). One of the two laser beams passed toward TMs and was partly tilted by three wedge prisms (WP). The other laser beam passed toward a reference mirror (RM). Each laser beam entered and was reflected by the target mirrors and the RM. Finally, the laser beam reflected from the target mirrors interferes with the laser beam reflected by the RM. When the MP moved, optical path difference between the two laser beams was varied. The displacements of the 6DOF motion errors of MP could be measured by counting the number of the interference fringes due to the two laser beams. The number of the interference fringes could be obtained by photo diodes (PD).

2.2 Displacement measurement in x direction

The schematic diagram of the principle of displacement measurement in x direction is shown in Figure 3. When MP moves to the x direction with a value of Δx, the relation of the displacement Δx and the number of the interference fringes which are caused by the interference

https://doi.org/10.1051/matecconf/201815306007

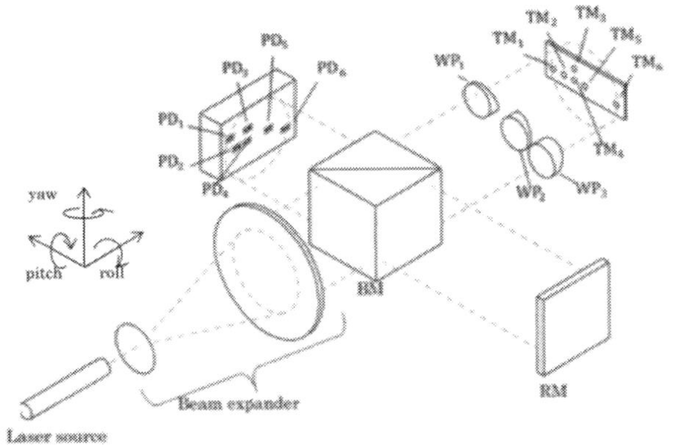

Figure 1. Configuration of optical system of 6 DOF laser interferometer.

of the laser beams reflected from the TM4 and the RM is shown as following equation (1).

$$\Delta x = \frac{\lambda}{2} n_4 \tag{1}$$

Here, λ denotes a wave length of the laser beam, n_4 denotes the number of interference fringe obtained by the PD_4 which is irradiated by the laser beam reflected from TM_4.

2.3 Displacement measurement in y direction

The schematic diagram of the principle of y direction displacement measurement is shown in Figure 4. A part of the laser beam is tilted by the WP_2 with the angle θ to the x axis in the x-y coordinate system. When MP moves in the y direction with the value of Δy, the optical path length of the tilted laser beam changes only $(\lambda/2)n_5$ where n_5 is regarded as the number of interference fringes obtained by the PD_5 which is irradiated by the laser beam reflected from TM5. The relation of the displacement Δy and the number of the interference fringes which is caused by the interference of laser beams reflected from the TM_5 and the RM is shown as following equation (2).

$$\Delta y = \frac{\lambda}{2\sin\theta} n_5 \tag{2}$$

2.4 Displacement measurement in z direction and roll angle measurement around x axis

The schematic diagram of the principle of z direction and roll angle measurement is shown in Figure 5 and Figure 6. TM_1, TM_6, WP_1 and WP_3 are used in the measurement. The displacement in z direction can be measured by the same method as that in y direction. The displacement in z direction is measured by the both laser beams tilted by the both WP1 and WP3 of which primary axes are rotated with 90 degree around x axis. Thus, the displacement Δz and the roll angle $\Delta\theta_x$ can be measured as the following equation (3) and equation (4).

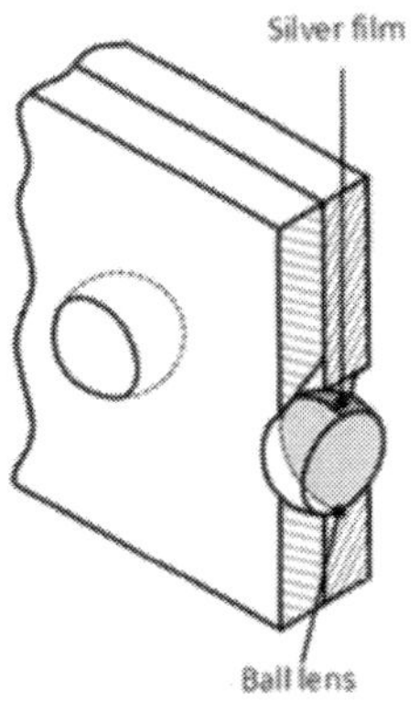

Figure 2. Moving plate(MP).

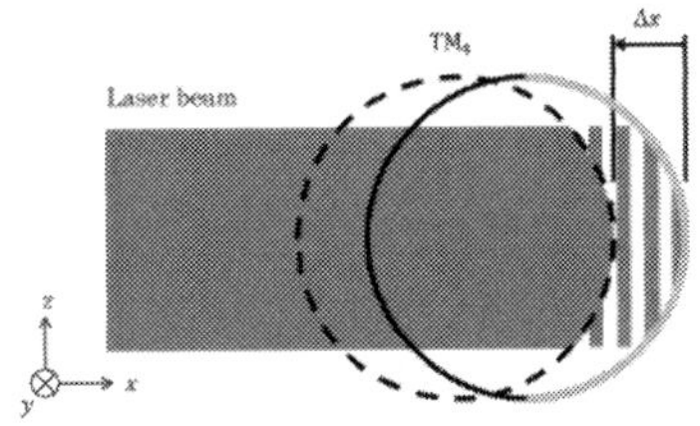

Figure 3. Measurement principle of x-displacement.

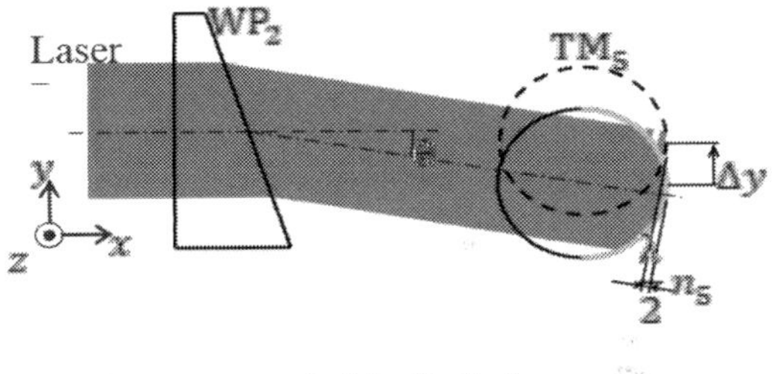

Figure 4. Measurement principle of y-displacement

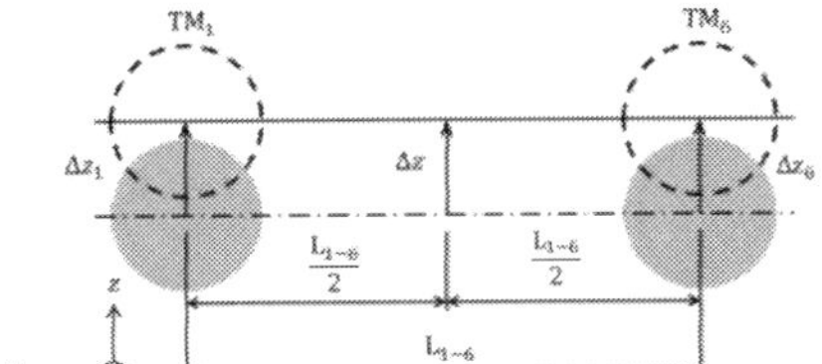

Figure 5. Measurement principle of z-displacement.

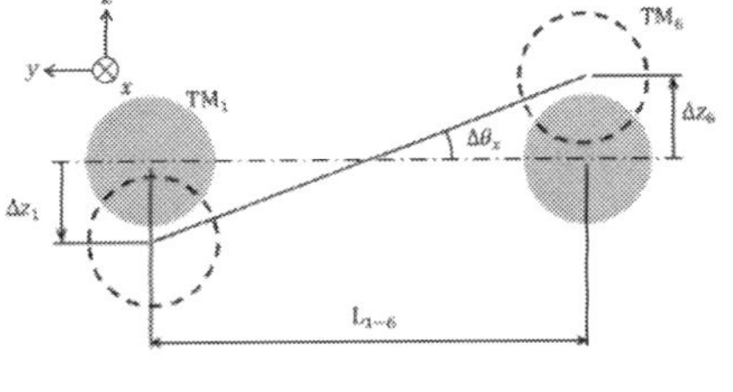

Figure 6. Measurement principle of roll angle around x-aixs.

 https://doi.org/10.1051/matecconf/201815306007

$$\Delta z = \frac{\lambda}{4\sin\theta}(n_1 + n_4) \quad (3)$$

$$\Delta\theta_x = \tan^{-1}\left(\frac{\lambda}{2L_{1-6}\sin\theta}(n_1 - n_6)\right) \quad (4)$$

Here, $L_{1\text{-}6}$ denotes the distance between TM_1 and TM_6.

2.5 Pitch and yaw angle measurement in the vertical plane to x axis

The schematic diagram of the principle of pitch angle measurement is shown in Figure 7. The pitch angle can be calculated by x direction displacement of TM_3 and TM_4 and the distance between TM_3 and TM_4. The x direction displacement of TM_3 and TM_4 can be measured by the same method of Δx mentioned in 2.2. The yaw angle can be measured as similar as the pitch angle using TM_2 ad TM_4 as shown in Figure 8. Thus, the pitch and yaw angle $\Delta\theta_y$ and $\Delta\theta_z$ are expressed as following equation (5) and equation (6).

$$\Delta\theta_y = \tan^{-1}\left(\frac{\lambda}{2L_{3-4}}(n_3 - n_4)\right) \quad (5)$$

$$\Delta\theta_z = \tan^{-1}\left(\frac{\lambda}{2L_{2-4}}(n_2 - n_4)\right) \quad (6)$$

- Here, $L_{3\text{-}4}$ denotes the distance between TM_3 and TM_4, $L_{2\text{-}4}$ denotes the distance between TM_2 and TM_4, n_2, n_3 and n_4 denotes the number of interference fringes obtained by the PDs which are irradiated by the laser beam reflected from TM_2, TM_3 and TM_4.

2.6 Measurement in bi-directional way

In the method mentioned above, the moving direction could not be distinguished. The photo diodes are replaced with the bi-photo diodes. By using the bi-photo diodes, the moving direction is able to be distinguished as shown in Figure 9.

3 Experiments

3.1 Experimental devices

The specifications of the optical devices are shown in Table 1. The MP was mounted on the xyz- and rotational stages, and driven by these stages. The displacement of the MP was measured 5530 Laser Calibration System (Agilent Technologies) as reference. Three axes piezo nanometer positioning stage (Physik Intrumente, P-733.3DD) and three rotation stages(Suruga Seiki, KGW06100, KGW06075 and KRW06360) were used.
The specifications of the used positioning stages are shown in Table 2.

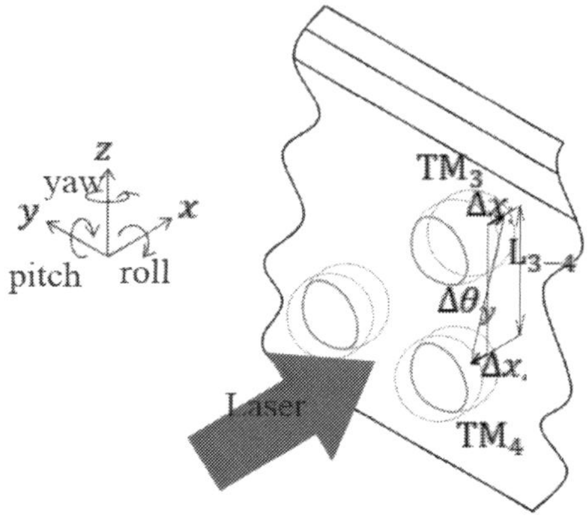

Figure 7. Measurement principle of pitch angle in the plane vertical to the optical axis.

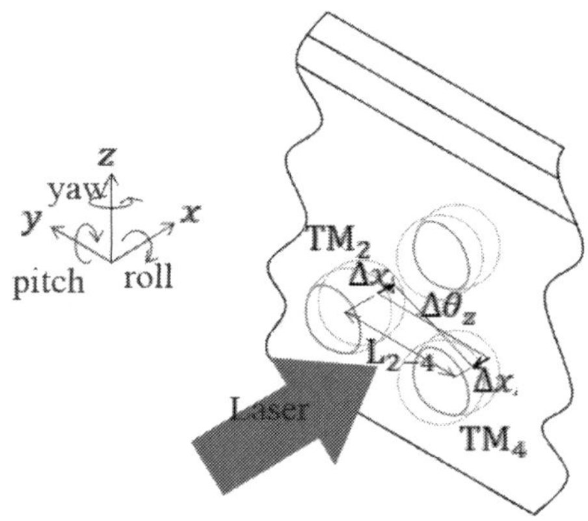

Figure 8. Measurement principle of yaw angle in the plane vertical to the optical axis.

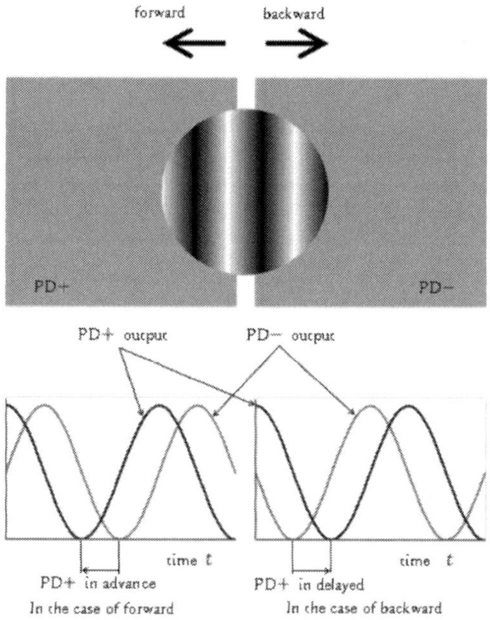

Figure 9. Distinguish of forward and backward direction.

https://doi.org/10.1051/matecconf/201815306007

Table 1. Specification of the optical system.

Item	value
λ:Wave length of the laser beam	632.57 nm
θ: Tilted angle of the laser beam	10 degree
Refractive indices of the ball lens	2.003
$L_{1\text{-}6}$: Distance between TM_1 and TM_6	48.06 mm
$L_{3\text{-}4}$: Distance between TM_3 and TM_4	7.92 mm
$L_{2\text{-}4}$: Distance between TM_2 and TM_4	8.19 mm

Table 2. Specification of stages.

Margin	Direction	Resolution	Repeatability
XYZ-Stage [P-733.3DD]	x,y and z	0.1nm	2nm, 2nm and 1nm
Gonio-Stage [KGW06100]	Roll	0.002466 degree	0.003 degree
Gonio-Stage [KGW06075]	Pitch	0.0032 degree	0.003 degree
θ -Stage [KRW06360]	Yaw	0.004 degree	0.01 degree

3.2 Experiment of detection of Bi-directional movement

In order to examine whether the bi-photo diodes can detect the moving direction, when the MP moves 5 μm along z-axis. The result is shown in Figure 10. Figure 10 a) shows a result when the MP moves forward along z-direction. Figure 10. b) shows a result when the MP moves backward along z-direction. It is proved that the moving directions are detected by comparing the phase of output signal of PD1+ and output signal of PD1-.

3.3 Experiment of movements

In this system, the principle of displacement measurement of y and z axis are similar, the principle of angle measurement of pitch and yaw are also similar. Therefore, the displacement in x and y axes and the angle roll and pitch are measured and compared with 5530 Laser Calibration System. Every measurement is iterated 5 times.

Figure 11. shows the result of displace measurement in x- and y- axis.

Figure 12. shows the result of angle measurement in roll and pitch.

In these measurements, the maximum displacement errors in x, y and z directions are 242nm, 179nm and 90nm. The maximum rotational errors (roll, pitch and yaw) are 1.75 arcsec, 2.35 arcsec and 1.67 arcsec. the results of this system meet with the reference instrument.

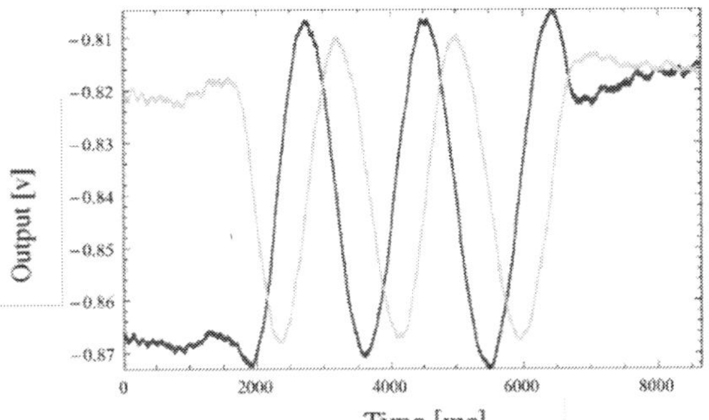

a) The output signal of PD1 in forward movement.

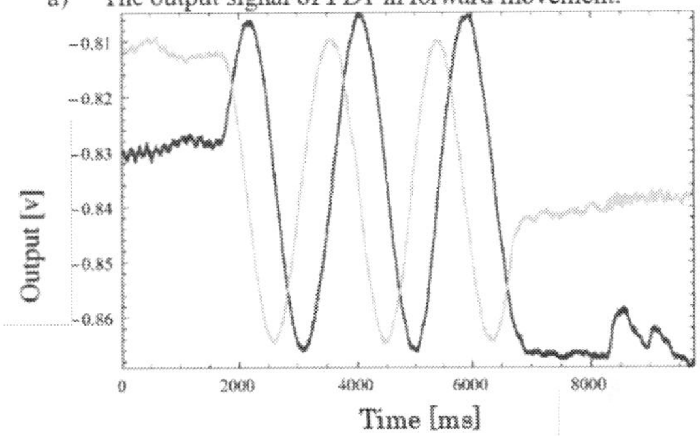

b) The output signal of PD1- in backward movement.

Figure 10. The output signal of PD1+(blue line) and PD-(green line) when the MP moves 5μm in z-direction.

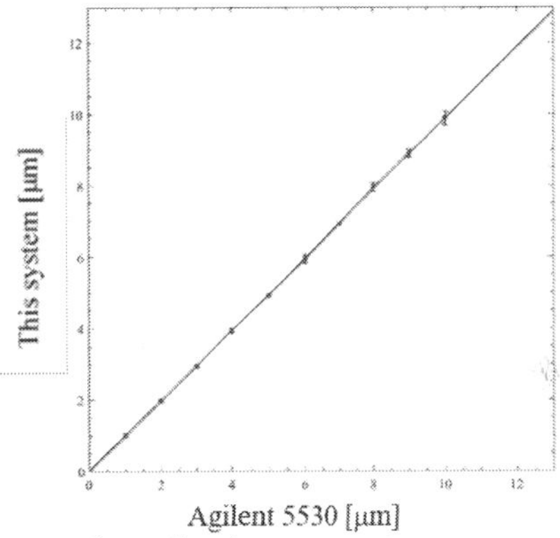

a) x-direction movement.

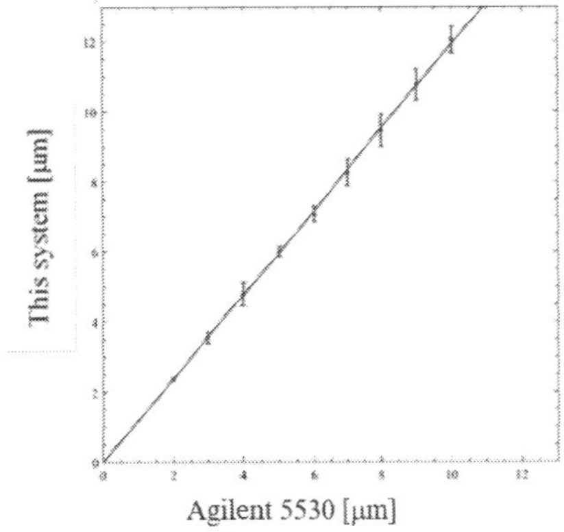

b) y-direction movement.

Figure 11. The relationship between the displacements of this system and Agilent 5530.

 https://doi.org/10.1051/matecconf/201815306007

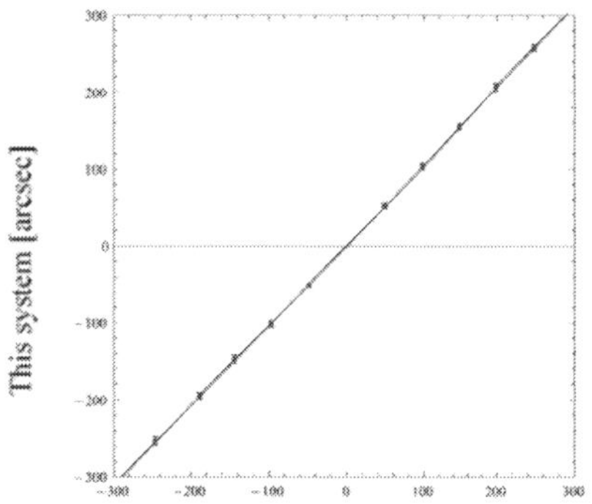

b) Roll angle movement.

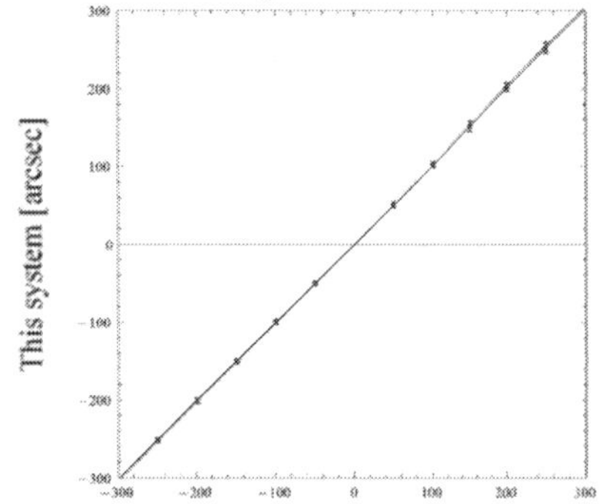

c) Pitch angle movement.

Figure 12. The relationship between the angle movement of this system and Agilent 5530.

4 Conclusion

The measurement system which can measure 6DOF of the stage was proposed. The proposed optical system was composed and the measurement result of this system were compared with 5530 Laser calibration system as reference. They are enough good to meet.

However, these experiments were performed independently. One displacement and/or one angle movement must affect the other displacements and angles. So two or more movements are composed and should be measured simultaneously in future work.

References

1. S. Shimizu, H. S. Lee and N. Imai, Int.J.Japan Soc. Prec. Eng. **28**, 273 (1994)
2. C.H.Liu, W.Y.Jywe, C.C.Hsu, Review of Scientific Instrument **76**,055110 (2005)
3. Q.Feng, B.Zhang, C.Cui, C.Kuang, Y.Zhai, F.You, Optics Express **22**,21,25805(2008)
4. K.C.Fan, M.J.Chen,W.M.Huang,Int.J.Machine Tools and Manufacture **38**, 155 (1998)
5. K.C.Fan, M.J.Chen, Precision Engineering **24**, 15 (2000)
6. J.Izawa, S.Kanno, M.Ozaki, R.Furutani, Proc. ISMTII 2015 (2015)
7. R.Furutani, Materials Science and Engineering **211** b012001 (2017)

Magnetic Microfluidic Linear Halbach Array Configuration for Cell Separation

Omer Saeed[1,2], lucienne Duru[1] and Deng Yulin[1]

[1]School of Life Science, Départment of Biomédical Engineering, Beijing Institute of Technology, Beijing 100081, China.
[2]Faculty of science and Technology, Department of Biomedical Physics, ALNeelain University, Khartoum 1121, Sudan.

Abstract. The different type of blood entities like red blood cells (RBCs), white blood cell (WBCs) and platelets counts is a critical point in routine medical tests. Circulating tumor cells (CTCs) counting and analysing in the blood is indicative of the stage and origin of cancer in particular time. Here we are describing a proposed microfluidic device, used a magnetic means for CTCs separation from a blood or buffer sample. The device characterizations were simulated using COMSOL MULTIPHYSICS software as well as practically validated. Using a known spiked number of CTCs in the sample, the purity and recovery rate were calculated to be 93% and respectively.

1 Introduction

Particle separations are actual particulate that are not dissolved in the sample but rather than particles are suspended in the sample, these samples would be referred as suspensions. There are a lot of different application of particle sorting and counting in blood analysis and environmental field. In fact one of the big areas in this field now is the ability to capture and sort circulating tumor cells (CTCs) in the blood. It turns out that the CTCs have slightly different properties than normal cells, they are more rigid and less squishy than the normal cells are[1].

Circulating tumor cells (CTCs) are thought to be cells that have been shed from a primary tumor and have transferred into the circulation system, those cells then have the potential to form a metastatic in the patient so the cells are very dangerous to the patient more so than once in the primary tumour because in most patients the mortality comes from metastatic cancer not from the primary tumour. These cells are really poorly defined as we really feel that we don't have the best definition of what a circulating tumour cell is, in order to capture a heterogeneous set of CTCs. the current definition most people agree is that given by cell search which is the only F.D.A. cleared method for enumerating. And they define a circulating tumor cell as a large nucleated cell, with a positive staining for EpCAM (Epithelial Cell Adhesive Molecules) and cytokeratin 8, 18 or 19 which is the epithelial cells intercellular marker and negative for leukocyte marker staining CD45. These cells are very difficult to study because they're in such low abundance, in one ml of whole blood there may be ten million white blood cells (WBCs) and five billion red blood cells (RBCs) while in a metastatic in a breast cancer patient may have on the order of one hundred circling tumour cells, so they're very difficult to separate them because of their extremely low abundance. We're interested in studying these because they hold such promise of prognostic value to the patient.

There are many researcher have studied the cells deformability and being able to sort different cells types based on those parameters[2][3]. Also there is some work that has been published earlier is well known now, that the cancer cells have slightly different size than normal cells[4] . So it's possible to separate cancer cells from noncancerous ones based on size with very high resolution.

Most recently, that have actually been shown that cancer cells have express different types of membrane protein than normal cells do, called surface markers, created pillar like structures functionalized antibodies groups on the wall will capture pushed cells and protein as they passed by those pillars. There are a lot of interests in counting and sorting the different types of particles in blood, either they are RBCs, WBCs cancerous and non-cancerous cells[5][6]. Cells counting are very important in medical diagnostics. Another example of cell counting in Human Aids Virus detection (HIV) diagnostic, where CD4 cells number in blood is an indicative of positive or negative diagnose[7]. All these techniques of particle and cell counting required methods to capture and sort particles in a liquid for sorting based on different properties. Besides blood analysis, also particle sorting has many applications in environmental and biology fields. Such in water monitoring, as a water sample may contains different types of organisms ranging from single cell organisms and bacteria, separating and counting these different life species will acts as environmental monitoring detector[8]. Cell sorting and focusing techniques are ultimately applied to flow cytometer concepts. A flow cytometer is a method for counting and differentiating cell types. The idea is that cells will flows in a single file through a window where they are detected by a laser beam. The cells suspension sample is combined with two other outsides liquid flows form a parallel streams through the length of the channel, called hydrodynamic techniques or sheath flows[9].

Basically we're interested in capturing a heterogeneous subset of circulating tumor cells, using a simple magnetic means technology that allows us to isolate a single circulating tumor cells that are appropriate for downstream analysis. We fabricated and test a microfluidic chip with permanent magnets linear array located at a distance from the channel wall to generate the magnetic field required to deflect and alter the magnetic particle and so magnetic particle labelled cell to the specific determined outlet. The collected cells sample has been characterized using flow cytometer and fluorescent microscope to determine the fabricated microdevice ability in cells separation.

2 Merterial and methods

The microfluidic device is simplify fabricated using PDMS with glass slide followed the standard photolithography process. The basic principle of the operation is relatively straight forward, part of particles could be magnetized and magnetophoresis is basically means a flow in the channel. In the figure .1 (A) below as the flow going from right to left, a magnetic field that's acting on the particles. The magnetic field can cause the particles to migrate at different rates depending on their size and magnetic properties. Here the sample mixture is going through the middle streamline. The sample is coming down the middle part of the channel and the others are just buffers. As continuous free flow electrophoresis, without magnetic field all the particles will just travel straight through the bottom, but with magnetic field different magnetized particles will travel at different speeds upwards depending on their properties. So the different types of particles would end up at different outlets.

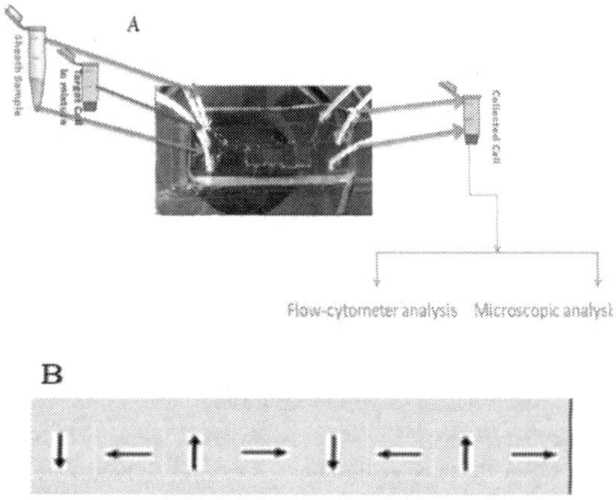

Figure 1. (A) Working and characterization steps of the proposed device. (B) The magnets orientation used.

Firstly, the magnetic field simulation of linear Halbach arrays is carried out by using COMSOL finite element software. The distribution of magnetic flux and the distribution of magnetic field are obtained. From these results, it can be seen that the Halbach array has the characteristics of one side magnetic field and the other side is very small. This feature lays a theoretical foundation for the development of various magnetic devices made up of Halbach arrays. The traditional linear Halbach arrays used her is shown in figure .1 (B). We used eight 3 mm3cubic permanent magnets, with specific orientation as in figure .1(B).

To represent what we would see in a patient sample we've done spiking experiments, where we spiked the chosen MCF7 Cells line into whole blood. Then ficoll gradient was performed to obtain a Buffy coat. Lastly we positively labelled the MCF7 Cell in the sample with anti-Epcam antibodies that have a magnet bead attached to it. The sample contained the labelled MCF7 is followed into our microfluidic device through the middle inlet using a syringe pump (Cole-Parmer, Vernon Hills, IL, USA) with a specific flow rate. The schematic graph of the sample preparation is shown in figure .2 below

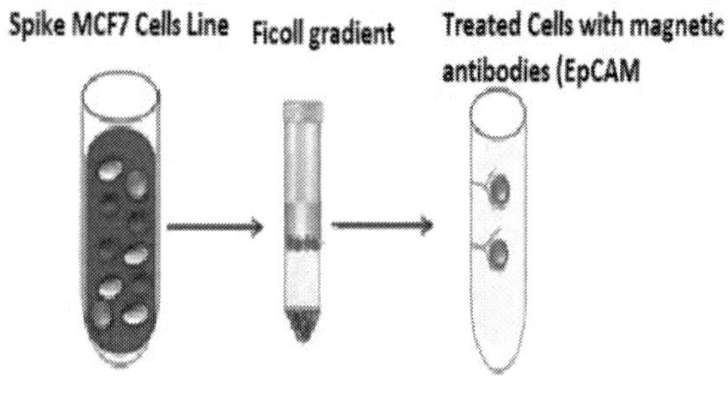

Figure 2. the sample preparations steps.

In order to solve the magnetic field distribution of the Halbach array, we have to solve the Maxwell equation by the basic theory of the electromagnetic field. The partial differential equation and boundary conditions of the magnetic field can be obtained from the Maxwell equation:

$$\nabla.(\mu_0\mu_r H) = 0 \quad (1)$$

$$56H = -\nabla V_m$$

Where μ0 and μr are the vacuum and material relative permeability respectively. H and V is the magnetic and potential field respectively.

3 Results and discussion

The designed three inlets and outlets microchannel,s width (w) and length (L), used here in this analysis were 200μm and 30 mm in y and x direction respectively.

 https://doi.org/10.1051/matecconf/201815306008

To determine the magnetic field distribution in the configuration above, a 2 D model was constricted and simulated using Finite Element Method (FEM) package COMSOL. The simulation model includes the polydimethylsiloxane (PDMS) channel, the permanent magnets array and the flow media, as shown in Figure. 1. In order to limit the simulation domain, outer circle was used. The array contains eight of 3 mm3cubic shape and 49600 A.m saturation magnetization NdFeB magnet was used. Theses magnets was placed at 2 mm away from the channel wall.

Magnetic field strengthH_y in both channel walls (at $y = 0$ and$H$) was computed, so as the magnetic field gradient in y direction $\frac{\partial H_y(x,y)}{\partial y}$ throughout the microchannel width has been estimated to be 592.813 A/m.

The resulted magnetic field distribution is shown in figure .3 below

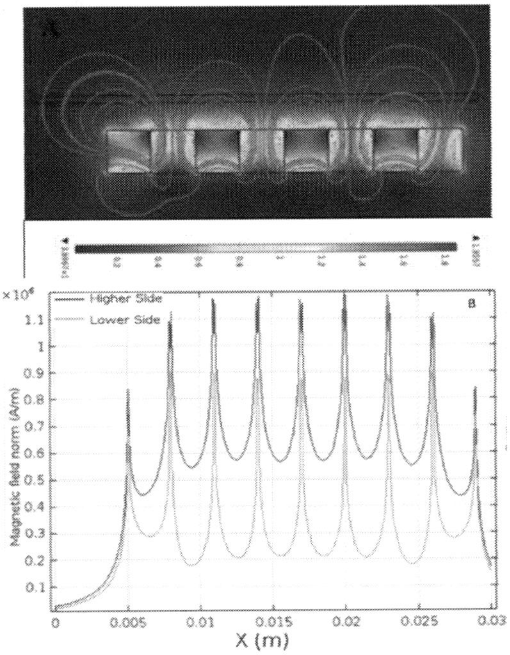

Figure 3. Magnetic field generated from the linear magnets array and the michrochannel location relative to the magnets. (A) Surface plot. (B) Magnetic field along the X axis at the surface of the magnets on higher and lower sides of the array.

Additionally the magnetic field gradient components X and Y have been calculated at different distance from the array surfaces, particularly, at 1, 2 and 2.250 mm and shown in figure .4 below.

The magnetic forces in both X and Y directions, that acting on the beads is given by[10]

$$F_{mag,\ x} = \frac{4\pi r_p{}^3 \Delta\chi}{3\mu_0} B_x \frac{\partial B_x(x,y)}{\partial x} \tag{2}$$

$$F_{mag,\ y} = \frac{4\pi r_p{}^3 \Delta\chi}{3\mu_0} B_y \frac{\partial B_y(x,y)}{\partial y} \tag{3}$$

Where B =the magnetic field, μ0 = the permeability of the space, Δx = magnetic susceptibility of the bead and that of the fluid difference and rP = radius of the bead magnetic core.

The magnetic beads acting force components X and Y were solved numerically and given in figure. 5. From there it's obviously that, the Y component of the force is much greater than the X one, with higher negativity, indicates the attractive force strength on the magnetic beads and affects their path with a displacement in y direction along the microchannel width, and end up through the desired exit.

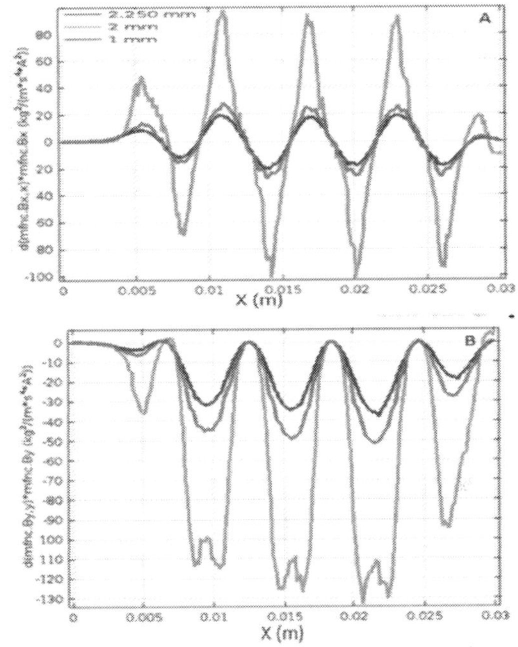

Figure 4. Magnetic field gradients. (A)X component. (B) Y component.

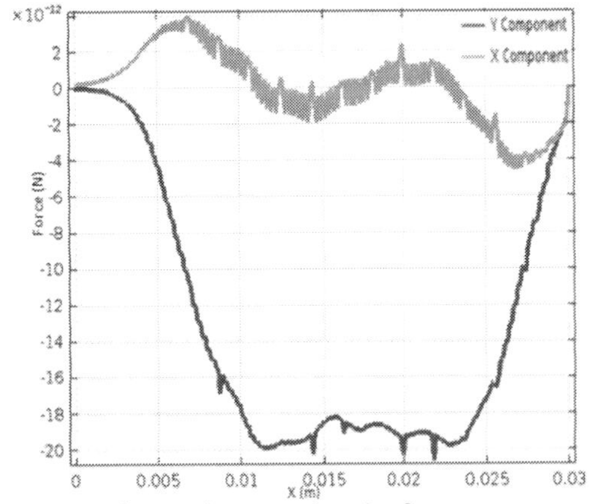

Figure 5. the X and Y magnetic acting force.

Magnetic particle trajectories for time saving, was simulated only in the main channel domain. This domain has the same dimension parameters of the channel (200 um and 30 mm) high and length respectively.

The magnetic particle trajectories under magnetic field effects were simulated numerically as shown in figure .6. It's clearly shown that, the magnetic beads followed a 15 mm distance inside the microchannel without suffering any displacement, but after this point (tip of the magnets array) they start to deflect from their original path as here the magnetic force is greater than the fluidic drag force.

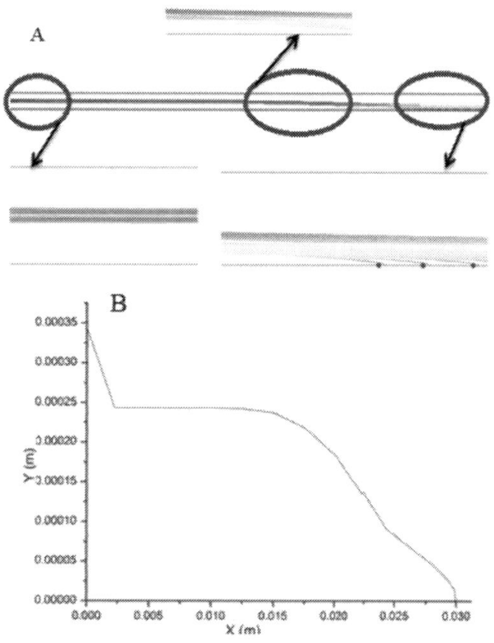

Figure 6. (A) snap shot images of the magnetic beads trajectories. (B) Beads trajectories across the channel width.

Experimentally, we tested the separation efficiency and specificity using a flow cytomer analysis. 10^2 to 10^5 labelled MCF7 cells in repeated assay were spiked in the prepared blood sample the results are shown below in figure. 7(A,B). We observed high separation efficiency in MCF7 cell separation from the sample. Around 83 % overall separation efficiency was achieved.

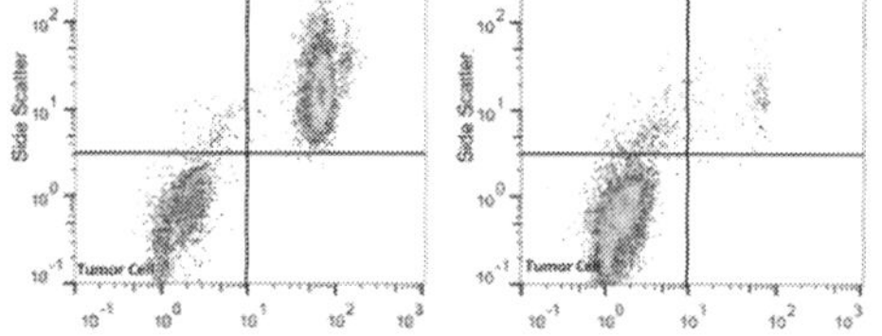

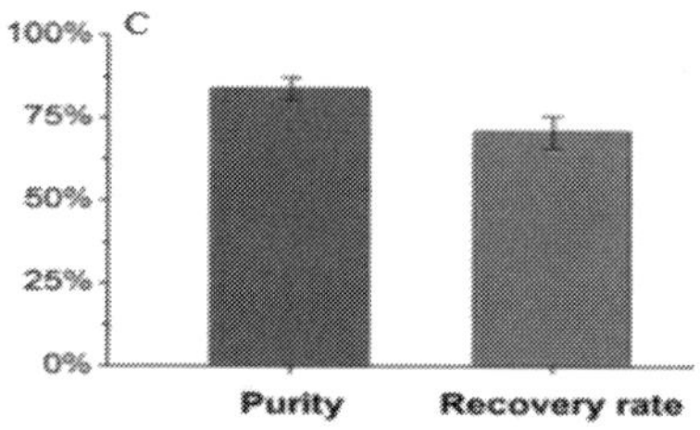

Figure 7. Flow cytometry results of (a) the mixture sample (WBCs mixed with tumor Cell), (b) separated tumor Cell sample collected through the lower outlet.

Additionally, the collected sample cells were stained for EpCAM, CD45, and DAPI are used to identify MCF-7 from leukocytes and for separated cell purity determination. The purity was almost about 87% as in figure .7(C).

4 Conclusion

Here we described the principles and experimental results of microfluidic technologies for cancer cell separation. The microfluidic devices are based on magnetic activated cell sorting. We fabricated and designed a simple low-cost chip for separation of CTCs from buffer /sample solution. We computationally simulated the external static magnetic intensity using modelling software COMSOL Multiphysics 4.3a (COMSOL, Inc.). The designed micro-device with permanent magnets array isolates the CTCs into the lateral -exit of the flow in the micro-channel. The micro- device was shown experimentally to separate around 87% of the spiked CTCs cancer cells (MCF-7). Microfluidic devices with an external permanent magnets array combined with a lab made fluidic control system are very useful for the precise and reliable microfluidic separation of CTCs as an integrated diagnostic instrument. Since the described designed micro-device protocols are quite simple and mass-producible, the microdevice can provide a useful and practical tool for allowing the micro-device to be costiveness producible as clinical tool for cancer diagnosis, prognosis, and personalized medicine.

Acknowledgements

This work in the school of life science, Beijing institute of technology and financially was was supported by the National Key Scientific Instrument and Equipment Development Project of China (No. 2012YQ040140).

References

1. W.-W. Sun, Z.-H. Xu, P. Lian, B.-L. Gao, and J.-A. Hu, "Characteristics of circulating tumor cells in organ metastases, prognosis, and T lymphocyte mediated immune response," Onco. Targets. Ther., vol. Volume 10, pp. 2413–2424, May 2017.

2. H. W. Hou, Q. S. Li, G. Y. H. Lee, A. P. Kumar, C. N. Ong, and C. T. Lim, "Deformability study of breast cancer cells using microfluidics," Biomed. Microdevices, vol. 11, no. 3, pp. 557–564, Jun. 2009.
3. M. Aghaamoo, Z. Zhang, X. Chen, and J. Xu, "Deformability-based circulating tumor cell separation with conical-shaped microfilters: concept, optimization and design criteria."
4. "Characteristics of circulating tumor cells (CTCs) in patients with lung cancer." [Online]. Available: https://www.mdlinx.com/oncology/conference-abstract.cfm/57200/?nonus=0&searchstring=&coverage_day=0&page=2. [Accessed: 03-Oct-2017].
5. C. W. Shields IV, J. L. Wang, K. A. Ohiri, E. D. Essoyan, B. B. Yellen, A. J. Armstrong, and G. P. López, "Magnetic separation of acoustically focused cancer cells from blood for magnetographic templating and analysis," Lab Chip, vol. 16, no. 19, pp. 3833–3844, Sep. 2016.
6. X. Mao, J. R. Waldeisen, and T. J. Huang, "'Microfluidic drifting'—implementing three-dimensional hydrodynamic focusing with a single-layer planar microfluidic device," Lab Chip, vol. 7, no. 10, p. 1260, Sep. 2007.
7. J. Darabi and C. Guo, "On-chip magnetophoretic isolation of CD4 + T cells from blood.," Biomicrofluidics, vol. 7, no. 5, p. 54106, 2013.
8. J. Bartram and R. Ballance, "Water Quality Monitoring -A Practical Guide to the Design and Implementation of Freshwater Quality Studies and Monitoring Programmes Edited Chapter 11 - BIOLOGICAL MONITORING," pp. 0–419.
9. S. F. Ibrahim and G. van den Engh, "Flow Cytometry and Cell Sorting," in Advances in biochemical engineering/biotechnology, vol. 106, 2007, pp. 19–39.
10. C.-L. Chang, C. A. Savran, S. Jalal, and D. E. Matei, "Micro-aperture chip system for high-throughput immunomagnetic cell detection," in 2012 IEEE Sensors, 2012, pp. 1–4.

Characteristic Analysis of longitudinal vibration Sucker Rod String With Variable Equivalent Stiffness

Jian Lv [1], Mingming Xing[2]

[1]*School of Mechanical and Vehicle Engineering, Linyi University, Linyi 276000,China*
[2]*School of Automation and Electrical Engineering, Linyi University, Linyi 276000,China*

Abstract. Considering the influence of variable equivalent stiffness on system response, the equivalent stiffness is defined as a step function, and a mathematical model of nonlinear longitudinal vibration of sucker rod string (SRS) is built. The dynamic response under displacement and load force excitation is solved by fourth-order Runge-Kutta method with zero initial condition. The results show the steady-state responses under the displacement and load force excitation of different function forms are different. The response curves of both displacement and velocity under the displacement and load force excitation of cosine function form have larger fluctuation than it under the displacement and load force excitation of sine function form. Therefore, the characteristic analysis of SRS plays an important role in understanding the influence of the excitation form and sensitive parameters on steady response.

1 Instruction

The SRS is a slender rod of several kilometres, which transmits the movement from mechanical rotary motion to plunger linear motion [1-2]. And, the characteristic analysis of longitudinal vibration of SRS is very important for the parameters optimization of sucker rod pumping system (SRPS) [3]. Therefore, the dynamic simulation of SRS is studied by domestic and foreign scholars. After wave equation of Gibbs [4], an improved model of SRPS is built by Dale Russel Doty and S.D.Lekia [5-6]. Recently, I.N.Shardakov, Luan Guo-hua and M.M Xing etal [7-9] analyse boundary conditions effecting rod law of motion and correct the excitation forces of rod longitudinal vibration. However, the most of aforementioned approaches have not been studied sufficiently. Relevant references [9-10] shows that the equivalent stiffness is a function of time, when the influence of the helical buckling of SRS on the equivalent stiffness is taken into consideration, which affects the characteristic of longitudinal vibration of SRS. Therefore, with non-constant equivalent stiffness, the dynamic Characteristic of SRS with Variable Equivalent Stiffness is necessary to be analysed.

In this paper, the dynamic model and nonlinear second-order differential equations are built, and the entire equilibrium equation is given in section 2. In section 3, the system equations of nonlinear longitudinal vibration of SRS are solved by fourth-order Runge-Kutta method with zero intimal conditional. In section 4, the dynamic responses under different function forms of load force excitation as well as the load force and displacement excitation are given, and the effects of different function forms of load force excitation as well as displacement excitation on the amplitude-frequency curves are analysed. In section 5, the significance of dynamic characteristic of SRS and conclusions are summarized.

2 Dynamic model and nonlinear second-order differential equations

In order to establish mechanical model of nonlinear longitudinal vibration for SRS, the fundamental assumptions are as follows: (1)The tubing string is anchored; (2)The viscous damping of liquid is concerned; (3) The tube- liquid column vibration is ignored; (4) Considering the deformation of helical buckling of SRS, the equivalent stiffness is a function of time.

Based on the assumptions, SRS is dispersed into a spring-mass-damper system and a mechanical model of nonlinear longitudinal vibration for SRS is established, in Fig.1. Fig.1 shows that SRS is a spring-mass-damper system with constant stiffness in 0~Li units and with non-constant stiffness in residual units. The direction from up-bottom is described as positive. The displacement, velocity and acceleration for polished rod are defined as u0, $\dot{u}_0$, $\ddot{u}_0$. Then a mathematical model of nonlinear longitudinal vibration of SRS is established, as follows.

 https://doi.org/10.1051/matecconf/201815306009

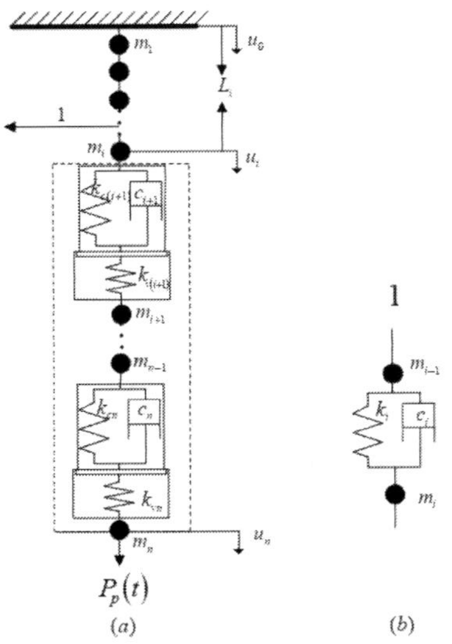

Figure 1. Mechanical model of nonlinear longitudinal vibration of SRS

$$\begin{cases} m_1\ddot{u}_1 + k_1(t)(u_1-u_0)+k_2(t)(u_1-u_2)+c_1(\dot{u}_1-\dot{u}_0)+c_2(\dot{u}_1-\dot{u}_2)=0 \\ m_2\ddot{u}_2 + k_2(t)(u_2-u_1)+k_3(t)(u_2-u_3)+c_2(\dot{u}_2-\dot{u}_1)+c_3(\dot{u}_2-\dot{u}_3)=0 \\ \vdots \qquad \cdots \qquad \vdots \qquad \vdots \\ m_{n-1}\ddot{u}_{n-1} + k_{n-1}(t)(u_{n-1}-u_{n-2})+k_n(t)(u_{n-1}-u_n)+c_{n-1}(\dot{u}_{n-1}-\dot{u}_{n-2})+c_n(\dot{u}_{n-1}-\dot{u}_n)=0 \\ m_n\ddot{u}_n + k_n(t)(u_n-u_{n-1})+c_n(\dot{u}_n-\dot{u}_{n-1})-P_p(t)=0 \end{cases} \tag{1}$$

Defining {u} = {u1, u2, …, un-1, un }T, a simplified mathematical model is given, as follows

$$[M]\{\ddot{u}\}+[K]\{u\}+[C]\{\dot{u}\}=\{F\} \tag{2}$$

Where,

$$m_i = \begin{cases} \dfrac{\rho_r g l_{0i}\pi D_i^2}{4} & i=1,2,3\cdots n-1 \\ \dfrac{\rho_r g l_{0i}\pi D_i^2}{8} & i=n \end{cases} \tag{3}$$

$$c_i = \frac{12\pi\mu m_i}{\rho_{ri}A_{ri}}\left(\frac{D_{ri}}{D_{ti}-D_{ri}}\right)\left[\left(0.20+0.39\frac{D_{ri}}{D_{ti}}\right)+\frac{2.197\times10^4}{25}\left(\frac{D_{ci}}{D_{ti}}-0.381\right)^{2.57}\frac{D_{ci}^{\ 2}-D_{ri}^{\ 2}}{l_{0i}D_{ri}}\right] \tag{4}$$

$$\begin{cases} u_0 = A_0\sin(\omega_f t) \\ P_p(t)=M_0+M_1\sin(\omega_f t) \end{cases} \tag{5}$$

$$k_x = \begin{cases} x=1,2,\cdots i \\ \dfrac{E_r A_{ri}}{l_{0i}} \\ x=i+1,i+2,\cdots n \\ \operatorname{sgn}(\dot{u})\varepsilon(\dot{u})\dfrac{E_r A_{ri}}{l_{0i}}+\operatorname{sgn}(-\dot{u})\varepsilon(-\dot{u})\dfrac{k_{ci}k_{vi}(t)}{k_{ci}+k_{vi}(t)} \end{cases} \tag{6}$$

Where mi is the element mass of SRS (kg); ρr is the element density of SRS (kg/m3); Di is the element diameter of SRS (m); l0i is the element length of SRS (m). Er is the Young's modulus of SRS (Pa); Ari is the element cross-sectional area of SRS (m2); kvi is the variable equivalent stiffness (N/m); Δγvi is the deformation of helical buckling section for SRS (m). μ is the kinematic viscosity of liquid (Pa.s), Dci is the outside diameter of sucker rod coupling (m), Dti is the outside diameter of tube (m), Dri is the element diameter of SRS (m).

For the researches on the deformation of helical buckling for SRS, Lubinski, Xiaojun, Mitchell and Guohua give theoretical analysis method and experiment [11-13]. And the damping coefficient is derived by Shimin, and the equation is expressed [2]. The main aim of this paper is to analyse the characteristic of SRPS. Therefore, the research on deformation of helical buckling for SRS is reported in Refs. [11-13] and need not be repeated here.

3 Solutions

The system equations of nonlinear longitudinal vibration of SRS are solved by fourth-order Runge-Kutta method. Therefore, the numerical model of nonlinear longitudinal vibration of SRS is obtained.

$$\begin{cases} dy(1)=y(2) \\ dy(2)=\dfrac{1}{m_1}\{k_1(t)u_0-[k_1(t)+k_2(t)]u_1+k_2(t)u_2+c_1\dot{u}_0-(c_1+c_2)\dot{u}_1+c_2\dot{u}_2\} \\ dy(3)=y(4) \\ dy(4)=\dfrac{1}{m_2}\{k_2(t)u_1-[k_2(t)+k_3(t)]u_2+k_3(t)u_3+c_2\dot{u}_1-(c_2+c_3)\dot{u}_2+c_3\dot{u}_3\} \\ \vdots \qquad \cdots \qquad \vdots \\ dy(2n-3)=y(2n-2) \\ dy(2n-2)=\dfrac{1}{m_{n-1}}\{k_{n-1}(t)u_{n-2}-[k_{n-1}(t)+k_n(t)]u_{n-1}+k_n(t)u_n+c_{n-1}\dot{u}_{n-2}- \\ \qquad (c_{n-1}+c_n)\dot{u}_{n-1}+c_n\dot{u}_n \\ dy(2n-1)=y(2n) \\ dy(2n)=\dfrac{1}{m_n}\{k_n(t)u_{n-1}-k_n(t)u_n+c_n\dot{u}_{n-1}-c_n\dot{u}_n+P_p(t)\} \end{cases} \tag{7}$$

In addition, for the first calculation, the initial conditions are the same for all the plots as given below

$$y(j)=0, \qquad j=1,2,3,\cdots,2n \tag{8}$$

4 Displacement and velocity response

The basic parameters are as follows: the diameter of SRS Dri=0.022m, the length of SRS Lr =1000m, the Young's modulus of the rod Er = 2.1×1011 Pa, the density of the rod ρr=7850 kg/m3, the deformation length of helical buckling section Δγvi = l0i, the load parameter M0=1000N and M1=9000N, the displacement parameter A0=2m, the excitation frequency ωf=1 rad/s.

4.1 Load excitation

When the load force is a cosine function and the amplitude of load force is not changed, Fig.2 gives the the steady-state response of displacement and velocity at the bottom of SRS under load force of cosine function form. In addition, when the load force is a sine function and the amplitude of load force is not changed, Fig.3 gives the steady-state response of both displacement and velocity at the bottom of SRS under load force of sine function form. In the details, when the load force is a cosine function, the amplitude of displacement response is bigger than the amplitude under load force of sine function form, and the velocity response of SRS has larger fluctuation. In addition, when initial value of load force starts from zero, the amplitudes of displacement and velocity response will be smaller.

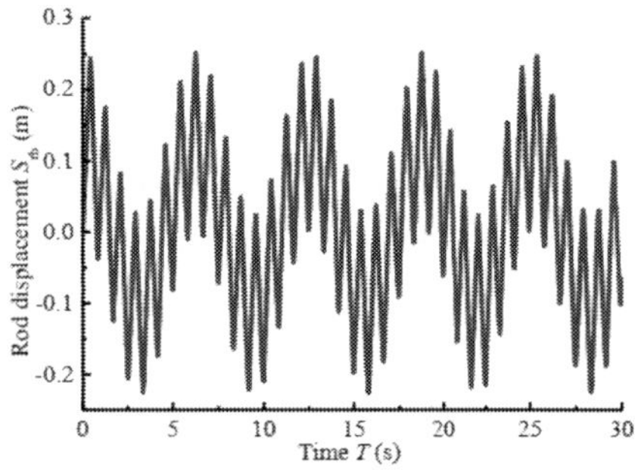

(a) Displacement of the bottom of SRS

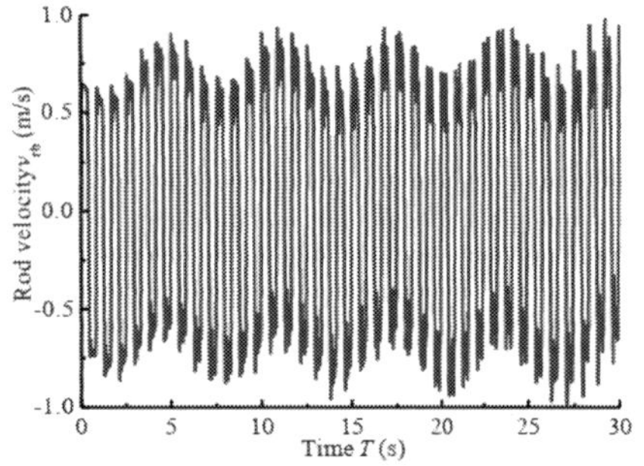

(b) Velocity of the bottom of SRS

Figure 2. The steady-state response of SRS under load force of cosine function form

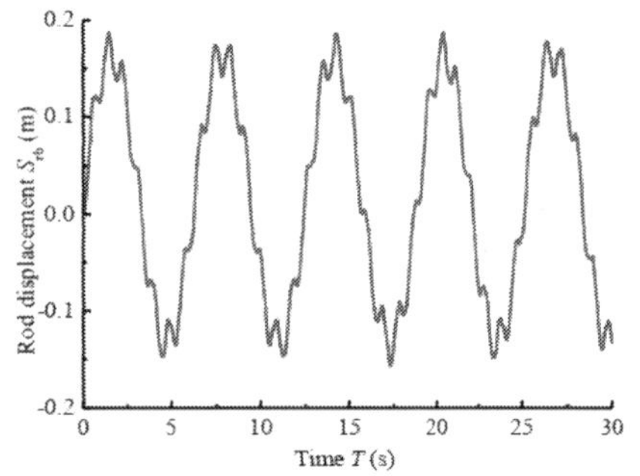

(a) Displacement of the bottom of SRS

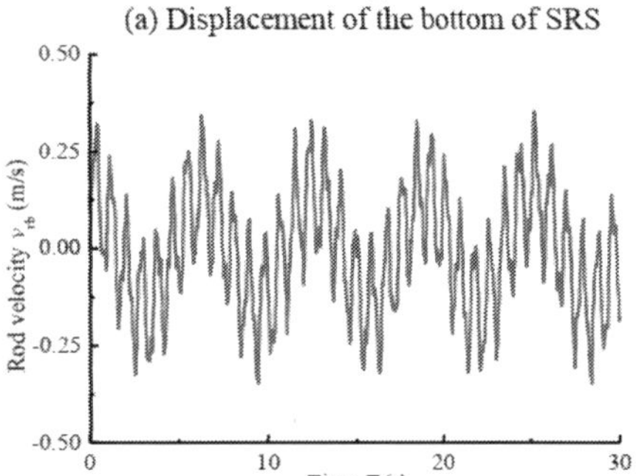

(b) Velocity of the bottom of SRS

Figure 3. The steady-state response of SRS under load force of sine function form

4.2 Load force and displacement excitation

When the load force is a cosine function and the displacement is a cosine or sine function, Fig.4 gives the steady-state response of both displacement and velocity at the bottom of SRS under load force of cosine function form. In addition, the load force is a sine function and the displacement is a cosine or sine function, Fig.5 gives the steady state response of both displacement and velocity at the bottom of SRS under load force of sine function form. According to the Figures, when the displacement excitation is a sine function, the amplitude response of both displacement and velocity are smaller than the response under displacement excitation of cosine function form in Fig.4. And the same phenomenon happens under load force of sine function form in Fig.5. Comparing Fig.4 with Fig.5, the amplitude of displacement response under load force of sine function form is bigger than the response under load force of cosine function form while the system is subjected to both load force and displacement excitation.

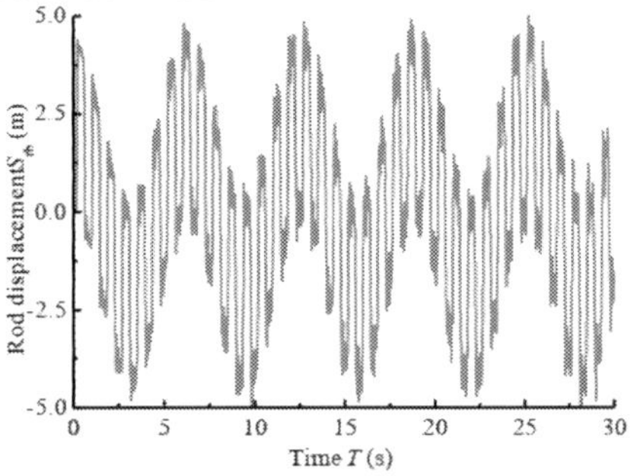

(a) Displacement of the bottom of SRS under displacement excitation of cosine function form

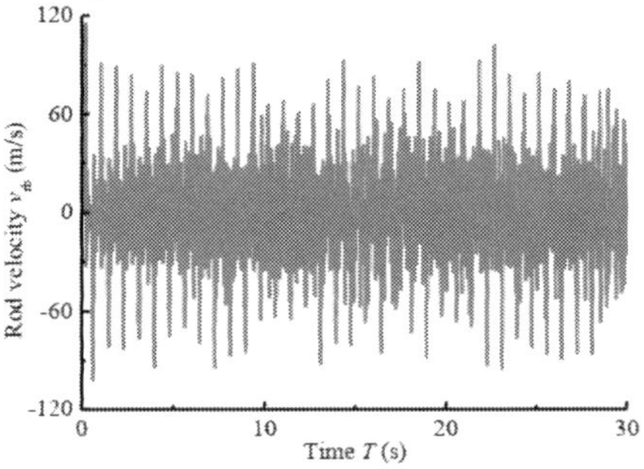

(b) Velocity of the bottom of SRS under displacement excitation of cosine function form

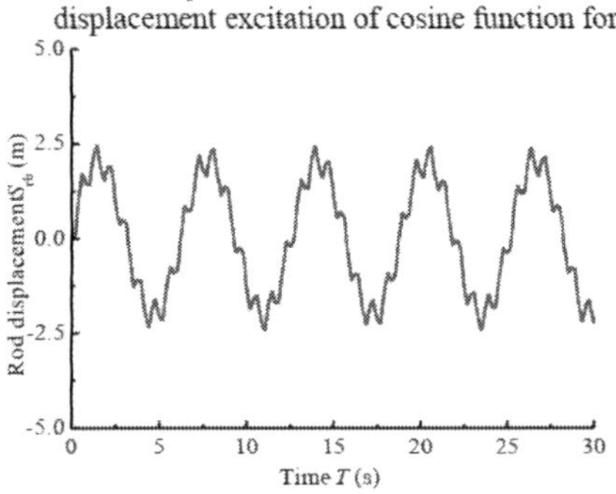

https://doi.org/10.1051/matecconf/201815306009

(c) Displacement of the bottom of SRS under displacement excitation of sine function form

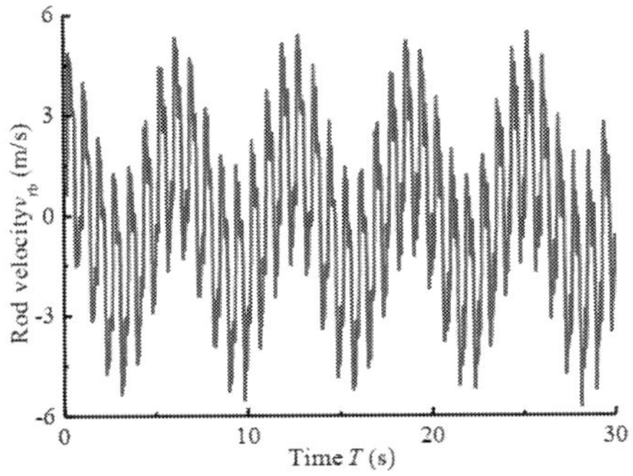

(d) Velocity of the bottom of SRS under displacement excitation of sine function form

Figure 4. The steady-state response of SRS under load force of cosine function form

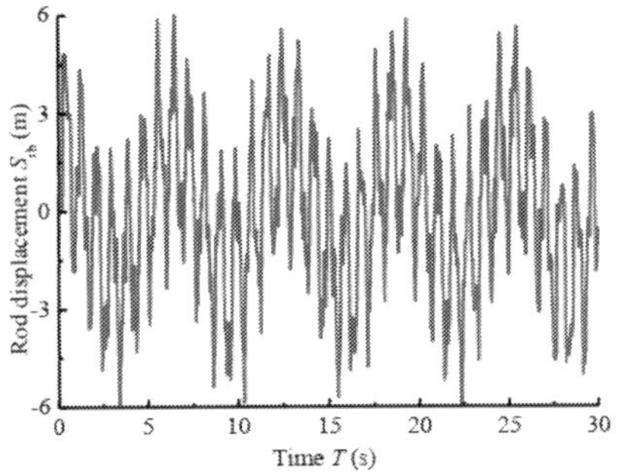

(a) Displacement of the bottom of SRS under displacement excitation of cosine function form

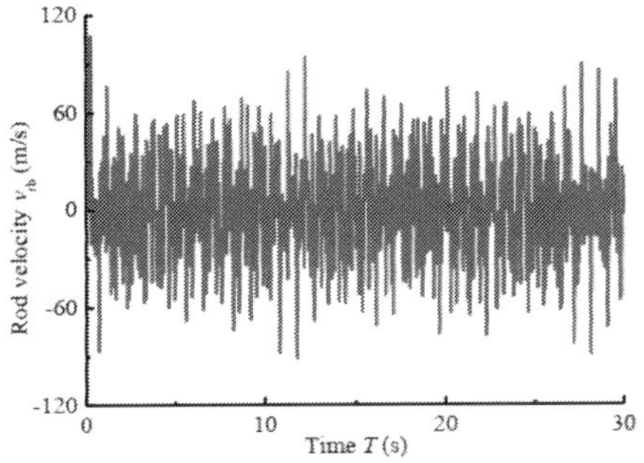

(b) Velocity of the bottom of SRS under displacement excitation of cosine function form

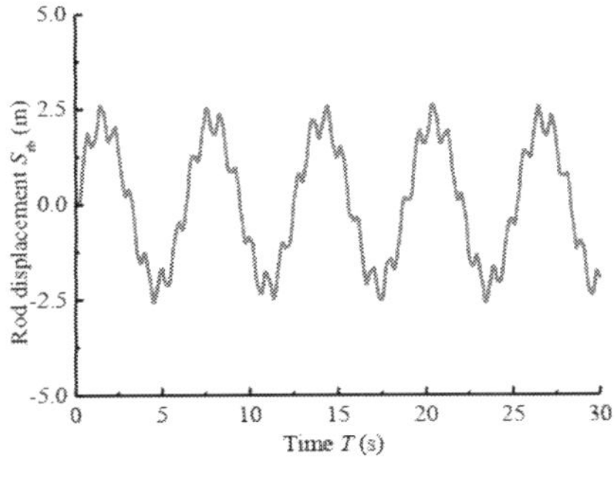

(c) Displacement of the bottom of SRS under displacement excitation of sine function form

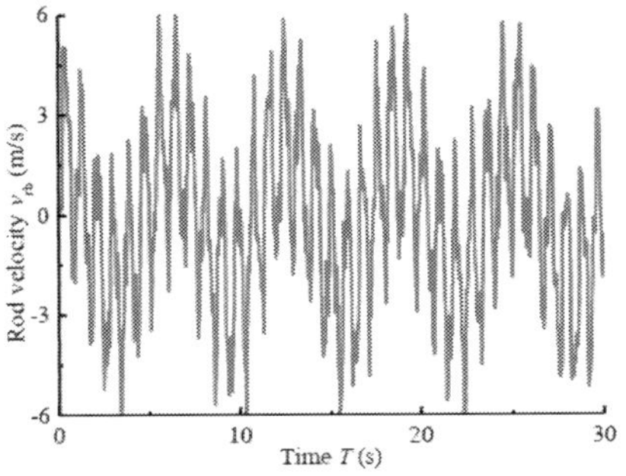

(d) Velocity of the bottom of SRS under displacement excitation of sine function form

Figure 5. The steady-state response of SRS under load force of sine function form

5 Conclusions

With the influence of the variable equivalent stiffness on steady state response, the sucker rod is dispersed into a spring-mass-damper mechanical system and a mathematic model of nonlinear longitudinal vibration of SRS is built. The dynamic response under displacement and load force excitation is solved by fourth-order Runge-Kutta method with zero initial condition.

The steady state responses under the displacement and load force excitation of different function forms are different. In the detail, the response curves of both displacement and velocity under the displacement and load force excitation of cosine function form have larger fluctuation than it under the displacement and load force excitation of sine function form. Therefore, the dynamic analysis of Sucker Rod String with variable equivalent stiffness plays an important role in understanding the influences of the excitation form and sensitive parameters on characteristic of SRS, which has important academic and practical significance for optimization of the performance parameters.

Acknowledgments

This research was financially supported by the Scientific Research Starting Foundation for Linyi University (Grant NO.LYDX2016BS032).

References

1. Xing M.M, Dong S.M. An improved longitudinal vibration model and dynamic characteristic of sucker rod string [J]. Journal of Vibroengineering, **16**(7),3432-3448(2014).
2. Shimin Dong, Computer simulation of dynamic parameters of rod pumping system optimization[M]. Petroleum Industry Press, Beijing(2003).
3. Xing M.M, Dong S.M. A new simulation model of beam pumping system applied in energy saving and

resource consumption reduction[J]. SPE Production & Operations, **30**(2):130-140(2015).
4. S.G. Gibbs, Predicting the behavior of sucker rod pumping systems, SPE Journal Paper 588-PA(1963).
5. S.G.Gibbs. and Neely, A.B, Computer Diagnosis of Down-Hole Conditions in Sucker Rod Pumping Wells, SPE Journal Paper, 1165-PA(1966).
6. S.G.Gibbs, A Review of Methods for Design and Analysis of Rod Pumping Installations, SPE Journal Paper, 9980-PA(1982).
7. I.N. Shardakov, I.N. Wasserman, Numerical modelling of longitudinal vibrations of a sucker rod string, Journal of Sound and Vibration 329(3, 1), 317-327 (2010).
8. DaCunha and Gibbs, Modeling a Finite-Length Sucker Rod Using the Semi-Infinite Wave Equation and a Proof to Gibbs' Conjecture, SPE Journal Paper, 108762-PA(2009).
9. Xing M.M. Response analysis of longitudinal vibration of sucker rod string considering rod buckling[J], Advances in Engineering Software, 99:49-58(2016).
10. S.A.M. Ghannadpour, B. Mohammadi, J. Fazilati, Bending, buckling and vibration problems of nonlocal Euler beams using Ritz method, Composite Structures 96, 584-589 (2013).
11. He, Xiaojun, Kyllingstad, Age, Helical Buckling and Lock-Up Conditions for Coiled Tubing in Curved Wells, SPE Journal Paper, 25370-PA (1995).
12. R.F. Mitchell, Buckling Behavior of Well Tubing: The Packer Effect, SPE Journal Paper, 9264-PA(1982).
13. Gao Guohua, Shell, and Stefan Miska, Effects of Friction on Post-Buckling Behavior and Axial Load Transfer in a Horizontal Well, SPE Journal Paper 120084-PA(2010).

https://doi.org/10.1051/matecconf/201815306010

Predictive Control Algorithm Based on Integral Action- Design and Implementation on a Conical Tank System

Vinayambika S Bhat[1], I. Thirunavukkarasu[1], S. Shanmuga Priya[2] and Shreesha C[1]

[1]*Dept. of Instrumentation and Control Engineering, Manipal Institute of Technology, Manipal University, Manipal- 576104, Karnataka, India*
[2]*Dept. of Chemical Engineering, Manipal Institute of Technology, Manipal University, Manipal- 576104, Karnataka, India*

Abstract. This article presents a Model Predictive Control (MPC) algorithm based on integral action. Level control in process industry is challenging because of nonlinearity presents in the shape of the tank, actuators etc. The conical tank system is taken as benchmark process in the present study. It is Single Input Single Output (SISO) nonlinear system whose cross-sectional area varies along the tank height. The control algorithm is simulated using MATLAB m-file environment. The effectiveness of the predictive algorithm is also presented by experimentally validating it on a conical tank system at different heights. The interfacing of m-file with the experimental setup is the challenging task faced during the initial stage of experimental validation.

1 Introduction

A basic problem in process industries is control of liquid level and flow in process tank. Conical tanks are extensively used in industries like petrochemical, food, wastewater treatment and hydrometallurgical industries [1]. Also, control of conical tank is a challenging problem because of its nonlinearity and constantly changing area of cross-section [2, 3]. Advanced and innovative algorithms are required for process control applications. Literature discusses that predictive control algorithms are effectively used in process industries [4, 5]. There are several methods within MPC algorithms that differ from each other based on the process model used [6]. However, these algorithms have some problems with offset i.e. the process output is not equal to the specified reference at steady state [7]. This offset problem can be solved by introducing the integral action [8]. The MPC with integral action algorithm is designed and implemented for conical tank system. The state space model of the conical tank system is considered for the study. The predictive algorithm is simulated in MATLAB environment. Also, the effectiveness of MPC with integral action has been validated by implementing it in real-time.

The article is outlined as follows. A brief introduction to predictive control algorithm is given section II. Section III presents details of the experimental setup. The methodology involved in developing the MPC integral action algorithm is discussed in section IV. Both the simulation and real-time results are presented and discussed in section V followed by conclusion in section VI. Also, in the following section, scope for future work in the reported research area is noted.

2 Fundamentals of predictive control algorithm

The basic structure of predictive control algorithm is shown in Figure 1. The main motive of predictive control algorithm is to find the input signal that best satisfies to a given performance criterion (cost function), it predicts how the system will behave if the signal is applied. It solves a control problem for an optimal control action to track the given input [5]. Models are used to predict the behaviour of dependent variables or outputs of the modelled dynamical system with respect to changes in the process independent variables or inputs. The main distinguishing factor of these predictive algorithms is their ability to deal with explicit constraints on system variables that are neglected in majority of the other control algorithms [7].

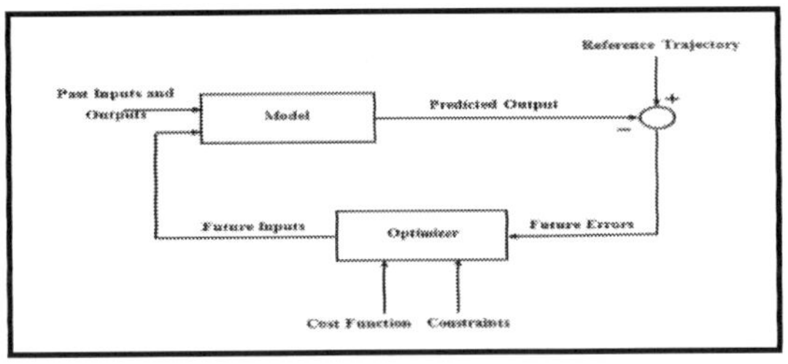

Figure 1. Basic structure of predictive control algorithm [5]

3 Experimental setup

The Conical Tank Level control system chosen for the study is shown in Figure 2. The Figure 2(a) represents the structure of conical tank followed by the experimental setup in Figure 2(b). The setup consists of level transmitter, pump, rotameter, conical tank, reservoir and power supply. The system descriptions are given in Table 1. The conical shape of the tank makes the level process nonlinear, since the area is directly proportional to the level. The tank is modelled by dividing the total height into four regions with 7cm for each region. The linearized model for the regions between 7cm- 14cm is obtained using the 2-point method. The VDAS-01 interfacing card is used for the real- time implementation of predictive control algorithm. The Data Acquisition Card (DAC) is used to interface the PC with the real-time system is shown in Figure 3. The MATLAB m-file code (Control algorithm developed) has been interfaced through conical tank system for the real-time implementation. The controller output is limited to 0-250. The First Order Plus Time Delay (FOPTD) model of the conical tank system for the second region (7-14cms) of locally lineralized model is given by Eq. (1) [2].

$$G(s) = \frac{0.925e^{-1.09s}}{25.05s+1} \tag{1}$$

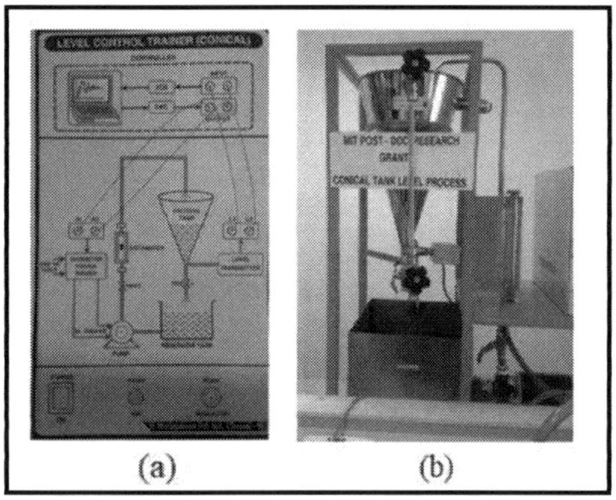

Figure 2. (a) Structure of the conical tank (b) Experimental setup of conical tank [2]

Table 1. Technical specification of conical tank setup [2]

Conical Tank	Motor
Height- 33cms	RPM 1500
Top Diameter-30cms	Head 24M
Bottom Diameter- 3cms	LPH 1600

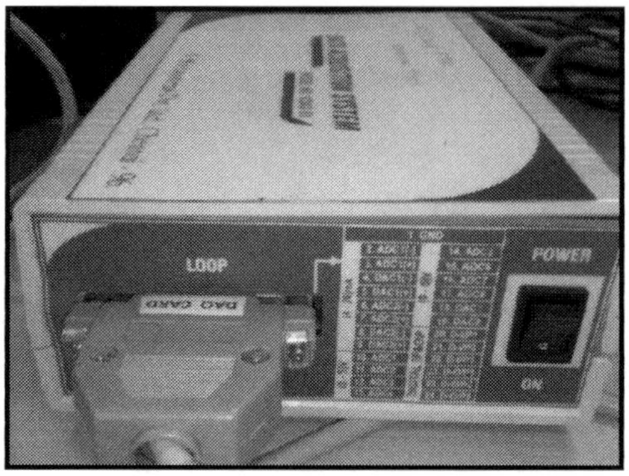

Figure 3. VDAS-01 Data acquisition card used for the real-time implementation

4. Methodology

A state space representation of the system has been obtained from the transfer function of the system. The process model can be described as follows.

$$x_{k+1} = Ax_k + Bu_k + v$$

$$y_k = Cx_k + w \tag{2}$$

Where A, B and C are the state space representation of the system of appropriate dimensions. The v and w are the disturbances which are unknown. That means, v may be an unknown constant or slowly varying process disturbance, and w may be an unknown constant or slowly varying noise vector. Variables y_k and u_k in the model Eq. (2) are the actual input and output variables respectively. To solve the MPC optimal control problem for Eq. (2) a model which is independent of the unknown disturbances has been considered [8]. That is Eq. (2) has been reframed as

$$\Delta x_{k+1} = A\Delta x_k + B\Delta u_k \tag{3}$$

$$y_k = y_{k-1} + C\Delta x_k \tag{4}$$

Where, $\Delta x_k = x_k - x_{k-1}$

Augmenting Eqs. (3) with (4) gives the state space model:

$$\begin{bmatrix} \Delta x_{k+1} \\ y_k \end{bmatrix} = \begin{bmatrix} A & 0_{n\times m} \\ C & I_{m\times m} \end{bmatrix} \begin{bmatrix} \Delta x_k \\ y_{k-1} \end{bmatrix} + \begin{bmatrix} B \\ 0_{m\times r} \end{bmatrix} \Delta u_k \tag{5}$$

$$y_k = \begin{bmatrix} C & I_{m\times m} \end{bmatrix} \begin{bmatrix} \Delta x_k \\ y_{k-1} \end{bmatrix} \tag{6}$$

The final state space model is as follows.

$$\Delta\bar{x}_{k+1} = \bar{A}\Delta\bar{x}_k + \bar{B}\Delta u_k \tag{7}$$

$$y_k = C\Delta\bar{x}_k \tag{8}$$

The state space model Eqs. (7) and (8) is used to define a Prediction Model (PM) of the form given by

$$y_{k+1/L} = p_L + F_L\Delta u_{k/L} \tag{9}$$

Where, $p_L = O_L\bar{A}\bar{x}_k$, $F_L = \left[O_L\bar{B} \quad H_L^d\right]$, OL is the extended observability matrix of the pair $\bar{A},\bar{C}$ and H_L^d is the Toeplitz matrix. The cost function along with control law is expressed as follows.

$$J = \Delta u_{k/L}^T H\Delta u_{k/L} + 2f_k^T \Delta_{k/L} \tag{10}$$

Where, $H = F_L^T Q F_L + P$ and $f_k = F_L^T Q\left(p_L - r_{k+1/L}\right)$. The P and Q are the weighting function selected by trial and error to achieve the stable closed loop response.

The unconstrained MPC control is given by

$$\Delta u_{k/L}^{*} = -H^{-1} f_k \tag{11}$$

The control law is given by

$$u_{k/L} = S\Delta u_{k/L} + c u_{k-1} \tag{12}$$

Here, $S \in R^{rL\times rL}$ and $c \in R^{rL\times r}$ are the matrices with ones and zeros. The MPC algorithm minimizes the cost function with respect to the constraints. This is a standard Quadratic Programming (QP) problem in terms of unknown future control increments and optimal solution is $\Delta u_{k/L}^{*}$ as defined in Eq. (11). A receding horizon strategy is used and only the first control increment Δu_k^{*} in the calculated $\Delta u_{k/L}^{*}$, is used for control. The actual control action to the process is then $u_k = \Delta u_k^{*} + u_{k-1}$ .The structure of linear quadratic controller is proportional to the integral form as in case of conventional PI control algorithm [9, 10]. Thus, the name predictive control algorithm with integral action [8].

5. Results and discussion

A state space equivalent model has been considered to develop the control algorithm in simulation. The corresponding A, B, C, D matrix in discrete domain for sampling period of 0.5 is given by

A= [0.9802], B= [0.125], C= [0.1439], D= [0.01502]

The MPC with integral action algorithm has been successfully implemented on the real-time system of SISO Conical Tank Level system and the output could track the reference signal. The output prediction and corresponding control input for setpoint of 50% is shown in Figure 4. Also, for the setpoint 70, the predicted output and the corresponding control input is shown in Figure 5. The plant output and manipulated variable response for setpoints 50% and 70% of tank level are shown in Figures 6 and 7 respectively.

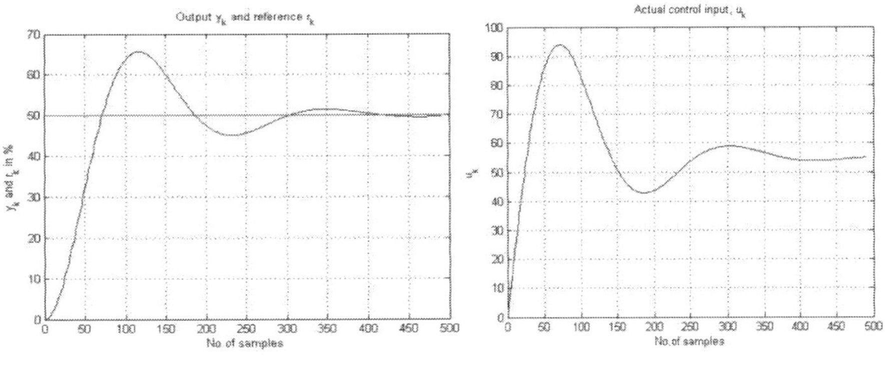

(a) (b)

Figure 4. (a) Output prediction vs number of samples at setpoint of 50 (b) Actual control input u_k vs number of samples

 https://doi.org/10.1051/matecconf/201815306010

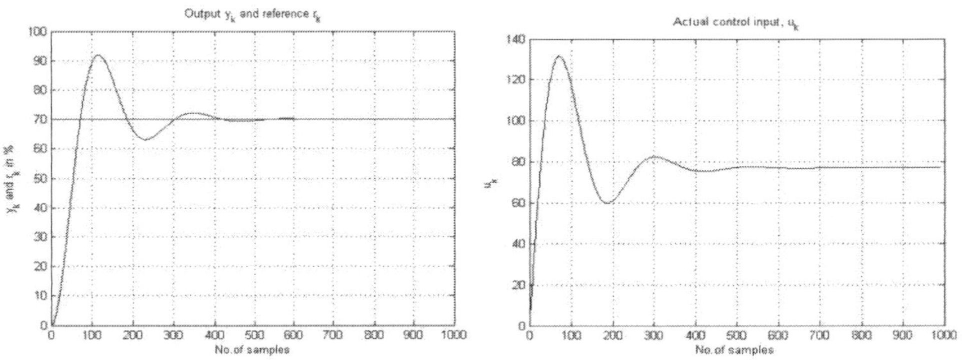

(a) (b)

Figure 5. (a) Output prediction vs number of samples at setpoint of 70 (b) Actual control input u_k vs number of samples

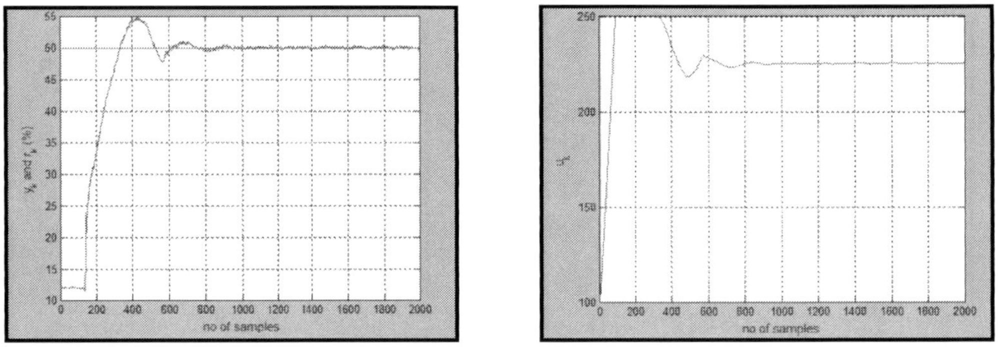

(a) (b)

Figure 6. (a) Experimental Process output and (b) Manipulated variable response for a setpoint of 50%

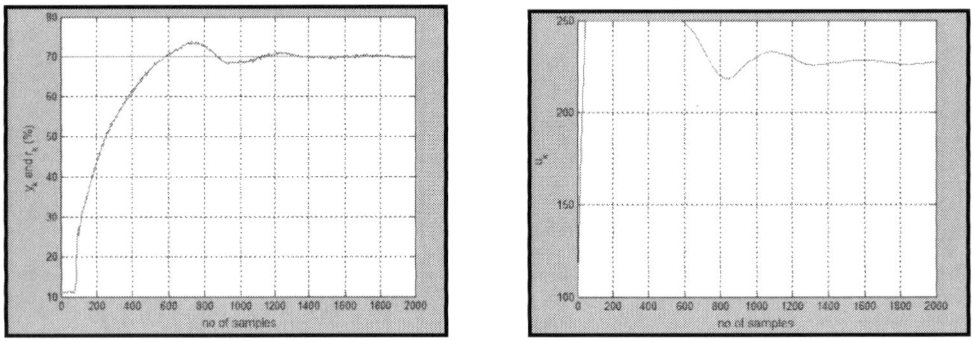

(a) (b)

Figure 7. (a) Experimental Process output and (b) Manipulated variable response for a setpoint of 70%

6. Conclusion

The predictive control algorithm along with integral action is considered for the design and experimental validation. The state space model of the system is obtained from the transfer function of the system. The effectiveness and practicality of the control algorithm is studied and simulated by using conical tank plant model. The closed loop servo response is obtained in simulation. The MATLAB m-file predictive control algorithm is interfaced to real-time using data acquisition systems VDAS-01. The same setpoint is maintained both in simulation and real-time for the comparison. The challenges faced during the m-file interface to the DAQ card and experimentation will be addressed during the presentation.

Future work

The MPC with integral action is extended to MIMO system. The control algorithm simulated and validated on a pilot plant binary distillation column. It was reported by the authors in the research article submitted to CHEMCON-2017, HIT, Haldia (Accepted for oral presentation) with experimental results of pilot plant binary distillation column. Few issues have been observed as drawbacks to retune the controller.

Acknowledgement

Vinayambika S. Bhat would like to acknowledge the Mangalore Institute of Technology and Engineering (MITE), Moodabidri, for sponsoring her Ph.D. programme. The authors would like to thank Department of Instrumentation and Control Engineering, MIT, Manipal University for providing the real-time experimental facility for carrying out the experimental work.

References

1. C. Lovaas, M. M. Seron, and G. C. Goodwin, Robust Output-Feedback MPC with Integral Action, *IEEE Transactions on Automatic Control*, **55(7)**, pp. 1531-1543, (2010).
2. M. K. Rakesh, R. S. Babu and I. Thirunavukkarasu, Sliding Model with Dead Time Compensation for Conical Tank Level Process, *IOSR Journal of Electrical and Electronics Engineering*, **9(2)**, pp. 59-64 (2014).
3. V. R. Ravi, T. Thyagarajan and G. Uma Maheshwaran, Dynamic Matrix Control of a Two Conical Tank Interacting Level System, *Proceedia Engineering*, **38**, pp. 2601-2610 (2012).
4. N. Sivakumaran and T. K. Radhakrishnan, Predictive Controller Design for Non-Linear Chemical Processes, *Indian Journal of Chemical Engineering*, **14**, pp. 341-349 (2007).
5. V. S. Bhat, I. Thirunavukkarasu, and S. Shanmuga Priya, Design of Unconstrained DMC to Improve the Distillate Product Purity of the Distillation Column, *MATEC Web of Conferences*, **77**, pp. 1-5, presented in ICPEME, Thailand, June (2016).
6. M. Manimaran., S. Malaisamy, M. M. Rafiq, V. Petchithai, V. S. Chitra, K. Kalanithi. and H. Abirami., Parameter Identification and Dynamic Matrix Control Design for a Nonlinear Pilot Distillation Column, *International Journal of ChemTech Research*, **7(1)**, pp. 382-388 (2015).
7. P. E. Orukpe, Model Predictive Control Fundamentals, *Nigerian Journal of Technology*, **31(2)**, pp. 139-148, (2012).
8. D. D. Ruscio, Model Predictive Control with Integral Action: A Simple MPC Algorithm, *Modeling, Identification and Control*, **34(3)**, pp. 119-129, (2013).
9. D. D. Ruscio, Discrete LQ Optimal Control with Integral Action: A Simple Controller on Incremental form for MIMO Systems, *Modeling, Identification and Control*, **33(2)**, pp. 35-44, (2012).
10. D. D. Ruscio, Model Predictive Control and Identification: A Linear State Space Model Approach, in *Proceedings of the* 36th *conference on Decision and Control*, USA, **4**, pp. 3202-3209, December (1997).

Design and Finite Element Analysis of Water Jet Energy Accumulator Barrel

Zhang Yuxian , Yang mengke, Wang Hong , Liu Binbin

Guilin University of Electronic Technology, Guilin 541004, China

Abstract. In order to solve the yielding failure problem of water jet energy accumulator barrel due to high pressure, the prestressed composite structure design and the theory of equal strength are used to determine the parameters of accumulator barrel, such as the internal cylinder of the composite cylinder, the radii of the inner and outer cylinders, the radius of sub-layer and the optimal interference. Then the ANSYS software is used to analyze the cylinder model with the finite element method, the equivalent stress of the effective node along the wall thickness of the simulation model is obtained by path extraction method and the distribution curves of the equivalent stress along the wall thickness with different internal pressure are also obtained. By analyzing the simulated results and curves, the design parameters of accumulator barrel meet the expected strength requirement.

1 Introduction

Water jet cutting is a kind of non-conventional machining method, which uses high speed and high kinetic energy fluid medium to working.The cutting process belongs to cold cutting, which does not affect the material structure, and has no phenomenon of mechanical stress deformation and warpage. Therefore, which is widely used in aviation, military industry, electronics and other fields.

Accumulator is an important auxiliary component of high pressure water in water jet cutting. It can save pressure, absorb shock and eliminate pulsation. The carrying capacity directly influences the cutting performance. In order to improve the actual strength of the barrel, in this paper, accumulator in the use of the process often due to high pressure and yield failure problem, the design of prestressed composite structure of accumulator cylinder, and make use of finite element method to valify, in order to improve the bearing capacity of the accumulator.

2 The failure form of accumulator barrel

Taking a company's WL-01025A water cut pressurization system (see Figure 1) as an example, to explain the function of the accumulator. After pressing the low-pressure water to the compound supercharger, it passes through the pipeline and enters the accumulator from the accumulator plug head. At this time, the accumulator barrel is subjected to approximate 400 mpa, it also absorbs shock and eliminates pulsation [1].

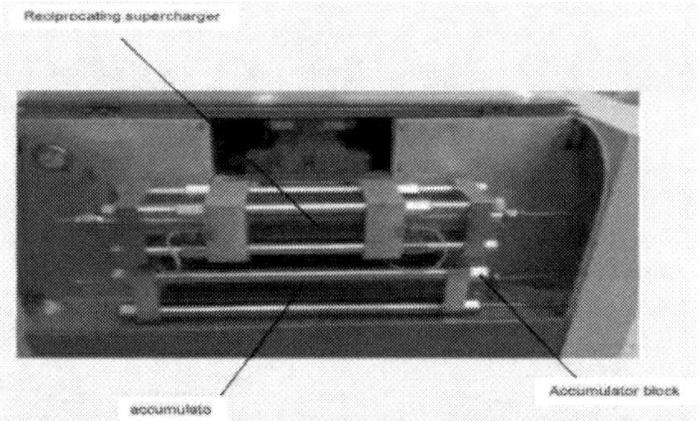

Figure 1. WL-01025A type water cutting supercharging system drawing

In the actual working process, the accumulator tube may have yield deformation, to make the plug seal failure, which resulting in high pressure water leakage, and water reducing rapidly in energy storage, it can't achieve the purpose of cutting [2]. The design of accumulator cylinder, using elastic failure criterion. When the cylinder maximum stress reaches the yield limit, the accumulator cylinder bring about yield failure.

2 Structure design of accumulator barrel

2.1 Problems existing in Single layer thick walled cylinder

If water cutting accumulator in the design of single layer thick wall cylinder, under high load, the stress distribution is extremely inhomogeneous, according to the fourth strength theory, when the equivalent s equal to the yield limit of material, $\sigma_{eq} = \sigma_s$ .accumulator cylinder start to yield, then the diameter ratio K = $\sqrt{\dfrac{0.577\sigma_s}{0.577\sigma_s - p_i}}$,thus when internal pressure p_i=0.577σ_s ,the diameter ratio is close to infinity, when the thick wall cylinder is yielding. From the above analysis, it is limited to increase the load-carrying capacity of the cylinder by increasing the thickness of the wall and using the high strength material. Therefore, it's necessary to change the structure of accumulator barre.

The prestress method can effectively improve the stress distribution of the cylinder and enhance the carrying capacity of the cylinder, in addition, it is simple and effective. In order to excavate the material potential further, the equal strength theory is adopted to design the prestressed cylinder.

2.2 Equal strength design of prestressed composite cylinder

2.2.1 Based on the equal strength theory of the cylinder diameter parameters

The relative structural parameters of the accumulator's prestressed composite cylinder are shown in figure 2. Figure 2 : r_a is inner barrel radius of the inner wall ; r_b is layered radius; r_c is inner wall radius of the outer cylinder; δ 1 is inner cylinder radial shrinkage; δ 2 is cylinder radial expansion tensor; δ max is the optimal interference of the inner cylinder which is relative to the outer cylinder; $p_{1,2}$ for shrunk interface pressure.

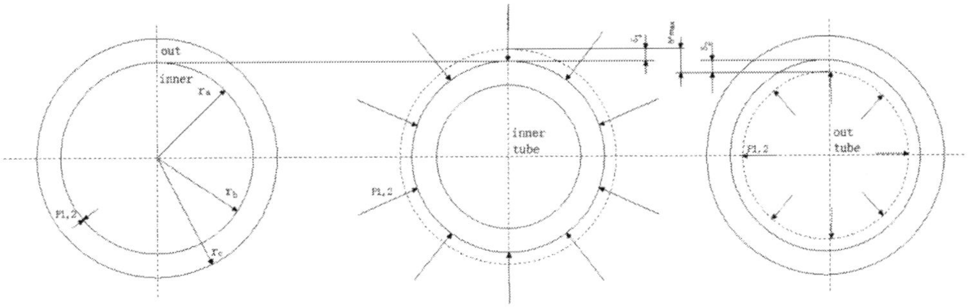

Figure 2. Structural parameters of prestressed composite cylinder sketch map

According to the elastoplastic theory of the cylinder,when maximum shear stress ,when maximum shear stress is $\tau_{max} = \dfrac{\sigma_s}{\sqrt{3}}$,the maximum permissible internal pressure p_i is

$$p_i = \frac{2\sigma_s}{\sqrt{3}}\left[1 - \frac{1}{2}\left(\frac{r_b^2}{r_c^2} - \frac{r_a^2}{r_b^2}\right)\right] \tag{1}$$

According to the equivalent strength design theory, it can be available[3].

$$p_i = \frac{2\sigma_s}{\sqrt{3}}\left[1 - \frac{1}{K}\right] \tag{2}$$

2.2.2 optimum interference calculation

Timoshenko deduction interface pressure formula of cylinder is

$$p_{1,2} = \frac{\delta E}{2r_b}\frac{\left(K^2 - K_1^2\right)\left(K_1^2 - 1\right)}{K_1^2(K-1)} \tag{3}$$

The optimum interference amount of the inner cylinder relative to the outer cylinder can be obtained[4] δ_{max}

$$\delta_{max} = \frac{2\sigma_s}{\sqrt{3}E}\left(\frac{\sqrt{r_a r_c}(r_c - r_a)}{r_c}\right) = \frac{2r_a\sigma_s}{\sqrt{3}KE}(K-1) = \delta_1 + \delta_2$$

2.3 instance calculation

A cylinder, radius of inner wall of cylinder $r_a = 23mm$, maximum allowable internal pressure $p = 500MPa$. Both inner and outer cylinder materials are selected 0Cr17Ni4CuNb, modulus of elasticity $E = 213GPa$,yield strength $\sigma_s = 856MPa$,Poisson ratio $\mu = 0.27$,tangential modulu $E_T = 210MPa$.

The concrete parameters of prestressed composite cylinder are as follows:inner radius of inner barrel $r_a = 23.0mm$,delamination radius $r_b = 32.7mm$,outer radius of outer cylinder $r_c = 46.5mm$,optimum interference $\delta_{max} = 0.0768mm$.

3 Finite element analysis of accumulator prestressed cylinder

3.1 The establishment of finite element model

Since the diameter of the accumulator is larger than the length and the material is nonlinear, the analysis belongs to the problem of plane strain and contact nonlinearity.

In addition, the model is symmetric and can be simplified to the two-dimensional 1 /4 model.According to the foregoing to establish model,the selection can be approximated by simulation with elastoplastic properties,two dimensional plane element of large deformation and large strain cylinder materials PLANE183,and Setting material mechanical parameters,the bilinear kinematic hardening model (BKIN) is adopted to simulate the material nonlinearity at the same time[5].

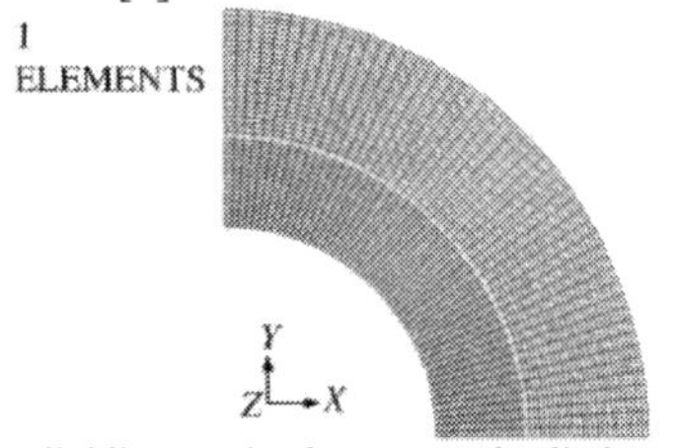

Figure 3. dividing mesh of prestressed cylinder model

As shown in Figure 3,the cell shape is set to a quadrilateral,the mesh is mapped by mapping,set the contact to simulate the external wall of the inner cylinder ectotheca and the outer cylinder wall,as to get preload effect. in order to obtain better convergence, the contact stiffness of the penalty function is set to 0.1; all nodes which X = 0 and Y = 0 ,applying 0 degrees of freedom for X, Y direction respectively . The entire loading process is set from 0 to 50 MPa to 600 MPa, and the solution process is divided into 13 load steps[6].

3.2 Finite element post processing and stress data analysis

The different load step equivalent stress nephogram is studied, When the internal pressure of the cylinder is less than 500 MPa, the equivalent stress in the cylinder is no more than the yield limit of the cylinder material,which the whole cylinder in the elastic stage does not yield; when the internal pressure is 500 MPa, the largest the value of equivalent stress of the cylinder wall and the inner wall of the barrel, and the outer tube, and the maximum equivalent stress is 856 MPa, just to reach the material yield limit, that under 500 MPa pressure, the inner cylinder wall and the inner wall of the outer cylinder to yield(After 1 /4 expansion, the 500 MPa stress distribution nephogram is shown in Figure 4); when the internal pressure is greater than 500 MPa, the inner wall of the cylinder inner wall and outer cylinder appear yield regions (under internal pressure of 600 MPa, prestressed cylinder equivalent stress distribution is showed in Figure 5).

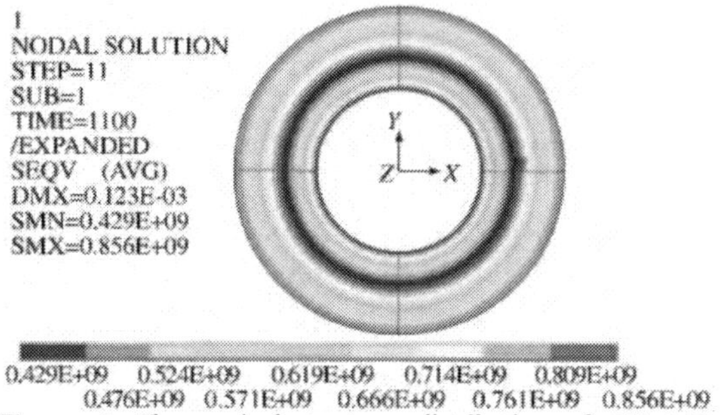

Figure 4. The equivalent stress distribution of 500 MPa prestressed cylinder

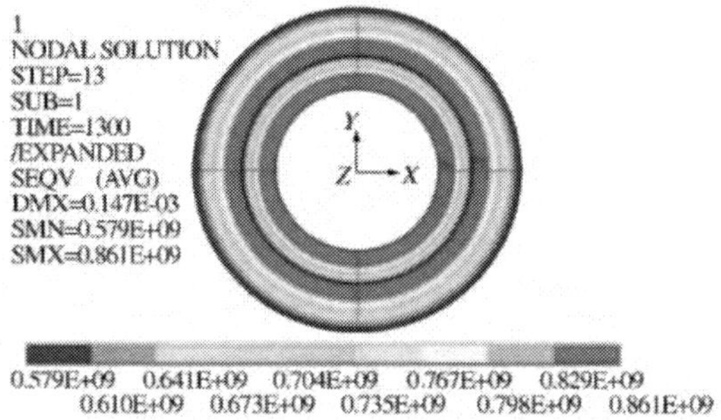

Figure 5. The equivalent stress distribution of 600 MPa prestressed cylinder

In order to further study on the equivalent stress distribution along the wall thickness at 100 MPa to 600 MPa pressure range, with 100 MPa steps and load in incremental manner , the cylinder continuous discretization compression process,along the wall thickness of the cylinder, the path is set up and the equivalent stress value of the effective node is extracted. The stress distribution curve along the wall thickness under different internal pressure is processed by MATLAB software[7].(See Figure 6)

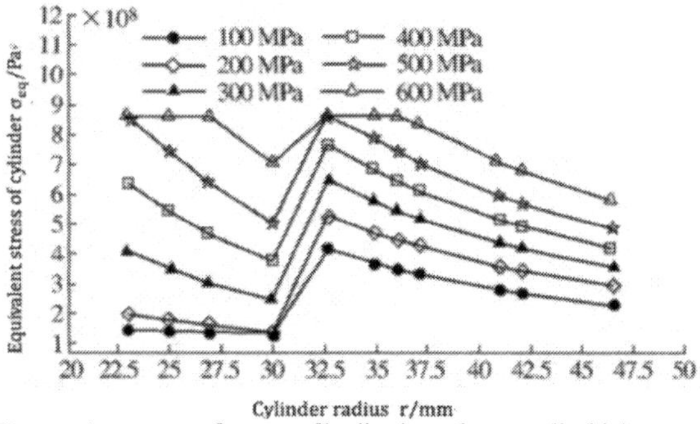

Figure 6. curves of stress distribution along wall thickness under different internal pressures

As can be seen from Figure 6, when the internal pressure is in the range of 100 MPa to 400 MPa, the equivalent stress within the cylinder does not reach the yield limit, and the variation tendency in equivalent stress along the wall thickness is similarly. The trends performance are: the inner cylinder, the equivalent stress along the wall thickness decreases, the interface to the inner cylinder and the outer cylinder stress increases suddenly, then along the wall thickness decreases until to the outer cylinder; Under internal pressure is 500 MPa, and the trend is similar to the before, but yield point along will be showed along the wall thickness near 23 mm and 33 mm; while the cylinder wall thickness of 33 mm to 36 mm the range is also a large yield.

4 Conclusion

1) the finite element analysis of the designed accumulator barrel is carried out by using ANSYS software, and the stress results show that the design parameters of the prestressed composite cylinder meet the requirements of the expected strength.

2) From the cylinder under different pressure along the wall thickness equivalent stress distribution curve, it can be seen: when the yield stress along the wall thickness isn't appeared, the stress distribution trend similar; Yielding, yield point first appeared in the inner wall of the inner tube and the inner wall of the outer cylinder, and with the increasing of pressure, the yield region will expand respectively from the inner wall of the inner barrel and the inner wall of the outer cylinder.

3) the finite element analysis method can effectively solve the problem of prestressed cylinder touching and nonlinear solution, and can simulate in complex compression environment accurately. It can provide the basis for improving the stress distribution of the extra high pressure cylinder.

References

1. Yang M G, Yu F, Kang C. et al. Study on moving characteristics of reciprocating intensifier [J] . Drainage and Irrigation Machinery,2009,27 (5) : 332-336 (in Chinese)
2. Guo Z F, Qin J P, Zheng Z Z. The research on superhigh pressure prestress composite cylinder of the water cutting Jet [J]. Chinese Hydraulics and Pneumatics, 2004(4) : 7-10 (in Chinese)
3. Yu T X, Xue P. Engineering elasticity and plasticity [M] . 2nd ed. Beijing: Higher Education Press, 2010: 123-135(in Chinese)
4. Zhang Y X, Wang H. The optimal design of shrunk cylinder super high pressure [J] . Machine Design and Research, 2006, 22(5) : 115-117 (in Chinese)
5. Xu C J, Li D S, Dou J P,et al. FEA of cylinder body for ultra-high pressure & shrink sleeve vessel [J]. China Metal forming Equipment & Manufacturing Technology, 2011, (6) : 83-86 (in Chinese)
6. Xiao Sa Studio, Zhang S Q. The new and classic course about ANSYS and workbench [M] . Beijing: Electronic Industry Press, 2004: 178-180 (in Chinese)
7. Wang Z L, Gong C, He Q. Master science computer of MATLAB [M] . Beijing: Electronic Industry Press, 2009: 221-239 (in Chinese)

https://doi.org/10.1051/matecconf/201815306012

Experimental Study on Friction of Hydraulic Cylinder in Different Sealing Systems

Ke Ma[1,2], Jiawei Wang[1], Le Gu[3]

[1]College of Mechanical& Electrical Engineering, Hohai University, Changzhou 213022, China
[2]Jiangsu Key Laboratory of Special Robot Technology, Changzhou 213022, China
[3]BOSCH Rexroth (Changzhou) Co., Ltd,China, Changzhou 213164, China

Abstract. The friction of the hydraulic cylinder mainly comes from the sealing part, the existence of friction leads the wear of the seal. In view of the serious impact of the damage of the seal on the hydraulic cylinder, a set of sealing test system was designed, for the purpose of investigation and analysis on the performance of different sealing kit configuration, in terms of friction with given hydraulic medium. In this paper, there will be the introduction of the principles, constitutions and functions of the testing system. The experimental results provide data support for the design and selection of sealing system of the hydraulic cylinder.

1 Introduction

As the actuator of hydraulic system, hydraulic cylinder is of wide application in kinds of engineering mechanics because of its simple structure and reliable operation[1]. While in the practical application of engineering mechanics, the problems such as crawling, insufficient output and poor pressure-holding causing by cylinder oil leakage seriously affect the stability, reliability and service life of the device[2]. In the process of the use of hydraulic cylinder, there exists some problems like oil waste, environmental pollution, service life shortening, machine downtime growth, reduction of the production efficiency, cost growth and production failure due to the damage to the seals[3-5].

Although the hydraulic cylinders of Bosch Rexroth have a common standard or professional range of products, but with the use of the field becomes wider, the use requirements become more different, in order to enable enterprises in the fierce market competition are still in a favorable position, and make the product can adapt to the huge market and own the high flexibility and adaptability, companies need to enhance the scientific design of the product, but also to ensure the quality of the case to reduce costs, with the lowest input to get the best results, the traditional choice can not meet these requirements, so exploratory of the experimental study of product components to grasp the performance is imperative[6-9].

2 Design of Test System

2.1 The design of hydraulic cylinder

The Methods section details the theoretical or experimental methods used. What justifies using a given method? What is special, unexpected, or different in your approach? If you use a standard or usual procedure, mention that upfront, too.

For the sealing performance test, a double rod hydraulic cylinder is designed as the test cylinder, because of the two rods on both sides of the piston, it is for two-way hydraulic drive. And at the meantime, the piston group was designed as the piston and piston rod one-piece, it aims to ensure the alignment of the piston and piston rod, guarantee the same working situation exactly in addition. The foot mounting is chosen to fix the hydraulic cylinder on the floor to process the horizontal test. And the cylinder head is connected with the cylinder tube by screw, the sealing flange are attached to both ends of the hydraulic cylinder. In addition, the displacement sensor is built in the piston rod.

As is shown in Figure 1, the piston and piston rod is one-piece, and the hydraulic cylinder seal is installed in the sealing flange, and the sealing flange is connected with the cylinder head by screw, the foot is threaded connection with the cylinder tube, at the meantime, to control the position of the piston rod, the displacement sensor is installed inside of the piston rod.

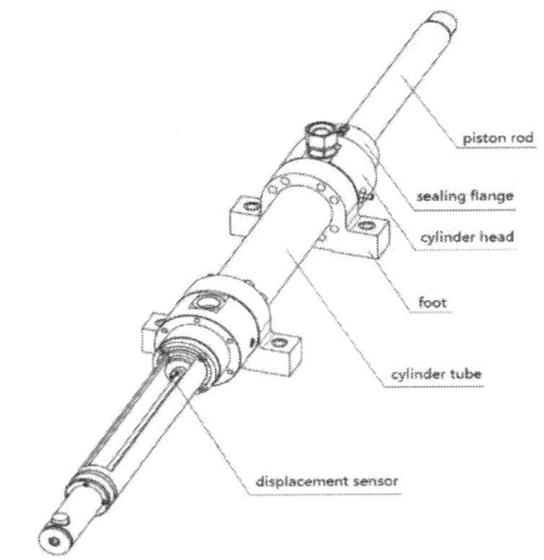

Figure 1. The double rod hydraulic cylinder

2.2 The design of the control system

The sealing system test requires to record the friction force caused by the test seal. Therefore, the pressure sensor in the two chambers of the device is installed to test the pressure change of the two chambers of the recording device. At the same time, the speed of the hydraulic cylinder is controlled and the stroke range of the movement can be adjusted, also need to achieve the precise positioning of the hydraulic cylinder movement control, so the displacement sensor signal needs to be sent to the test device, as a reference, control the high frequency alarm valve input signal, through the high frequency alarm valve signal adjustment to achieve the hydraulic cylinder movement position and speed control. Bosch Rexroth's electro-hydraulic controller HNC is chosen to meet the requirements of motion control and high-speed data acquisition and storage, according to the type and quantity of signal acquisition and the requirements of motion control. The control system frame is shown in Figure 2 .In the control system, the electrical components information is as the Table 1.

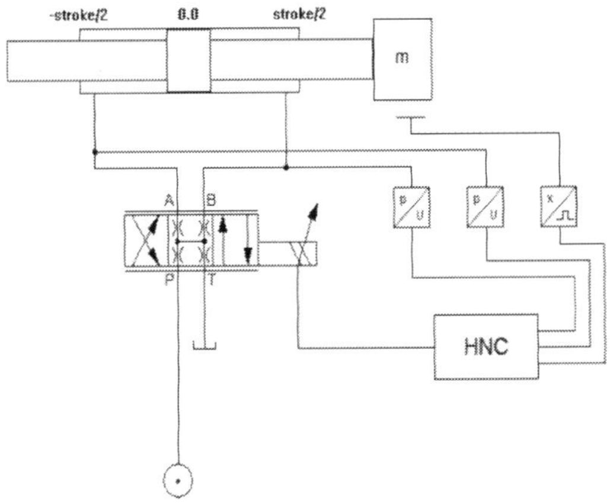

Figure 2. The control system frame

Table 1. The Selection electrical components

Electrical Components	Model	Range
Pressure sensor	HM20	0~50Bar, 4~20mA Output
Displacement sensor	RH-M-0500-M-D60-1-A01	500mm, 4~20mA Output
HNC	VT-HNC-2-3X/P-I-E0/G02	Two-axis HNC with TCP/IP interface
High frequency Response Valve	4WRTEV125L-4X/6EG24K31A1M	Input Signal -10~10V

2.3 Introduction of the two kinds of sealing system

At present, in hydraulic cylinder design, we usually take the series seal. In the present industrial applications, the detailed and suitable application that using the different sealing systems is not clear. In this paper, the common piston rod seal system includes M-sealing system and T-sealing system are taken test to compare the friction force in different speed with the same test condition. As shown in the Figure 3, M-sealing system is usually constituted with a stepseal, a u-ring and a wiper scraper seal[10-12]. As shown in the Figure 4, T-sealing system is usually constituted with two stepseals and a wiper scraper seal. And because of the limitation of time and experimental conditions, the V-sealing system can be tested in later study.

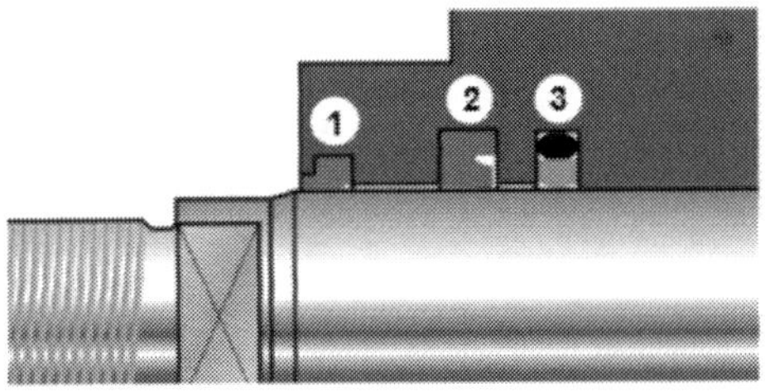

1-Wiper scraper seal, 2-Lip ring, 3-stepseal
Figure 3. The M-sealing system

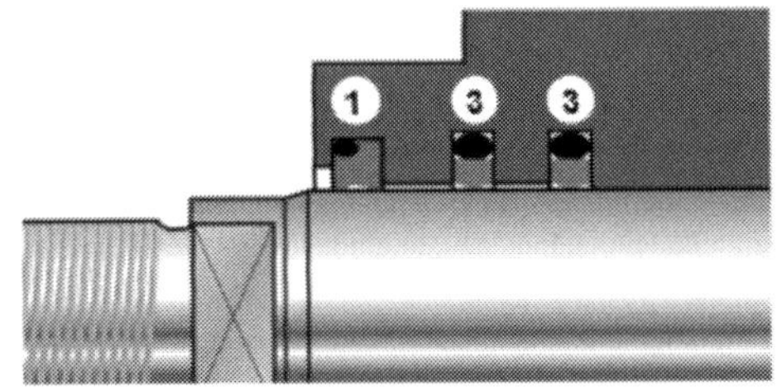

1-Wiper scraper seal, 2-Lip ring, 3-stepseal
Figure 4. The T-sealing system

3 Experimental

Aimed at the different working conditions, such as stable working condition, high frequency working condition, working condition with pressure-holding requirement and the working condition with friction requirement, now, do research on the fiction and sealing performance of the M-sealing system and the T-sealing system to find out the most appropriate occasion.

The research done on fiction at the specified sealing system, mainly test the friction force caused by seals of hydraulic cylinder. To do this experiment, we install the pressure sensors in the two cavity pressure joints of the double rod hydraulic cylinder and write the programs in HNC. Predict the fiction value by analyzing the friction curve in different speed, and layout them in the same cure, compared the curves in the two sealing systems in the last. Considering the actual situation of the hydraulic power unit, we choose the test pressure as 50Bar, and control the speed at 150mm/s, 150mm/s, 200mm/s and 250mm/s to get the curves.And the actual experimental equipment is shown as Figure 5.

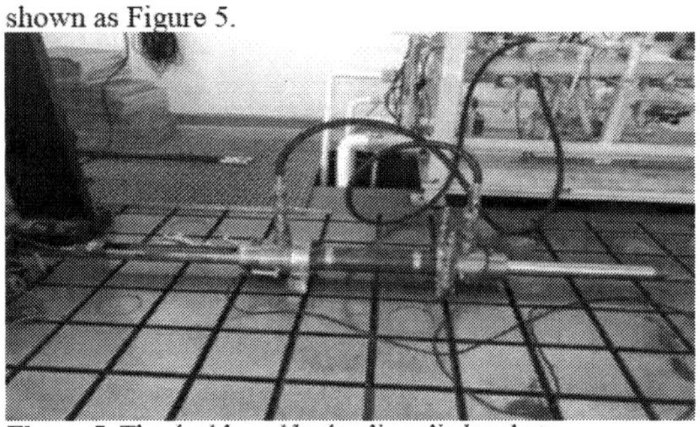

Figure 5. The double rod hydraulic cylinder photo

4 Results and discussion

In this experiment, we get the pressure valve in different speed by the pressure sensor installed in the pressure joints of the sealing flange, and through the scale conversion, we can finally achieve the real-time friction force.

4.1 The Friction of M-sealing system

When the pressure is 50 Bar, and the speed is 100mm/s, the friction curve is shown as Figure 6.

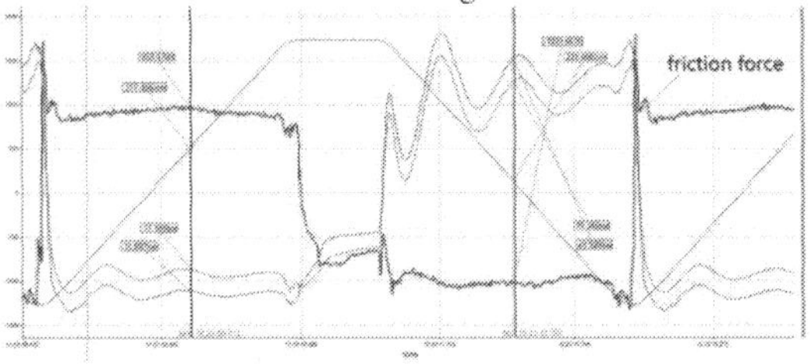

Figure 6. The friction curve in 100mm/s of M-sealing system

In above picture, the purple curve represents the change of friction force. In the process and return, take a symmetric position relatively.

From the friction force curve, the process friction is 953.470N, the return friction is 1033.863N, we can estimate the comprehensive friction is 993.667N.

With the speed of 150mm/s, 200mm/s and 250mm/s, the friction curves are shown as Figure 7, Figure 8 and Figure 9.

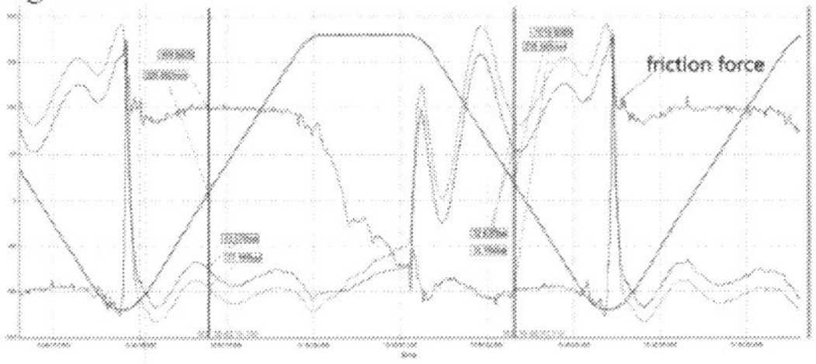

Figure 7. The friction curve in 150mm/s of M-sealing system

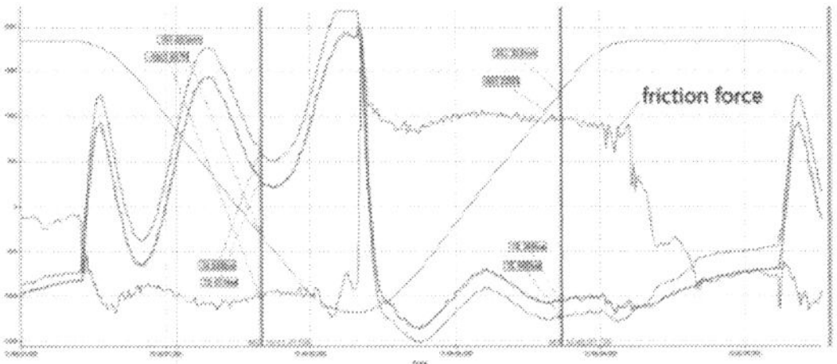

Figure 8. The friction curve in 200mm/s of M-sealing system

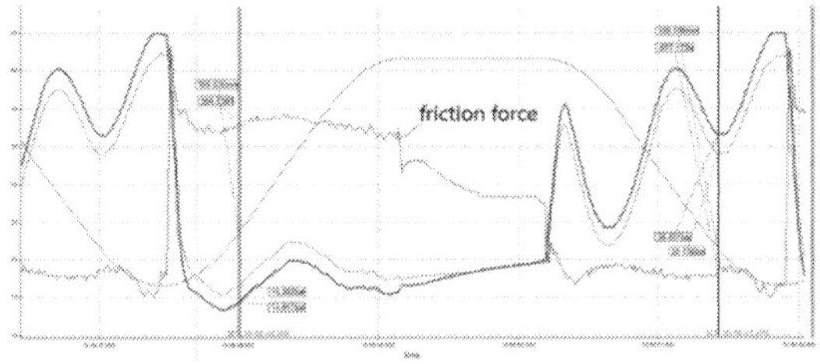

Figure 9. The friction curve in 250mm/s of M-sealing system

For ease of comparing the shape of the curves in different speed, put the average friction force in one curve by data analysis software, the detailed shape is shown as Figure 10.

In the shape of friction force in different speeds, with the higher speed, the friction force is generally a downward trend, but in the speed of 150mm/s, it presents a rising friction value. Maybe it can attribute to the accuracy of the pressure sensor, in addition, the whole test system is designed artificially, and it also exists system error.

From the above curves, we know that the higher speed, the lower friction force. Meanwhile, in the M-sealing system, the average friction force is 983.251N.

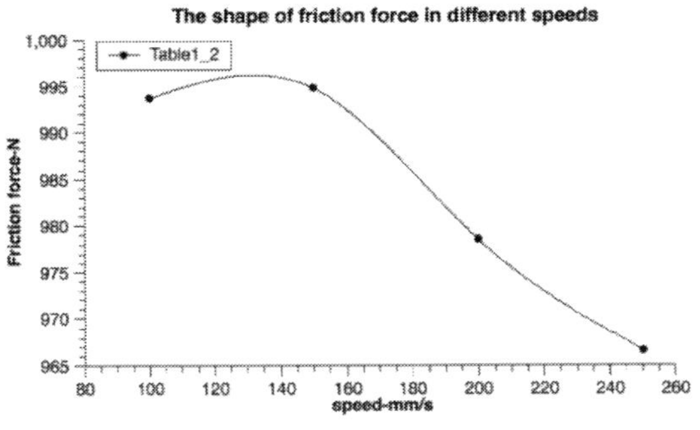

Figure 10. The shape of friction force in different speeds in M-sealing system

4.2 The Friction of T-sealing system

In the T-sealing system, the friction curves in the speed of 100mm/s, 150mm/s, 200mm/s and 250mm/s are shown as Figure 11, Figure 12, Figure 13 and Figure 14.

Put the average friction force in one curve by data analysis software, the detailed shape is shown as Figure 15.

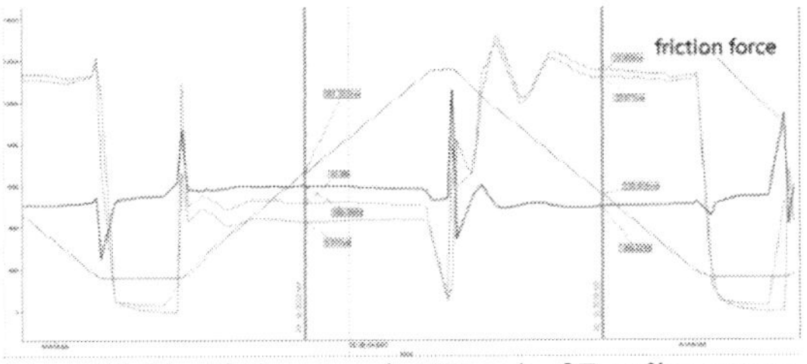

Figure 11. The friction curve in 100mm/s of T-sealing system

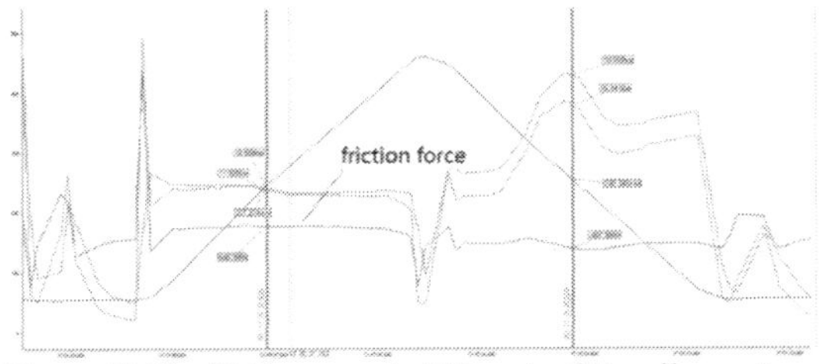

Figure 12. The friction curve in 150mm/s of T-sealing system

From the above curve, the general trend in the T-sealing system is as the trend in the M-sealing system. The higher speed, the lower friction force. Meanwhile, in the M-sealing system, the average friction force is 476.702N.

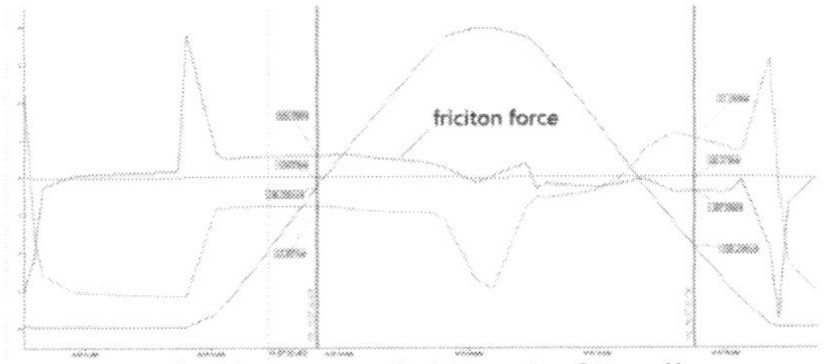

Figure 13. The friction curve in 200mm/s of T-sealing system

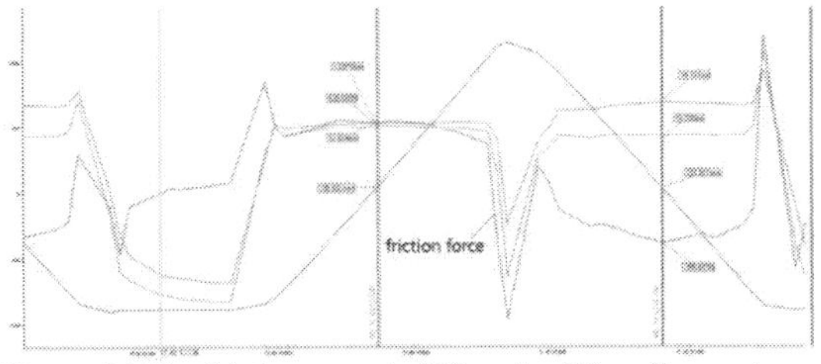

Figure 14. The friction curve in 250mm/s of T-sealing system

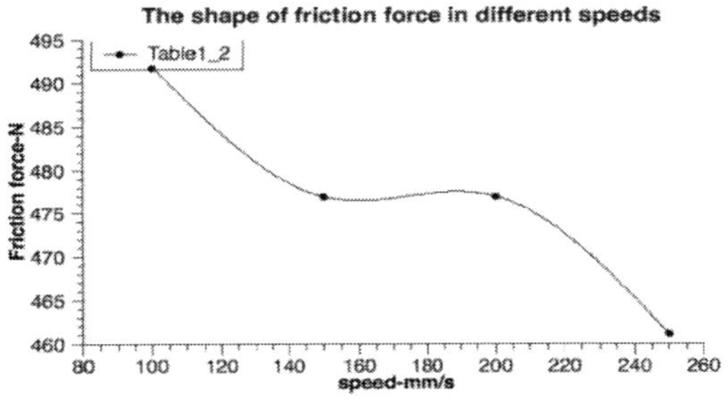

Figure 15. The shape of friction force in different speeds in M-sealing system

By comparing the friction force curves in the two kinds of sealing system, the characteristics of the M-sealing system and the T-sealing system are displayed. The M-sealing system has the following characteristics: durability, high strength, sensitive dust and well dust prevention. And the T-sealing system has an obvious advantage, that is low friction (with no crawling).

5 Conclusions

With the increasing application of hydraulic cylinder in industrial engineering, the design of cylinder is more reasonable, and the process technology is better perfect. The difficulty in design of hydraulic cylinder is the sealing system, because the failure of hydraulic cylinder mostly attributed to the failure of the seals. Therefore, take a study in sealing system and own a technical standard in selection of the sealing system is feasible.

Through the experiment, the most suitable applying situation of the different sealing systems is clear and definite. In the design of hydraulic cylinder, the M-sealing system should be chosen in the steady conditions, and in the working condition of high frequency, the T-sealing system should be considered. The research results provide the technical support in seals choosing, and meanwhile, with the requirement of friction predict, the experimental study can give the data support.

Acknowledgements

This research was supported by the the Natural Science Foundation of Jiangsu Province(Project No. BK20140247)and the Postgraduate Research & Practice Innovation Program of Jiangsu Province (Project No. SJCX17_0161).

References

1. X.L. Xu, W.S. Wei . *Hydraulic Cylinder and Its Design*. National Defense Industry Press, Beijing. (2011)
2. X. Zhou, L. Jiang, J. Gao. Simulation analysis for inner leakage of hydraulic cylinder based on amesim. *Machine Tool & Hydraulic*, **40**, 23, 147-150 (2012)
3. S.J. Shen, Q.H. Yang. Analysis and solution of hydraulic cylinder's leakage problem. *Advanced Materials Research*, **189-193**, 664-667 (2011)
4. X.X. Zhao, S.S. Zhang, C.L. Zhou, Z.M. Hua, R. Li, J.H. Jiang. Experimental study of hydraulic cylinder leakage and fault feature extraction based on wavelet packet analysis. *Computers & Fluids*, **106**, 33-40 (2015)
5. W. Wang. Analysis on the side leakage amount of the friction between piston and cylinder block in axial piston pump. *Applied Mechanics and Materials*, **635-637**, 341-345 (2014)
6. C.Y. Liu, F. Li. The design of seal style for the long journey hydraulic cylinder. *Lubrication Engineering*, no. 2, 167-169 (2005)
7. Y.D. Tang. *Hydraulic Cylinder Seal Technology and Application*. Machinery Industry Press, Beijing.(2016)
8. E.Z. Yang. Prospect of Modern Hydraulic Transmission Technology. *Hydraulics Pneumatics & Seals*, no. 4, 1-6 (2003)
9. X. Zhang, Y. Bai, F. Zhang, P. Zhao. The research of performance testing system of sealing washer worked upright column for hydraulic support. *Hydraulics Pneumatics & Seals*, no. 5, 39-42 (2012)
10. Z.D. Hu, J.H. Jiang. Friction torque of rotating hydraulic cylinder with deep circular groove oil film face seal. *Advanced Materials Research*, **97-101**, 1937-1941 (2010)
11. G.B. Yu, B. Dai, L. Huang, J.F. Nie, Y. Zhao. No-contact hydraulic dynamic seal characteristic study base on cfd. *Applied Mechanics & Materials*, **274**, 612-615 (2013)
12. G.D. Buyalich, K.G. Buyalich. Comparative analysis of the lip seal in hydraulic power cylinder. *Applied Mechanics & Materials*, **770**, 402-406 (2015)

Analysis of the jet pipe electro-hydraulic servo valve with finite element methods

Zhao Kaiyu[1], Wu Qingxun[1] , Zhang Lijian[1] , Yuan Zhaohui[2] and Yang jiaqiang[2]

[1]*Beijing Institute of Mechanical Equipment, 100854 Haiding district Beijing China*
[2]*Northwestern Polytechnical University, 710129 Chang'an District of Xi'an, China*

Abstract. The dynamic characteristics analysis about the jet pipe electro-hydraulic servo valve based on experience and mathematical derivation was difficult and not so precise. So we have analysed the armature feedback components, torque motor and jet pipe receiver in electrohydraulic servo valve by sophisticated finite element analysis tools respectively and have got physical meaning data on these parts. Then the data were fitted by Matlab and the mathematical relationships among them were calculated. We have done the dynamic multi-physical fields' Simulink co-simulation using above mathematical relationship, and have got the input-output relationship of the overall valve, the frequency response and step response. This work can show the actual working condition accurately. At the same time, we have considered the materials and the impact of the critical design dimensions in the finite element analysis process. It provides some new ideas to the overall design of jet pipe electro-hydraulic servo valve.

1 Introduction

Electro-hydraulic servo valve is used in many modern control systems. Due to the working oil pollution is a major factor that causes the servo valve does not working properly, the anti-pollution ability of the servo valve is very important. This servo valve driven by the pre-stage jet pipe nozzle, and the harm caused by the debris and impurity particles is no too bad, conversely, it's very easy to maintenance management, and its anti-pollution ability is very efficient.

The jet pipe electro-hydraulic servo valve which was driven by jet pre-stage has been applied in many prospects, so its characteristics has been put in-depth research. The jet pipe servo valve itself contains a magnetic, electrical, hydraulic multi-fields coupling analysis, every part is very complex, and the coupled fluids analysis cannot be carry out in any simple way. So the study of it has become more and more difficulty. Many of the existing products have a lot of problems, for example, the dead zone is too large, poor linearity, connection instability, poor dynamic characteristics etc. Especially when the valve constitutes a closed-loop system with steering gear, it is prone to instability. Further study of jet pipe electrohydraulic servo valve is necessary [1].

2 The overall structure and principles of a jet pipe valve

The total composition of jet pipe hydraulic servo valve can be divided into four parts: Moving iron permanent magnet torque motor, jet pipe amplifier, slide valve and feedback components. The coordinating relations among the various components were established through force feedback principle. The overall actions of the jet pipe electro-hydraulic hydraulic servo valve were described as below.

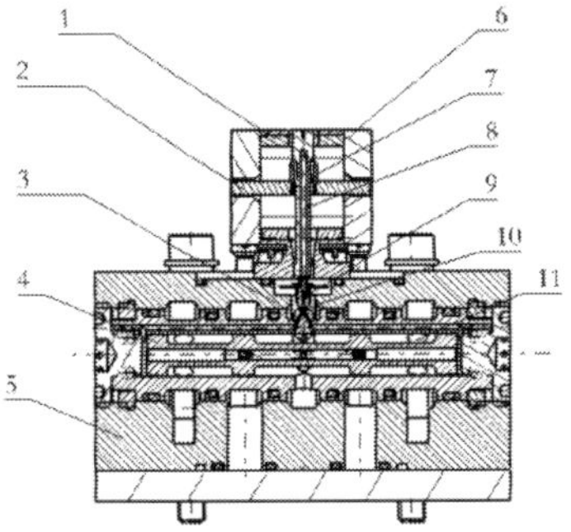

Figure 1. Front view of jet-tube hydraulic servo valve

In this figure, 1: Weak permeable material, 2: Armature and coil, 3: receiving hole, 4. Spool, 5: valve sleeve, 6: Strong permeable material, 7: The spring tube, 8: Jet pipe, 9: nozzle, 10: Overflow chamber 11:O-ring.

When we input a control current i to the torque motor, the armature will produce a deflection angle θ , so the jet pipe is deflected from the neutral position, the nozzle displacement was x^p. The fluid flow accepted by the two receiving holes of the nozzle below is different; the pressure differential would be$\Delta p = p_1 - p_2$ formed in the nozzle receiving chamber. The pressure difference drives the spool and feedback stick move to one end aboutx_v. When feedback stick continue deformation, the feedback force from stick and fluid pressure effects on ends of spool reaches dynamic equilibrium. In this case, the torque from feedback lever, spring tube, electromagnet torque motor, inlet pipe as well as the safety wire and leaf spring reach equilibrium.

3 Feedback spring assembly stiffness calculation

Feedback spring assembly was composed with the lever and connects pieces. The movement of the spool affecting the nozzle angle of deflection through feedback spring assembly. Use the Catia static analysis, put armature a rotating electromagnetic torque to analog actual circumstances, a direct force applied to the end of the feedback lever, the jet tube will be rotated a certain angle, the nozzle in response to a slight displacement. When put a Rotational torque on the armature, it is more obvious to cause the displacement of the jet pipe fuel injector end, the inlet pipe and twisted wire. So, fuel tube and twisted wire will produce a reaction torque on jet pipe. Furthermore, the lower part of the feedback spring tube assembly also have deformation accompanied the jet pipe nozzle [2].

If we give electromagnetic torque on the armature and force on the lower end of the feedback lever, we will get deflection displacement of the jet pipe. Record the relationship among them; we can analyze stiffness of the feedback spring assembly about the jet pipe. The various parts of the stress cloud as follows:

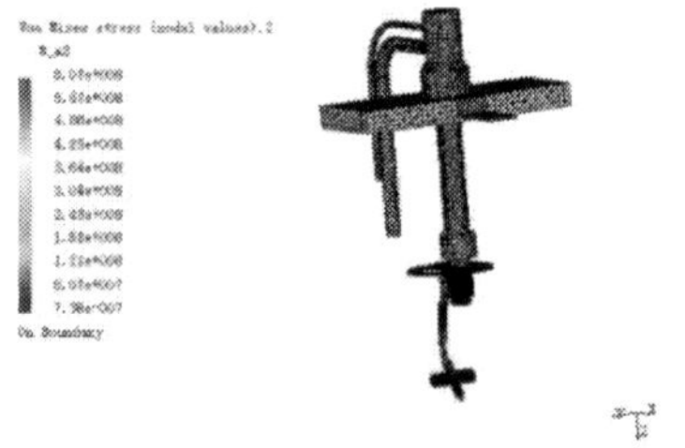

Figure 2. Armature feedback component stress cloud

Throughout the jet pipe, armature and feedback spring assembly, the force is mainly concentrated in the parts of armature and feedback spring where binding with the jet pipe. In addition, the parts of inlet pipe and twisted wire where connect with jet pipe were also the larger force area. As written in Eq.(1),M_1 indicates reaction torque of the leaf spring. M_2 indicates reaction torque of the spring tube. M_3 indicates reaction torque of the twisted wire. M_4 indicates reaction torque of the oil conduit. M_f indicates reaction torque of the feedback stick.

$$M_0 = M_1 + M_2 + M_3 + M_4 + M_f \quad (1)$$

It is not only related to the deflection angle of the jet pipe, but also the motion displacement of the spool. Then we take use of the Catia finite element simulation and get the relationship data.The mathematical relationships between them were fitted as follows:

$$M_f = 27.28\theta + 311.85x_v + 3.53\times10^{-5} \quad (2)$$

$$f = 10754x_v + 277.98\theta + 4.59\times10^{-5} \quad (3)$$

4 Torque motor characteristics analysis

The torque motor in jet pipe electrohydraulic servo valve was an iron moving torque motor, which armature rotation direction is parallel to the working air gap flux density. This torque motor was constituted with permanent magnet, the upper guide magnet, the lower guide magnet, armature and the control coil. Permanent magnet magnetized the upper and lower guide magnet as the Arctic and the Antarctic. Two control coils trap around the armature. There will constitute four working air gap between the armature and the guide magnet. The magnetic permeability of the working air gap was showned respectively:

$$G_1 = \frac{\mu_0 A_g}{g - x} \quad (4)$$

$$G_2 = \frac{\mu_0 A_g}{g + x} \quad (5)$$

Where $A_g$is the area of the air gap, g is the thickness of the air gap when armature is at the neutral position. x is the displacement of ends of the armature (pole face center) from the neutral position. We have plused the magnetic drop in the working air gap which were generated by fixed magnetic force and control magnetic potential. Then we will get the total magnetic pressure drop in working air gap:

$$U_1 = \frac{B_p g}{\mu_0} + \frac{B_c g}{\mu_0} \quad (6)$$

$$U_2 = \frac{B_p g}{\mu_0} - \frac{B_c g}{\mu_0} \quad (7)$$

Where $B_p$is the fixed magnetic flux density in the air gap when armature was at the neutral position. B_c is equivalent to the control magnetic flux density when the armature in the neutral position. Now we can calculate the electromagnetic suction of each working air gap by the formula (4), (5), (6), (7) which were enumerated above. Then we can obtain the electromagnetic torque as follows[3]:

$$M_d = \frac{20.4\times10^{-8} B_p A_g a N_c}{g} i_c + \frac{40.8\times10^{-8} B_p^2 A_g a^2}{\mu_0 g}\theta \quad (8)$$

Models and magnetic field distribution was shown in the following figure:

 https://doi.org/10.1051/matecconf/201815306013

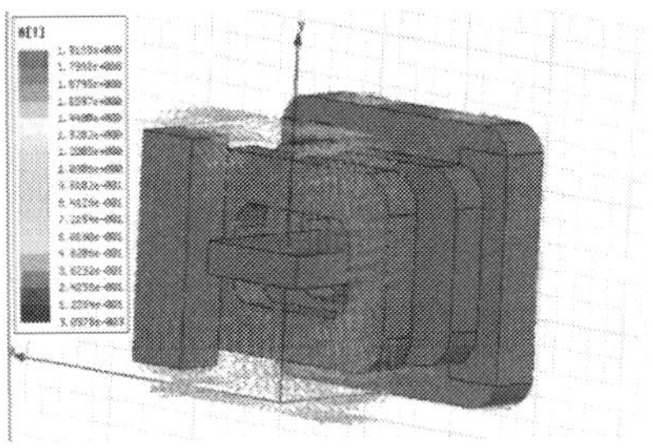

Figure 3. Magnetic field distribution of torque motor

The electromagnetic torque generated by the motor was ralated to the coil current and the armature deflection angle. Changing the deflection angle of the armature and the current respectively, we will obtain the relationship among them:

$$Md = -0.002328 + 2.838ic + 16.79\theta \quad (9)$$

This is the relationship among electromagnetic torque, current and the armature declination obtained by the finite element analysis methods. It is very important for the analysis of whole servo valve system.

5 Analysis of the flow field between the jet pipe and receiver

We select the ansys workbench13 as the tool of fluid simulation. The total process Include pre-processing (modeling, material properties, meshing, etc.), load, and solve (applying the boundary conditions, determine the solution method, etc.), post-processing (view and display)[4]. We meshed different parts of the model by tetrahedral and hexahedral use ANSYS ICEM CFD, and made 2 scale local grid refinements of the imports. The mesh size is 0.1 or 0.2 and the number of units is about 447,000. We use the SIMPLE method to solve the simulation. Fluid model use the k-epsilon two-equation model, When the nozzle deflection is 0.2 deg, convergence precision is about 10 ^ -5. The partial enlarged velocity vector diagram of nozzle - receiving hole was shown below:

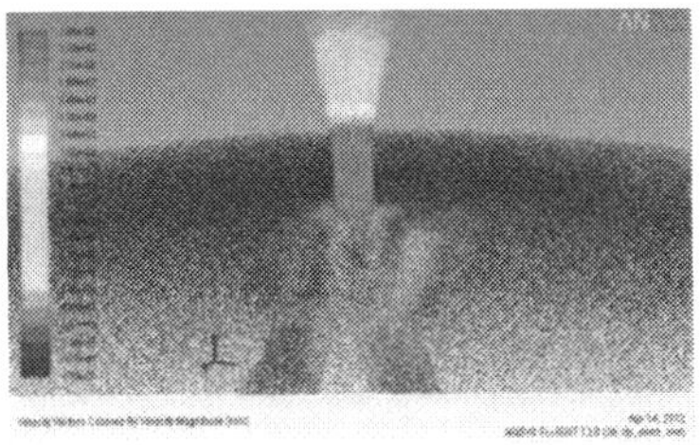

Figure 4. The velocity vector partial enlarged view of 0.2 degrees nozzle declination

We simulate the pressure difference at the two sides of spool by jet pipe CAD models in different declination. We consider the jet pipe is deflected in one direction only due to the spool movement and jet pipe deflection is symmetrical. The calculation results was shown in Fig.5:

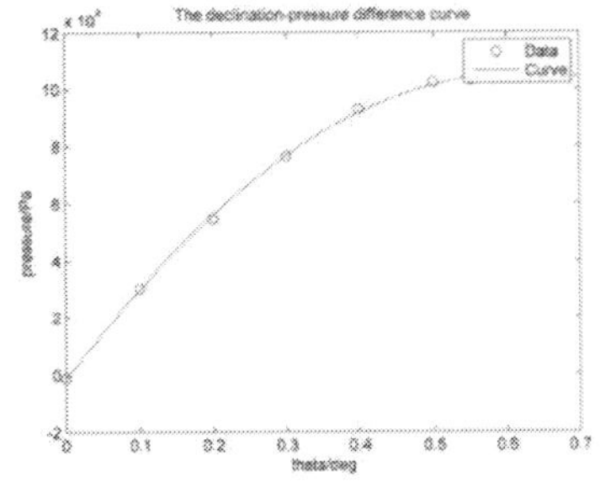

Figure 5. Pressure differential curve of different jet pipe declination

It is visible that the flow rate increases into a cavity and decreases into the other when the jet pipe is offset to one side. So the pressure difference increases gradually, this relationship can be approximated by a equation:

$$\Delta p = -1.878\times10^{11}\theta^2 + 3.175\times10^{9}\theta + 4.388\times10^{5} \quad (10)$$

This is the relationship between the jet pipe declination and pressure differential at both ends of spool.

6 Analysis of jet pipe electrohydraulic servo valve performance

6.1 mathematical models

The stiffness of feedback lever and spring tube was confirmed by torque balance of armature assembly. The torque balance equation of armature assembly was [5]:

$$K_t i_c(S) + K_m\theta(S) = J_a\ddot{\theta} + B_a\dot{\theta} + K_a\theta + rA_p p_p - r^2 8\pi C_p^2 psx_a\theta + (r+b)K_f[(r+b)\theta + x_v] \quad (11)$$

The left side of equation was electromagnetic torque of the torque motor, while the right was the anti-torque on the armature, the last anti-torque was came from the deformation of feedback lever. The pressure differential of spool ends was$\Delta p = p_1 - p_2$, one forceF_f applied to the spool came from the feedback lever, the other force $\Delta p \cdot A$came from fluid pressure differential of the spool ends. The kinetic equilibrium equation can be listed:

$$\Delta P \cdot A - F_f = m\cdot\ddot{x}_v + \beta\cdot\dot{x}_v + F_{ff}(\dot{x}_v) \quad (12)$$

We can get the finite element model of the whole valve through the analysis builded before.

6.2 Simulation and Modeling

The armature feedback components, torque motor and the jet pipe amplifier section has been simulated by the finite element software. We need to connect the various parts to build the overall model. The armature assembly damping is very important for the torque balance equation of torque motor. In addition, when we came to the force balance equation of spool parts, the valve internal loss along the tubule way should also be considered, as well as the steady-state fluid power. Given input current is sinusoidal, the amplitude is 0.043 and frequency is 200Hz. The relationship between the output displacement of spool and input signal was shown as below:

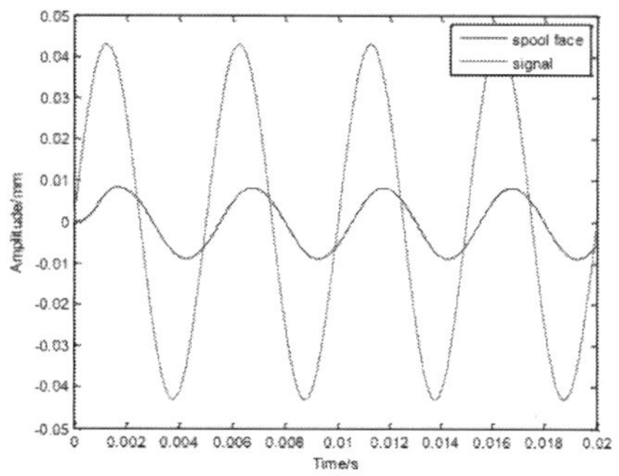

Figure 6. The relationship between Input signal and output displacement of spool

It can be seen that the spool displacement has a phase lag relative to the signal, but it can follow well. The frequency response characteristic of the whole system was shown as follows:

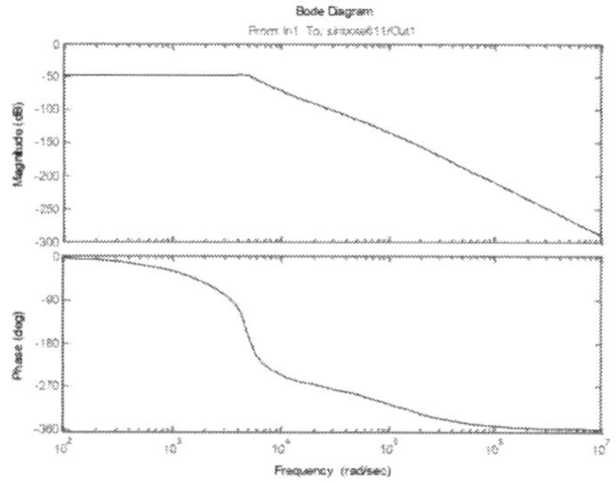

Figure 7. The frequency response curve of jet pipe electrohydraulic servo valve

The frequency response is the essential characteristics of the servo valve in the frequency domain, related to the geometrical structure, material, external working conditions and other factors, determined by the torque motors, slide valve and jet tube amplifier. Jet-tube hydraulic servovalve frequency responses have second-order characteristics related to the power sliding valve.

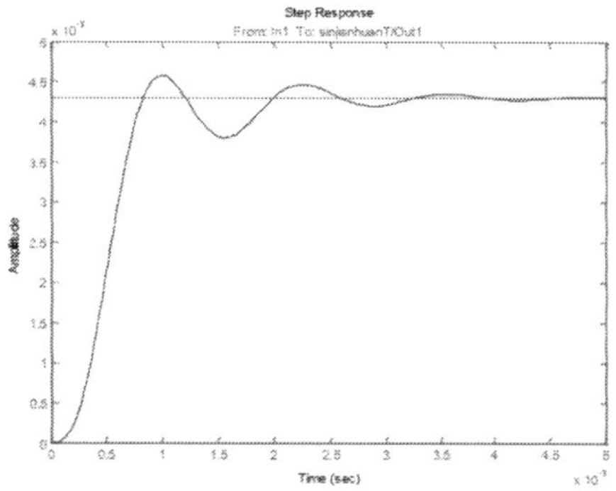

Figure 8. The step response curve of jet pipe electrohydraulic servo valve

Step response related to the oil viscosity, friction characteristics and dynamic characteristics of the spool primarily. As can be seen from the figure, the adjust time of servo valve is 0.005s, the overshoot is 12.5%, which is obviously an ideal state, the adjusting time will be longer slightly in actual test.

7 Conclusions

We obtained the relationship of the key structures about jet pipe electrohydraulic servo valve by finite element simulation calculation, and summed up the dynamic characteristics of the jet pipe servo valve.

(1) We got the overall stiffness of armature feedback components by the method of finite element simulation, and obtained the relationship of the jet pipe deflection angle and electromagnetic torque in the case of certain feedback force.

(2) We got the relationship about electromagnetic torque, current and armature deflection angle by the finite element electromagnetic analysis.

(3) Analyze the relationship between the jet pipe declination and spool pressure at both ends through the ansys Fluent module.

(4) Analyze the overall dynamic characteristics of the valve with the mathematical simulation tools Matlab Simulink, We have got the relationship between input and output signal, the frequency response curve and the step response curve.

(5) We can got more reasonable and accurate results in simulation of the jet pipe valve through finite element analysis methods. We should take full advantage of the advanced simulation tools to obtain satisfactory results in the design process.

References

1. AV Mesropyan, RR Sharipov, Mathematical Modeling of Transient Processes in the Jet Pipe Servoactuator with a Dual-Mode Controller [J]. Procedia Engineering, (2016)
2. Xuan Hong Son Pham, Yao Bao Yin, Xi Zhang. Optimal Design for Torque Motor of Jet Pipe Electro-hydraulic Servo-valve Based on Dynamic Stiffness [C]. Intelligent Human-Machine Systems and Cybernetics (IHMSC), (2012)
3. M Jelali, A Kroll. Hydraulic servo-systems: modelling, identification and control [M].P. 62-68 (2012)
4. G.M. Ivanov, V.K Sveshnikov. Power efficient improving of digital hydraulic servo actuator Engineer Vestnik,7. P. 29-37. (2015)
5. AV. Mesropyan, RR. Sharipov. The simulation of complex digital control systems in the hydraulic servoactuater, RAS, Moscow. (2013)

Application of Video Recognition Technology in Landslide Monitoring System

Meng Qingjia[1, 2], Wang Chenhui[2]

[1]*Key Laboratory of Geological Environment Monitoring Technology, MLR, 1305 Qiyii Zhong Road Baoding 071051, China*
[2]*Center for Hydrogeology and Environmental Geology Survey, CGS, 1305 Qiyi Zhong Road Baoding 071051, China*

Abstract. The video recognition technology is applied to the landslide emergency remote monitoring system. The trajectories of the landslide are identified by this system in this paper. The system of geological disaster monitoring is applied synthetically to realize the analysis of landslide monitoring data and the combination of video recognition technology. Landslide video monitoring system will video image information, time point, network signal strength, power supply through the 4G network transmission to the server. The data is comprehensively analysed though the remote man-machine interface to conduct to achieve the threshold or manual control to determine the front-end video surveillance system. The system is used to identify the target landslide video for intelligent identification. The algorithm is embedded in the intelligent analysis module, and the video frame is identified, detected, analysed, filtered, and morphological treatment. The algorithm based on artificial intelligence and pattern recognition is used to mark the target landslide in the video screen and confirm whether the landslide is normal. The landslide video monitoring system realizes the remote monitoring and control of the mobile side, and provides a quick and easy monitoring technology.

1 Introduction

Landslide Geological Hazards is a geological hazard caused by geological disasters in China. Now more advanced monitoring technologies are 3D laser, BOTDR, GNSS, osmotic pressure, inclinometer and so on. A variety of monitoring methods are proposed for different landslide types: displacement monitoring, physical field monitoring, and groundwater monitoring and external induced factor monitoring to monitor, prevent, treat and evaluate the landslide bodies. With the development of video surveillance technology, the artificial intelligence is promoted. The video for automatic analysis, identification and retrieval of intelligent video surveillance system are achieved.

Intelligent video recognition technology is applied to landslide monitoring in this paper. The target trajectory tracking is carried out by using the difference chart method to realize the image and data, and the dynamic trajectory recognition system is established. The system is established in order to facilitate the timely transmission of image information to the receiving terminal, and retain valuable image and data information which method can achieve low cost, low power consumption, intelligent, network and remote. In the field of geological disaster monitoring, the system starts remotely and has sleep function. The system can be controlled through the 4G network by the terminal man-machine interface for intelligent identification processing to send and receive orders. The application of landslide video monitoring technology provides intuitive data and image information for post-level expert analysis and decision-making, and it basically realizes disaster information uniformity, integrity and real-time efficient reporting. For the geological disaster group, the system provides more reliable technical support.

2 Intelligent Video Recognition Technologies

Intelligent video recognition system refers to the use of intelligent video analysis algorithm. The system detects and identifies automatically to set the scope of the target which is an integrated system. The computer is used instead of the monitor to achieve the monitoring of the target body. The computer can do a series of calculation, interpretation, and classification etc., it can also maintain the image sequence database, detect abnormal actions, storage image sequence and send the data image information to the monitoring center.

Landslide video recognition process is affected by the surrounding natural environment, for example trees, light shadows, seasonal weather changes, etc... In the extraction of the background trajectory, the edge of the landslide reference frame is needed to constantly select repeatedly. Using three frame image difference methods, the pseudo-motion information caused by the background

 https://doi.org/10.1051/matecconf/201815307001

movement and green plant shaking is removes. The target body is separated from the complex background.

3 Foreground and background discrimination

The segmentation between the foreground and the background needs to compare the current frame with the background colour and brightness difference. If a pixel colour distortion value is close to the pixel coding of a coded word and its brightness is in the encoding of the word brightness range. The pixel may be classified as a foreground, and vice versa as a background. After the self-improvement of the colour class library, the edge feature of the target body and the background is extracted, and the target is tracked and the target model is established.

3.1 Colour clustering and weight index

In the separation process of foreground and background, the video block color clustering and color weight index data structure is established. The visual field RGB color space is converted to YUV space by using the average method that the color and brightness is separated. This will directly record each pixel changes; each pixel block is managed in units. The weight of each piece of color clustering is calculated, that is, the color weight. The color weight is the reduction of the pixel class. Thus the extraction of the target body can be avoided the impact of light.

Color weight:

$$W = (1 - \alpha) \times W + \alpha \times M$$

α is adaptive learning parameter(0.5~0.3), adjust weight is W;

M is the foreground mask, getting pixel values that match color clustering M is 1, Otherwise it is 0.

3.2 Geometric feature recognition

First, the target body geometric feature is recognized, the distance between the various parts of the target body are accurately measured, These distances are identified as features to identify.

3.3 Pseudo-motion removal and selection of key frames

A new method of moving target detection in complex scene is used to extract the motion information of the monitoring target. Using motion compensation technology, the key frame of the three frames as eliminate the camera, leaves and other pseudo-motion are selected. The coordinates of the pixels in three frames are set respectively as $f_1(x_1, y_1)$ 、 $f_2(x_2, y_2)$ 、 $f_3(x_3, y_3)$. The motion parameter from the first frame to the second frame is (dx_1, dy_1). The motion parameter from the second frame to the third frame is (dx_1, dy_1):

$$\begin{cases} f_1(x_1, y_1) = f_1(x_1 - (dx_1 + dx_2), y_1 - (dy_1 + dy_2)) \\ f_2(x_2, y_2) = f_2(x_2 - dx_2, y_2 - dy_2) \\ d(x, y) = |f_1(x, y) - f_2(x, y)| \times |f_2(x, y) - f_{3(x,y)}| \end{cases}$$

When $d(x, y)$ is not zero, the key frame of the target is the pixel. But this information may still have a small amount of pseudo-motion information. Then the clustering analysis techniques are used to reduce the error in the follow-up work, and extract the target key frame. This way is to avoid overlapping two adjacent objects and remove the pseudo target point.

4 Identification and locking of the target

After the foreground target is determined, the target is detected in the environment which the foreground target is located in, and the target body is selected in the static background. For the monitoring of landslides, it is targeted, fixed and non-random changes. Camera position, focal length and other parameters are fixed. The relative area of the target body is relatively stable. The color, texture and shape of the reference material need to be extracted which is mainly to avoid the influence of factors such as light and shadow, seasonal variation of trees and so on, and to reduce the variation of landslide displacement.

After the image recognition target is locked, the point cloud data will be concatenated through the CAD image processing software to construct the surface of the target body, so that the dynamic variation surface model of the landslide body is obtained. During the process of landslide, the shape of the landslide is changing constantly. In the process of video recognition, it is necessary to record all the shapes in the course of the movement of the target. Once the relative motion is generated, the basic characteristics of the landslide movement can be clearly documented and provides a basic data base for geological research. (Figure 1)

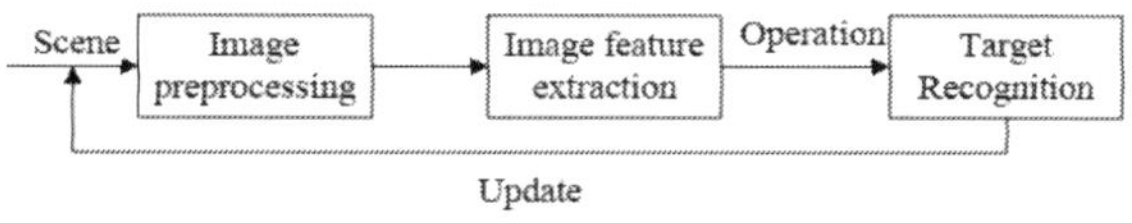

Figure 1. Image processing system.

4.1 Image acquisition equipment

Through the laser scanning system, Space point information can quickly access high density and high precision in real-time.

Through the DS-2DF8823IV-A800 camera, the high-definition images are obtained. This kind of camera can control light, through the fog, 3D noise reduction, and intelligent movement tracking.

4.2 Target characteristics and signal processing

CAD software will be used to pretreat the digital signal acquisition, extract points, lines, and surface features. the point cloud data is used to build surface model.

4.3 Image recognition processing

The selected image frame is set color clustering, recognition and other operations to select the key frame. The part of the dummy data will be removed.

4.4 Artificial intelligence

Target selection: The shape of the target body will change during the course of the movement, and the background will vary with the movement of the target. The real-time image and the original reference image of the record will change slowly. This process maybe bending, or tilting, or stretching, etc., the edge of the target body is slowly changing, on the basis of image recognition, the surface changes are increased, the background of the background continue to filter out.

Morphological processing: Before the target body moves, there is a comparison of the initial template. Through the expansion or corrosion of the image manipulation method, the target part is really complemented to avoid the monitoring of the object of the incomplete.

5 Target body tracking processing

For the dynamic study of the landslide disaster in this paper, it is difficult to track the video target recognition. Once the relative movement of the landslide occurs, its shape, movement speed and surrounding background are likely to change. Therefore, it is necessary to carry out motion simulation during the tracking process of the target body, and records a variety of possible changes, that is, the possible trajectory Variety. To a certain extent, this process can reduce the amount of calculation and expand the scope of application.

The Gabor filter is used to extract the feature, and the tracking problem of the target is transformed into the classification problem of target and background. A variety of morphological parameters may occur in landslide bodies. Gabor filters can be high frequency and maintain progressive stability under the specific structure and size of perturbation Calculation.

After the target body is determined, the target body is marked with a special mark in the system, and the possible movement pattern in the system is estimated in advance and the edge of the target body is set. The relative displacement of the target body can automatically record the trajectory of the target violation.

6 Landslide video surveillance

6.1 Landslide foreground and background separation

Three frame image difference map method is used to separate the landslide body from the surrounding natural environment, and remove the background and green plant shaking and other pseudo-motion information.

6.2 Landslide recognition

The edge of the landslide is treated by morphological corrosion and expansion to eliminate the shadow. The form of the entire landslide is completeness.

6.3 Landslide tracking

Gabor filter is used to extract the features of the landslide, and the landslide body is marked with special markings. The possible motion patterns are preliminarily estimated in the system. The landslide Edge boundaries, in the event of landslide displacement, can automatically record the landslide violation of the boundaries of the trajectory.

6.4 Track analysis

After the target tracking of landslide foreground and background is completed, the position information of the recorded position is analysed and calculated, and the change of the relative position is calculated, and the meshing is simulated in the simulated 3D graphics. (Figure 2)

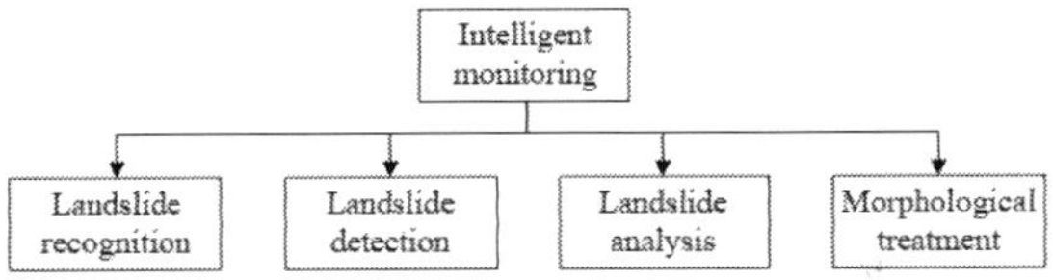

Figure 2. Intelligent analysis

7 Design of video surveillance software for landslide

Landslide video monitoring center is the terminal server to achieve data image reception, quality inspection, query display, statistical analysis and other fixed functions. With the development of science and technology, only limited to use the terminal server to control the operation of monitoring equipment has been unable to meet the needs of the public.

Mobile phone is widely used for monitoring and provides a more convenient and economical platform. Fixed control terminals have been unable to meet today's needs. In the network under the cover, anytime, anywhere a variety of information can be accessed. Through the Internet, the system links to the cloud and receives monitoring image information.

Landslide video surveillance software is designed to be able to access anytime, anywhere monitoring information. In the smart phone, the system will be able to access the terminal server through the network to require information, and to achieve remote monitoring of the monitoring equipment. (Figure 3)

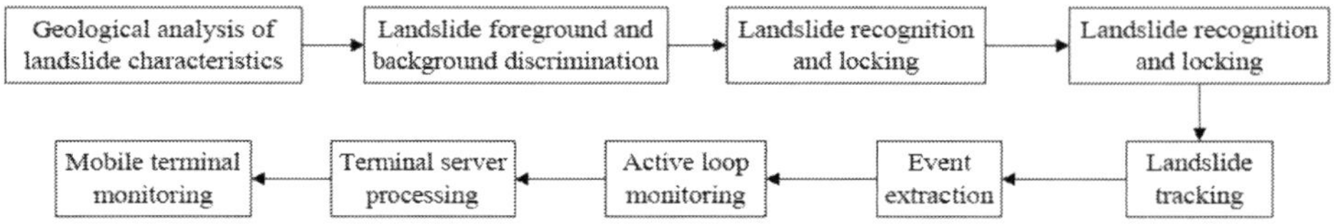

Figure 3. Monitoring process.

8 Summaries

Intelligent video recognition system has developed a mature self-contained system. The computer vision is applied to the landslide monitoring to achieve landslide monitoring image data visualization monitoring.

1) Image pre-processing: The colour clustering and weight index and pseudo-motion processing algorithms are used to outline the foreground and background interval.

2) Target detection: The colour, texture and shape of the landslide and surrounding environment are relatively stable, which mainly avoid the influence of light and shadow, seasonal variation of trees and so on. This way can reduce the error of landslide displacement.

3) Target tracking: a variety of dynamic characteristics of the landslide is estimated to reduce the amount of calculation. The system records the dotted line feature beyond the boundary, and draws the three-dimensional map.

4) Post-processing: the information will be returned to amend the use of open operation to remove the tiny noise in the image which corrects the overall image contour and ensures the integrity of the image.

In the application of landslide disaster monitoring is still relatively simple, the overall design is only from the dynamic simulation point. It does not carry out practical application testing. In the field application, there are still many immeasurable factors; the system is still constantly improving.

References

1. Tang Yaming, Zhang Maosheng, Xue Qiang, Bi Junbo, Landslide Monitoring and Early-warning: an Overview. GEOLOGICAL REVIEW, 58 (2012)
2. Zhang Xiaoyu, Research on Foreground Extraction and Action Recognition for Intelligent Video Surveillance, Zhejiang University, (2012)
3. Huang Dong, Abnormal behavior detection in intelligent video recognition system. Beijing University of Posts and Telecommunications. (2014)
4. Tu Tianyi, Fast Detection and Recognition of Human Face in Video Surveillance System, Xiangtan University, (2005)
5. Yang Wei, Zhang Tianwen, A new method for the detection of moving targets in complex scenes. COMPUTER RESEARCH & DEVELOPMENT, 35 (1998)
6. Chen Xiao, The research for feature-based algorithm of video tracking, XIDIAN UNIVERSITY, (2008)
7. Zhang Guilin, Zhang Tianxu, Wei Luogang, Xie Xianming, An edge-feature-based approach to the extraction and tracking of a moving object, J.Huazhong Univ. of Sci & Tech. 22 (1994)
8. Zhu Hui, LI Zaiming, Moving Objects Detection Algorithm in Video Sequences, SIGNAL PRCESSING, 18 (2002)
9. Hu Xianming, Yan Echuan, Zhou Yu, Zhang Ting-ting, Fractal Characteristics of Landslide Monitoring Point Movement Trace and Its Application, Chinese Journal of Rock Mechanics and Engineering, 31 (2012)
10. Li Xiaojun, Feng Xiating, Yang Chengxiang, HUANG He, Landslide Monitoring and Intelligent Analysis On Deformation Prediction Based On Three Dimensional Geographic, Chinese Journal of Rock Mechanics and Engineering, 23 (2004)
11. Xu Jinjun, Wang Haicheng, Luo Yuzhen, Wang Shangqing, Yan Xueqing, Deformation monitoring and data processing of landslide based on 3D laser scanning, Rock and Soil Mechanics. 31 (2010)
12. Chen Xianhua, Study on Image Processing and Recognition Techniques for Intelligent Video Monitoring System, Ocean University of China, (2004)

Research on Ground Calibration Technology for Geomagnetism Sensors

Zhang Hui , Li Chuanjun and Liu Yukuan

School of Aerospace Engineering, Beijing Institute of Technology, Beijing 100081, China

Abstract. The measurements of geomagnetism sensors are often affected by self-factors, install-factors and environment factors, therefore they often exist error and cause measurement accuracy reduction. Focus on these problems, this paper analyse existing ground calibration algorithms and their merits and demerits respectively. Besides, a comprehensive geomagnetism sensor error model is presented. Based on this model, a least square method based ellipsoid fitting calibration and compensation algorithm is presented. The experiment results show that this novel calibration algorithm can effectively restrain and compensate geomagnetism signal measurement error.

1 Introduction

Geomagnetism filed is elementary physical field of earth, it has characteristics such as all-time, all-weather, all-region etc. Therefore geomagnetism filed has abundant parameter information. Particularly, geomagnetism total filed, three components of geomagnetism, magnetism inclination, magnetism declination and geomagnetism gradient can be used for magnetism navigation [1]. Geomagnetism sensors have many advantages such as small volume, low cost, high precision and so on. Furthermore, they have high ability to anti-impact or overload. Hence geomagnetism sensors have widely used in commercial and military filed. The purpose of this paper is to calibrate and compensate of geomagnetism sensors and finally implement geomagnetism navigation through calibrated geomagnetism information [2].

Existing ground calibration algorithms include follows: 1) Ellipsoid fitting method, this method is based on one assumption. That is, under the impact of magnetic sensor measurement error, the measurement trajectory of magnetic field can be approximate to an ellipse trajectory. The essence of Least Squares ellipsoid fitting method algorithm is to find a group of elliptic parameter, which can minimize the distance between the measurement data and the fitting data in a sense. The advantage of this method is easy to calculate, but the compensation effect for three-axis magnetic sensors is limited [3]. 2) Magnetic variation calibration method, this method trying to calculate rotate, stretch and translation factors to correction ellipsoid trajectory into circle trajectory. Then using this model to filter the abnormal signal. This method is also easy to perform, but the precision of compensation calibration is also limited [4]. 3) Kalman filter method. Kalman filter is a common parameter estimation method for linear system. Extend Kalman filter (EKF) and Unscented Kalman filter (UKF) can be used to estimate and compensate model parameters in real time, but these methods need higher precision reference to be standard [5]. 4) Neural network calibration method. Using the magnetometer's output as independent variable, with the taylor expansion of objective function, this algorithm deduce the optimized objective function. Then using this function as network node to build network, and perform functional approximation for magnetometer's signal. This method can calibrate three-axis magnetic sensors error simultaneously, which can enhance performance of magnetic sensors. However, this method also needs higher precision reference to be standard [6]. 5) Ellipsoid fitting method. Within the region that magnetic field strength is constant and know, the measurement data under ideal situation should be a standard sphere. While because of measurement error, the standard sphere is aberration and becomes an ellipsoid. And using the property of ellipsoid, the algorithm can estimate error correction parameters of magnetic sensors, and performing calibration compensation. Ellipsoid fitting method usually using iteration method or least square method to calculate. However, iteration based ellipsoid fitting is vulnerable to initial value and noise, it's easy to divergence and the computation burden is increase [7]. On the other hand, tradition least square based ellipsoid fitting method gets higher precision error compensation, but it also turn into unstable when constraint matrix is singular [8].

2 Geomagnetic sensor error model

In an invariable measure system and constant testing environment, the measurements of earth magnetic field exist error. Generally speaking, the measurement data are often constituted by random error and system error. The former error are normally distributed, which can be noted as $e \sim N(0,\sigma_e^2)$. The latter error are fixed(or changed at a certain regular), and it often include instrument error, installation error and environment error. We analyze these error respectively, and finally build geomagnetic sensor system composition error mathematical model.

Owing to the wide frequency bandwidth of geomagnetic signal, the measurement signal is easy to interference by other magnetism signal. Besides, be limited in machinery manufacturing equipment, production technology level and material characteristics, magnetic sensor exist measurement error, which can be classified as bias error, sensitivity error and non-orthogonal error [9]. The definition of these error are shown as follows.

Bias error: This kind of error means when the external magnetic field is zero, the output measurement data is not zero. Bias error can be modeled by a fixed matrix such as $b_o = \begin{bmatrix} b_{ox} & b_{oy} & b_{oz} \end{bmatrix}^T$.

Sensitivity error: Sensitivity error are often caused by the different sensitivity or amplification gain between three-axis sensors. Sensitivity error can be modeled as follow:

$$K_s = \begin{bmatrix} s_x & 0 & 0 \\ 0 & s_y & 0 \\ 0 & 0 & s_z \end{bmatrix} \tag{1}$$

Non-orthogonal error: When the three-axis are set non-orthogonal, just like figure 1, it often cause measurement error, that is, non-orthogonal error. Non-orthogonal also can be modeled as:

$$K_p = \begin{bmatrix} \sqrt{1-\cos^2\beta-\cos^2\gamma} & \cos\beta & \cos\gamma \\ 0 & \cos\alpha & \sin\alpha \\ 0 & 0 & 1 \end{bmatrix} \tag{2}$$

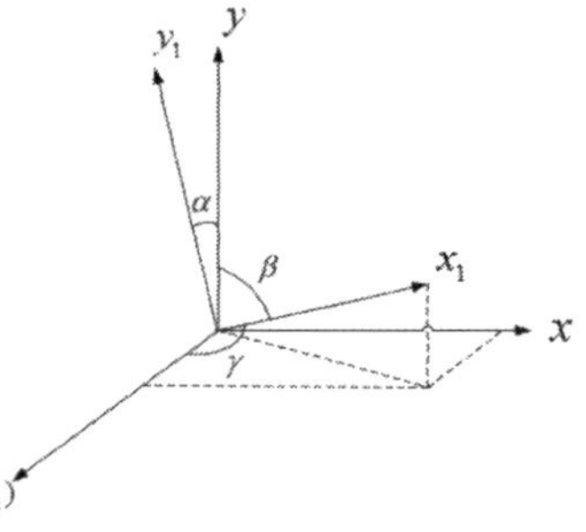

Figure 1. The non-orthogonal error model of geomagnetic sensor

Installation error: Installation error is caused by the misalignment between actual installation position and ideal installation position. Installation error is shown as figure 2 and its mathematic model can be expressed as:

$$K_n = \begin{bmatrix} 1 & 0 & 0 \\ 0 & \cos\varepsilon_3 & \sin\varepsilon_3 \\ 0 & -\sin\varepsilon_3 & \cos\varepsilon_3 \end{bmatrix} \begin{bmatrix} \cos\varepsilon_2 & \sin\varepsilon_2 & 0 \\ -\sin\varepsilon_2 & \cos\varepsilon_2 & 0 \\ 0 & 0 & 1 \end{bmatrix} \begin{bmatrix} \cos\varepsilon_1 & 0 & -\sin\varepsilon_1 \\ 0 & 1 & 0 \\ \sin\varepsilon_1 & 0 & \cos\varepsilon_1 \end{bmatrix}$$

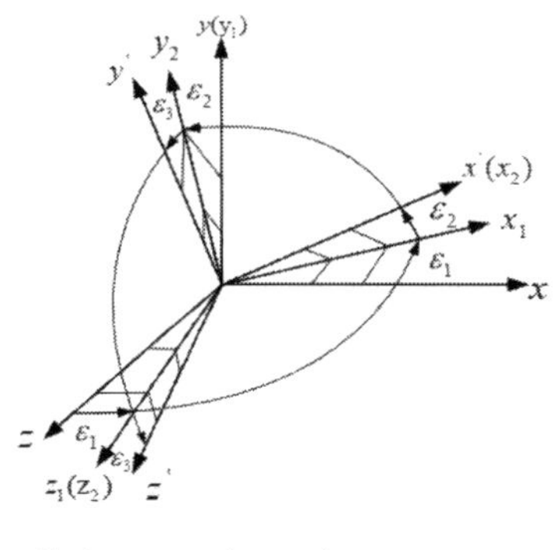

Figure 2. Installation non-orthogonal error

Environment error means when using magnetic sensors in iron-magnetism environment, the error is caused by these magnetic materials. It can also be classified as hard magnetic error and soft magnetic error. Just as their name imply, the hard magnetic error is caused by hard magnetic materials' magnetic field, or interfered by electronic equipment. Hard magnetic error can be expressed as $b_h = \begin{bmatrix} b_{hx} & b_{hy} & b_{hz} \end{bmatrix}^T$. The soft magnetic error is caused by soft magnetic materials' induced magnetic field when they setting in geomagnetic field. Soft magnetic error can be expressed as

$$K_m = \begin{bmatrix} a_{11} & a_{12} & a_{13} \\ a_{21} & a_{22} & a_{23} \\ a_{31} & a_{32} & a_{33} \end{bmatrix} \tag{3}$$

As described previously, if we assume the original magnetic field is B, and the measurement magnetic field with various error is B_m, then we can deduce system error mathematic model as follow:

$$B_m = K_n K_p K_s (K_m B + b_h) + b_o + e \tag{4}$$

It can also be simplified as:

$$B_m = KB + b + e \tag{5}$$

Where $K = K_n K_p K_s K_m B$ is total transformation matrix, $b = K_n K_p K_s b_h$ is total offset vector. K is invertible matrix, hence the sensor error mathematic model can be deduced from equation (5) as follow:

$$B = K^{-1}(B_m - b - e) \tag{6}$$

Where random error can be compensated by hardware, hence e can be ignored, and sensor error mathematic model is:

$$B = K^{-1}(B_m - b) \tag{7}$$

3 The Principle of ellipsoid fitting and error compensation

If the position of geomagnetic sensor is fixed, the measurement data of magnetic sensor should be one constant value theoretically. That is, no matter what direction the vehicle heading to, the measurement data is same. Hence the magnetic data can be modeled as a standard sphere in space. However, because of the existence of various error, the standard sphere is distortion to ellipsoid. The principle of ellipsoid fitting is to using ellipsoid fitting to actual testing data, and estimating the central and shape parameters of ellipsoid. Finally, the parameters of sensors error model can be calculated.

Ellipsoid is one kind of quadric surface, the equation of ellipsoid is:

$$\begin{aligned} F(\xi,v) &= \xi^T v \\ &= ax^2+by^2+cz^2 \\ &\quad +2fxy+2gxz+2hyz \\ &\quad +2px+2qy+2rz+d \\ &= 0 \end{aligned} \tag{8}$$

Where $\xi=[a,b,c,f,g,h,p,q,r,d]^T$ is shape parameters of quadric surface, $v=[x^2,y^2,z^2,2xy,2xz,2yz,x,y,z,1]^T$ is measurement data vector. Now if measurement is performed for n times, we can define $D=[v_1 \quad v_2 \quad \dots \quad v_n]^T$ as measurement batch measurement data. Secondly, using fitting ellipsoid to minimize the sum of squared distances from measurement data to fitting ellipsoid. This process can be expressed as follow:

$$\min\left|F(\xi,v)\right|^2 = \min(\xi^T D^T D\xi) \tag{9}$$

Now we build function

$I=a+b+c, J=ab+bc+ac-f^2-g^2-h^2$, then the least square solution under the constraint equation $\alpha J - I^2 > 0$ is $\min(\xi^T D^T D\xi)$. When $\alpha = 4$, the shape parameters of fitting ellipsoid is obtained.

The error model of geomagnetic sensor can be rewritten as follow:

$$B^2-(B_m-b)^T(K^{-1})^T K^{-1}(B_m-b)=0 \tag{10}$$

Let $A=(K^{-1})^T K^{-1}$, the equation (7) can be written as:

$$(B_m-b)^T A(B_m-b)=B^2 \tag{11}$$

Besides, equation (8) can be also written as:

$$(B_m-B_0)^T R(B_m-B_0)=1 \tag{12}$$

Expand the above equation, we can get next equation:

$$B_m^T R B_m - 2B_0^T R B_m + B_0^T B_0 = 1 \tag{13}$$

Where $B_m=[x \quad y \quad z]^T$, $B_0=-R^{-1}[p \quad q \quad r]^T$ is the coordinate of optimal fitting ellipsoid's center point. The shape parameters matrix of fitting ellipsoid notes as R

$$R=\begin{bmatrix} a & f & g \\ f & b & h \\ g & h & c \end{bmatrix} \tag{14}$$

From all above equations, we can get

$$b=B_0, A=RB^2$$

The singular value decomposition (SVD) of A is

$$A=U\Sigma U^T$$

Now set

$$Q=U\sqrt{\Sigma}U^T$$

then $A=U\Sigma U^T=U\sqrt{\Sigma}U^T U\sqrt{\Sigma}U^T=Q^T Q$.

Besides, $A=(K^{-1})^T K^{-1}$, hence $B=U\sqrt{\Sigma}U^T(B_m-b)$.

4 Experiments and Results

To verify the precision of Ellipsoid fitting method, a rotation experiment has been performed. Just like figure 3. Three axis geomagnetic sensor system is put on a rotary table. The frame of rotation system is constructed by aluminum material, which can decrease magnetic interference. The geomagnetic sensor system is constructed by three axis TMR linear sensor—TMR2305 and three axis magneto-resistive sensor HMC1043.

Figure 3. The rotary table used in experiment.

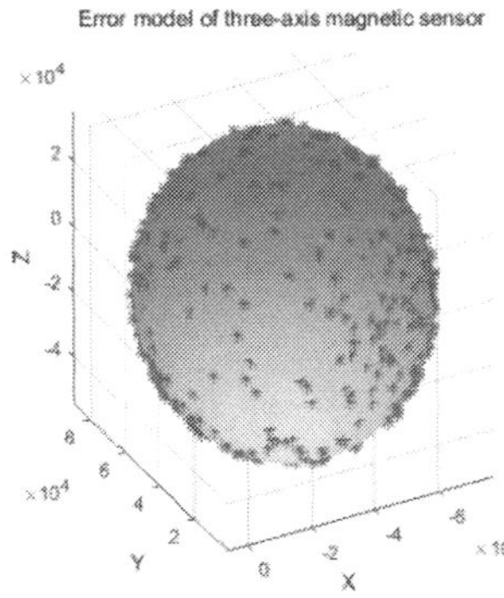

Figure 4. Measurement results for HMC1043.

Figure 4. shows the ellipsoid fitting result for HMC1043, as what we discussed before, the measurement results is distortion to ellipsoid. Figure 5 shows the residuals error of ellipsoid fitting.

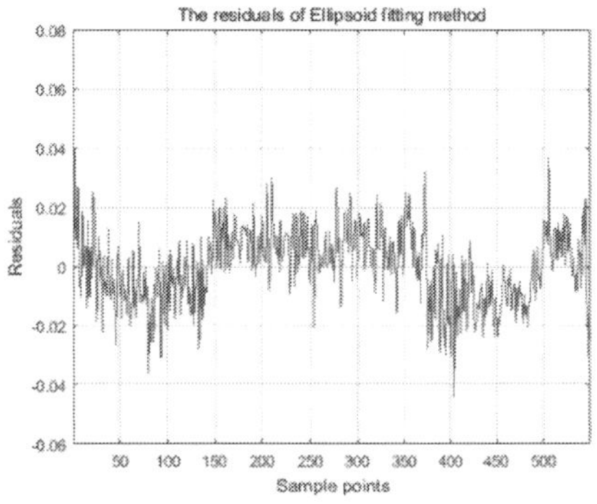

Figure 5. The residuals error of ellipsoid fitting.

Based on these data, the shape parameters of fitting ellipsoid and sensor error calibration model can be expressed as follows:

$$R=\begin{bmatrix} a & f & g \\ f & b & h \\ g & h & c \end{bmatrix}$$
$$=10^{-9}\begin{bmatrix} 0.5897 & 0.0050 & 0.0059 \\ 0.0050 & 0.5103 & -0.0133 \\ 0.0059 & -0.0133 & 0.4934 \end{bmatrix}$$

$$K^{-1}=\begin{bmatrix} 1.3220 & 0.0059 & 0.0069 \\ 0.0059 & 1.2287 & -0.0162 \\ 0.0069 & -0.0162 & 1.2083 \end{bmatrix}$$

$$b=[-35318.1397 \quad 45038.2748 \quad -9220.9663]^T$$

Figure 6 shows the calibration results for HMC1043 magnetic sensor.

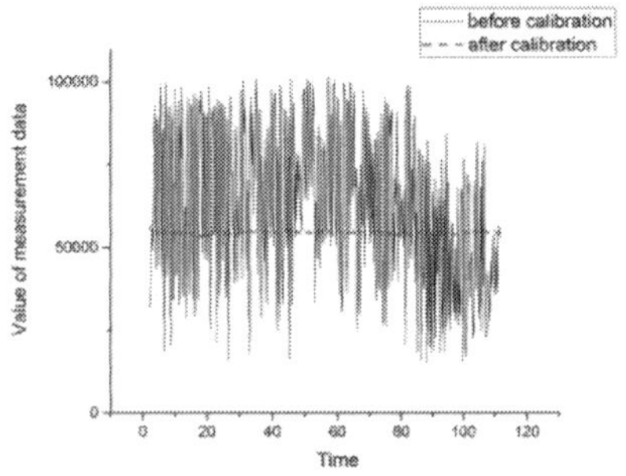

Figure 6. The measurement data and calibration data.

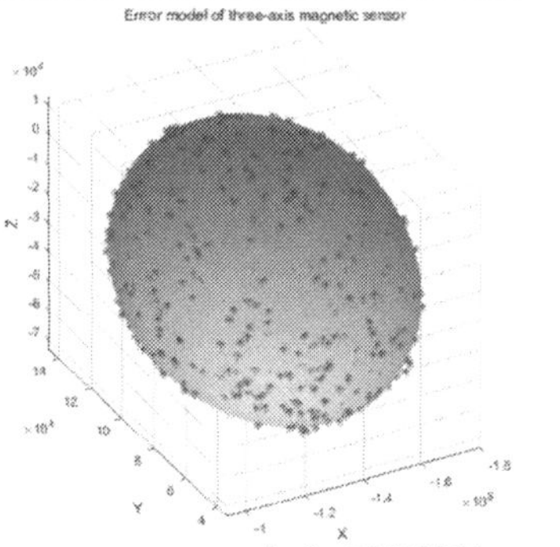

Figure 7. The measurement results for TMR2305

Similarly, figure 7 shows the measurement results for TMR2305 magnetic sensor.

Figure 8 shows the residuals of ellipsoid fitting. Based on these testing data, the shape parameters of fitting ellipsoid and sensor error calibration model can also be calculated. Finally, figure 9 shows the measurement data after calibration.

$$R=\begin{bmatrix} a & f & g \\ f & b & h \\ g & h & c \end{bmatrix}$$
$$=10^{-9}\begin{bmatrix} 0.5353 & -0.0147 & -0.0189 \\ --0.0147 & 0.3320 & -0.0219 \\ 0.0189 & -0.0219 & 0.5329 \end{bmatrix}$$

$$K^{-1}=\begin{bmatrix} 1.2583 & -0.0196 & -0.0225 \\ -0.0196 & 0.9915 & -0.0290 \\ -0.0225 & -0.0290 & 1.2553 \end{bmatrix}$$

$$b=[-136987.0139 \quad 90713.6502 \quad -29807.8842]^T$$

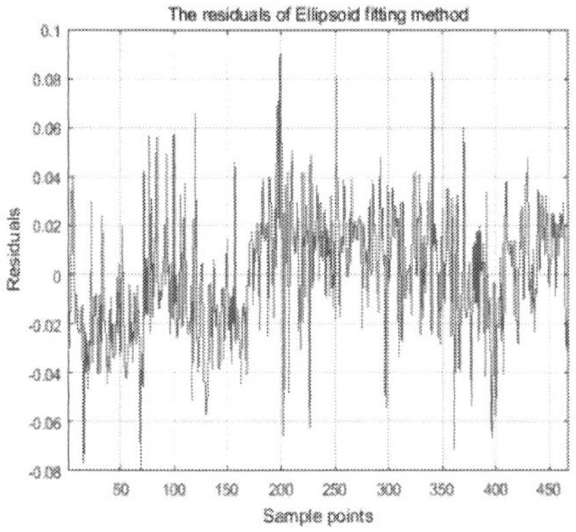

Figure 8. The residuals of ellipsoid fitting.

It can be seen that the transfer matrix of these two kinds of geomagnetic sensor is similarity, while the bias vector of TMR2305 is bigger than HMC1043. When performing least squares based ellipsoid fitting, the measurement data of all two kinds of sensors have been compensated effectively.

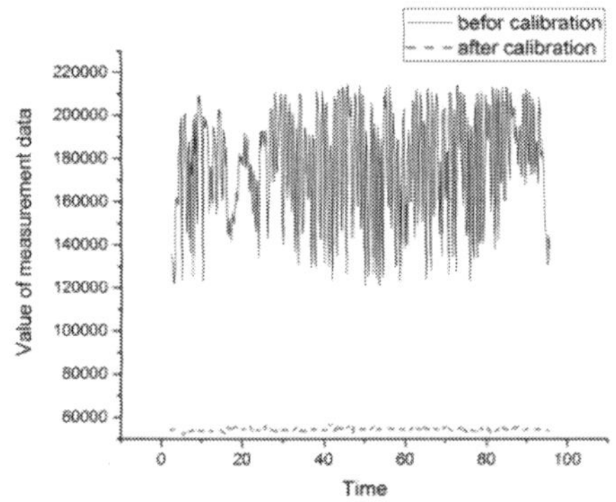

Figure 9. The measurement data and calibration data.

5 Conclusion

This paper summarizes the common methods of geomagnetic sensors' calibration and compensation methods, meanwhile compared their advantages and disadvantages. Besides, this paper presents a novel least squares based ellipsoid fitting calibration and compensation method. The error types and error model have been analysed. Finally, real test experiments have been performed for common sensor HMC1043 and TRM2305 respectively, and the results prove that this method is more accuracy and effective. Of course, because of the influence of various factors in the actual environment, magnetic field compared with an ideal value exists errors. The aim of our further work is to shield the existing magnetic field and to build a certain magnetic field.

Acknowledgment

This paper is supported by National Natural Science Foundation of China (61473039) and the key laboratory of dynamics and control of flight vehicle, Ministry of education, school of aerospace engineering, Beijing Institute of Technology.

References

1. Goldenberg F. Geomagnetic navigation beyond the magnetic compass. *IEEE ION Position, Location and Navigation Symposium*. Coronado, CA. 2006: 684-694.
2. Li, X. P., Fan, J., Ding, J., & Qian, X. B. (2006). Multi-core orthogonal fluxgate sensor. *Journal of Magnetism & Magnetic Materials*, 300(1), e98-e103.
3. Gebreegziabher, D. (2007). Magnetometer autocalibration leveraging measurement locus constraints. *Journal of Aircraft, 44*(4), 1361-1368.
4. Yang, X., & Huang, S. (2004). Magnetic deviation analysis and verification for magnetic compass. *Journal of University of Electronic Science & Technology of China, 33*(5), 547-550.
5. Soken, H. E., & Hajiyev, C. (2011). In flight magnetometer calibration via unscented Kalman filter. *International Conference on Recent Advances in Space Technologies* (pp.885-890). IEEE.
6. Huang, Y., & Hao, Y. (2012). Error correction of magnetic field component gradiometer based on flann and least-squares. *Chinese Journal of Scientific Instrument, 33*(4), 911-917.
7. Wang, M., Wu, Y., Yu, B., & Automation, S. O. (2015). Research on tri-axial magnetic sensor calibration and measurement error compensation based on iterative algorithm. *Ordnance Industry Automation*.
8. Zhang, Q., Pan, M., Chen, D., & Pang, H. (2012). Scalar calibration of tri-axial magnetometer with linearized parameter model. *Chinese Journal of Sensors & Actuators, 25*(2), 215-219.
9. Huang, X. G., & Wang, J. (2011). Error analysis and compensation methods for geomagnetic signal detection system. *Acta Armamentarii,32*(1), 33-36.
10. Yang, X., & Huang, S. (2004). Magnetic deviation analysis and verification for magnetic compass. *Journal of University of Electronic Science & Technology of China, 33*(5), 547-550.

Calculation for Hard Measured Loads on the Travelling Mechanism of Bridge Inspection Vehicle

Meng Han[1], Yongneng Lu[2], Hongfeng Ma[2]

[1]*State Key Laboratory of Mechanics and Control of Mechanical Structures, Nanjing University of Aeronautics and Astronautics, NanJing, 210016, China*
[2]*State Key Laboratory of High-end Construction Machinery Intelligent Manufacturing, Research Institute of Xu Zhou Construction Machinery Group, XuZhou,221001, China*

Abstract. In allusion to the recently problem of some loads to be calculated and hard measured of travel mechanism of bridge inspection vehicle, based on the principle of superposition in the elastic mechanics, the FEA simulation and test methods have been used to research the hard-measured loads on the travel mechanism of bridge inspection vehicle. The calculation case of the driving force in the driving case and the side load in the turning case of traveling mechanism of bridge inspection vehicle have been solved with this method, the results are used to be applied reference for the property optimization and model selection of mechanical bearing inside of travel mechanism of the bridge inspection vehicle. The research method is of well-reference value for the calculation of the unknown loads of whole machine and parts of construction machinery.

1 Introduction

In recent years, with the fast construction of bridge and the development of the urban viaduct construction in the world [1], the bridge inspection trade is of the well development potentiality, and the bridge inspection vehicle will be of the well market prospect, but it will be of the hard market competition at the same time [2], so it is very important to improve the performance and technology research of the bridge inspection vehicle.

Figure 1. The bridge inspection vehicle.

The travel mechanism is the key bearing component of bridge inspection vehicle [3]. When the bridge is inspected, the travel mechanism not only is used to kept the whole machine stable, but also it takes the important responsibility to keep travelling. So whether the strength calculation of parts and components and the driving force were correct, not only it would influence its performance, but also it would impact the safety of the inspectors.

For the strength calculation of the travel mechanism, the calculation method has been very professional, it is widely used to calculate the stress of parts and components with FEA simulation, and based on the results, the structure will be improved, in order to make the mechanism be satisfied with the strength and stiffness and standards.

Recently, the research methods of the driving force, turning force and friction which is very peculiar, but it is very unrealized to measure them [4-7], it is widely used to calculate and design with the analog method, but because of which results are very conservative, it is easily caused by wasting the resource and increasing the cost.

The test and FEA simulation are combined with together in this paper, the driving force and turning force of the driving mechanism has been calculated, and the results would be used to be applied reference for the optimization of driving mechanism.

2 Method of calculation

The model of hydraulic reducer of bridge inspection vehicle used to be selected with experience after generally calculated, when the bridge is inspected, the driving force is hard to be measured which is exported by hydraulic reducer of driving mechanism, so it is not easy to be judged whether the selected model is correct or not，at the same time, when the bridge is inspected on the curve bridge, it is unrealized to measure the turning force and the research of which is very peculiar.

Recently, the research of driving force and turning force of driving mechanism of bridge inspection vehicle is very peculiar. In allusion to the characteristic of driving

https://doi.org/10.1051/matecconf/201815307003

mechanism of bridge inspection vehicle, the calculation method of unknown force has been summarized in this paper: based on the principle of superposition in the elastic mechanics, through the straightly driving test and the turning test, the stress of driving mechanism had been collected, and all of the FEA simulation model of driving mechanism in all of the case, and the force was loaded in order to be calculated the stress of driving mechanism parts, then the value of loaded force was adjusted to reduce the error between the test stress and the simulation stress, when the error would be controlled in the scope of 10% of the design , the loaded force would be regarded as the force which hardly to be measured.

2.1 Principle of superposition [8-12]

Principle of superposition means that the stress、strain and displacement of the elastic structure loaded complex force could be added through the stress、strain and displacement of the elastic structure loaded simple force under the precondition of infinitesimal deformation and linear strain.

2.2 Driving case force solution of bridge inspection vehicle

When the bridge inspection vehicle is worked, the driving mechanism supports on the ground, which bears the supporting force F_y from the ground; when the driving mechanism is worked in the straightly driving case ,which bears the supporting force F_y from ground, the driving force F_x from the hydraulic reducer and the friction force f from ground; when the driving mechanism is worked in the turning case, which bears the centripetal force F_R which caused by the factor such as turning centrifugal case and the relative displacement(Fig.2).

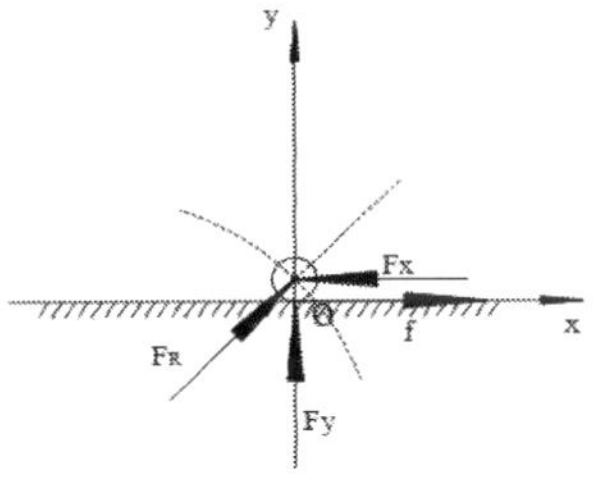

Figure 2. The loads of bridge inspection vehicle.

The parts of driving mechanism is in the elastic transformation scope, according to the principle of superposition in the elastic mechanics, one stress state is $\vec{\sigma}_{CASE1} = \vec{\sigma}_y$ in the static supporting case, one stress state is $\vec{\sigma}_{CASE2} = \vec{\sigma}_y + \vec{\sigma}_{F_x}$ in the straightly driving case and the other stress state is $\vec{\sigma}_{CASE3} = \vec{\sigma}_y + \vec{\sigma}_{F_x} + \vec{\sigma}_{F_R}$ in the turning case. So the driving force F_x and centripetal force F_R of outrigger can be calculated according to the three different cases.

3 Calculation case

3.1 Force calculation in the straightly driving case

3.1.1 Test in the straightly driving case

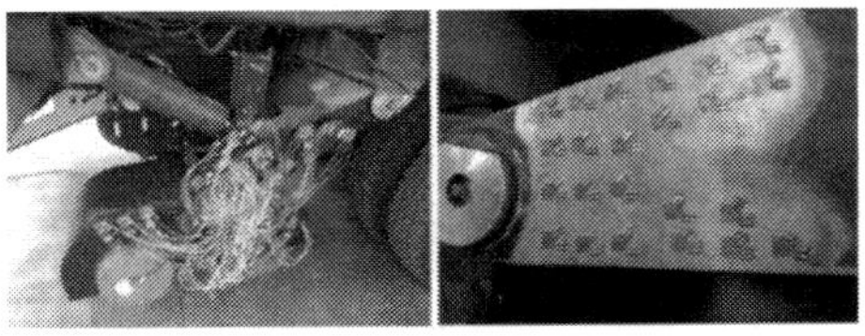

Figure 3. The strain sensors arrangement of travelling mechanism of bridge inspection vehicle.

The strain sensors are pasted on the parts of driving mechanism of bridge inspection vehicle, in order to monitor the stress of driving mechanism in the straightly driving case, when the outrigger is out of the ground, the values of sensors are cleared to zero, then it is supported and worked, the sensors are controlled to collect the stress.

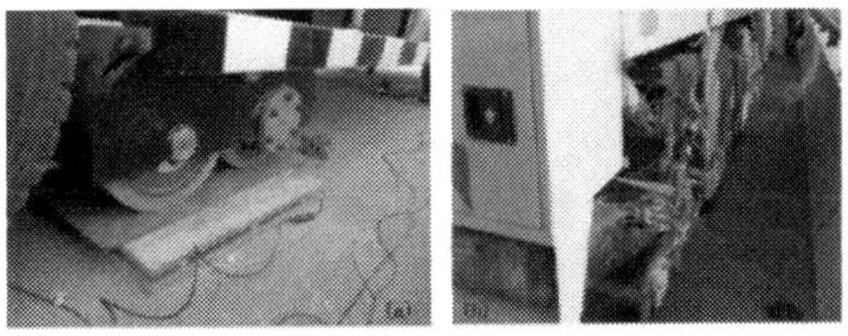

Figure 4. (a) load collecting test in the supporting case (b) strain collecting test in the straightly driving case

3.1.2 Simulation in supporting case and in straightly driving case

The FEA simulation model of driving mechanism has been established, and the driving force has been loaded, then the solved results are contacted with the test results, the loaded driving force would be adjusted according to the error between the solved stress and the test stress, it would be controlled in the scope of 10%, then the loaded driving force would be thought as the force output from the vehicle. After contacted, the error is in the 10% scope of the results, and the driving force is 30kN, which being considered as the practical driving force value.

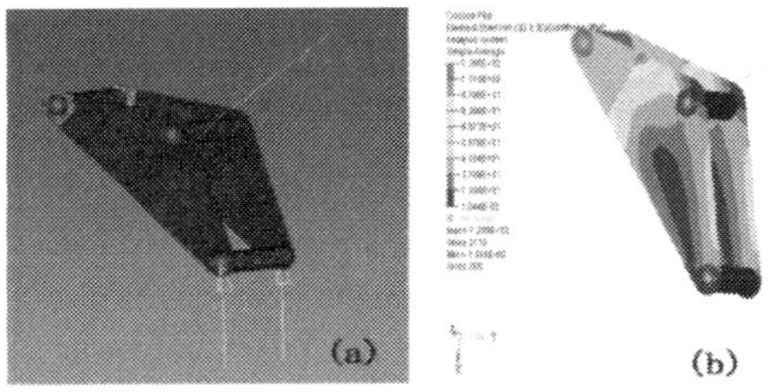

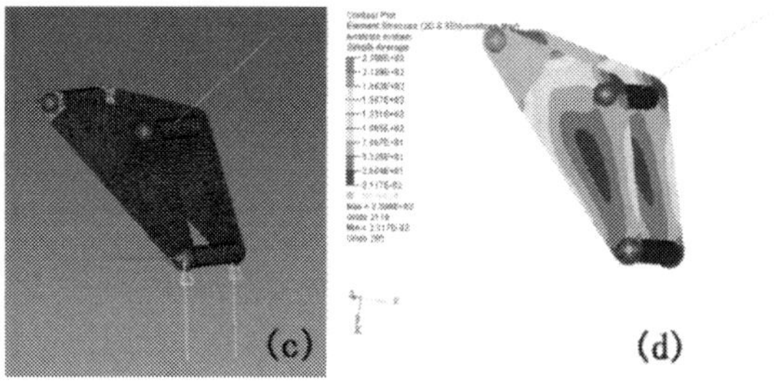

Figure 5. (a) simulation model in the supporting case (b) simulation results in the supporting case (c) simulation model in the straightly driving case (d) simulation results in the straightly driving case

3.2 Force calculation in the turning case

When the bridge inspection vehicle is worked on the turning case, it causes the side force and changes the bearing force situation, the selected bearing model is decided from side force FR 、 driving force FX and vertical force FY together. The vertical force FY could be collected by being assigned sensor under the driving mechanism, the driving force FX has been solved with the above-mentioned method. According to the above-mentioned method, the side force FR would be solved in this paper, it can supply some data for the model selected of driving mechanism later.

3.2.1 Test of turning force

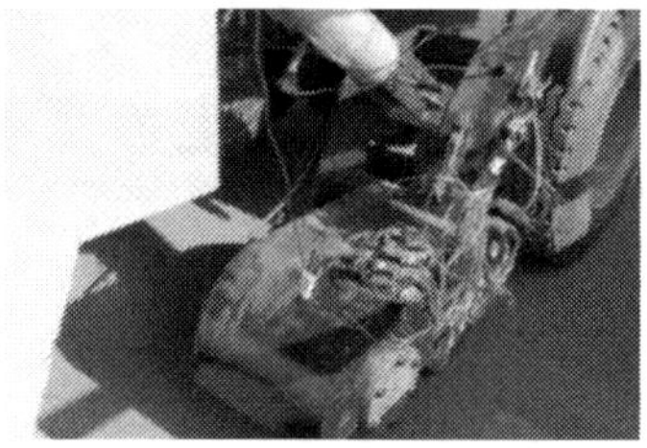

Figure 6. The strain test in turning case.

According to the above-mentioned method, in order to calculate the turning force of driving force, it is necessary to establish the FEA simulation model of turning case, the side force should be load, according to the actual stress, the side force should be adopted to keep the error between the test stress and the solved stress being in the scope of 10%, then the turning force satisfy requested is the force needed to solved.

3.2.2 Simulation in turning case

The simulation model of side force of driving mechanism has been established, the side force has been loaded, the parts stress of driving mechanism has been calculated, and the result has been compared with test stress, then the loaded force value has been adjusted to keep the error between them being in the scope of 10%, then the side force should be solved is 21.7kN, it is considered as the actual side force.

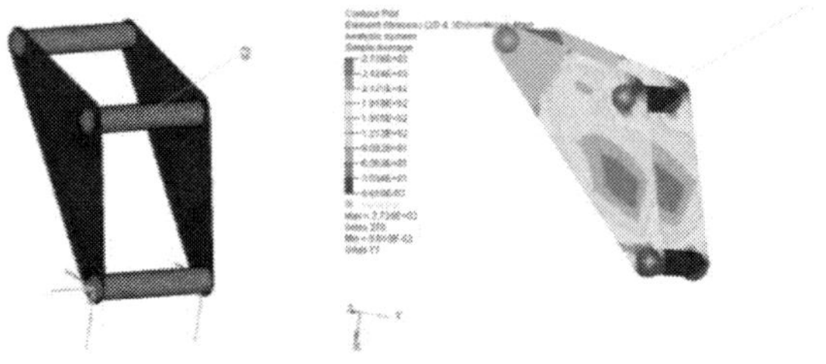

Figure 7. The simulation model in turning case.

The test results of driving mechanism in the all case as follows:

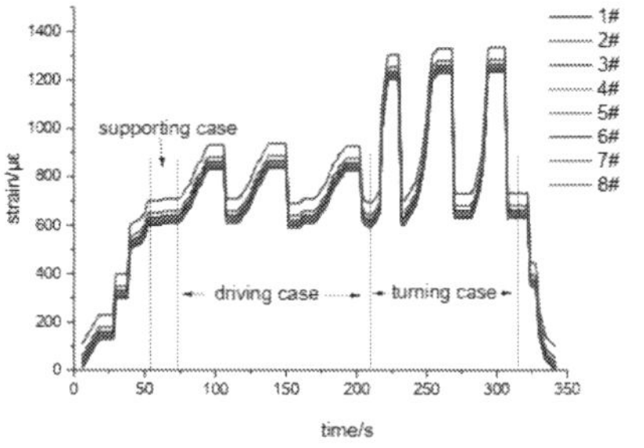

Figure 8. The curves between time and strain of travelling mechanism of bridge inspection vehicle

4 Conclusion

(1)Based on the principle of superposition in the elastic mechanics, it forms a force solved method by combined the simulation with test, and the driving force and side force has been solved of driving mechanism of bridge inspection vehicle, the results are used to be applied reference for the property optimization and model selection of mechanical bearing inside of travel mechanism of the bridge inspection vehicle;

(2)The calculation method combined the FEA simulation with test can be used to solve other hard measured loads of the construction machine , and it can be widely used in the force calculation of the other parts , so it is of very important value in the project.

Acknowledgement

This work is sponsored by the National Natural Science Foundation of China (11272147,10772078), Chinese Aviation Science Foundation(2013ZF52074), Foundation of State Key Laboratory of Mechanical Structural Mechanics and Control(0214G02), SKL Open Fund (IZD13001-1353, IZD150021556), Priority Academic Program Development of Jiangsu Higher Education Institutions, Foundation of National Scholarship of China (201706830073).

References

1. G.H. Tan, G.H. Brameld, D.P. Thambiratnam, Development of an analytical model for treating vehicle–bridge interaction, Eng. Struct. 1998(20):54–61.
2. Y.B.Yang, C.H. Chang, J.D.Yau, An element for analyzing vehicle bridge systems considering vehicle's pitching effect, Int. J. Numer. Methods Eng. 1999(46):1031–1047.
3. Reza Akhavian,Amir H. Behzadan. Construction equipment activity recognition for simulation input modeling using mobile sensors and machine learning classifiers, Advanced Engineering Informatics,2015,29(4):867-877.
4. Jacek Karliński,Eugeniusz Rusiński,Tadeusz Smolnicki. Protective structures for construction and mining machine operators, Automation in Construction,2008,17(3):232-244.
5. Csaba Raduly-Baka, Timo Knuutila, Mika Johnsson, Olli S. Nevalainen. Construction of component tapes for radial placement machines, Computers & Operations Research,2010,37(8):1488-1499.
6. T.D. Tran, H. Schlegel, R. Neugebauer. Application of Modal Analysis for Commissioning of Drives on Machine Tools, Procedia CIRP,2016,40:115-120.
7. Je-Keun Oh, Giho Jang, Semin Oh, et.al. Bridge inspection robot system with machine vision, Automation in Construction,2009,18(7):929-941.
8. Yi-Chu Chen, Chi-En Yang, Shih-Chung Kang. A lightweight bridge inspection system using a dual-cable suspension mechanism.Automation in Construction,2014,46:52-63.
9. N.S.S. Mar, P.K.D.V. Yarlagadda, C. Fookes. Design and development of automatic visual inspection system for PCB manufacturing. Robotics and Computer-Integrated Manufacturing, 2011,27(5):949-962.
10. Jung-Fa Hsieh, Psang Dain Lin. Application of homogenous transformation matrix to measurement of cam profiles on coordinate measuring machines,nternational Journal of Machine Tools and Manufacture,2007,47(10):1593-1606.
11. O.A. Zambrano, J.J. Coronado, S.A. Rodriguez. Failure analysis of a bridge crane shaft.Case Studies in Engineering Failure Analysis,2014,2(1):25-32.
12. Gunter Mennig, Klaus Stoeckhert. Manufacturing and Machining Methods.Mold-making Handbook,2013,471-606.
13. M. Ogle. Structural Integrity Issues Associated with Bridges,Comprehensive Structural Integrity,2003,1:319-336.

Measurement of Global Solar Radiation data using Raspberry Pi and its estimation using Genetic Algorithm

S.Shanmuga Priya[1] , Arunabh Borkataky[2], Sneha Reddy[3], I.Thirunavukkarasu[4]

[1]Associate Professor,[2,3]Student, [4]Associate Professor (Senior Scale)
[1,2,3]Department of Chemical Engineering/ [4]Department of ICE
Manipal Institute of Technology, Manipal Academy of Higher Education (MAHE)
Manipal.Karnataka. India.

Abstract. The demand for more efficient and environmentally benign, non-conventional sources of energy came into picture due to increasing demands for human comforts. Solar energy is now the ultimate option. In this paper, the instruments used to measure the solar radiation at Innovation Centre, MIT Manipal were connected to a Raspberry Pi to access the data remotely. Genetic Algorithms were formulated, so that the monthly mean global solar radiation in Manipal can be effectively estimated. Meteorological data such as humidity, temperature, wind speed, etc. were used as inputs to train the networks. A successful network was made between the data loggers and the Raspberry Pi. The data collected by the data loggers from the devices are transmitted to the Raspberry Pi which in turn sends the data to an internal server. The Raspberry Pi can be accessed using any SSH client such as PuTTY. The meteorological data was collected for the years 2010-2014 in order to formulate the Artificial Intelligence models. The validity of the formulated models were checked by comparing the measured data with the estimated data using tools such as RMSE, correlation coefficient, etc. The modelling of solar radiation using GA was carried out in GeneXpro tools version 5.0.

1 Introduction

Sources of renewable energy have been vital for humans from the start of civilization. The sun is a working fusion reactor that as of now supplies more energy than humans could perhaps require. Solar power is a standout amongst the most encouraging renewable energy sources for being more predictable than wind energy and less vulnerable against changes in regular climate designs than hydropower.

For the advancement of any solar energy project, site-oriented and long term solar radiation information is required right from asset appraisal to design of the system to evaluation and optimization of its performance and transient forecast of solar radiation for operational achievability. Such information is to a great degree fundamental in various field applications, such as climatology, solar energy technologies, architectural and energy saving building designs, illumination, simulation of solar power plant and agricultural research.

Due to high maintenance and calibration costs of solar measurement instruments, the solar radiation data are constricted in many meteorological stations across the globe. This has led to the development of innumerable models for the estimation of solar radiation.

Manipal, located in south-west India has a hot and humid weather condition during the whole year. The obtained solar radiation data can be used for various solar based applications.

The objective of the work is to integrate the existing solar radiation apparatus (pyranometer, pyrheliometer, sunshine duration sensor and UV meter) with a Raspberry Pi so that the data can be retrieved tenuously. to make the assembly water-proof to withstand the rainfall and humidity in Manipal, to utilise historical meteorological data to formulate different models for monthly mean global solar radiation estimation and to validate the models with the real time data obtained from Innovation Center, MIT Manipal. Genetic Algorithm has been relatively less utilised in India for the estimation of solar radiation; it's most utilisation being in Gulf countries. A prediction of the monthly mean solar radiation in Manipal can be made using the formulated Artificial Intelligence models. Genetic Algorithm is used to solve both constrained and unconstrained optimization problems. The genetic algorithm works on the principle of reproduction system by which the data set is repeatedly modified on the basis of the best fit to evolve into best reproduced offspring or data set. This finally leads to the best outcome of data set producing the optimal solution.

Aybar-Ruiz et. al [1] used grouping genetic algorithm to answer optimal solution of features and extreme learning machine algorithm to carry out the prediction and merged them into a novel scheme which uses input of the system and output of the weather meso-scale model.

Haydar and Yasemin et.al [2] researched on the estimation models of horizontal global solar radiation using the weather data of Turkey. They used vigorous coplot technique which lessens the number of covariates and then utilized the monthly and yearly mean daily horizontal global solar radiation prediction models using genetic algorithm. Hongzing et. al [3] proposed an optimal sizing technique made using genetic algorithm, whereby the configurations of a hybrid solar-wind system using battery banks is optimized. The method proved to give good optimization performance and hence achieved minimum annualized cost of system and the required loss of power supply probability.

Will et. al [4] solved the issue of selecting variables to estimate solar radiation through the use of Niching Genetic Algorithm. The algorithm provides a way to predict the input variables for a desired seasonal variable prediction problem and also approximate a given seasonal variable using databases with the aid of missing data. Qin et. al [5] offered a compared the performances of four different shortwave solar radiation models in different climatic conditions and terrain in China. The Yang's Hybrid model was found the most efficient in terms of performance in predicting solar radiation in China than the other three models namely: hourly solar radiation model, efficient physically based model and neural network model.

Ji Wu et. al [6] developed a novel approach to predict solar radiation time series by a genetic way by scrutinizing multi model framework. This model estimates the optimal fragment for the series of solar radiation of Singapore and hence proves to have greater consistency and accuracy than other models such as TDNN, ARMA and hybrid model. The clustering accuracy of the model is to be improved with an improvised capability of obtaining the relationship of data with certain configurations.

Jianyuan et. al[7] performed a critical review on the existing solar radiation estimation models and compared and analysed the models on estimation type as well as from the viewpoint of time scale. By this study they concluded that the artificial neural networks and the sunshine duration fraction models have analogous performances.

Nwokolo et.al [8] made a comprehensive and quantitative review on Africa's solar radiation estimating empirical models and concluded that due to lack of meteorological stations in sites of Africa, models using longitude, latitude, maximum sunshine duration, solar declination, extra-terrestrial radiation and day number of the year inputs for exact measurement should be established. Cano et. al [9] researched on a statistical procedure to evaluate the global solar radiation at ground level with the help of the data extracted from meteorological satellites that provide sufficient ground resolutions and extensive coverage as well. Great precision is obtained for two different time periods using this model for the data obtained by a ground level pyranometers.

Yongbin and Ronghuai et.al[10] created a low cost and portable weather monitoring system for technology rich classroom which is based on raspberry pi. This system allows in the collection, storage and transmission of data. Prediction of global solar radiation should be made as a target for this equipment using soft computing models.

2 Methodology

2.1 Solar radiation monitoring equipment

Solar radiation equipment's record solar radiation data which can utilized further for analysis of solar based applications. Data collection is done using solar radiation instruments such as pyranometer, pyrheliometer and sunshine duration as shown in Figure 1 and Figure 2..

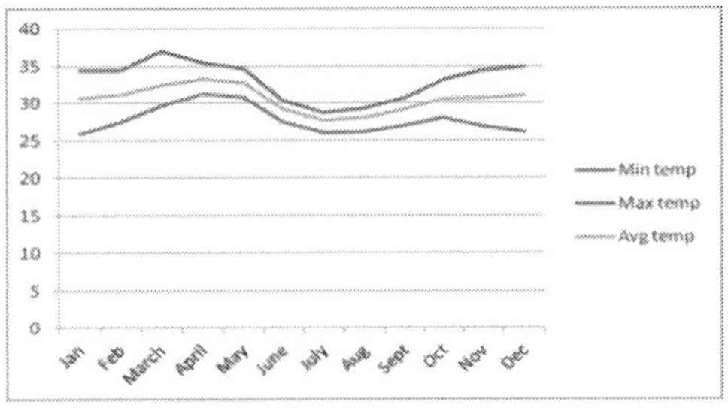

Figure 1. Average Monthly temperature history in Manipal from 2010-2014.

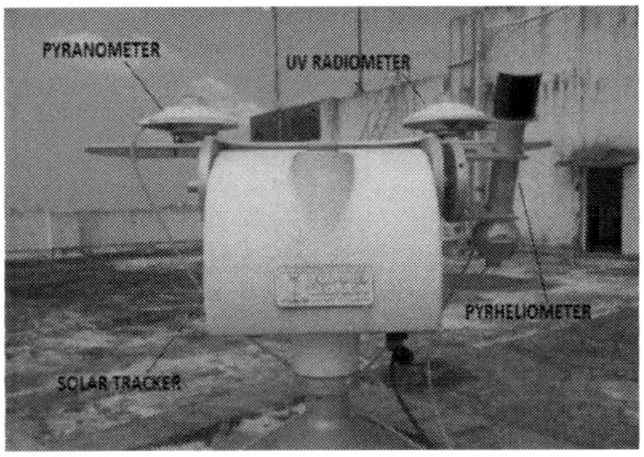

Figure 2. Sun tracker with Pyranometer, Pyrheliometer and Global UV Radiometer installed at Innovation Centre, Manipal.

2.2 Raspberry Pi

The Raspberry Pi as shown in Figure 3, costs only 35$ and about Rs 3000 in India.

https://doi.org/10.1051/matecconf/201815307004

Figure 3. A Raspberry Pi board

The Raspberry Pi is a series of credit card–sized single-board computers developed in England, United Kingdom by the Raspberry Pi Foundation with the intent to promote the teaching of basic computer science in schools and developing countries. The original Raspberry Pi and Raspberry Pi 2 are manufactured in several board configurations through licensed manufacturing agreements with Newark element14. The hardware is the same across all manufacturers. In February 2016, the Raspberry Pi Foundation announced that they had sold eight million devices, making it the best-selling UK personal computer, ahead of the Amstrad PCW.

2.2.1Modifications in the system

- To protect the unit from extreme weather conditions we have replaced with external grid powered voltage adapters.
- The box is placed some centimetres above the ground for better air circulation (cooling) and to prevent direct contact with water in case of heavy rain.
- Silica gel has been kept in the box to prevent humidity.

2.2.2 Working of the Raspberry Pi system

- The data collected by the data loggers are written to RS232 and also to the SD cards inside the loggers(Set interval 1 min).
- Two RS232 to USB adapters are used to connect to the Raspberry pi unit.
- It collects the data and sends it to an internal server. The data can be found at local network.
- To access the data, one must be connected to the institute LAN.
- The Raspberry Pi can be remotely accessed using an SSH client such as Putty.

Figure 4, Figure 5 shows the integration of solar radiation monitoring equipment with Raspberry Pi.

Figure 6 shows the flowchart for the connections of data loggers to the raspberry Pi.

Figure 4. Plastic container integrated with Raspberry Pi and data loggers

Figure 5. Connections inside the box

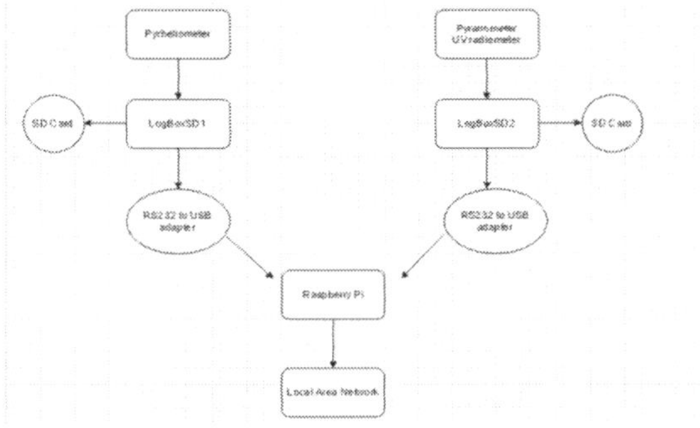

Figure 6. Flowchart deplicting the connections of data loggers to the Raspberry Pi

Table 1. Monthly mean variation in input parameters (for a period 2010-2014)

Month	Humidity	Wind Speed (m/s)	Pressure (Kpa)	Sunshine Duration (hrs)
Jan	51.73	1.91	100.48	9
Feb	44.46	1.66	100.38	9
March	44.95	1.59	100.34	10
April	59.50	1.76	100.23	11
May	72.15	2.46	100.11	11
June	81.49	3.72	100.03	8
July	85.09	4.15	100.07	9
August	85.23	3.64	100.12	9
September	82.86	3.17	100.18	10
October	78.59	2.10	100.25	10
November	73.51	1.94	100.33	10
December	61.81	2.01	100.36	9

2.3 Genetic Algorithm

The Genetic algorithm is based on Darwin's Theory of "Survival of the Fittest". Individuals are created and eliminated based on a fitness function which is determined from the input output relationship.GA used different set of meteorological constraints for the prediction of solar radiation. Different researchers have used different parameters for the prediction of radiation. Based on the literature survey it is found that prediction accuracy of these models get changed with the geographical and meteorological variables as input parameters as shown in Table 1. To select the relevant input parameters one has to use different combinations of input parameters to evaluate prediction accuracy of these models which requires large computational analysis.

Table 2 Input Parameters for different developed models

Model	Input Parameters
GA-1	Minimum Temperature, Average Temperature, Maximum Temperature, Humidity, Pressure and Wind Speed
GA-2	Average Temperature, Maximum Temperature, Humidity and Wind Speed
GA-3	Average Temperature, Wind Speed and Sunshine Duration
GA-4	Wind Speed and Sunshine Duration

3 GSR estimation using GA models

Four Genetic Algorithm models were formulated by using various combinations of input parameters as mentioned in Table 2. Out of the three models which were created, the one with least statistical errors will be chosen as the best model to estimate the monthly mean global solar radiation in Manipal. The modelling was gone in GeneXproTools version 5.0. It can be used to produce dominant predictive models for Time Series Prediction and Regression Predictions, Classification & Logistic Regression and explore Logic Synthesis. The parameter settings for the modelling process is shown in Table 3. 70% of the statistics was used for training the algorithm and remaining 30% was used for testing and validation.

Table 3. Parameter settings

Type	Parameter	Value
General	Chromosomes	30
	Genes	2
	Head Size	3
	Linking Function	Addition
Complexity	Generation w/o change	2000
	No of tries	3
	Max complexity	5
Genetic Operators	Mutation Rate	0.044
	Inversion rate	0.1
	IS transposition rate	0.1
	RIS transposition rate	0.1
	One-point recombination rate	0.3
	Two-point recombination rate	0.3
	Gene recombination rate	0.1
	Gene transposition rate	0.1
Numeric Constants	Constants per gene	2
	Data type	Integer
	Lower Bound	-10
	Upper Bound	10

3.1 GA model I

For this model minimum temperature, humidity, maximum temperature, average temperature, pressure and wind speed were taken as the input parameters. Figure 7 shows the regression plot of the model. The expression tree of the model is shown in Figure 8.

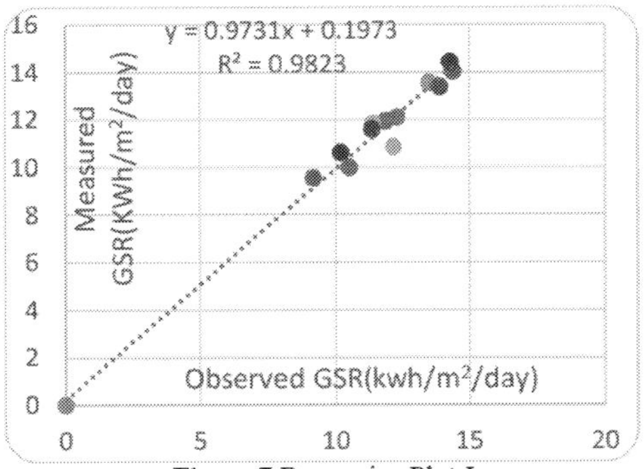

Figure 7.Regression Plot-I

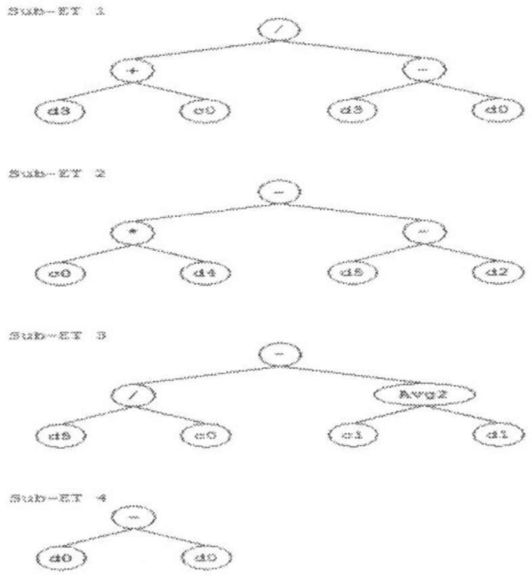

Figure.8 Expression Tree-I

From the expression tree in Figure 8, the model can be explicitly written as:

$$GSR = \frac{d3+c0}{d3-d0} + (c0*d4) - (d5-d2) + \frac{d5}{c0} - \frac{c1+d1}{2}$$

Where,
d0 = Min Temp
d1 = Max Temp
d2 =Avg Temp
d3 = Humidity
d4 = Pressure
d5 = Wind Speed
G1C0 = 9.0
G2C0 = -3.674e-04
G3C0 = 4.03596
G3C1 = 4.39120

 https://doi.org/10.1051/matecconf/201815307004

3.2 GA model II

For this model maximum temperature, average temperature, humidity and wind speed were taken as the input parameters. Figure 9 shows the regression plot of the model. The expression tree of the model is shown in Figure 10.

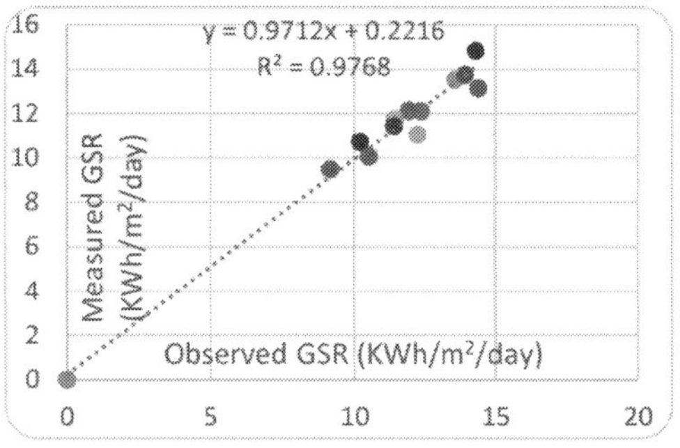

Figure 9. Regression Plot-II

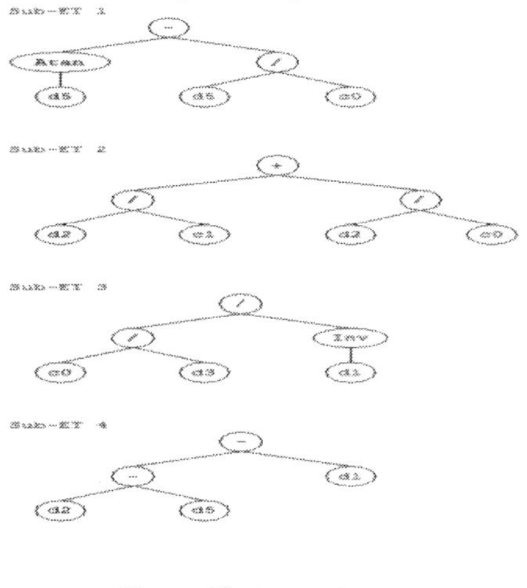

Figure 10. Expression Tree – II

From the expression tree in Figure 10, the model can be explicitly written as:

$$GSR = (a\tan(d5) - \frac{d5}{c0}) + (\frac{d2}{c1} + \frac{d2}{c0}) + \frac{c0*d2}{d3} + (d2 - d5) - d1$$

Where,
d0 = Min Temp
d1 = Max Temp
d2 = Avg Temp
d3 = Humidity
d4 = Pressure
d5 = Wind Speed
G1C0 = 3.5964
G2C1 = 4.3429
G2C0 = 4.1010
G3C0 = 4.1010

3.3 GA model III

For this model average temperature, sunshine duration and wind speed were taken as the input parameters. Figure 11 shows the regression plot of the model. The expression tree of the model is shown in Figure 12.

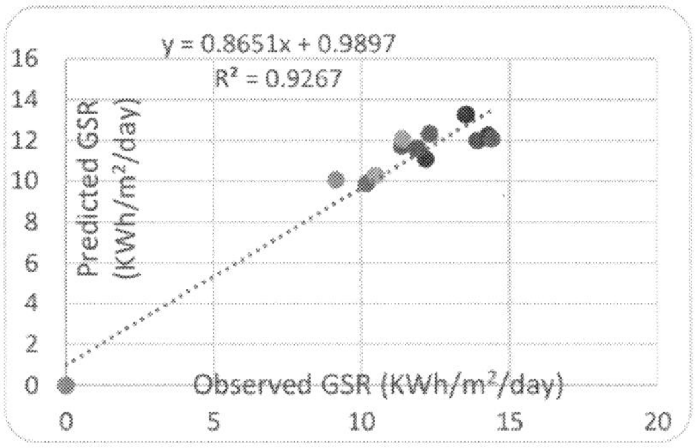

Figure 11. Regression Plot – III

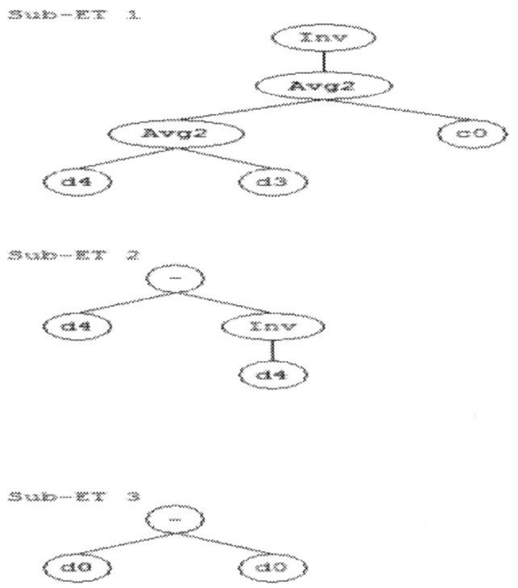

Figure 12. Expression Tree – III

From the expression tree in Figure 12, the model can be explicitly written as:

$$GSR = \frac{4}{d2+d3+2c0} + d2 - \frac{1}{d4}$$

Where,

d0 = Avg Temp
d3 = Wind Speed
d4 = Sunshine duration
G1C0 = -4.870418

3.4 GA model IV

For this model average temperature, sunshine duration and wind speed were taken as the input parameters. Figure 13 shows the regression plot of the model. The expression tree of the model is shown in Figure 14.

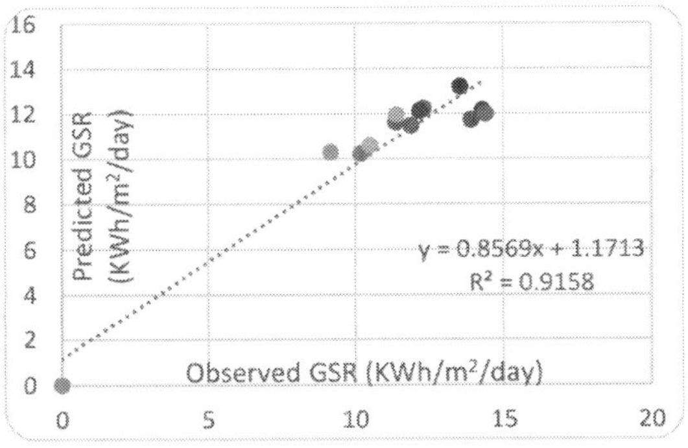

Figure 13. Regression Plot – IV

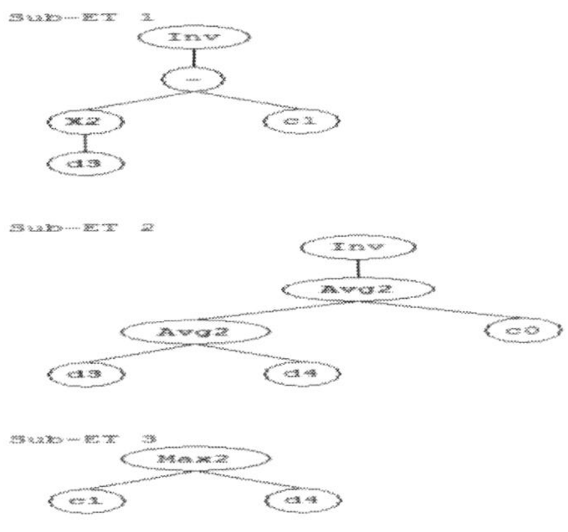

Figure 14. Expression Tree –IV

From the expression tree in Figure 14, the model can be explicitly written as:

$$GSR = \frac{1}{d4^2 - c1} + \frac{4}{d3 + d4 + 2c0} - \max(c1, d5)$$

Where,
d3 = Wind Speed
d4 = Sunshine duration

G1C1 = 9.0

G2C0 = -4.870418

G3C1 = -8.0

Table 4 Validation of the developed models

Model	Parameter	Training	Testing
Model I	MSE	0.297	0.177
	RMSE	0.545	0.42
	MAE	0.384	0.386
	Correl.	0.9	0.99
Model II	MSE	0.237	0.525
	RMSE	0.486	0.724
	MAE	0.353	0.602
	Correl.	0.922	0.932
Model III	MSE	0.35	3.386
	RMSE	0.592	1.84
	Correl.	0.887	0.945
Model IV	MSE	0.219	3.81
	RMSE	0.468	1.95
	Correl.	0.962	0.886

Of the four GA models which were developed using GeneXproTools 5.0, the one which is best suited for Manipal will be determined based on statistical parameters like correlation coefficient, Mean Absolute Error, Root Mean Square Error, etc.From Table 4, it is evident that Model I is the best explicit model to estimate the monthly mean global solar radiation in Manipal. Thus the monthly mean global solar radiation in Manipal can be explicitly expressed as :

GSR = Global Solar Radiation

$$GSR = \frac{H+9}{H-T1} + ((-3.67 * 10^{-4} * P) - (W - T3)) + (\frac{W}{4.036} - \frac{4.391 + T2}{2})$$

on ($KWh/m^2/day$)
H = Humidity
T1 = Minimum Temperature (°C)
T2 = Maximum Temperature (°C)
T3 = Average Temperature (°C)
P = Pressure (KPa)
W = Wind Speed (m/s)

4 Conclusion

The unavailability of solar radiation statistics in many meteorological stations worldwide is mainly due to high cost of maintenance and calibration of the solar radiation measuring devices. The hurdles involved to measure the solar radiation has led to the formulation of various models for its prediction. Four genetic algorithm models were developed in order to predict the monthly mean global solar radiation in Manipal. Solar radiation prediction was made using the developed models and it was found similar to the measured data. The advantage of the GA model is that the output can be explicitly written in terms of the input parameters which is not possible in the ANN model.

Acknowledgement

The authors would like to prompt their gratitude and indebtness to the Innovation Centre, Manipal Institute of Technology, Manipal Academy of Higher Education (MAHE), Manipal by entertaining our necessities and enabling all the provisions utilized for the project.

References

1. Aybar-Ruiz, A., et al. "A novel Grouping Genetic Algorithm–Extreme Learning Machine approach for global solar radiation prediction from numerical weather models inputs." *Solar Energy* 132 (2016): 129-142.
2. Demirhan, Haydar, and Yasemin Kayhan Atilgan. "New horizontal global solar radiation estimation models for Turkey based on robust coplot supported genetic programming technique." *Energy Conversion and Management* 106 (2015): 1013-1023.
3. Yang, Hongxing, et al. "Optimal sizing method for stand-alone hybrid solar–wind system with LPSP technology by using genetic algorithm." *Solar energy*82.4 (2008): 354-367.
4. Will, A., et al. "On the use of niching genetic algorithms for variable selection in solar radiation estimation." *Renewable energy* 50 (2013): 168-176.
5. Qin, Wenmin, et al. "Comparison of deterministic and data-driven models for solar radiation estimation in China." *Renewable and Sustainable Energy Reviews* 81 (2018): 579-594.
6. Wu, Ji, et al. "Prediction of solar radiation with genetic approach combing multi-model framework." *Renewable Energy* 66 (2014): 132-139.
7. Zhang, Jianyuan, et al. "A critical review of the models used to estimate solar radiation." *Renewable and Sustainable Energy Reviews* 70 (2017): 314-329.
8. Chukwujindu, Nwokolo Samuel. "A comprehensive review of empirical models for estimating global solar radiation in Africa." *Renewable and Sustainable Energy Reviews* 78 (2017): 955-995.
9. Cano, Daniel, et al. "A method for the determination of the global solar radiation from meteorological satellite data." *Solar Energy* 37.1 (1986): 31-39.
10. Hu, Yongbin, and Ronghuai Huang. "Development of Weather Monitoring System based on Raspberry Pi for Technology Rich Classroom." *Emerging Issues in Smart Learning*. Springer, Berlin, Heidelberg, 2015. 123-129.

Dynamic Behavior Analysis of Non-Contacting Hydrodynamic Finger Seal Based on Fluid-Solid-Interaction Method

Hua Su[1], and Jiru Wang[2]

[1] *School of Mechanical Engineering, Northwestern Polytechnical University,710072 Xi'an, China*
[2] *School of Electrical Engineering, Xi'an Jiaotong University,710049, Xi'an, China*

Abstract. Finger seal is an advanced compliant seal and can be utilized to separate high (HP) and low pressure (LP) zones in high speed rotating shaft environment. The work to be presented concerns the dynamic behavior of a repetitive section of a two-layer finger seal with high- and padded low-pressure laminates. The dynamic performance of the finger seal are analyzed by the coupled fluid-solid-interaction (FSI) simulations. By using the commercial software ANSYS-CFX, the numerical simulation results of interactions between the gas flow and fingers structural deformation are described when the radial periodic excitation from the shaft applies to the finger seal. And the gas film loading capacity, gas film stiffness and leakage varied with time are put forward in different working conditions. Compared with the dynamic performance analysis results based on equivalent dynamic method, the FSI dynamic analysis shows some different characteristics which are more accordance with actual circumstance. Moreover, it is shown that under low pressure differential and high rotation speed the non-contacting finger seal with advance features both in sealing effectiveness and potential unlimited life span can be obtained by rational structure design. But for the non-contacting finger seal with circumferential convergent pad working in high pressure and low rotating speed conditions, it is difficult to improve the sealing performance by the way of changing the structure parameters of finger seal. It is because the high pressure plays a major role on this sealing situation.

1 Introduction

Finger seal is new type seal presented after labyrinth seal and brush seal which is regarded as a revolutionary new technology in air-to-air sealing for secondary flow control and gas path sealing in gas turbine engines. It is given more attention in past two decades because of its excited performances and relative low manufacture costs. The first generation finger seal is a kind of contacting seal, which is composed of multiple staggered laminates with each having a large number of flexible fingers around the rotor surface[1]. It is found that the finger seal has demonstrated leakage considerably lower than that shown by conventional labyrinth seal, and at a cost much lower than that of the commonly used brush seal. However, when the radial run-out of rotor exist unavoidably, the flexible finger feet may contact with the rotor surface and result in heating and wear especially in high-speed and high-temperature operation conditions. Therefore, non-contacting finger seal has been came up with in recent years. The main significant structural feature of non-contacting finger seal is aerodynamic self-acting lift pads added at the bottom of the low pressure fingers. And certain clearance is reserved between the pads and the rotor surface. Under the differential pressure and rotary motion of rotor, hydrodynamic pressure will generate in the flow film between rotor and pads to lift the finger seal and maintain the non-contacting operation condition. This non-contacting finger seal are expected to be used in high-speed and high-temperature sealing conditions with better performance.

The researches on non-contacting finger seal at present mainly include the following fields: firstly, the different structural design of lifting pad including circumferential converged[2], axial and circumferential double converged[3], step-shaped, herringbone grooved[4] and micro-textured [5]etc. Secondly, analysis on static and dynamic performances of finger seal under different working conditions. Non-contacting finger seal is generally serviced in dynamic condition, so it is crucial to concern about its dynamic performance and working stability. Most current analysis on dynamic performance of finger seal is based on mass-spring-damper equivalent dynamic model. Though it can get the tacking characteristics of seal to the stimulus from rotor and vibration performance of seal, there are still some disparities from practice due to some assumptions applied in the equivalent model. In order to further understand the dynamic performance of the hydrodynamic finger seal and the flow characteristic of the fluid, the fluid-solid coupling analysis method based on the prototype of the sealing system is used in presented work to concern about the dynamic performance of the non-contacting finger seal.

This paper introduces the seal geometry in second part. And the fluid-solid coupling analysis method is discussed in third part. Then the dynamic performances of non-contacting finger seal in different working conditions and with different structures are analyzed in the fourth part, and it ends with the conclusions.

2 Finger seal geometry

The finger seal with circumferential converged pad consists of front plate, spacer, high-pressure (HP) finger, low-pressure (LP) finger with pad and back plate, as shown in Figure 1. The finger stick profile is traced by means of arc. The detailed geometry parameters in Figure 1 are shown in Table 1.

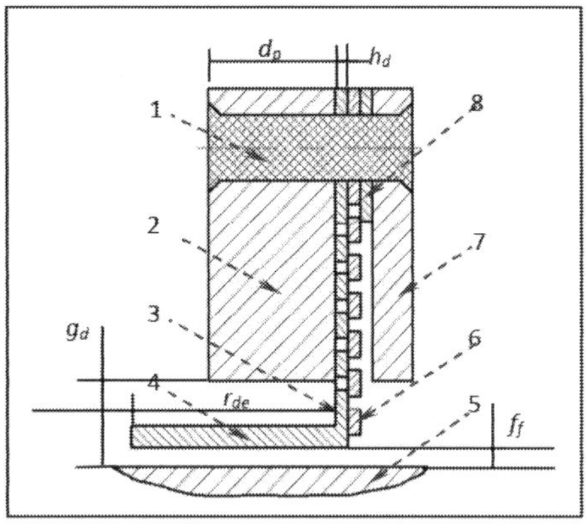

1-vivet 2-back plate 3-LP finger 4-pad 5-rotor
6-HP finger 7-front plate 8-spacer
(a) Axial cross section

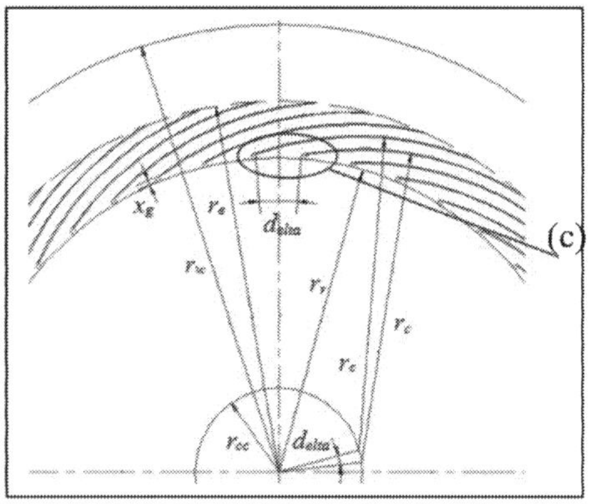

(b) Head face

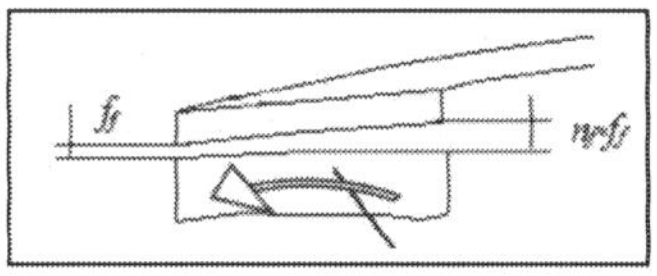

(c) Partial enlarged view

Figure 1. Geometry of finger seal with circumferential converged pad

Table 1. Geometric Values of the Basic Finger Seal

Finger Geometry	Symbol	Value
Outer radius of finger (mm)	r_w	122.65
Radius of finger root (mm)	r_e	116.45
Rotor Radius (mm)	r_r	108
Stick Arc Radius (mm)	r_c	114.6
Radius of Center Circle (mm)	r_{cc}	20
Finger Repeat Angle(°)	d_{elta}	4.444
Radial height of HP foot	x_g	1.22
Minimum radial clearance (mm)	f_f	0.00635
wedge ratio of clearance	n_f	3
Axial width of LP pad (mm)	r_{de}	6.35
Radial gap of the back plate(mm)	g_d	1
Thickness of back plate(mm)	d_p	5
Thickness of finger seal (mm)	h_d	0.762

3 Fluid-solid interaction analysis on dynamic behaviors of finger seal

The dynamic behavior of non-contacting finger seal is analyzed based on the displacement response of finger pad to the rotor stimulus in given working conditions. The rotor produce radial run-out because of its unbalanced mass, which will indirectly impact on pad by gas film and results in radial movement of the fingers. Since the high compliance of the fingers, the force from the gas film move the finger seal and results in a change of flow field profile. These in turn contribute to further change the load distribution and flowing condition of gas film. Thus the dynamic analysis of non-contacting finger seal is based on transient fluid-structure interaction(FSI) analysis of finger seal.

3.1 Finite element model of fluid-solid field

Considering the circumferential convergence finger seal is a cyclic symmetry structure, to simplify the dynamic analysis procedure, a circular unit including a LP finger stick and a two-half HP finger stick are chosen to establish the solid structure model (shown in Figure 2), and the corresponding flow field between the finger pad and rotor is as gas flow field model (shown in figure 3). Here the solid and gas film models are established based on commercial software ANSYS-CFX.

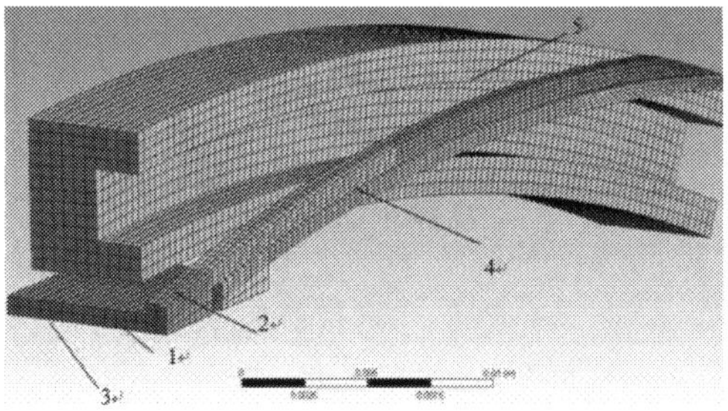

1-half HP finger 2-half HP finge 3-hydrodynamic pad
4-back plate 5-LP finger
Figure 2. Solid model of non-contacting finger seal

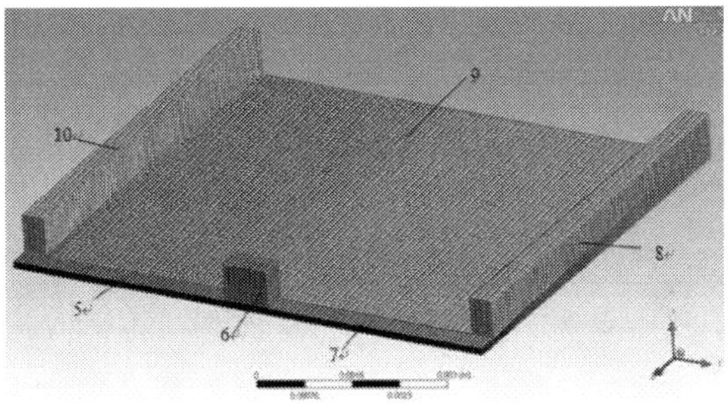

Figure 3. Gas film model of non-contacting finger seal

3.2 Boundary conditions

The mechanical boundary conditions in solid model of circumferential convergence finger seal on dynamic analysis including:

1)The outer faces of HP and LP are fixed as well as the back plate.2)The surface in HP finger face to high pressure is applied to pressure differential.3) The surface under the pad is set to be fluid-solid intersection surface. 4) The contact surface between LP plate and back plate is set to be friction contact-pair and the friction coefficient μ is 0.2. 5) The contact surface between HP and LP is set to be friction contact-pair to demonstrate the relative movement between HP and LP slice. Because there is only dynamic pad in LP stick which results in different lift movement between HP and LP.

According to Figure 3, the boundary conditions in fluid flow filed are as follows:1) The inlet (front surfaces 5, 6 and 7) are set opening boundaries and the high pressure are given. 2) Top surfaces 8 and 10 and the back surface 9 are outlet of the gas field, which are set opening boundaries and the low pressure are given.3) The two cut sides of 8 and 10 are set to be cyclical boundaries.4) The lower surface of 9 is contacting with rotor surface, which is set moving wall and the velocity and dynamic stimulus are given.5) The upper surface of 9 is interface of solid and fluid, which is set to be wall.

3.3 Forms of stimulus

The unbalanced force of rotor is a harmonic exciting force. Considering the phase and frequency of the radial run-out of rotor are same as those of the unbalanced force of rotor, the rotor displacement stimulus may substitute for unbalanced force of the rotor, and the amplitude the displacement stimulus is the radial runout of the rotor. Then the rotor displacement stimulus can be expressed as

$$y(t) = \alpha \sin(\frac{n\pi}{30}t) \quad (1)$$

Where, α is the runout of the rotor, n is the rotate speed of the rotor.

3.4 Parameters of dynamic performance

Here we use leakage M, film load capacity F, and gas film stiffness K to describe the dynamic behavior of the non-contacting finger seal. A finger seal with lower leakage, higher load capacity and higher stiffness simultaneously exhibits good dynamic performances.

3.4.1 Leakage

The seal leakage is relative with the area and velocity of the flow at outlet. These two factors are all vary with time. The mass flow leakage of the seal at a moment can be calculated by the following formula

$$M = \int \rho \cdot v ds \quad (2)$$

Where, $\int ds$ is the area of outlet, v is the velocity of the gas in outlet, ρ is the density of the gas.

The average leakage of the seal in a rotor rotating cycle is defined as

$$M_{av} = \frac{\sum_{t=1}^{n} M}{n} \quad (3)$$

The variable coefficient of the leakage at every moment is defined as

$$M_{cv} = \frac{M - M_{av}}{M_{av}} \quad (4)$$

3.4.2 Load capacity

The load capacity of the gas film of non-contacting seal is the radial film force applied to the finger pad. The load force is calculated as

$$F = \left|\int PdA\right| \quad (5)$$

Where, $\int dA$ is the area of the coupling surface, P is the pressure at the coupling surface.

3.4.3 Gas film stiffness

Gas film stiffness is the derivative of film load with respect to film thickness, which is defined as

$$K = -\frac{dF}{d\delta} \quad (6)$$

Where, F is film loading force, δ is film thickness. In a periodic excitation the film thickness of finger seal can be calculated approximately as

$$\delta = \overline{Y}(t) - y(t) \quad (7)$$

Where, $\overline{Y}(t)$ is average position of the pad bottom, y(t) is the rotor displacement.

3.5 Calculating method

Based on SST model and K-ω equations the finger seal flow is simulated as turbulent model. The convergence is set as 10^{-4}. The original position of rotor and finger are set according to the installation. In a periodic motion of rotor the stimulation is discretized to compute the fluid flow and solid deformation at a moment, and its result is set as the initial value for the next compute point till the end of the whole computing period.

 https://doi.org/10.1051/matecconf/201815307005

4 Results and discussing

4.1 Compared with the results from equivalent dynamic model

Using the structural and working conditions from Reference[6],the finger seal displacement responses are calculated by the FSI method from the current paper, and the results are compared with the method of equivalent dynamic model from reference[6]. The comparing results are shown in Figure 4.

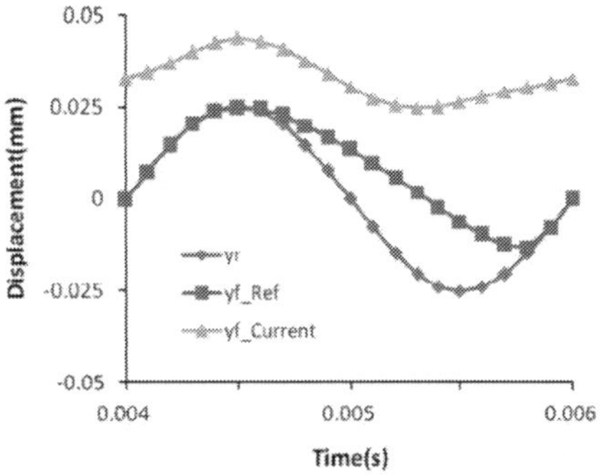

Figure 4. Finger's displacement response of the two methods

From Figure 4 it is found that the two results are quite different. Firstly, in Reference[6] the initial gap is set to the initial minimum film thickness,6.35μm, then the initial displacement response is zero because the initial minimum film alteration due to the hydrodynamic effect is not considered. In current computation the initial film thickness will increase to about 30μm at pressure of 6.9KPa and rotating speed of 30000rpm. Secondly, in the first 1/4 increasing period, or the equivalent dynamic model, the finger followed the rotor closely because the initial film thickness is very small with greater film stiffness. And the finger's curvature is not considered in the equivalent dynamic model, so the finger stick stiffness is constant. However, in current paper the finger response is a little slower because the initial film thickness is bigger and the film thickness is decrease gradually hence to the film stiffness increase subsequently. And at the same time the finger's curvature is decreased resulting the increasing of finger's stiffness. Thirdly, in the period from the highest position to the lowest position, the finger's response lags behind the rotor because of the hysteresis of finger seal. Though this phenomenon can be seen in the two results, but the displacement decreasing value for dynamic model is greater than that of the current method because the film stiffness in the equivalent dynamic model is bigger. Lastly, in the last 1/4 period the finger's rising position is different. For dynamic model when the rotor rise from the lowest position the fingers move down due to the inertia. But actually when the film stiffness is small and the finger stiffness is bigger, the finger will stop at its lowest position due to the frictional hysteresis before the rotor arrives at the lowest position. And the finger will rise up following the rotor due to the increasing film load capacity.

4.2 Effect of pad parameters on the finger seal dynamic performance

The load capacity of dynamic finger seal is mainly caused by the dynamic pad. Here the varieties of dynamic pad parameters are discussed on the performance of finger seal.

4.2.1 Wedge ratio of clearance

Figure 5 and Figure 6 show the different wedge ratio effects on the leakage and film stiffness respectively.

For the case of lower pressure(1psi,6.9kPa) -higher velocity(30krpm),when the minimum film thickness is given the clearance of the film increase with the wedge ratio, as well as enhance the dynamic effect which increase the film load capacity. So it is found that *M*av is increasing and *M*cv is decreasing. Meanwhile the enlarged film thickness will decrease the film stiffness especially in the condition of lower pressure drop and higher velocity, which is shown in Figure 6. Hence in low pressure and high velocity working conditions the wedge ratio should be selected not only to avoid the contact between the rotor and finger pad but also to maintain lower leakage.

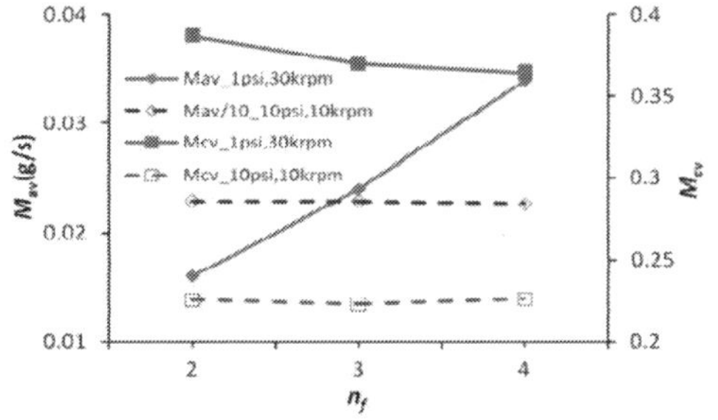

Figure 5. Effects of wedge ration on leakage

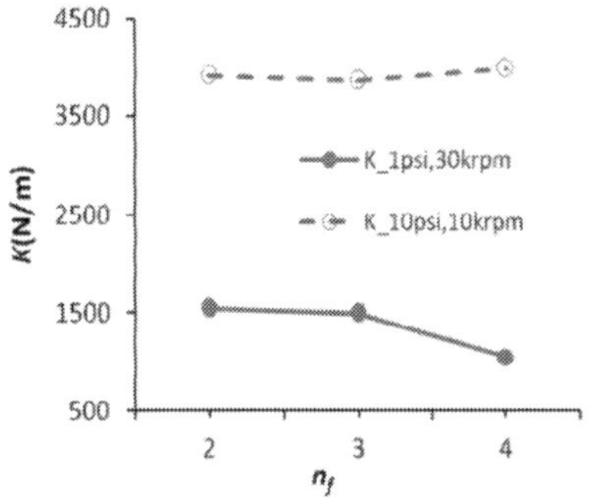

Figure 6. Effects of wedge ration on film stiffness

For the case of high pressure (10psi, 69kPa) -lower velocity(10krpm), the leakage and film stiffness do not vary obviously. Because the pad lifting displacement

(80μm) is much greater than the original clearance (10μm) which make the film clearance increase largely. Then the effect of the variation of the film due to the wedge ratio is relatively small. In higher pressure conditions the film form have little effect to the finger seal performance. So in later parts we just discuss the performance in lower pressure -higher velocity conditions.

4.2.2 Minimum film thickness

Figure 7 and Figure 8 show the leakage and film stiffness with different minimum film thickness with 1psi pressure drop and 30krpm velocity. When the wedge ratio of film is keep constant, the film thickness will enlarge with the minimum film thickness increase, which will result in increasing of leakage and fluid velocity. Then the load capacity is enhanced. However with the film thickness increasing the film stiffness will decrease. So for lower pressure-higher velocity conditions, thinner film thickness can keep high film stiffness and low leakage.

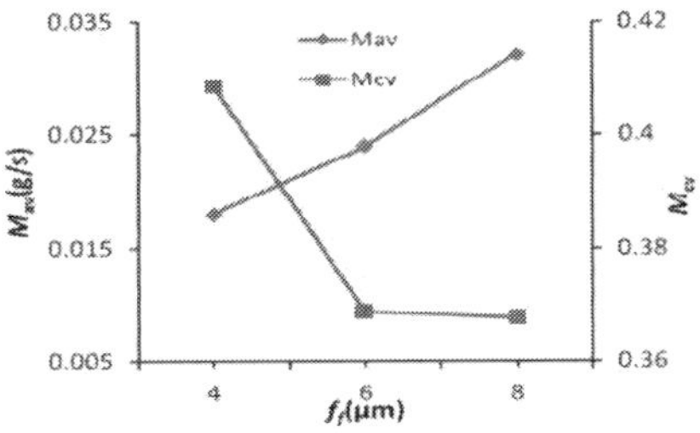

Figure 7. Effects of minnmum film thickness on leakage (1psi,30krpm)

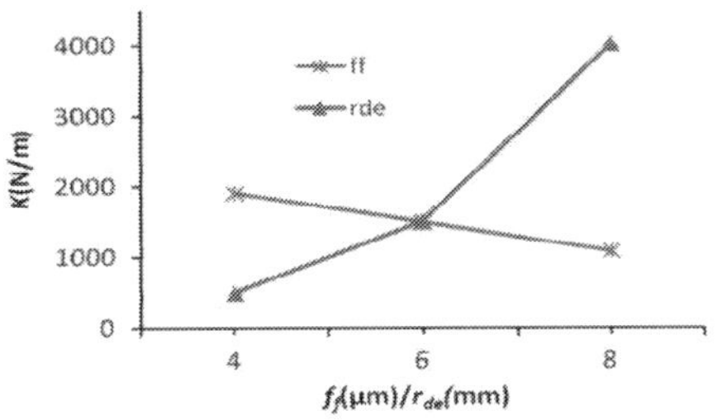

Figure 8. Effects of minnmum film thickness and axial length of pad on film stiffness(1psi,30krpm)

4.2.3 Axial length of the dynamic pad

Figure 9 and Figure 8 show the effects of axial length of pad on the leakage and film stiffness respectively.When the axial length of pad increase the fluid flow resistance increase, then the leakage decrease and the film stiffness increase. The pad can follow the rotor well when rotor has radial runout and the leakage variation is small too. For lower pressure-high velocity case it is better to lengthen the pad axial length to keep high film stiffness and low leakage.

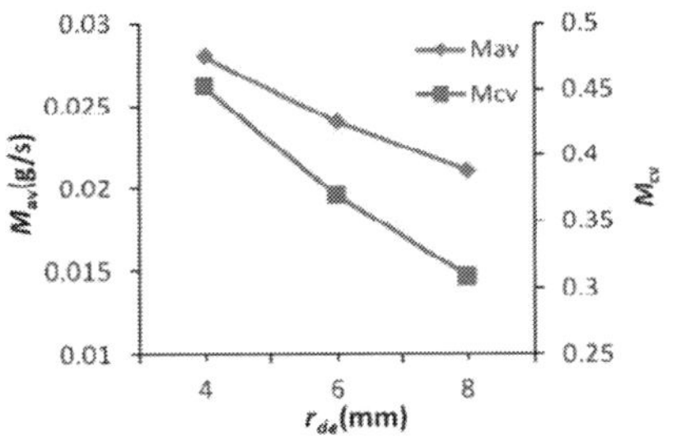

Figure 9. Effects of axial length of pad on leakage(1psi,30krpm)

5 Conclusions

The dynamic performance of the non-contacting finger seal is analyzed based on 3D FSI method, which can describe the movement between the rotor and dynamic pad much accurately. However, it cost more computing time than solving the equivalent dynamic model.

It is found when the non-contacting finger seal working in higher pressure drop the pad will lift obviously, which results in high leakage and the dynamic potential may not be developed. It is necessary to improve the performance of the finger seal adapting to high pressure conditions.

For lower pressure and high velocity conditions the pad parameters will influence the leakage and film stiffness of the seal definitely. For the cases studied in this paper, when the wedge ratio is 3, the minimum film thickness is 6 to 7 μm, and the longer pad axial length, the film stiffness keep higher and leakage is lower as well as the pad can avoid contact with the rotor, which maintain better dynamic performance.

Acknowledgment

The authors would like to extend their appreciation to the National Natural Science Foundation of China for supporting this work under Grant 51575445.

References

1. G K. Arora , M.P.Proctor, B. M.Steinetz. and I. R.Delgado, AIAA-**99-2686**, 1-18(1999)
2. M.J. Braun, V.V. Kudriavtsev, B.M. Steinetz, and M.P. Proctor, The 9th International Symposium on Transport Phenomena and Dynamics of Rotating Machinery, Honolulu, Hawaii, February 10-14, 1-8(2002)
3. H. Marie, 42nd AIAA/ASME/SAE/ASEE Joint Propulsion Conference & Exhibit 9-12 July 2006, Sacramento, California. **AIAA 2006-4931**.1-18(2006).
4. D.P. Fleming,"Gas Seal Pad With Herringbone-Grooved Rotor—Stiffness and Load Capacity,"**NASA/TM—2006-214333**,pp.1-13 (2006)
5. D.X Lang, H. Su, Acta Aeronautica et Astronatica Sinica, **33**,, 1540-1546(2012)
6. M.J.Braun , H.M Pierson. , D.Deng, Tribology Transactions, **48**, 531-547(2005)

Optimization Design and Calculation of Mooring System Parameters based on Genetic Algorithm

Liang-yu Xia, Li Zhi , Ling Chen and Yue Cheng

College of Electrical Engineering & New Energy, CTGU, Yichang 443002, China

Abstract. Based on the static equilibrium model of mooring system parameters, the multi - objective optimization model is established and the genetic algorithm is introduced as the optimization algorithm, which improves the crossover operator. Taking the example of a single point mooring system, concerned optimization design and calculation has been done. Compared with the general genetic algorithm, the result of this algorithm is superior, significantly improving the stability of the mooring system.

1 Introduction

The single point mooring system has been widely used in both ship navigation and the comprehensive utilization and development of Marine resources. The small changes of mooring system parameters can greatly affect the stability of the whole system, so a large number of experts have carried out relevant researches. Ming-wu Wang [1] established a two-dimensional model of the underwater buoy system and calculated relevant parameters according to static equilibrium; John [2] used catenary method and polygon approximation method for analysis and calculation; Xing-ping Xu [3] studied single point mooring system from the perspective of dynamics.

Genetic algorithm is a kind of random searching method which is developed by the genetic mechanism of biology. It regards the possible solution in the problem as an individual in the group, and encodes it as a chromosome. Then simulating the biological evolution process, the group receives repeated hybrid operation and chooses the better one continuously, according to the evolution of the biological sector. At the same time, the optimal individual in the optimal group can be found, which is the optimal solution for satisfying the requirements.

2 Improvement of the genetic algorithm of crossover operators

The basic genetic algorithm also has many shortcomings, which often can only approach the optimal solution rather than calculate the optimal solution, and sometimes find the local optimal solution. In order to solve these problems, generally beginning with genetic algorithm three basic genetic operators (select operator, crossover operator and mutation operator) of genetic algorithm. In this paper, improved crossover operator's genetic algorithm [4 ~ 5] is used to solve the multi-objective optimization problem.

To enhance the diversity of population in the early evolution period, cross-operation between the parent individuals should be expanded to generate a variety of new individuals, which can choose many excellent individuals by copying the operator. However, at the late stage of evolution, with the high frequency of evolution, the solution group gradually approaches the optimal solution. At this point, if we still use a high crossover rate, it will produce a lot of new individuals, which distribute divergently in the entire search space, leading to the low proportion of good individuals in the population. The high rate will delay the convergence process. Therefore, the crossover rate is modified as follows:

$$p_c = p_{c0} - \frac{d(p_{c0} - p_{c\min})}{D} \qquad (1)$$

Where: p_{c0} -a initial crossover rate; d - the current evolutionary algebra; D -the total evolutionary number. It is not difficult to see that with the increase of the iteration times, the crossover rate decreases and the convergence process is accelerated. Genetic algorithm is the search process of every generation based on the adaptability, which is mainly composed of coding mechanism, control parameters, fitness function and genetic operators. The main calculation process of solving the multi-objective optimization problem is as follows: The process is shown in Fig. 1:

a. Initialize the population: randomly generate pop_size feasible chromosomes.
b. Calculate the population fitness: fitness function is used to evaluate the adaptability of each chromosome. In this paper we use the objective function as fitness function, the maximum level of

fitness means the best individual.

c. Selection process: select one chromosome at a time as a new subgroup. Repeat this process pop_size times to get a new generation.

d. Variation operation: supposing the probability of cross operation is P_m. Randomly select a chromosome C_i as the father of the mutation and randomly generate real numbers which belong to $[0,1]$. If $s < P_m$, a mutation in one of the genes of the chromosome is selected randomly.

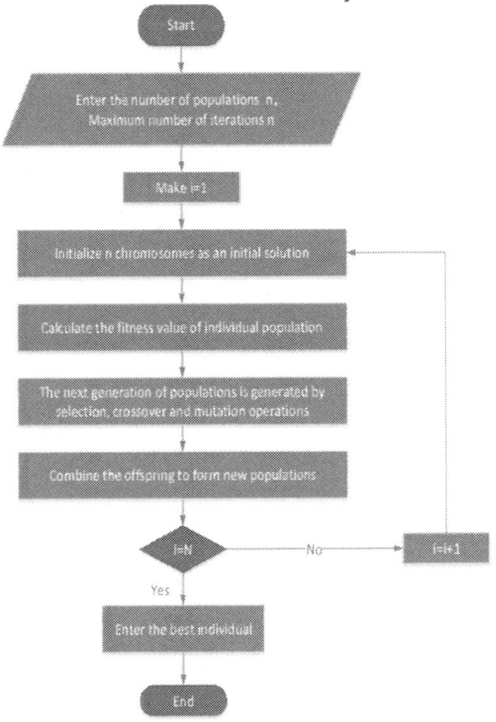

Figure 1: Flow chart of genetic algorithm for improved crossover operator

3 Static analysis of mooring system

The single point mooring system consists of steel pipe, steel bucket, heavy ball, electric welding anchor chain and special anti-drag and shifting anchor. Its schematic diagram is shown in Figure 2. This paper uses a polygon approximation method of concentrated mass [3] to conduct two-dimensional static analysis of mooring system.

3.1. Buoy force analysis

The equation of equilibrium is as follows:

$$\begin{cases} \textit{horizontal direction}: \ T\cos\theta_i = f \\ \textit{vertical direction}: \ T\sin\theta_i = F_{buoyancy} - G_{buoy} \end{cases}$$

Where: $F_{buoyancy}$ -the buoyancy of the buoy in the sea, G_{buoy} -the gravity of the buoy itself, T -the tension of the steel tube, f -the resistance of the float when the buoy is swimming. The buoyancy is $F_{buoyancy} = \rho_{sea} g \times \pi (D/2)^2 h$, resistance is $f = (1/2)\rho_{sea} C_D S V |V|$, θ is the angle between the tension T and the horizontal direction, D is the diameter of the buoy, C_D is the resistance coefficient of the fluid to the buoy, S is the horizontal wetting area of the buoy, V is the unidirectional flow rate that represents the fluid at the buoy.

3.2. Force Analysis of Steel Pipe and Steel

The equation can be:

$$\begin{cases} \textit{horizontal direction}: \ T_i\sin\theta_i = F_{buoyancy\ i} - G_i + T_{i-1}\sin\theta_{i-1} \\ \textit{vertical direction}: \ T_i\cos\theta_i = f + T_{i-1}\cos\theta_{i-1} \end{cases}$$

Where: $F_{buoyancy\ i}$ -buoyancy of the steel pipe in the water, f -the water resistance, G_i -the gravity of the steel tube, T_i -the force of the previous steel tube, T_{i-1} -the force of the next steel pipe. And we can get $T_i = \sqrt{(F_{buoyancy\ i} - G_i + T_{i-1}\sin\theta_{i-1})^2 + (f + T_{i-1}\cos\theta_{i-1})^2}$, $\theta_i = \arctan\dfrac{F_{buoyancy\ i} - G_i + T_{i-1}\sin\theta_{i-1}}{f + T_{i-1}\cos\theta_{i-1}}$ by derivation. It can be seen from the derivation formula that if the former tension T_{i-1} is given, the following tension T_i can be calculated, and the steps in the middle can also be calculated by this method. The force analysis of the steel drums is similar to that of steel pipes, and the difference is that the gravity of the steel drum will be combined with the gravity of the steel ball.

3.3. Stress analysis of the link

As the volume of the link is very small, it can be considered without buoyancy and hydrodynamic effect, that is, only bears its own gravity and the force of the previous link and the chain, we can get the equilibrium equation:

$$\begin{cases} \textit{horizontal direction}: \ T_i\sin\theta_i = G_i + T_{i-1}\sin\theta_{i-1} \\ \textit{vertical direction}: \ T_i\cos\theta_i = T_{i-1}\cos\theta_{i-1} \end{cases}$$

3.4. Analysis of the relevance of each component`

The total length of the projection of each member in the vertical direction should equal the depth of the seabed when the system is stable, which is $h + \sum_{i=1}^{N} l_i\cos\theta_i = H$. The radius of the floating circle of the buoy is $r = \sum_{i=1}^{N} l_i\sin\theta_i$, θ can be derived from the stress analysis in the previous text.

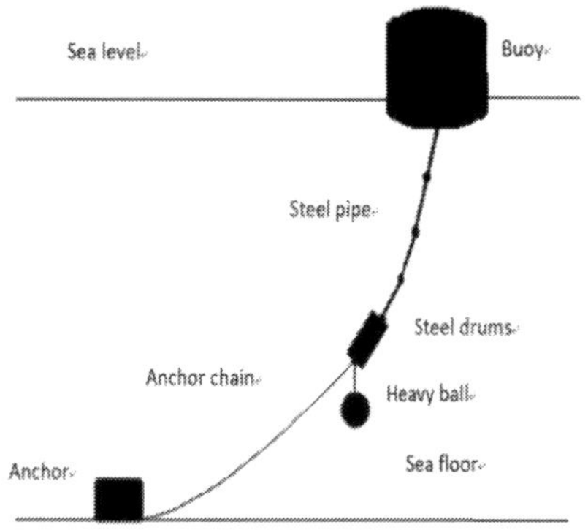

Figure 2. Schematic diagram of mooring system.

4-Establishment of multi-objective model

Taking the complex external conditions consideration, mooring system can still work properly, so properly designed mooring system parameters is indispensable, including the type of anchor k and the heavy ball quality m . Therefore, a multi-objective optimization model is established. Optimization Goal 1: The depth d of the buoy is affected only by the type k of the chain, the length s and the mass m of the heavy ball, so it can be expressed as: $\min d(m,k,s)$. Optimization Goal 2: The area r of the buoy is affected only by the type k of the chain, the length s and the mass m of the heavy ball, so it can be expressed as: $\min r(m,k,s)$.

Create the following two constraints:

(1) the quality of heavy ball is indefinite and variable in a range, there are:

$$M_{\min} < m < M_{\max}$$

(2) The steel drum tilt angle can not exceed a certain degree, and the smaller the better, there are:

$$0 < \theta < \theta_{\max}$$

5-Mooring system parameter optimization case analysis

Select a type of mooring system [6] as an example of multi-objective optimization calculation. The basic parameters of the mooring system are as follows: Buoy system can be simplified as a cylinder, the bottom diameter and high of which both are 2 m. The quality of the buoy is 1000 kg. The quality of the anchor is 600 kg. Common model and its parameters of near shallow sea observation are listed in table 1. There are four sections of steel pipe, one meter of each length, and the diameter is 50mm, and the quality of each steel tube is 10kg. The weight of the heavy ball is 1000kg to 5000kg. The environmental parameters of the mooring system are as follows: the measured water depth of the sea area is 20m, the wind speed of the laying point is 36m / s and the sea water velocity is 1.5m / s.

Table 1. Common anchor chain models and parameters

Type	Length (mm)	Mass per unit length (kg/m)
I	78	3.2
II	105	7
III	120	12.5
IV	150	19.5
V	180	28.12

The main control parameters of genetic algorithm are population number, times of iterations, hybrid rate, recombination rate and mutation rate. The final result and the efficiency of the genetic algorithm are affected by the population size. The small population will lead to the missed solution. The large population will increase the times of calculation of the adaptive function. The higher the crossover rate and the recombination rate are, the faster the individual updates in the population, but if too high, it will lead to the destruction of the outstanding individual. If too low, the quality of the population can not be further improved. The mutation rate cannot be too high, and if too high, the algorithm can continuously search the solution space and influence its convergence. When optimizing the design of mooring system parameters, the values of appropriate adjustment parameters are calculated through multiple trial calculations, and the relevant control parameters of genetic algorithm are shown in table2:

Table 2. genetic algorithm related parameters

Parameter	The numerical
Population quantity	200
Number of iterations	100
Crossover rate	0.9
Number of objective functions	2
The mutation rate	0.1
Recombination rate	0.7

When using genetic algorithm to optimize the parameters of mooring system, a feasible solution can be obtained. The results of the improved crossover operator genetic algorithm and the results of the original genetic algorithm are shown in table 3.

Table 3. Results comparison

	Improved genetic algorithm	Primitive genetic algorithm
Number of links	161	230
type	IV	II
Ball weight	4601	4792
Steel drum Angle	4	3.4
Winding radius	19.71	22.43
depth of immersion	0.3	0.4

Through the comparison of the above results,choose the type IV anchor chain. When the mass of the weight of the ball is 4601kg, the length of the chain buoy of the draft depth and the swimming area and the weight of the heavy ball can reach a lower value. The smaller the buoyancy depth is, the more obvious the buoy is, the smaller the swimming area is, the higher the stability of the whole system is. In this paper, the optimal design value of mooring system parameters obtained by improved crossover operator genetic algorithm has better effect on improving the stability of mooring system.

6-Mooring system parameter optimization case analysis

The parameter optimization design of the mooring system is a multi - objective optimization problem with constraints. Using the genetic algorithm of improved crossover operator, the multi - objective optimization design of each parameter is solved. The result of the example shows that the optimized parameters obtained by this method can meet the constraints and improve the stability of mooring system. Therefore, this method can be carried out in multi-parameter optimization design effectively and provides a approach for other types of mooring system parameter optimization design.

References

1. Wang Mingwu. The static analysis and attitude calculation of the ocean submersible system [J]. Marine technology, 2001, (04) : 41-47.
2. Berteaux, H.O. Buoy Engineering. John Wiley & Sons, New York, 1976
3. Li Zhihai, Xu Xingping, Wang Huili. Development of mooring system of offshore platform [J]. Oil field machinery, 2010, (05) : 75-78.
4. Xie Yanli, Xu Qinglin, Liang Wenchao. A genetic algorithm based on crossover and mutation operators [J]. Computer technology and development, 2014, (04) : 80-83.
5. Lu houqing, Chen liang, Song Yisheng, Wu Zhimin, Zou Yunbo. An improved algorithm for the crossover operator of genetic algorithm [J]. Journal of PLA university of science and technology (natural science edition), 2007, (03) : 250-253.
6. national mathematical modeling contest for college students in 2016, question A, www.mcm.edu.cn/

Research on Multioperator silk screen printer Based on Communication Protocol Macro

Zhang Zhiming, Sun Jun, Lu Binbin, Duan Yaoshuai

School of Electrical & Mechanical Engineering, Wuhan Textile University, 430073 Wuhan City, China

Abstract. Multioperator silk screen printer is automatic silk screen printing equipment. There are many advantages, such as energy saving, reduce the labor intensity and so on. The control system consists of an intermittent movement main motor and eight color working station printing cycle motion motors. The system controls nine frequency converters through protocol macro communication by using PLC. This paper studies the principle and process characteristics of multioperator silk screen printer. Besides, this paper adopts protocol macro sequence generation method for communication between PLC and frequency converter. This method can control the programming of parameters, such as frequency, operation and monitoring. It also can realize the printing of eight color working stations and monitor the printing status in real time. The practical application proves that the method meets the technical requirements and the reliability is good.

1 Introduction

The silk screen printing technology is to make silk screen plate with various patterns place on the fabric. Each colour needs a silk screen plate. The active colour paste is added on the silk screen plate when printing. Then the silk screen plate is covered on the fabric. Under the action of the printing scraper and the slurry scraper, the colour paste is applied to the fabric through the gap of silk screen plate. Through the chromatography, a variety of colour patterns become bright colour and rich patterns of finished products [1]. Traditional silk screen printing technology is manual and semi-automatic. The shortcomings are labour intensity, pollution, waste of resources and low printing quality. Therefore, it restricts the development of printing industry in our country more and more [2].

With the development of automation technology, automatic silk screen printing technology began to develop into a cutting edge trend. Automatic silk screen printing technology has the advantages of energy saving, environmental protection, labour intensity reduction and so on. In order to meet the needs of multiple colour printing, independent asynchronous motors are used at each working station. Frequency converters are adopted to adjust speed. The drive uses protocol macros to communicate with multiple frequency converters to control per station printing.

2 The system principle of multi working stations silk screen printing

Multi working station screen printing consists of cloth feeding device, centering device, printing device, platen driving device, printing body lifting device and drying and cooling device and so on. The fabric is attached to a plate moving along the radial circulation by a cloth feeding device. Then the fabric is imported printing unit by centering device. When printing, the plate is stationary, the floral board descends and the scraper is scraped. When the printing is completed, the floral board is raised. The fabric moves to next station with the plate and enters the next printing unit.

Multi working station screen printing is controlled by nine asynchronous motors. Eight of them are colour working stations. The working principle is shown in figure 1.

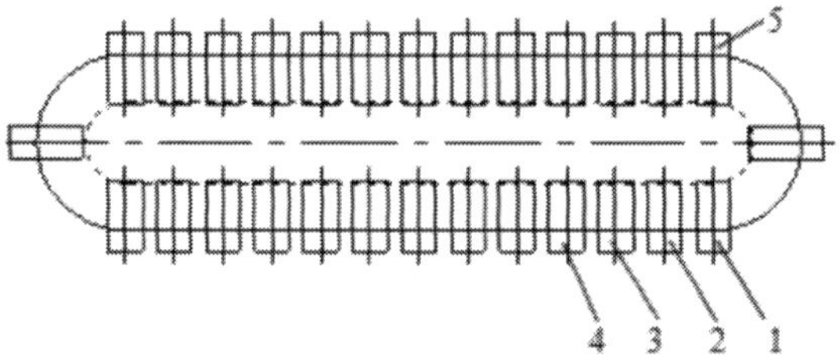

1- Paving position 2- Printing position 3-Drying 4-Cooling
5- Collecting cloth

Figure 1. Working principle of multi working stations silk screen printing

In figure 1, 2, 3, and 4 are a combination. They are arranged in eight printing positions in a clockwise order. Each printing position is controlled by an asynchronous motor and frequency converter with its speed and steering. Therefore, this system realizes the reciprocating movement of the printing scraper and the slurry scraper to achieve silk screen printing[3]. The actual device is shown in figure 2.

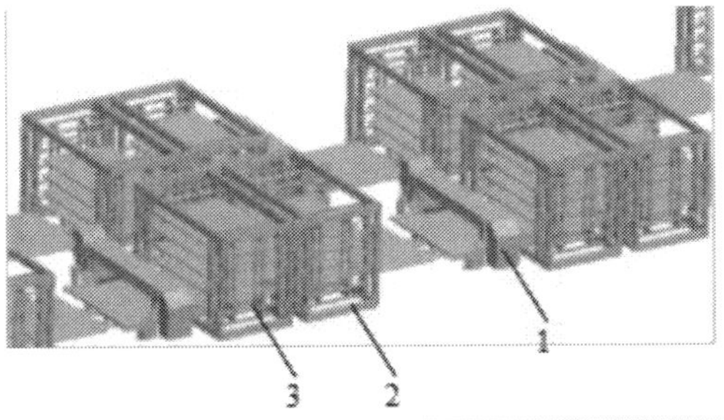

Figure 2. Equipment of multi working stations silk screen printing

3 The control system of multi working stations silk screen printing

Multi working stations silk screen printing control system is shown in figure 3. PLC uses CP1H series CPU of OMRON. The CPU is equipped with communication module. It has RS-485 communication function. The frequency converter adopts Mitsubishi FR-E500 series, supporting protocol macro programming with PLC. RS-485 bus connection is used between slave stations. The use of pneumatic electromagnetic valve can realize the up and down movement of the scraper. So the slurry can be printed on the fabric[4].

Protocol macro is a communication method. It is used to control the PLC and general equipment equipped with RS-232C or RS-422/485 port for data exchange. Through the support of communication protocol macro, a variety of communication equipment for communication protocol can be customized. Therefore, the instructions of the communication protocol macro can be set according to the user's requirements. It also implements an independent definition to send and receive data sequences and communicate with peripheral devices[5].

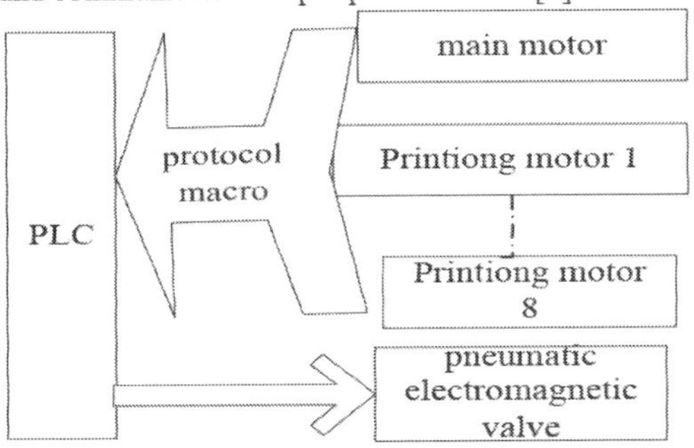

Figure 3. Controlling system diagram of multioperator silk screen printer

4 The protocol macro system of multi working stations silk screen printing

The multi working stations silk screen printing protocol macro system mainly consists of three movements: platform intermittent movement, scraper reciprocating printing movement and electromagnetic valve pneumatic movement. Each movement is required to operate in accordance with the specified process. The key point of the control system is how to realize the communication between PLC and frequency converter, send frequency and operation instructions to nine frequency converters and monitor the printing status.

4.1 Protocol macro communication settings

In order to make protocol macro communication with PLC, coincident parameters are required to be set up. These parameters include baud rate, stop bit, even-odd check and data length and so on. Only the parameters are consistent, communication can be done. In PLC, they are stored in a special register DM6555 and DM6556 by bit combination. Set the data length to 1. No even-odd check. The stop bit length is 1. The baud rate is 9600.

4.2 Communication protocol

Frequency converters and PLC communications use sixteen hexadecimal numbers. Data are automatically transmitted between them using ASCII code. Table 1 is the control code for communication.

Table 1. Controlling code.

Signal	Code	Function
ENQ	05h	Communication request
ACK	06h	Correct response
NAK	15h	Error response

The definition of data frames stipulates the configuration of data blocks during communication. The communication protocol data request from PLC to frequency converter is shown in figure 4. ENQ is a communication request. The frequency command code is HED. The data account for 4 bits. The sum check code is the 2 ASCII characters represented by the lowest byte of the checked ASCII data sum. The waiting time is the waiting time between the data received from the PLC and the transmission response data.

ENQ	Frequency converter station number	Instruction code	Waiting time	Data	Sum check
1	2 3	4 5	6	7 8 9 10	11 12

Figure 4. Communication protocol format

4.3 Create protocol macros

According to the frequency data frame format of figure 4, a communication protocol macro can be created in the host computer configuration software. First, set up

SendMessage and ReceiveMessage information. According to the communication request, check code, data length and data, the structure of sending information and receiving information is generally shown in Table 2.

Table 2. Data structure of protocol macro

Identification code	address	length	data	Error checking	Terminator

The identification code is the data that indicates the beginning of the information. The address is an identifier that sets the node number or indicates that the information sent to the destination. The length is automatically attached to the data length. The data is set for the content of the information. The error check code is setting the SUM, LRC or CRC as the error check code. When sending, the specified error code is automatically appended. When receiving, the error control is automatically carried out according to the error checking code specified in the information. At the same time, the specified amount of data is received. The terminator is the data that sets the end of the information.

For example, in the protocol macro communication sequence, the data for the transmit frequency is <h>+ "00ED07D0" +<c>. 00 indicates the frequency converter station number. The ED indicates that the setting frequency is written to the frequency converter RAM. 07D0 stands for that the frequency is hexadecimal. That is, decimal 2000. The frequency is 20Hz (2000*0.01=20). <h> represents the information data header. <c> represents the data check code.

4.4 Step settings

4.4.1 Repeating counter

Setting the number of repeated steps, when each step is performed, the counter will add 1. There are two ways to set up. One way is to reset and the other way is to keep it. If set to reset, once the step is interrupted for some reason, the counter will start counting at the time of next execution. That is, it will still start from 0. If set to keep it, once the step is interrupted, the counter will still keep the last count and continue to count at the time of next execution.

4.4.2 Retrial times

The number of retries can be set to 0-9. It is the number of retries that can be made when an error occurs while sending a command. It is used for the number of retries that can occur after an error has occurred in the transmission.

4.4.3 Response notices

If the setting is YES, it means to allow depositing the received data in the third operand of the PMCR instruction or software specified receiving area. The simple point is to allow writing the received data.

4.4.4 Next step

There are 4 options: the end, the next step, the direct jump to a certain step and the neglect (That is, END, NEXT, GOTO, ABORT).

4.4.5 Error handling

Same as above, there are also 4 options. When an error occurs, it will repeat the work first and retry. If there is still an error, the error will be handled.

4.4.6 Receive array

Receive array is another way to receive. It has the function of processing the received data separately. In total, 16 desired reception values can be set. Then when the data is received, the next step can be done according to its requirements. That is, to turn to execute a certain step of the set. When the received data does not meet any of the conditions, the next step will also can be set when the condition is not satisfied.

4.4.7 Read word (R)

The "address" or "data" in the sending and receiving information can be set to read word data.

4.4.8 Wild card (*) and word

When receiving the data, the amount of data can be set to wild card (*) and word.

4.4.9 The first-order equation using the variable N

The first-order equation containing the variable N is used for importing the address and data. The variable N will add 1 whenever repeated counters are repeated by a repeated counter specified in the step of a communication sequence.

4.5 Protocol macro control system implementation

The protocol macro sets the user communication protocol through the software. Each step can be a send and receive process. In each step, the sending step, the receiving step, the sending and receiving steps can be defined. The data sent and received can be defined completely by the user. Each sequence can perform 16 steps. That is, each sequence can communicate with different devices using different communication formats. In the ladder diagram program, the use of PMCR instructions can easily call different sequences to achieve different communication tasks. Figure 5 is the protocol macro communication program.

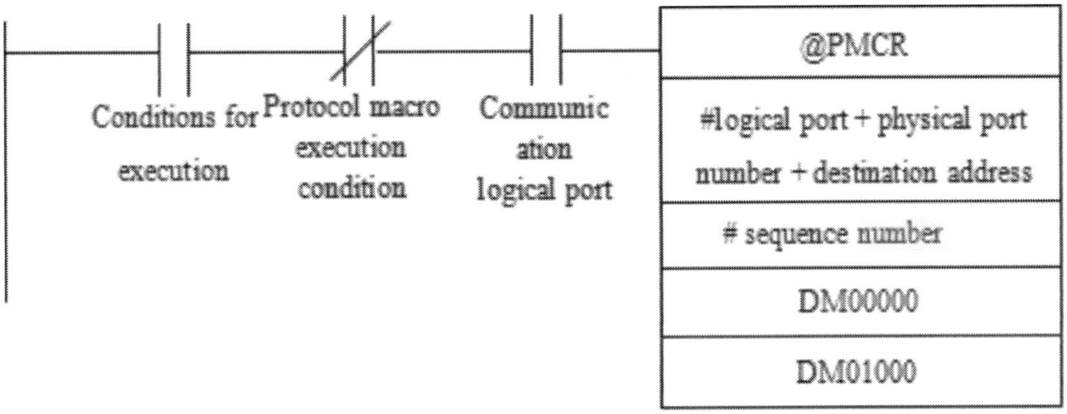

Figure 5. The instruction of protocol micro

The first and second operands of the protocol macro execution instruction are the control word, specified communication port and communication sequence (number 000-999). DM00000 is the first word of the sending data, which is used to store the words of the current send frame. The sending data is stored in the memory area at address DM00000+1. If no data is sent, the number is #0000. DM01000 is the first word of the received data, which is used to store the words of the received frame. The received data is stored in the memory area at address DM01000 + 1. If no data is received, one memory addresses can be specified at random. The execution of the communication instruction does not overwrite the data of the address.

4 Conclusions

According to the requirements of multi working station silk screen printing machine process to the control system, the control system principle based on protocol macro is researched by configuration software of host computer. The communication programs are also developed. Moreover, multicolor printing is achieved. The equipment diagram is shown in figure 6. The device can simplify the automatic silk screen printing control program, shorten the development time and improve the versatility. The research and practice show that the communication protocol macro realizes multi frequency converter control, simplifies the field wiring and has high versatility and high utilization. The communication protocol macro also creates the conditions for realizing the automation and intelligence of silk screen printing machine, improves work efficiency and reduces labour intensity.

1- Printing mechanism 2-Platform mechanism

Figure 6. The finished equipment

References

1. Z. M. Zhang, Q. Xu, S. Q. Mei. Journal of Wuhan Textile University, **26**, 71,(2013)
2. P. Liu. Journal of Wuhan Textile University,(2011)
3. Y. Q. Liu. Silk screen printing,**8**,24,(2013)
4. Z. W. Wang, S. Q. Mei, X. Du. Light Industry Machinery,**29**,53,(2011)
5. H. Y. Wu, X. N. He, X. D. Lv. Electric Drive, **31**,26,(2001)

Design and Fabrication of Integrated Microchannel and Peristaltic Micropump System for Inertial Particle Separation

Utku Sönmez, Muhammed Bekin and Levent Trabzon

Department of Mechanical Engineering, Istanbul Technical University, Istanbul, Turkey
MEMS Research Center (www.mems.itu.edu.tr)

Abstract. In particle separation applications, conventional syringe pumps are widely used to supply fluid flow into microchannels at a controlled flow rate. However, their bulky structures lack the development of compact particle separation systems which is essential for all LoC (Lab on a Chip) systems. In this study, we designed and fabricated a peristaltic micropump which can be integrated into an inertial particle separation microchannel at the same layer with a compact design. Since inertial particle separation can be done without a need for an external force field, we aimed to develop a μTAS (Micro Total Analysis Systems) system which is able to realize particle separation in an integrated micropump-microchannel system. The circular micropump channel made of two PDMS layers and its width is optimized. The 3D-Printed micropump is actuated by a stepper motor, and the rate of pumped fluid is monitored by an LCD screen connected and programmed to system according to the system parameters. Micropump has a theoretical capacity of supplying particle carrying fluid at the flow rate of 25.47 ml/min when the stepper motor is rotated at 330 rpm.

1 Introduction

Micropumps are widely used to produce microscale fluid flows for the application fields ranging from biological and chemical assays like drug delivery systems, micro-dosing control mechanisms and particle separation devices to aerospace and microelectronics industry [1]. There is plenty of micropump designs that can be found in literature dedicated to carry out different tasks [2]. They are generally classified with respect to the working principles of their actuator mechanisms. From this point of view, SMA (Sheet Metal Alloy) actuated micropumps [3], piezoelectric micropumps [4], ICPF (Ionic Conductive Polymer Film) based micropumps [5], peristaltic micropumps [6] and MHD (magnetohydrodynamic) micropumps [7] are developed in the last decade for variety of applications. SMA actuated micropumps, piezoelectric micropumps, and ICPF micropumps contain a deformable diaphragm, which pumps the fluid to a certain direction with the aid of one-way valves. However, the maximum flow rate that they can supply is not sufficient for applications like inertial particle separation which requires extraordinarily high flow rates [8] since the maximum achievable frequency limit for their membranes is not high enough. On the other hand, they lack on biocompatibility either due to the high electric voltage need to be used, or the materials required to be employed [9]. In this study, we aimed to develop a compact micropump system that can be used for inertial particle separation applications, and integrate it to the *Sunflower* microchannel geometry that we developed before.

2 Micropump design

The micropump that would be suitable for the inertial particle separation should supply particle carrying fluids with relatively high flow rates without fluctuations, and must be integrated to another microchannel geometry in which inertial particle separation will be realized. Furthermore, the mechanism should not be harmful for the biological cells (e.g. high voltage, high working temperature, physical impact, high accelerations) to not to affect their viability to allow post processing of the separated cells. Given these conditions, peristaltic micropumps are the most appropriate type of micropump since they meet all these conditions with a rather simple design and low cost. In literature, there are two similar configurations of peristaltic micropumps which can be integrated to an inertial separation microchannels. Both of these positive displacement pump configurations trap a determined amount of fluid body between two restrictive elements, and force it to move together with them to a certain direction as they displace. The first configuration uses magnetically attracted metal balls that squeeze the deformable micropump channels and sweep them by means of rotating magnets that are connected to a motor [10]. The second configuration, uses rotating ball bearings that are directly connected to the motor, and creates the squeezing effect by the preload applied on the micropump channel. The latter one has a line contact between the roller bearings and the deformable PDMS surface instead of a point contact, which in turn deforms the PDMS layer in a more uniform trend, and closes the gaps formed near the microchannel side walls (Figure 1). This reduces the backflow significantly, allowing the micropump to pump the fluid at higher flow rates for the given rotational speed. On the other hand, metal balls in the first configuration are fixed to their circular orbit by means of magnetic forces. It is reported that the metal balls draw apart from their circular orbit and spread away randomly at higher rotational speeds due to the increasing centripetal acceleration acting on them, in a way that the mechanism loses its function [11]. In the rotating ball bearing configuration, this problem is not observed since the rotating ball bearings are fixed to their circular orbit by means of rigid bars. Therefore, the maximum achievable rotational speed of the micropump, thus the maximum achievable flow rate, only depends on the maximum rotational speed of the motor that is used to rotate the ball bearings. Apart from that, the width of the micropump channel should not exceed the width of the squeezed cross section to have proper peristaltic effect. At the same time, the amount of fluid that can be pumped in one revolution linearly increases by the increasing channel width. In the rotating metal balls configuration, the diameter of the balls can be increased to work with wider micropump channels to have higher flow rates.

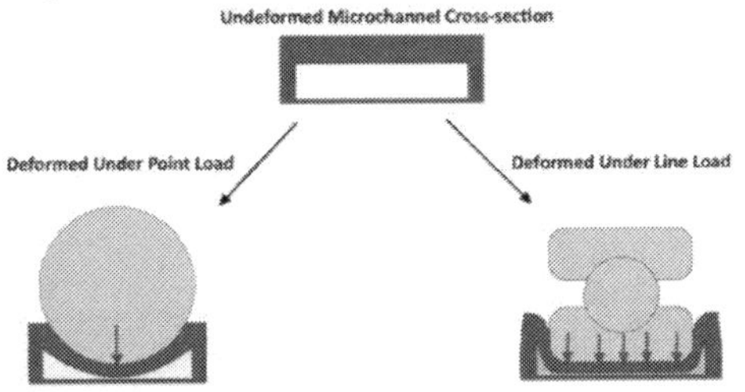

Figure 1. The differences between point load applied by metal balls and the line load applied by the roller bearing itself, in terms of the ability to clog the gap in a uniform manner.

However, in this case, maximum achievable rotational speed decreases since the centripetal acceleration acting on the metal balls increases for the given rotational speed and leads to the breakdown of the mechanism due to the reason described above. In the rotating ball-bearings design, the width of the micropump channel can be tailored freely since there is plenty of ball-bearing with different widths in the market, and the centripetal acceleration acting on the ball bearing is balanced by the rod connected to ball-bearings. Therefore, the fluid pumping capacity of the micropump can be increased tremendously by optimizing the micropump channel width. Due to the advantages described above, we decided to use the peristaltic micropump configuration with rotating ball bearings in our integrated microchannel-micropump system. The concept design described by Zhang et al. [12] for this configuration was used to pump the fluid between nl/min and μl/min range. To meet our system requirements this mechanism was modified to have volumetric flow rates at the range of ml/min with less pulsation. Also, it is integrated to a microchannel geometry that we developed in which passive particle separation occurs by means of inertial lift forces and Dean forces. In the following sections, details regarding to the design and the fabrication of such a micropump system is given together with its integration to cell-separating microchannel in a way that it forms a versatile microfluidic platform.

2.1 Micropump Channel Design

After examining various micropumps and selecting the most appropriate micropump configuration for our application, we carried out further analysis for the determination of the optimum design variables of the selected peristaltic micropump configuration. The micropump is designed to be manufactured at the same layer width of the *Sunflower* microchannel [13]. Due to the soft lithography technique that is used to fabricate the microchannel layer of the micropump, it is obligatory for the microchannel and the micropump channel to have the same height. Since the optimum *Sunflower*-microchannel height is determined to be 60 μm, we used this value as the height of the micropump channel. On the other hand, the width of the micropump channel can be tailored to increase the amount of fluid pumped per revolution. However, in previous works, various researchers stated that at very high aspect ratios (i.e., width >> height) the roof of the microchannel tends to collapse during the oxygen plasma bonding step [14] [15], sticking on the bottom layer permanently, which, in turn, prevents the fluid partially from flowing through the micropump channel. To find the maximum micropump channel width that can be manufactured without roof collapse, we prepared various micropump channel samples with different channel widths ranging from 2 mm to 5.5 mm on a single wafer (Figure 2).

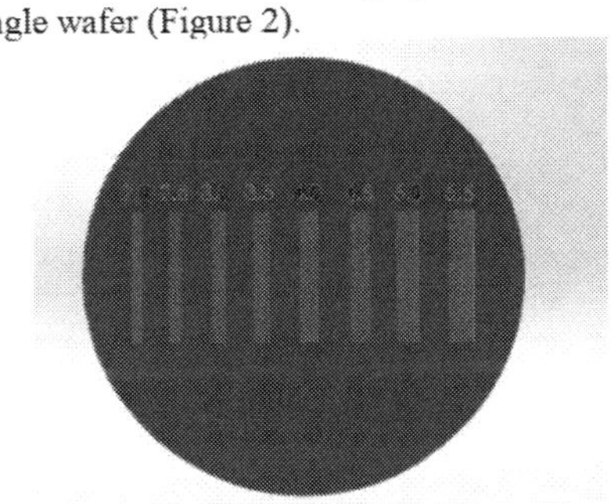

Figure 2. Micropump channel samples with different channel widths are prepared on a single wafer to investigate the maximum channel width that can be achieved without PDMS roof collapse.

We detected that at the width of 5.5 mm, PDMS starts to collapse down and clog the micropump channel permanently (Figure 3). Given the experimental results, we decided to have a micropump channel width of 4.5 mm to be on the safe side.

Figure 3. As the channel aspect ratio increases, the roof of the PDMS channel tends to collapse. Experiments done using channel samples with different channel widths show that roof starts to collapse (marked area with red circle) at channel width of 5.5 mm for the given channel height of 70 μm.

2.2 Rotating Bearing Bar and the Ball Bearings

The number of rods was selected to be 3 to further decrease the pulsation, and the length of each was set to 40 mm which is the radius of curvature of the micropump channel. The angle between the rods was set to 120° in such a way that the volume of fluid trapped between each rod is the same. The width of the ball bearings was selected to be 5mm, which is sufficient to squeeze the micropump channel entirely. Ball bearings were selected from the catalogue of the bearing manufacturer BRC (Washington, United States). Between the ball bearings which have the width of 5 mm, the one which has the smallest outer diameter (d_o =11 mm) was selected to have a circular motion without locking. The rods were designed to have a decreasing diameter in two steps starting from 8 mm at the centre, to 4 mm at the edge, which is the inner diameter of the selected ball bearings. We added the stepped decrement of the diameter to ensure that the ball bearing can still rotate around their own axis freely after the assembly and the rods are deflected less for the given bending force, which rises due to the force applied on the ball bearing while squeezing the micropump channel. Furthermore, we designed the rods with a circular hollow at the centre to nail steel wires, which would further increase their strength against bending moment (Figure 4). Due to the intrinsic properties of the microchannel that are described in the first section of the paper, it only functions well if it is designed to separate particular particles, which means that different PDMS microchannels should be produced to carry out different separation tasks. Furthermore, PDMS microchannel should be disposable for biological applications to eliminate the risk of cross-contamination. For this reason, we designed our micropump in such a way that the PDMS layer can be easily removed and a new one can be replaced without affecting the components of the micropump. Therefore, a hole with 4 mm diameter drilled at the centre of the bars to connect them to the axis of the motor by means of a screw and a nut, which can be removed and reassembled to replace the microchannel layer. This assembly configuration is also crucial to create the axial force, which would push the ball bearings along the micropump channel.

2.3 Sunflower Microchannel and Micropump Channel Integration

The micropump channel and *Sunflower* microchannel which will be at the same PDMS layer should be manufactured together in the same pattern to accomplish the particle separation task. The pump channel is located around the *Sunflower* microchannel in such a way that the outlet of the micropump channel is the inlet of the *Sunflower* microchannel. Since the maximum radius of curvature of the *Sunflower* microchannel is 2.5 mm, the radius of curvature of the pump channel is selected to be 4 mm which is sufficiently high to allow the roller bearings to turn about the microchannel, and not excessively high to increase the footprint of the integrated system. Furthermore, two serial reservoirs are located between the inlet of the *Sunflower* microchannel and the outlet of the micropump channel to decrease the fluctuations intrinsically occurring in peristaltic pumps. Consequently, the integrated form of microchannel-micropump channel is established, as shown in Figure 4. PDMS will be cut. It does not play any functional or structural role in the system. By adding the roller bearings and the bearing rod on top of the system, and rotating them via an electric motor, the final configuration of the integration is obtained.

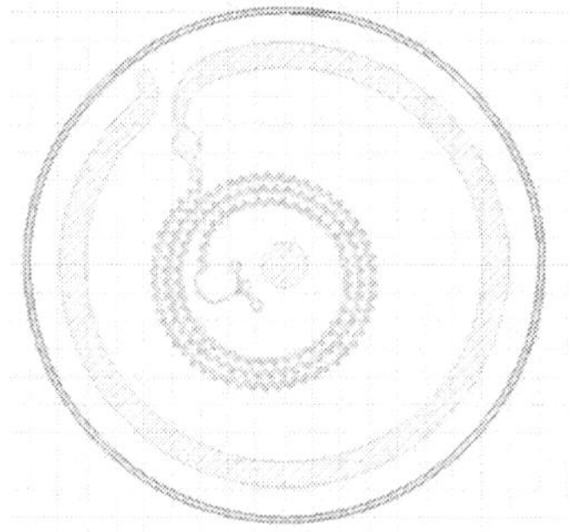

Figure 4. The integrated micropump channel and the *Sunflower* microchannel with two subsequent reservoirs.

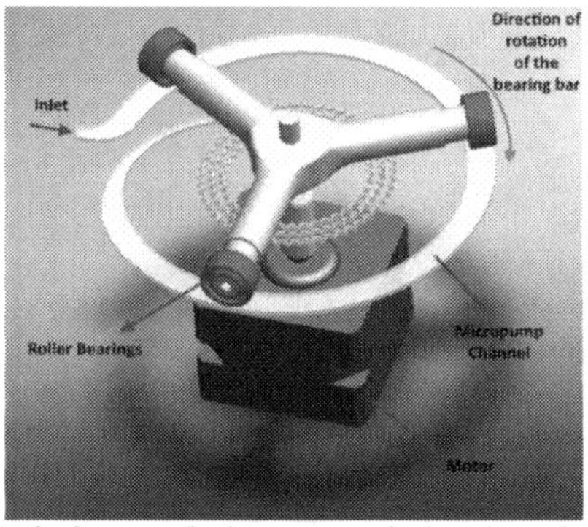

Figure 5. The integrated micropump and microchannel system.

2.4 Micropump Case

Micropump case is designed to carry the microchannel and micropump channel, and to house all the electronics used in the mechanism. Furthermore, inlet and outlet tanks are located into the micropump case by means of a movable door, which allows us to fill them with the suspension that contains mixed particles, and to collect the fluid from the outlet of the microchannel, which contains separated particles. Plus, a housing for the motor is located at the centre of the case, and other required housings for power switch, LCD monitor and Arduino exits are located with proper dimensions. The case is then designed to be mounted to a bottom plate, which is fixed to the case with screws. The final design of the micropump case can be seen in Figure 6.

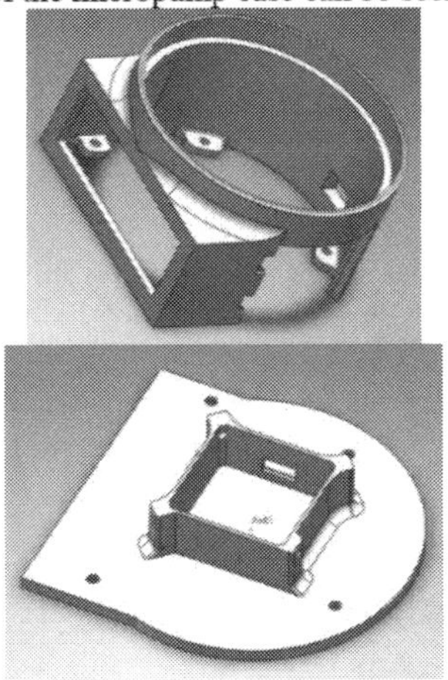

Figure 6. The micropump case (left) and bottom base (right) designed to be able to contain all the electronic and mechanic parts necessary for the system.

Further, inlet and outlet tanks are located on the movable gate on one side of the micropump case. The inlet tank is designed to have a higher inner volume than the outlet tanks since fluid will be separated into two at the exit of the microchannel. The inlet has a 9 ml inner volume which enables us to use the micropump at the optimum flow rate of the *Sunflower* microchannels for 6 minutes, which is sufficiently high for biological assays, since the specimen volume is much smaller under normal circumstances. Also, a small hole with a diameter of 1.8 mm, which is the diameter of the tygon tubings that we used, is located at the bottom of the inlet tank to maintain a continuous fluid flow from the inlet tank to the micropump channel.

2.5 Motor and Control Mechanism

Stepper motor was selected to be used to create the necessary rotational movement since it is easy to control it precisely. For this purpose, Nema-14 stepper motor (National Electrical Manufacturers Association, Virginia-USA) was purchased. The selected stepper motor has 200 steps per revolution which corresponds to 1.8° of step angle that is sufficient to eliminate fluctuations which may rise due to the discontinuous rotation of the stepper motor. Also, it has 7.4 V of operating voltage that is also suitable for our application. To drive the stepper motor, EasyDriver ROB-12779 (SparkFun Electronics, Colorado-USA) stepper motor driver was purchased. Furthermore, a 16x2 IIC/I2C/TWI serial LCD display (Surenoo Technology, Guangdong, China) was purchased and integrated to the stepper motor, then it was programmed to show the volumetric flow rate that is pumped by the micropump to the user. Moreover, Arduino Nano microcontroller was purchased to control and program the stepper motor driver and the LCD display via a computer. Lastly, an on/off switch is acquired and connected between the 9V Battery power source and the stepper motor to start and shut down the system. All the electronic equipment is connected to each other by using jumper wires.

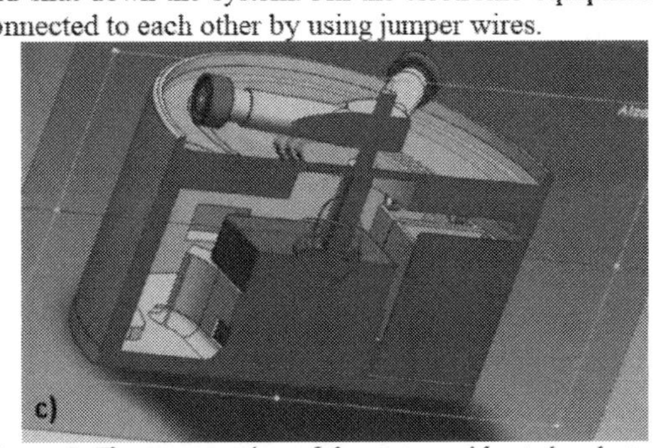

Figure 7. The cross-section of the system without the electronic equipment, wirings and tubing.

3 Fabrication of Micropump

Micropump base, micropump case, bottom plate and Y-shaped bearing bar were manufactured by a Ultimaker-2 3D Printer (Geldermalsen-Netherlands) from the CAD documents produced in Solidworks (Figure 8). All the parts were produced by using a PLA filament with a tolerance of 0.2 mm.

Figure 8. Micropump parts produced by 3D Printer before the assembly.

Supporting parts used by the 3D Printer were then removed by using appropriate tools, and fine-tuning of the parts was done via a sandpaper disk. Three roller bearings (685 Z, BRC Bearing Company) were close fitted to the edges of the bearing bar. The step motor is placed in its housing on the bottom plate and fixed by screws to its centre.

The PDMS layer of the microseparation system that is composed of micropump channel and *Sunflower* microchannel is first constructed by using CAD software, Solidworks. Then, the CAD geometry is transferred to Heidelberg DWL-66FS laser writer (Heidelberg, Germany) for the mask fabrication. A chrome mask was produced by using DWL. Before the photolithography process, single-side polished Si-wafer was cleansed with Deionized (DI) water and dried with N2 gas to remove any deposits that might have been on the surface. This was followed by spreading Su-8 3000 negative photoresist on the Si-wafer. The PR spreading process was done by spin coating in a Laurell WS 400 B Spinner (North Wales, USA). By using the the Su-8 3000 datasheet from MicroChem , necessary process information is found and the necessary rotational speed for the desired photoresist layer thickness is provided. Later on, soft baking is done for 45 minutes at 95°C on a hotplate (Prazitherm, Düsseldorf, Germany). Then, using the previously prepared mask carrying the desired pattern, the Su-8 coated Si-wafer was subjected to contact lithography under UV light, in an OAI Model 200 Mask Aligner (San Jose, USA). Sequentially, a post-exposure bake was performed by putting the Si-wafer back on the hot plate on two levels to ensure the solidification of the parts affected by the UV light. The Si-wafer was then immersed in a developer (1-methoxy-2- propanol acetate) medium for 12-15 minutes in which the soluble regions of the PR would dissolve. The wafer was later rinsed with isopropanol and DI water, and dried with N2 gas. After the fabrication process was done, the master template was characterized under the microscope to check for any errors that would cause structural failure in the microchannel manufacturing. After preparing the master template, the PDMS replica molding process started with preparing the PDMS (Sylgard 184, Dow Corning) by mixing it with a curing agent at a weight ratio of 10:1. The pre-polymer was then degassed in a vacuum chamber (Sheldon Manufacturing, Inc.) to remove air bubbles present within the homogeneous mixture. After all bubbles were released, the PDMS prepolymer was slowly poured on the master template, and baked at 90°C for more than 3 hours to solidify. The PDMS was later detached from the master template, and fitting holes were punched into the PDMS mould with a 2 mm biopsy punch. Afterwards, the detached PDMS layer and another PDMS layer that has no pattern on it, were placed inside a Harrick Plasma Cleaner (New York, USA) for the plasma bonding process. After the plasma bonding process inlet and outlet tubings are fit into the punched areas, and the whole PDMS layer is located under the bearing bar.

4 Results and Discussion

After the fabrication of each part and the integration of the electronics, assembly of the micropump is done (Figure 10). Given the defined micropump channel dimensions, this peristaltic micropump has the capacity to pump 79 mm^3 of fluid per revolution into the microchannel. Under the given loads on the stepper motor, we experimentally found that it is possible to reach 330 rpm by reducing the micro delay of the stepper motor down to 60 μs.

Figure 9. The final form of the micropump with a coffee cup for scaling purposes.

Therefore, by using the stepper motor at its maximum capacity, it is possible to pump the fluid at the flow rate of 23.75 [ml/min] which is quite higher than the peristaltic micropumps that can be found in literature [16]. However, with respect to the backflows and inefficiencies in terms of trapping the fluid mass, the efficiency of the peristaltic pumps are prone to decrease down to 0.5. Even in this case, the volumetric flow rate capacity of 11.7 [ml/min] will be sufficiently high for our inertial cell separation application. Our work on the characterization of micropump is still in progress. As a continuation for this research, we want to test our micropump in cell separation experiments, and we also would like to compare the separation efficiencies of the syringe pump and our peristaltic pump. Furthermore, viability of the cells that pass through the micropump channel should be another consideration. One should bear in mind that the squeezing and dragging motion of the peristaltic pump may affect the viability of the cells. If it really happens, this design would not be an ideal solution for the future LOC devices for inertial separation. Our work on these issues are currently in progress.

5 Conclusions

An integrated peristaltic micropump mechanism which is capable of supplying fluid flow at ml/min scale is designed, optimized, and fabricated. The portable final product is directly integrated to a passive cell separation microchannel which is previously developed by the authors, in which biological cell separation and enrichment tasks can be done. Together with the inlet and outlet tanks, actuator mechanism, programmable LCD screen, tubings, integrated power source, the micropump system can be considered as a self-contained μ-TAS device with a low fabrication cost.

References

1. D J Laser and J G Santiago. "A review of micropumps," J. Micromech. Microeng. 14, R35–R64, 2004.
2. Nan-Chyuan Tsai, Chung-Yang Sue. "Review of MEMS-based drug delivery and dosing systems," Sensors and Actuators A 134, 555–564, 2007.
3. D. Xu, L. Wang, G. Ding, Y. Zhou, A. Yu, B. Cai. "Characteristics and fabrication of NiTi/Si

diaphragm micropump," Sensors and Actuators A: Phys. 93, 87–92, 2001.
4. M. Koch, N. Harris, A.G.R. Evans, N.M. White, A. Brunnschweiler. "A novel micromachined pump based on thick-film piezoelectric actuation," Sensors and Actuators A: Phys. 70, 98–103, 1998.
5. S. Guo, T. Fukuda. "Development of the micro pump using ICPF actua- tor," Proceedings of IEEE International Conference on Robotics and Automation, vol. 1, pp. 266–271, 1997.
6. Insoo Leea, Pyohwan Hong, Chanseob Choa, Byeungleul Leec, Kyunghan Chund, Bonghwan Kimd. "Four-electrode micropump with peristaltic motion," Sensors and Actuators. 245, 19–25, 2016.
7. A.V. Lemoff, A.P. Lee. "An AC magnetohydrodynamic micropump," Sensors and Actuators B: Chem., 63, 178–185, 2000.
8. Zhou, J., Papautsky I. "Fundamentals of inertial focusing in microchannels, " Lab on a Chip, 13, 1121-1132, 2013.
9. G. Shuxiang, T. Fukuda. "SMA actuator based novel type of micropump for biomedical application, " Proceedings of IEEE International Conference, vol. 2, pp. 1616–1621, 2004.
10. Yobas L., Tang K., Yong S., and Ong E. K. "A disposable planar peristaltic pump for lab-on-a-chip, " Lab on a Chip, 8, 660–662, 2008.
11. Bekin M., Say U., Ozer, A. "Polimer tabanlı kendinden tahrikli entegre mikroparçacık ayrıştırma sistemleri tasarımı, üretimi ve karakterizasyonu [Turkish]," B.S. Thesis, Dept. of Mechanical Engineering, Istanbul Technical University, Istanbul, Turkey, 2014.
12. Zhang X., Chen Z., Huang Y. "A valve-less peristaltic pumping method, " Biomicrofluidics. 9, 014118, 2015.
13. Sonmez U., Jaber S., Trabzon L. " Investigation of superposition in microchannel geometry for inertial particle separation, " Proceedings of SB3C International Conference, Maryland, USA, Vol. 1, pp. 1079-1080, June 2016.
14. Lee et. al. "PDMS Nanoslits without Roof Collapse, " Bull. Korean Chem. Soc., Vol. 30, No. 8, 2009.
15. Huang et al. "Stamp collapse in soft lithography, " Langmuir, 21, 8058- 8068, 2005.
16. Du M., Ye X., Wu K., Zhou Z. "A peristaltic micro pump driven by a rotating motor with magnetically attracted steel balls, " Sensors, 9, 2611- 2620, 2009.

Study on High energy efficiency photovoltaic facility agricultural system in tropical area of China

Zhiwu Ge[1], Tiantian He[2], XiaoYanyang[1], JianbiaoDai[3] XuefeiJiang[4]

[1]*Hainan normal University, Physics Department, No. 99, Long Kun south road, Haikou, China*
[2]*Hainan normal University, Horticulture Department , No. 99, Long Kun south road, Haikou, China*
[3]*Shanghai JiaoTong University, Physics Department, No. 800 Dongchuan RD. Minhang District, Shanghai, China*
[4]*Hainan University, Horticulture Department, No. 58 People's RD, Haikou, China*

Abstract. The photovoltaic facility agriculture is developing rapidly in recent years, but there are many problems brought out, even in some important demonstration projects, due to the lack of standards. In order to solve some of these problems, we set up a photovoltaic facilities agricultural system in Guilinyang University City, Haikou, China and make an in-depth study on the photovoltaic facility agricultural system and its related problems. In this paper we disclose some of the experimental results. We plant corianders under two kinds of solar cell panels and general double glass assembly already sold on the market. Experiments showed that the square format cell panels are much better than row type, and the next one is general double glass assembly sold on the market, the last is the case without any shelter. 30 days after planting, the height of coriander plants are 50mm, 30mm, 23mm and 20mm correspondingly. The two typical solar cell panels have gaps between cells, and can save much more energy and improve power generation efficiency, we arrange the panels at optimum tilted angle, and design the system as open structure to save more energy. The photovoltaic facilities agricultural system we set up in Guilinyang University City can achieve much high solar energy efficiency than others and has broad application prospects.

1 Preface

With the growing development of people's living standard, the material demand is growing also. The resource consumption will greatly increase. But as the earth's non-renewable resources is becoming less and less, it is possible to exhaust the non-renewable resources in the near feature. In the exploration process of new energy, it is found that the light energy from the sun is almost inexhaustible, and it is the first choice to take the place of the non-renewable resources[1-3].

In recent years, under national policy support, solar energy has become a fast growth energy application form all over the world. China also spare no effort in the development and applications of solar energy[1-3].

These days photovoltaic facility agriculture is developing rapidly, Because of the common and complementary aspect of photovoltaic (PV) and agriculture on the land use, the industry has been seeking a combination for breakthrough, making it one of the important developing direction currently[4-9].

But because the lack of national standards, many problem occurred, even in some important demonstration projects. The main problem need to solve is how to consider both the photovoltaic power generation and plant yield above the same field. In this paper, we will try to solve this problem[4-9].

2 High-efficiency photovoltaic facilities agricultural system

In order to study the solar panels effects on plants below, we set up a photovoltaic facilities agricultural system in Guilinyang University City, Haikou, China. Fig.1 shows the photo of this system and Fig.2 shows the typical solar panels of different structure.

Figure1. Photo of tropical high-efficiency photovoltaic facilities agricultural system in Guilinyang University City, Haikou, China

There are gaps between the photovoltaic cells, and sunlight can cast directly on the ground without absorption. For typical double glass solar components, the theoretical sunlight transmittance μ is

$$\mu = \mu_1 \times \mu_2 \times \mu_3 = 92\% \times 91.5\% \times 92\% = 77.4\% \quad (1)$$

μ_1 and μ_3 is the total solar transmittance of glasses, μ_2 is is the solar transmittance of EVA film.

So we can get that the loss of sunlight ratio μ_L is

$$\mu_L = 1 - 77.4\% = 22.6\%$$

(a) (b)

Figure.2 solar panels of different structure

3 Study on the effects of corianders planted under different solar panels

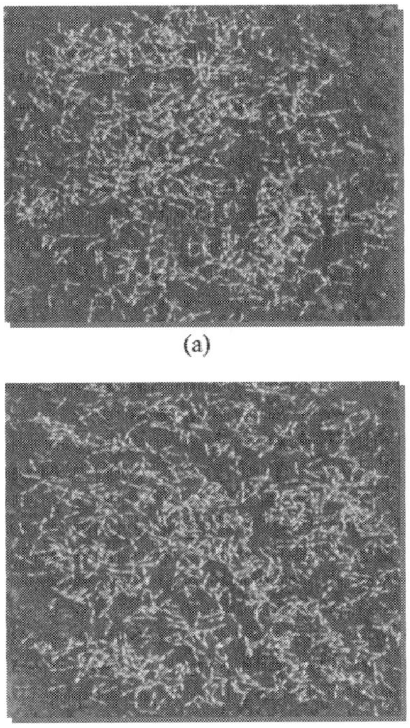

(a)

(b)

Figure 3. Photos of coriander taken in March 3rd

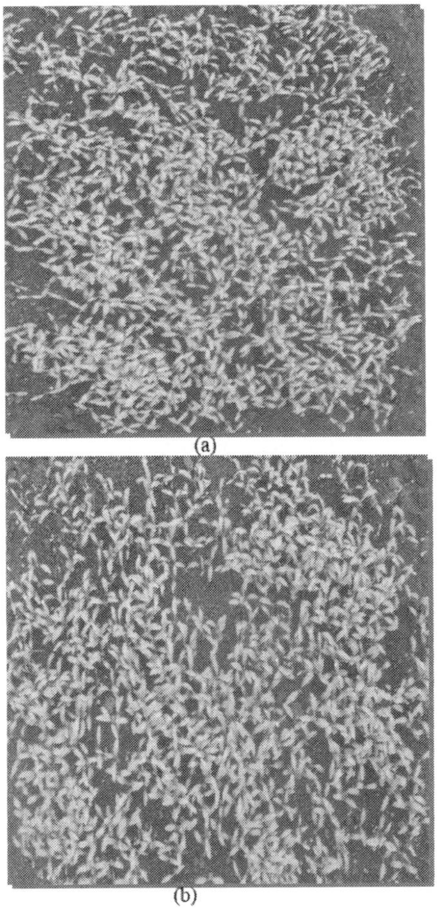

(a)

(b)

Figure 4. Photos of coriander taken in March 7th

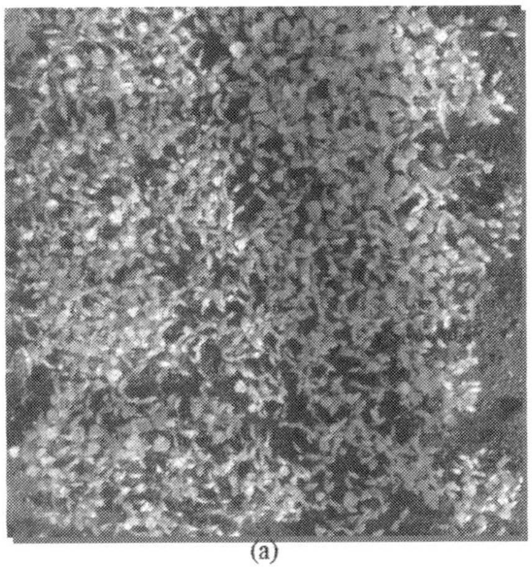

(a)

https://doi.org/10.1051/matecconf/201815308003

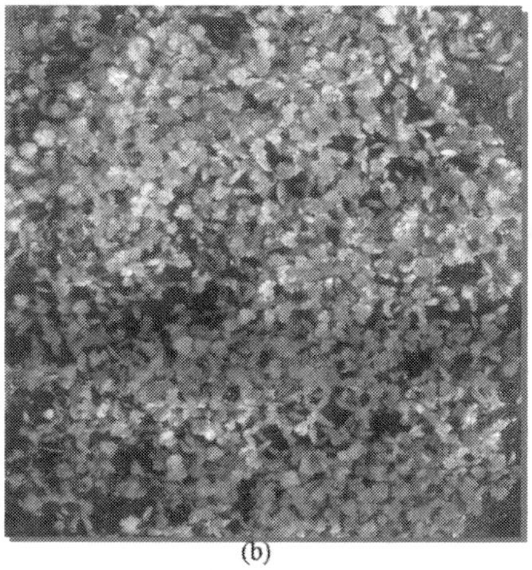

(b)

Figure 5. Photos of coriander taken in March 13th

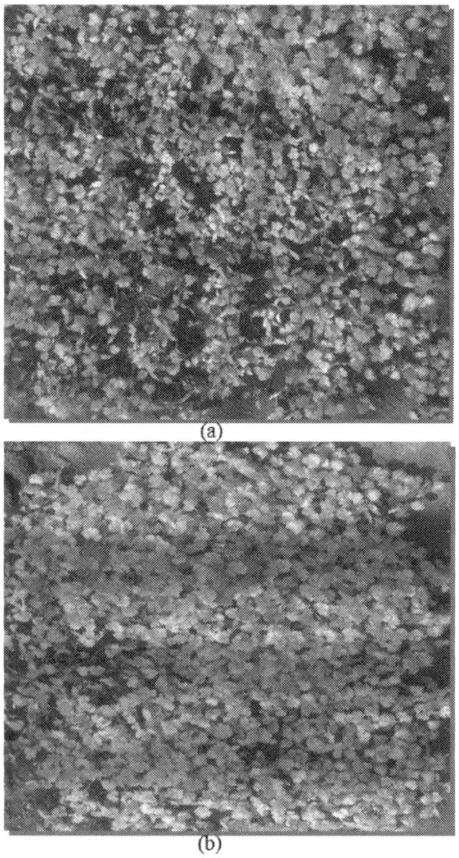

(a)

(b)

Figure 6. Photos of coriander taken in March 21st

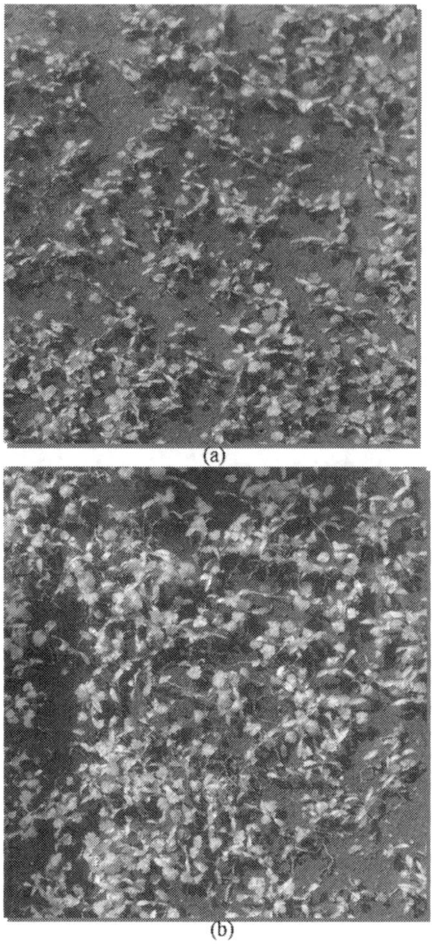

(a)

(b)

Figure7. Photos of the other coriander taken in March 21st

The corianders were planted in February 19th，2017, under different solar panels.

Fig. 3(a) to Fig. 6(a) were photos of corianders planted under solar panels as shown in Fig.2(a), and Fig. 3(b) to Fig. 6(b) were photos of corianders planted under solar panels as shown in Fig.2(b). We water each plant as often as require, and all the other conditions are the same.

From February 19th to March 3rd, 10 days after planting, corianders speared out of the earth one after another. Fig.3 shows the photos of planted corianders taken in March 3rd, 12 days after planting. Fig.4 shows the photos of coriander taken in March 7th, 16 days after planting. Fig. 5 shows the photos of coriander taken in March 13th, 22 days after planting. Fig. 6 shows the photos of coriander taken in March 21st, 30 days after planting. In each figure, compare the photo(a) to photo(b), It can be seen that over time, the difference becomes more and more obvious. Corianders planted under solar panels as shown inFig.2 (b) growth much better and yield highly compare with that under Fig.2 (a). Fig. 7 shows the photos of the coriander under other conditions taken in March 21st,

Fig. 7(a) shows the photos of coriander plants grow naturally, without any shelter, and Fig. 7 (b) shows the photos of coriander plants grow under general double glass assembly sold on the market.

We measured the height of coriander plants at March 21st, 30 days after planting, and got the quantitative results, the average height $H_1$of plants in Fig.6(b) is50mm, while H_2 in Fig.6(a) is 30mm, H_3 in Fig.7(a) is 20mm , and H_4 in Fig.7(b) is 23mm. Fig.7(a) shows the case without any shelter, so it a can be set as comparative benchmarks for height comparison.

$$R_{13} = \frac{H_1 - H_3}{H_3} = \frac{50-20}{20} = 150\% \quad (3)$$

$$R_{23} = \frac{H_2 - H_3}{H_3} = \frac{30-20}{20} = 50\% \quad (4)$$

$$R_{43} = \frac{H4 - H_3}{H_3} = \frac{23-20}{20} = 15\% \quad (5)$$

R_{13}, R_{23} and R_{43} are the height comparison ratio of case shows in Fig.6 and Fig.7(b) compared with that in Fig.7(a).

The reasons of above results are that solar panels in Fig.2(a) and (b) have much higher sunlight transmittance than general double glass assembly sold on the market, and also just proper for plants grows, compared with that in Fig.7(a) and (b), there are too much sunlight in Fig.7(a) and too less in Fig.7(b). The reason of coriander plants shown in Fig.3(b) to Fig.6(b) yield much than Fig.3(a) to Fig.6(a) is that solar panels in Fig.2(b) can cast fairly well-distributed sunshine than that under Fig.2(b) [13-17].

4 Conclusion

For typical double glass solar components, the theoretical sunlight transmittance is 77.4%, a lot of solar energy has been lost, this caused because of the absorption of EVA film and glasses, field tests showed that the real solar energy loss is more than 26%. Using the special solar panels whit gaps, the direct solar energy utilization ratio increased by 26%, and by proper arrangement, the crop yield can also increase significantly, in our experiment, the crop height increased by more than 50%, other parameters have been detected and will be analyzed in detail and publish later.

In this photovoltaic facility agricultural system, the wind and plant under solar panels can reduce the temperature of silicon solar cells, resulting in improved power efficiency[10], and all the solar panels were arranged at optimum tilted angle to get the most solar energy[11-12]. The open structure caused the system to maximize the use of local natural climatic conditions, and does not require other electrical energy consumption equipment to adjust.

So our photovoltaic facilities agricultural system set up in Guilinyang University City has much high efficiency than others.

Acknowledgement

This work was financially supported by the Hainan Natural Science Foundation in 2016(Grant Number 20165198) and 2015 key projects fund of Haikou National University Science and Technology Park.

References

1. 1.http://xueshu.baidu.com/s?wd=paperuri:(c8a0b8d51472cede3d00a617f9c147c)&filter=sc_long_sign&sc_ks_para=q%3D%E6%B5%B7%E5%8D%97%E7%9C%81%E5%A4%AA%E9%98%B3%E8%83%BD%E5%88%A9%E7%94%A8%E7%8A%B6%E5%86%B5%E8%B0%83%E6%9F%A5&tn=SE_baiduxueshu_c1gjeupa&ie=utf8&sc_us=7327178283693853975
2. Xufei. zhang. Renpeng. Xiao, Science and Technology and Equipment, 12, 35-37,(2010).
3. Yi .shen, Thesis of Lanzhou University, 1-13,(2014)
4. Zhiwu Ge, Yue Li, Chengling Bian， Study of a typical photovoltaic greenhouse in Hainan tropical island ， 2016_5th International Conference on Energy and Environmental Protection， Shenzhen, China. September 17-18, 2016， Accepted
5. Zhiwu Ge, RuipingWang, The survey of photovoltaic agricultural development in Hainan, Science and Technology Economic Herald, 2017, (3): 110-111
6. GeZhiwu, Design of Small scale Independent photovoltaic facilities above farm land ， 2014 International Conference on Environmental Protection and Human Health, Wuhan, Hubei, China， December 13-14, 2014,236-239
7. Xu Yong, discussion on some problems of solar greenhouse, modern agricultural science and technology, 2015, (11): 29-32
8. http://solar.ofweek.com/2014-06/ART-260009-11000-28835847.html
9. http://guangfu.bjx.com.cn/news/20161201/793211.shtml
10. YanXialin, Zhiwu Ge, etc., Study on temperature and power generation efficiency of crystalline silicon solar cells, 2017, (11): 42-43
11. XiaoYanyang, Zhiwu Ge, etc., Structure and wind resistance design of photovoltaic facilities agriculture system in Guilin University Town, 2017, (11): 87,68
12. Study on the optimum tilted angle of solar cells in Hainan tropical photovoltaic facility agricultural system, The 2017 3rd International Conference on Energy Materials and Environment Engineering， Bangkok， Thailand， March 10-12, 2017 :012073
13. Gezhiwu*,Bianchengling, Liujie, Research on the effect of photovoltaic solar cells setting up above farmland, 2015 International Conference on New Energy Science and Research, Changsha， Hunan， China， June 20-21， 2015. 52-56.
14. Gezhiwu*,Zhouyu, Analysis of photovoltaic Cell Arrangement in the Independent solar Power generating System Installed above Farm Land, 2015 Information Technology and Mechatronics Engineering Conference, Chongqing, China, March 28-29, 2015 , vol.7, 87-89.

15. Zhi Wu Ge*,Xue Fei Jiang, Research on the solar cell arrangement and its effect on the photovoltaic protected agriculture systems in the Hainan tropical region, 2015 International Conference on Architectural , Energy and Information Engineering, Hong Kang, China, July 15-16,2015, 253-256
16. Gezhiwu*, Bianxiufen , Weijiande , Design of Small scale Independent photovoltaic facilities above farm land , 2014 International Conference on Environmental Protection and Human Health, Wuhan, Hubei, China, December 13-14, 2014,236-239
17. Gezhiwu*, XiaoRixin, WangRuiping, Influence of solar Panels in distributed photovoltaic Power generating System above Farm Land on Field and Crops,The2014 International Forum on Applied Energy and Environment, Shenzhen, Guangdong,, China, November 28-29, 2014. Applied Mechanics and Materials Vol.737,20-23

Structural Characterization of Hexagonal Braiding Architecture Aided by 3D Printing

Zhengning Li[1,3], Ge Chen[1], Haichen Lyu[1], Chenwang Yuan[2] and Frank Ko[3]

[1]*College of Mechanical Engineering, Donghua University, 201620 Shanghai, China*
[2]*Department of Mechanical Engineering, University of Saskatchewan, S7N 5A9 Saskatoon, Canada*
[3]*Department of Materials Engineering, University of British Columbia, V6T 1Z4 Vancouver, Canada*

Abstract. Hexagonal braiding method has the advantages of high shape compatibility, interlacing density and high volume fraction. Based on hexagonal braiding method, a hexagonal preform was braided. Then, by following the characteristics of repeatability and concentricity of hexagonal braided preform, the printed geometry structure was got in order to understand and optimize geometric structure to make it more compact like the braided geometric structure. Finally, the unit cells were defined with hexagonal prism to analyze the micro-geometric structure of hexagonal braided preform.

1 Introduction

Braiding is a traditional textile process that has found wide application for composite manufacturing for the unique characteristics of braided geometry structures such as high damage tolerance, high strength and stiffness, low-cost, reliability and reproducible net-shaped preforms. Generally, braided geometry structures are formed by Cartesian braiding and rotary braiding. However, the limitations of these two braiding techniques, such as low packing density, shear deformation and limited preform geometry structure availability constrained further application of braiding geometry structures.

In order to significantly increase the packing density and yarn carrier density of the braiding loom, the concept of hexagonal braiding was developed at AFML (Advanced Fibrous Materials Laboratory) of University of British Columbia. After three generations of optimization, a computer controlled modular hexagonal

Figure 1. The Hexagonal Machine in UBC.

braiding machine was designed and fabricated jointly by AFML at University of British Columbia (UBC) and ITA at RWTH Aachen University. The hexagonal braiding loom consists of a series of hexagonal horngears and can be easily expanded to variable braiding requirements, as shown in Figure 1 [1].

In hexagonal braiding machine, each hexagonal horngear can drive 6 yarn carriers, comparing to the orthogonal gears in Tuzuki looms with 4 yarn carriers per horngear; The 18% more yarn carriers can be placed in approximately the same machine footprint area. Due to the universal nature of loom design, the hexagonal braiding allows for a large family of complex 3D polygon shapes such as triangle, hexagon, star, bifurcation, etc., by simply rearranging the yarn carriers as the cross-section shape of preforms and controlling the yarn carriers moving along specific paths, as is shown in Figure 2. More importantly, the horngears in hexagonal loom can rotate 60°, 120°, 180° or 240°, which meets the braiding requirement flexibly and the traditional braiding technology, like Cartesian braiding and rotary braiding, cannot do.

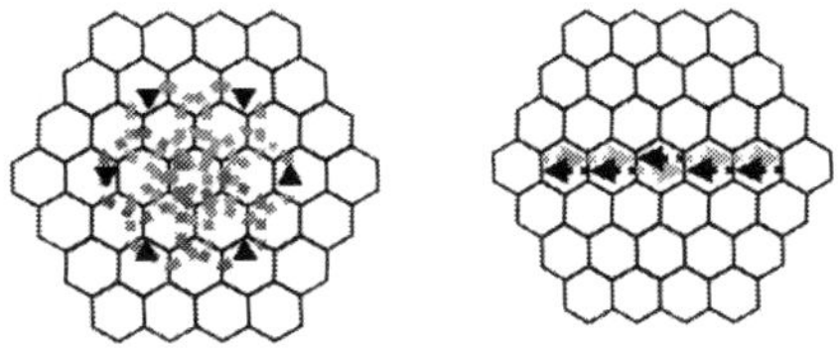

Figure 2. The variable paths of yarn carriers.

The hexagonal geometry structure varies according to the yarn carrier paths and shapes of preforms. For each unique geometry structure, the micro-geometry structure needs to be well defined in order to understand its geometry structure and predict the mechanical property. Instead of going through the laborious loom set-up and braiding process, 3D printing, typically Fused Filament Fabrication (FFF), is a promising method to explore various design options and fabrication methods in the formation of complex objects, As shown in Figure 3, after designing the model in CAD software, the model is sliced into a number of layers in slicing software, and then the 3D printer adds the materials layer by layer to fabricate the designed model. This will significantly reduce capital investment and production cost [5]. Recognizing these benefits 3D printing technology has attracted increasing attention as prototyping equipment for 3D textile preforms, including the topological design. For printed braiding geometry structure, the most excellent characteristic is that the geometry structure is intuitive and can be easy to get, compared to the costly and complex braiding process on braiding looms.

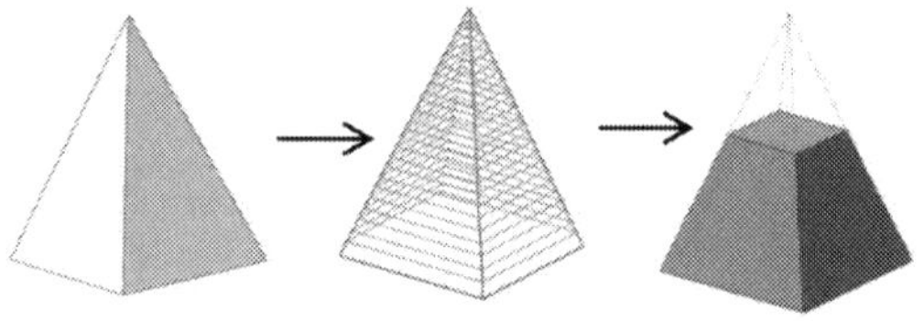

Figure 3. The procedure of 3D printing

2 Hexagonal process and parameters

2.1 Yarn carriers' motion principles

As it is mentioned above that hexagonal braiding geometry structures vary according to different movement paths of yarn carriers and yarn carriers' displacement as well, in order to articulate the hexagonal braiding process and hexagonal geometry structure, a kind of solid circular hexagonal geometry structure is created, which contains 30 yarns in the braided preform. In this braiding process, all the 30 yarns are braiding yarns and located at the flanges of seven hexagonal horngears, as shown in Figure 4, there are no axial yarns in this preform. The central horngear 0-1 keeps stationary all the time during braiding process, and the horngears with odd numbers, like 1-1, 1-3, 1-5, rotate clockwise at the same time, likewise, the horngears with even numbers, like 1-2, 1-4, 1-6, rotate counterclockwise at the same time; for each rotation of horngears, the rotation angle is 60°, as shown in Figure 5. These two groups of horngears rotate in turn, and yarn carriers are passed from one horngear to another horngear to lead the yarns intertwining in space, and then the hexagonal preform is produced. The motion sequence and rotation directions of horngears are shown in Table 1. The "+" means horngear rotates clockwise by 60° and the "-" means the horngear rotates counterclockwise by 60°.

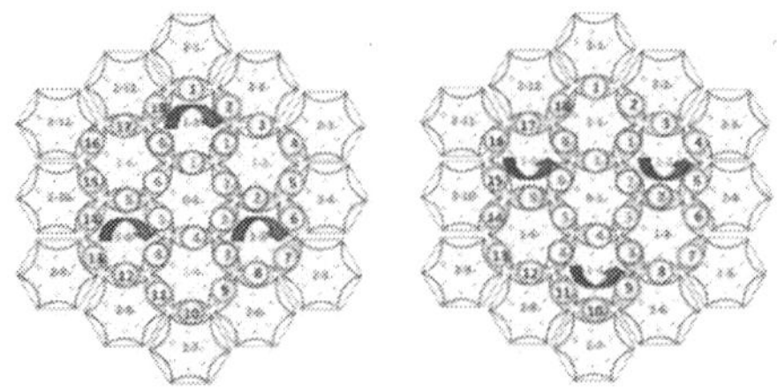

Figure 4. The movement principles of hexagonal horngears

Table 1. The motion sequence and direction of horngears

Horngear number	Step 1	Step 2
1-1	+	
1-2		-
1-3	+	
1-4		-
1-5	+	
1-6		-

2.2 Yarn carriers' path and trajectories

If the yarn carriers' positions are recorded after every step and the positions of one single yarn carrier are connected with lines, the lines will create a concave polygon path in X-Y plane (red lines); for every single yarn, their paths will be like concave polygons (red lines and blue lines), as shown in Figure 6. Following the motion principles mentioned above, every yarn carrier will return their original positions after 30 steps; actually, because of sequence of horngears rotation, the position of every yarn carrier has changed 15 times.

It seems that the paths of 30 yarns preform is the same as maypole braiding, but their braiding processes are totally different, because in maypole braiding process, only half of the intersection points among horngears can be occupied by yarn carriers in order to preventing yarn carrier collision, on the contrary, in hexagonal braiding process, all of the intersection points can be occupied by yarn carriers.

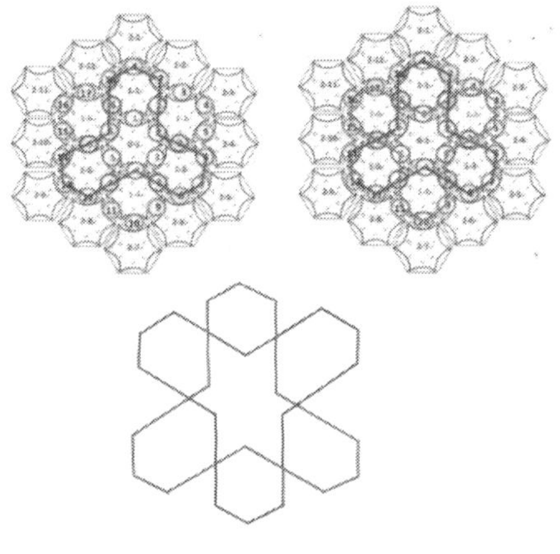

Figure 5. The paths of yarn carriers

During braiding process, braiding yarns are driven by yarn carriers into interlacing in space to produce preforms, and the take-up mechanism will take the preform from the forming plane. If the yarn carriers' positions of every

https://doi.org/10.1051/matecconf/201815308004

step on the braiding bed are connect along the Z-axis, the general trajectories of yarns carriers can be described, the trajectories of yarn carriers can be considered as the yarn trajectories. The yarn carriers' positions are defined as spatial coordinates, and then imported into Matlab to create the trajectories, as show in Figure 6, it seems that the top view of yarn trajectories in Matlab have the same paths as that of yarn carriers in Figure 5. Obviously, the trajectories described in Matlab are not real yarn trajectories, because in braided preform, all the yarns are stretched and cross the preform along the Z-axis in the shortest distance, but as for the yarn trajectories in Matlab, the coordinates are connected by lines, not even curves, so the trajectories should be optimized.

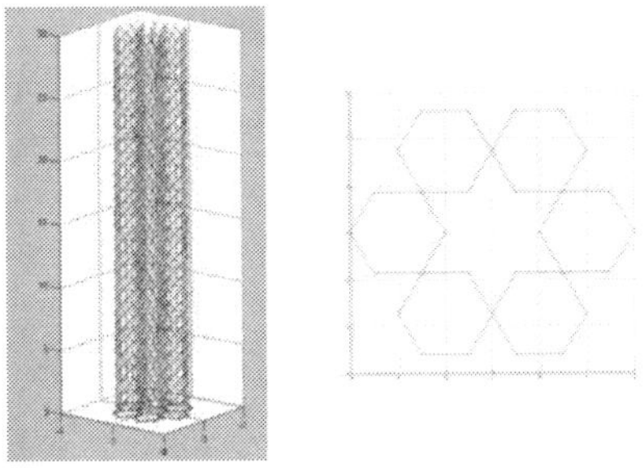

Figure 6. The trajectories of yarns described in Matlab

3 The establishment of hexagonal geometric structure and optimization by 3D printing

For the indirect description of yarns' trajectories built in Matlab, the hexagonal geometry structure and micro-structure are not easy to understand. If a real 3D model with these trajectories can be got, it would be helpful to clarify and optimize the hexagonal geometry structure. Previous research papers have explored the potential and possibility of printed woven and Cartesian geometry structures, which are based on 3D printing or additive manufacturing, specifically fused filament fabrication (FFF) technology, in that case, the hexagonal geometry structure can be fabricated in the similar way.

3.1 The 3D model built in CAD

In order to print 3D geometry structures, the 3D CAD drawing should be built. In this paper the CAD software, SolidWorks, is used to generate the 3D model. It is easy to import the coordinates of every yarn into SolidWorks to create yarn curves. By using spline tool, the coordinates are connected by splines and the hexagonal geometry structure has been built, as shown in Figure 8. Compared with the yarn trajectories described in Matlab, the yarn curves in SolidWorks are smoother.

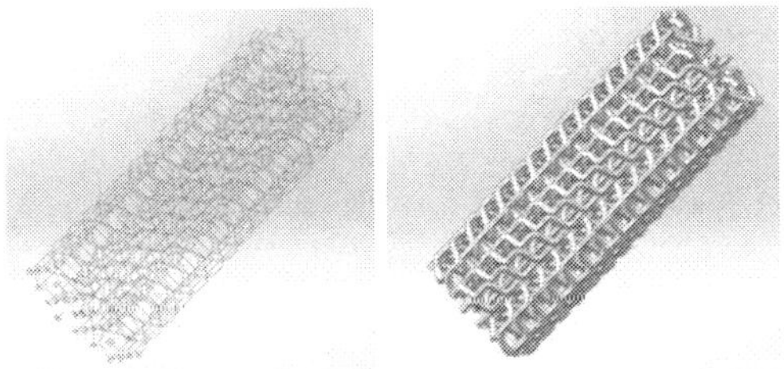

Figure 7. Line drawing and piped structure of yarn trajectories in CAD

A round circle is used to extrude along the spline to create the piped 3D model entity, likewise, all the yarns' entities are built to get the 3D hexagonal piped model, as shown in Figure 7, it seems that all the yarns has the similar tendency along Z-axis in 3D hexagonal piped model. According to the surface of the piped model, yarns have obvious buckling angles at every changing position periodically, the reason for the buckling is that the piped lines go through the positions step by step, not connect the neighbouring coordinates with the shortest line, but it can not be observed in the 3D piped model that whether the internal parts of yarns have the same tendency or not. Hence, the 3D printed model is required.

3.2 The fabrication of 3D printing model

Before the fabrication of 3D printed model, the 3D piped model needs to be sliced into layers in the slicing software to generate G-code, which can control the nozzles' movement according to the shape of model's cross-section to extrude and add the molten filament materials on the bed of 3D printer.

The printing parameters about 3D printer are shown in Table 2. Limited by the maximum printing size of the 3D printer, the hexagonal geometry structure to be printed was 1/3 part of the all geometry structure, however, the same geometry structure repeats three times during the 30 steps, namely, the printed geometry structure can reveal the yarn trajectories.

Table 2. Parameters to print hexagonal geometry structure

Parameter	Value
Filament Materials	ABS
Nozzle Diameter	400μm
Layer Thickness	210μm
Nozzle Temperature	220°C
Bedplate Temperature	105°C
Printing Speed	30mm/s

Because of the huge gaps among yarns, the support materials are required to add into the model in order to let the printing materials stick together layer by layer during printing process. More importantly, the nozzles' diameter determines printed model's resolution and fidelity, so the yarns diameter in hexagonal geometry structure should be larger than 1000 μm, namely, 1.00 mm, in which the nozzle's diameter is 400 μm. For the printed model is

https://doi.org/10.1051/matecconf/201815308004

only used to understand hexagonal geometry structure, not to test the mechanical property, the printing model's infill density is not a requirement and can be set at random.

After the model printing was finished, the support material was removed form the model and the printed hexagonal geometry structure was got, as shown in Figure 8, the printed hexagonal geometry geometry structure has the same exterior looking as the 3D CAD model, and the yarns interlacing in the center can be observed clearly. However, obviously, the gaps among yarns, longitutionally and radially, are extraordinarily large, it means that the hexagonal geometry geometry structure need to be optimized. Through the tracing yarn (red) in the model, it reveals that every yarn went from the edge.

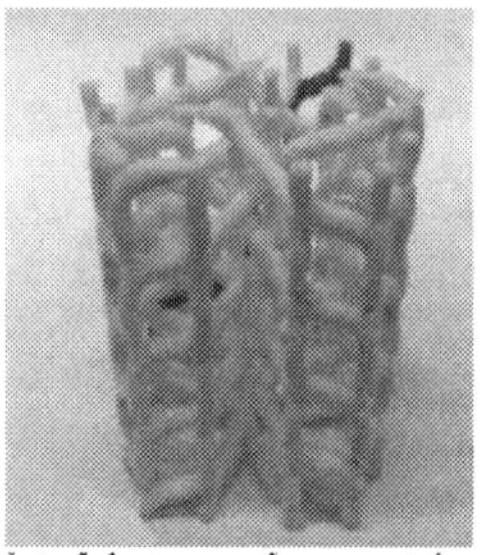

Figure 8. The printed hexagonal geometric structure without optimization

3.3 The optimization of hexagonal structure

All the positions of the yarn carriers are concentric, and the yarns are symmetric to the z-axis, the positions are defined as coordinates (x_i, y_i, z_i), i is positive integer. According to the group thoery, the space lattice which consists of the positions can be converted to new ones by matrixing to make the geometry structure more compact. For the conversion, the conversion matrix is defined as W_i.

$$W_i \begin{bmatrix} x \\ y \\ z \end{bmatrix} = \begin{bmatrix} \rho_{i1} & 0 & 0 \\ 0 & \rho_{i2} & 0 \\ 0 & 0 & \rho_{i3} \end{bmatrix} \begin{bmatrix} x \\ y \\ z \end{bmatrix} = \begin{bmatrix} x' \\ y' \\ z' \end{bmatrix} \tag{1}$$

The conversion coefficients, ρ_{i1}, ρ_{i2}, ρ_{i3}, have some relationship with the positions' distance to the center and the jamming condition, and $\rho_{i1}, \rho_{i2}, \rho_{i3} < 1$. Let us assume that the positions move along the radius to the center, also made assumptions for geometric geometry structure of hexagonal braiding process:

(1) All the braiding yarns have circular cross section, same linear density and constant fiber packing density;

(2) During the braiding process yarns tension is high enough to make a non-crimp yarn path in hexagonal braiding geometry structure.

So, the coordinates' conversion will follow the relationship below,

$$\begin{cases} x' = \rho_{i1}, \cdot \sqrt{x^2+y^2} \cdot \cos\left(\arctan\left(\frac{y}{x}\right)\right) \\ y' = \rho_{i2} \cdot \sqrt{x^2+y^2} \cdot \sin\left(\arctan\left(\frac{y}{x}\right)\right) \\ z' = \rho_{i3} z \end{cases} \tag{2}$$

If the rotation angle of yarn jamming is taken into account, function (2) can be revised as,

$$\begin{cases} x' = \rho_{i1}, \cdot \sqrt{x^2+y^2} \cdot \cos\left(\arctan\left(\frac{y}{x}\right) \pm \varphi_i\right) \\ y' = \rho_{i2} \cdot \sqrt{x^2+y^2} \cdot \sin\left(\arctan\left(\frac{y}{x}\right) \pm \varphi_i\right) \\ z' = \rho_{i3} z \end{cases} \tag{3}$$

Imported the converted coordinate into SolidWorks and use spline to create the trajectories, the piped model was got, as show in Figure 9.

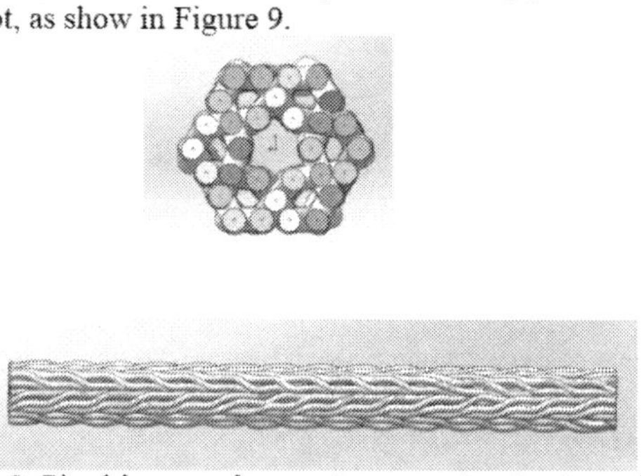

Figure 9. Piped hexagonal geometry geometry structure with optimization

The optimized hexagonal geometry structure has obvious tube in the center of cross-section; for one hand, the yarns deformation and jamming are not considered in this structure, for the other hand, the tubes actually exist in the braided hexagonal preform. As shown in Figure 10, the braided hexagonal preform was cured with resin and cut into pieces; there was a tube in the center of cross-section. Without considering the deformation of yarns, the optimized geometry structure is close to the braided hexagonal preform.

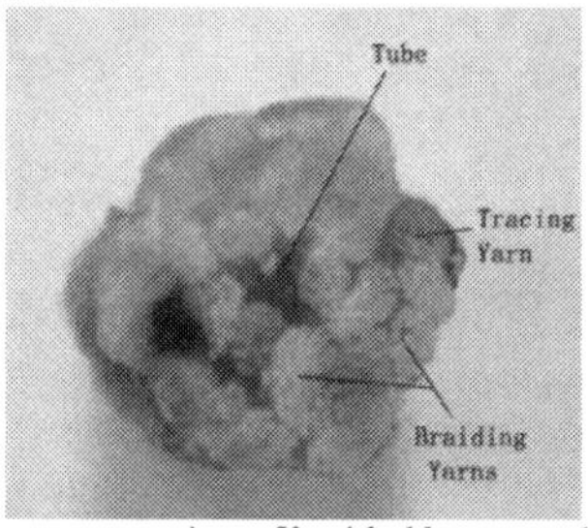

Figure 10. The cross-section of braided hexagonal preform with cured resin

Then, the optimized hexagonal geometry structure was printed by the 3D printer. Compared with the hexagonal geometry structure without optimization, the optimized hexagonal geometry is more compact, hence it is no need to add support materials during printing to make the geometry structure generated. The printed hexagonal geometry structure with optimization is shown in Figure 11.

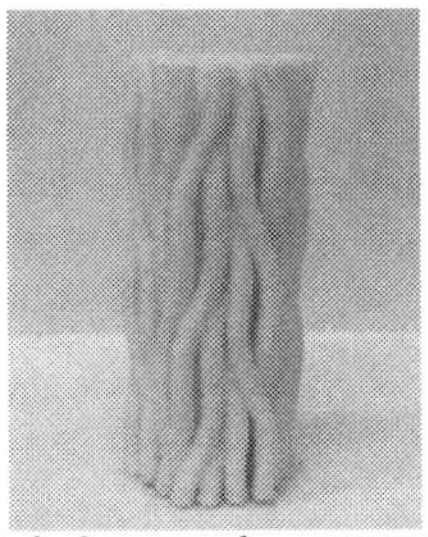

Figure 11. Printed hexagonal geometric structure with optimization

4 The unit cells of hexagonal structure

Traditionally, researchers tend to define unit cell geometry to model 3D braided composites. The approach used in geometric modeling of textile geometry structures is to first determine the dimension, shape and fiber geometry structure of the unit cell based on process and structural analysis; using the unit cell geometry identified, the relationship between processing variables and key geometric parameters can readily be established [5]. For traditional braided geometry structures, three types of unit cells, interior cell, corner cell, surface cell, are used to describe the braiding micro-geometry structure. However, for hexagonal geometry structures, because of the repeatability and concentricity, the hexagonal prism can be used to define the unit cell, in that case, there will be two types of unit cells, as show in Figure 12 and Figure 13. In both types of unit cells, they contain 6 bundles of yarns, for one unit cell the yarns have 3 directions, for the other unit cell, the yarns have 6 directions.

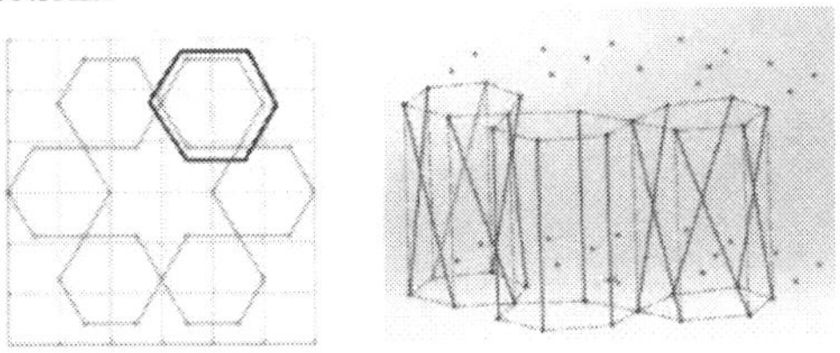

Figure 12. The definition of hexagonal unit cell

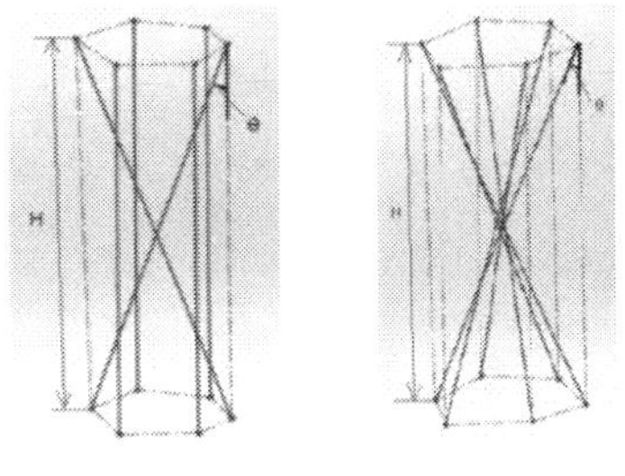

Figure 13. Two types of hexagonal unit cells

5 Results

The paper has demonstrated the possibility of using 3D printing technology to aid the geometry structure establishment of hexagonal geometry structure. An optimized hexagonal braiding geometry structure was built through coordinates' conversion and printed geometry structure. Based on micro-geometry structure analysis, two types of unit cells were defined. It reveals that 3D printing offers us a direct and fast way to establish the hexagonal geometry structure and fabricate the braiding geometry structure. Moreover, in this paper only one kind of circular hexagonal geometry structure was discussed and for the variable hexagonal geometry structures they may have other micro-geometry structure and unit cells, hence, in the future the research emphasis will put on them.

Acknowledgement

Zhengning Li acknowledges the support of China Scholarship Council (CSC)

References

1. Schreiber, F., K. Theelen, and S. Schulte, 18th international conference on composite materials, Jeju Island, Korea (2011)
2. Tsuzuki, M., Kimbara, M., Fukuta, K., Machii, A., US Patent 5,067,525 (1991)
3. Schreiber, F., et al, 17th International Conference on Composite Materials (2009)
4. Quan Z, Larimore Z, Wu A, et al, Composites Science and Technology, E **126**, 39-148 (2016)
5. Du G W, Ko F K. Unit Cell Geometry of 3-D Braided Geometry structures. Journal of Reinforced Plastics and Composites, E **12**, 752-768 (1994)

Design of Intelligent Manufacturing Big Data Cloud Service Platform

Danlin Cai[1,2,3,4], Mingyu Chen[1,2,3,4], Daxin zhu[1,3,4,5], Junjie Liu[1,2,3]

[1] *Quanzhou Normal University,China*
[2] *Fujian Provincial Key Laboratory of Data-Intensive Computing,China*
[3] *Fujian University Laboratory of Intelligent Computing and Information Processing,China*
[4] *Fujian Provincial Big Data Research Institute of Intelligent Manufacturing,China*

Abstract. With the coming of the intelligent manufacturing, the technology and application of industrial big data will be popular in the future. The productivity, competitiveness and innovation of the manufacturing industries will be improved through the integrated innovation of big data technology and industries. Besides, products, production process, management, services, new form and new models will be more intellectualized. They will support the transformation and upgrading of manufacturing industry and the construction of an open, shared and collaborative ecological environment for intelligent manufacturing industry.

1 Background

In recent years, due to the popularity of Internet and smart phones, data size tends to have an explosive growth. According to the prediction of International Data Corporation IDC, world's data may double every two years and till 2020, the data volume stored in electronic forms will reach 35ZB. Humans have entered the age of big data. With the deep integration of informatization and industrialization, the information technology has permeated into all aspects of industrial chains, bringing abundant data owned by industrial enterprises. Industrial big data are data produced in the informationization application process of industrial fields and they are featured by large scale, multi-sources, continuous sampling, low value density and strong dynamic nature. Big industrial data has become an important driving force for the new round of industrial revolution.

Industry is the foundation and pillar of the national economy as well as an important symbol of a country's economic strength and competitiveness. With the development of emerging technologies such as cloud computing, big data and Internet of Things, a new round of industrial revolution has taken shape in the world with manufacturing transformation and upgrading as the most important task. The major industrialized developed countries in the world have formulated industrial re-development strategies. In February 2012, the United States released the "National Strategic Plan for Advanced Manufacturing", which will promote the development of advanced manufacturing industries to the national strategic level. The National Strategic Plan for Advanced Manufacturing provides three major principles, five major goals and corresponding countermeasures to promote the development of the advanced manufacturing industry in the United States from the aspects of investment, labor force and innovation. In April 2015, Germany proposed the "Industry 4.0" strategy, emphasizing that through the full integration of information network and industrial production system, the products and production equipment, the vertical and inner parts of the factory, the factories and factories can all be integrated through the CPS (Physical Information System) to realize the production intellectualization, enhance manufacturing flexibility and engineering efficiency. In 2015, France launched the "New Industrial Policy in France" with an overall layout of "one core and nine major fulcrums."

One core refers to "future industry". The main content is to achieve the transformation from industrialization to digitization and intellectualization as well as the business model transformation driven by transformation and upgrading of production tools. Nine major fulcrums include new resource development, sustainable cities, environmentally friendly vehicles, internet technology, big data technology and new medicine etc. In May 2015, the State Council formally issued the "Made in China 2025" program. The program put forward to focus on promoting the integration of information technology and industry with the intelligent manufacturing as the main direction, developing intelligent equipment and products, promoting the production intellectualization of the process, cultivating new production methods and realizing intelligent production and intelligent manufacturing through the construction of intelligent factories.

2 Problem of Intelligent Manufacturing Big Data Urgently Needs to Be Solved

China's economic development has entered a period of new normal. The state vigorously implements innovation-driven strategy with manufacturing industry as the mainstay of the national economy. Industry 4.0 is the fourth industrial revolution led by intelligent manufacturing. "Made in China 2025" is a strategic program at the national level, which aims to speed up the integration of a new generation of information technology and manufacturing industries as the main line and promote intelligent manufacturing as the main direction with the goal of meeting the needs of major technological equipment for economic and social development and national defense construction so as to realize the historic transition of manufacturing industry from big to strong by strengthening industrial fundamental ability, enhancing the level of comprehensive integration, improving multi-level and multi-type personnel training system, promoting industrial transformation and upgrading and cultivating manufacturing culture with Chinese characteristics.

The key issues facing by intelligent manufacturing industry are the transformation from mechanization and automation to digitization and intelligent manufacturing, from focusing on growth rate to focusing on quality and efficiency, and from production-oriented manufacturing to service-oriented manufacturing. In this process, urgent solutions are needed for three important issues.

(1) How to promote the application of big data in all aspects of the industrial chain so as to achieve networking and intellectualization in the manufacturing industry, and to enhance the production technology and efficiency of industrial enterprises?

(2) How to promote the integration and development of a new generation of information technologies such as big data, Internet of Things, and cloud computing so as to promote the transformation and upgrading of traditional industries and the acceleration of the development of new industries?

(3) How to make use of big data to promote the deep integration of informatization and industrialization, propel the gathering of industry data and carry out the application of big data fusion in the industry?

Solving these three major issues will have a direct bearing on exploring a new mode of advanced manufacturing in China and enhancing our ability to provide typical experiences and exemplary leadership for the "Made in China 2025" to accelerate its implementation nationwide.

3 Main Service Contents of Intelligent Manufacturing Big Data Platform

The problem of effective management of huge data resources and the management of massive data mining has become a common issue that needs to be faced and solved by academia and industry. As shown in Figure 1, we conducted a three-level big data application for the intelligent manufacturing industry.

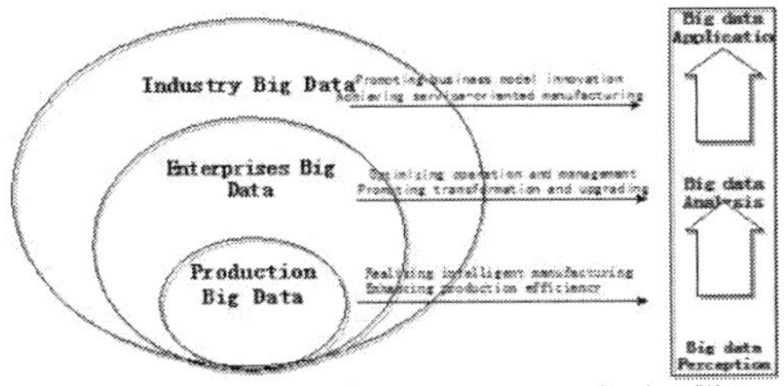

Figure 1. Main service contents of intelligent manufacturing big data platform

(1) Intelligent manufacturing and production big data can be used to achieve networking and intellectualization of manufacturing industry and establish advanced production system to realize intelligent production so as to enhance enterprise production technology and efficiency.

The workshop management level needs to be improved. Production line sensors can be used to collect temperature, pressure, heat, vibration, noise and other data so as to achieve several results, including equipment diagnosis, electricity consumption analysis, energy consumption analysis and quality accident analysis. Big data can be used in the production process to analyze and optimize the entire work flow. It integrates and aggregates the data of all aspects of production and manufacturing, establishes a virtual model for the production process of industrial products, simulates and optimizes the production work flow, and makes an integrated analysis of the manufacturing data of various links, which is conductive to improve and optimize the production process and promote the establishment of a modern production system. Through automated and intelligent control of the entire manufacturing process, information sharing, system integration and business collaboration are promoted. Scientific decision-making in the production work flow, and the automation, personalization, flexibility and self-optimization of the production process at the greatest extent will lead to the realization of intelligent manufacturing with precise, high-end and agile manufacturing capabilities.

(2) Intelligent manufacturing enterprise big data can be used to achieve lean management and promote transformation and upgrading of enterprise by optimizing the operation and management system as well as the industrial supply chain. RFID and other electronic identification technology, Internet of Things technology and mobile Internet technology can help industrial enterprises to obtain a complete big data of product supply chain. By analyzing these data, the efficiency of storage, distribution and marketing will be significantly improved and the cost will be sharply declined, which will optimize the supply chain and promote the convergence and optimization of the whole process of operation and management of intelligent manufacturing enterprises. The integration of resources such as production data, financial data, management data, purchasing data, sales data and consumer behavior data will help enterprises to find the best investment proportion of production factors through data mining analysis, and to realize research and development, supply

and marketing, operation and management, production control, seamless integration of business and financial processes as well as business collaboration so as to promote the integration, reorganization and optimization of business, decision-making and operational processes.

(3) Intelligent manufacturing industry big data can be used to promote in-depth integration of information technology and industrialization, propel enterprises to gather industry data and accelerate business model innovation so as to realize service-oriented manufacturing. The use of big data can also help intelligent manufacturing enterprises to innovate their products and services by developing new business model. The real-time monitoring of product operating status can be conducted through embedded sensors in products and the product sales data and customer data can be obtained through the business platform. By analyzing and predicting these data, intelligent manufacturing enterprises can develop various online value-added services, including fault early warning, remote monitoring, remote operation and maintenance and quality diagnostics etc. The personalized, online and convenient value-added services will expand the product value space and realize the transformation of operation model from product-oriented model to "manufacturing + service" model as well as provide relevant services and decision-making basis for other industries.

4 Design of Intelligent Manufacturing Big Data Platform

For big data application in intelligent manufacturing from workshop to enterprise to industry, from point to surface, it still follows the general rules of the big data system, that is, big data perception (acquisition), big data analysis and big data application. The goal of intelligent manufacturing big data application is to build a data chain which covers the whole procedure, process of industry and the whole life cycle of the product. Figure 2 shows the main link of intelligent manufacturing big data in practical application, which includes data perception, data analysis, data application and other three levels.

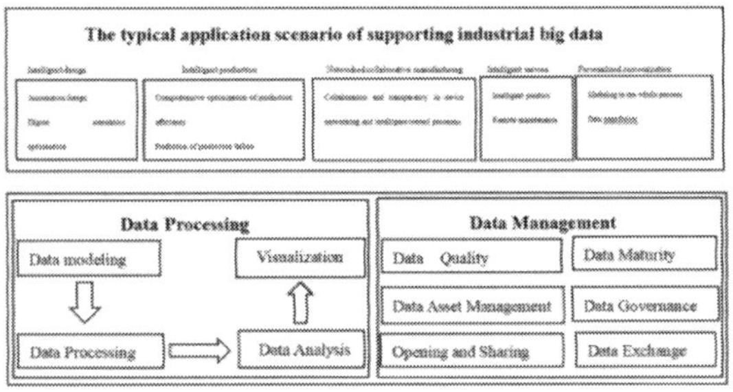

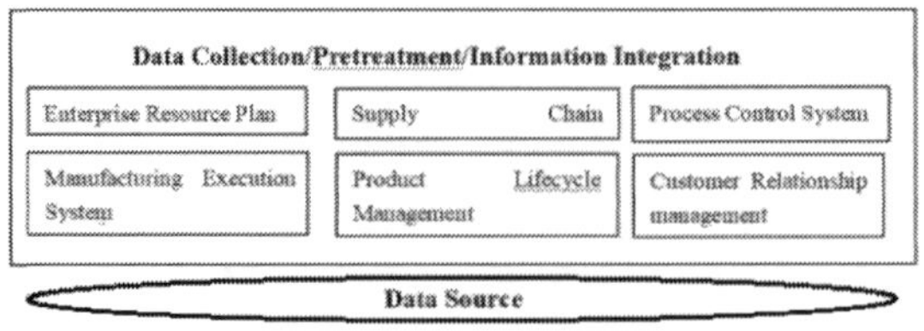

Figure 2. Three-tier architecture of intelligent manufacturing industry

Data sensing layer mainly realizes the collection and integration of data in all parts of the industry and gets through the data connection of the existing information system. The data sources include three categories. The first one is business data related to the production and operation of enterprises, which mainly includes lots of product research and development data ,customer information data, production data, logistics supply data and environmental data accumulated from enterprise information system. The second category is device object data, which refers to the equipment operation parameters, operating condition parameters, operating environment parameters and other data which can be used to evaluate the operation status of the production and equipment under the internet operation mode of industrial production equipment. The third kind is external data, which refers to the data of external Internet source of relevant enterprises related to industrial production activities.

Data analysis layer is the core of industrial big data. Its key target is to realize the industrial production data for intelligent process, intelligent products, intelligent new mode of new forms, intelligent management and intelligent service in the field of data processing and data management. Through data modeling and data processing, we can realize the visualization of data result and 3D industrial scene. And we can realize data management in data quality, capability maturity, data asset management, data open sharing and trading, etc.

Data application layer mainly generates visual description, control, decision making and other applications based on the results of data processing and data management in order to realize intelligent design, intelligent production, networked collaborative manufacturing, intelligent service, personalized customization and other typical intelligent manufacturing modes.

5. Application prospect of intelligent manufacturing big data platform

The result of industrial big data mining and analysis can be used in research, development and design, complex production process, product demand forecast, industrial supply chain optimization and industrial green development.

5.1 Applications in the design field

Traditional research, development and design got finished products by experiencing trail-manufacture, revise and iteration repeatedly in the lifetime of products, while industrial big data provides new advantages to research, development and design. From the perspective of traditional process, at an early stage, it is necessary for designers to start from preliminary investigation and put customer demands into consideration, and then find out loopholes in the market and relevant linkages by divergent thinking. However, under the influence of industrial big data, designers can publish their own products online. Users do not only pay for their favorites through the platform but also can form personalized customize products adding their thoughts.

5.2 Applications in the optimization of complex production process

Facing complex production process in the modern industrial manufacturing production lines, the industrial Internet of things is equipped with plenty of sensor devices. Real-time collected data can complete analysis in various forms including device diagnostics, electrical power consumption, energy analysis, quality accident analysis (violation of production regulations and failure parts), etc. Firstly, as for production process improvement, using big data in the production process can help to analyze the whole production flow chart and get to know how each link works. Once some process deviates from the standard one, there will be an alarm signal, so errors or bottlenecks will be more quickly detected and problems will be easier solved. In addition, big data technology can be used to build virtual model, simulate and optimize production process. When all processes and performance data can be rebuilt in the system, this kind of transparency will help manufacturers improve their production process. Secondly, in terms of energy consumption, by using sensors to supervise whole procedures during the process of equipment production, abnormal energy consumption or peak can be found timely to optimize the manufacturing process and reduce energy consumption.

5.3 Applications in product demand forecast

For product development, by analyzing current changes in demand and combining forms, the adjustments in the function and performance of the products can be made through focuses of consumers. First, web crawler technology and web-service technology are utilized to obtain related basic data, enterprise international data, user's behavior data and third-party data. And with persona, target user attributes can be described objectively and accurately and statistics in preferences and function demands can be made, which contributes to more acceptable products with core needs and providing more potential sales opportunities. And the persona is smart with system grouping to gain various target-user groups and then push targeted marketing. In addition, through historical data of multidimensional combination, the occupation and changes in regional demand, market popularity and ordinary combination and the lever of consumers can be observed to adjust product and distribution strategies.

5.4 Applications in the optimization of industrial supply chain

The application of industrial big data in supply chain is mainly embodied in the chain optimization. In other words, through information integration in the whole industry chain, the collaborative optimization and dynamic equilibrium in the production system are achieved, which further enhances productivity and reduces costs. The primary application includes optimization of supply chain distribution system and quick response to user requirements. Optimization of supply chain distribution system can help industrial enterprises obtain big data of supply chain by RFID electronic identity technology, networking technology and mobile Internet technology. Utilizing data in sales, production sensor and database from suppliers, industrial manufacturing enterprises can analyze and predict the needs in various regions of the world accurately to encourage efficiency of distribution and storage. Utilizing data produced by sensors in the product to analyze failure parts and confirm requirements, those enterprises can predict when and where integrating parts will in need; this largely improves timeliness of production, reduces repertory and optimizes the supply chain.

5.5 Applications in industrial green development

The target of green manufacturing is to realize least consumption, minimized or even no environmental pollution in the process of design, manufacturing, use and disposition. It emphasizes that information related to resource consumption and environmental information should be handled and considered integrally. Information flow, material flow and energy flow used to manufacture system should also be combined to integrate and optimize systematically. By monitoring and management in the process of recipe, crafts, purchases of materials, manufacturing, storage, transportation, usage, overhaul and scrap, taking data acquisition as the front and data analysis and mining as service, NGIT provides strong guarantee for manufacturing enterprise. The 2015 Industrial Green Development Special Action Implementation Plan officially released by Ministry of Industry and Information put "Organization and Implementation of Digital Energy Efficiency Promotion Plan" as an important task, which includes pilot project of green data center of key industries and the construction of national energy conservation monitoring and analysis platform. The new technology such as Internet, cloud computing, big data and industrial robots will guide and promote industrial equipment towards green development. It will certainly have a positive impact on industrial transformation, upgrading and green development through developing and perfecting manufacturing improvement program in equipment industry based on big data.

6 Big Data Service Circle of Intelligent Manufacturing Industry Cluster

Intelligent manufacturing enterprises, in the help of electronic identification technology like RFID, Internet of Things technology and mobile Internet technology, obtain a complete big data of supply chain. The analysis on these data helps to achieve warehousing, distribution, sales efficiency and substantial cost reduction, thus optimize the supply chain. It integrates resources such as data of production, finance, management, purchasing, sales and consumer behaviour of intelligent manufacturing enterprises to help them find the best investment proportion of production factors through data analysis, realize the seamless convergence and business collaboration in the research and development of supply and marketing, management, production control and all business and financial procedures, as well as promote the integration, restructuring and optimization of the process of business, decision-making and management. The big data and cloud computing technology can further tap production relevance in intelligent manufacturing enterprises cluster to provide customized, temple, real-time business information analysis services, and to enhance strategic planning, access to competitive intelligence and industrial information sharing capabilities. After using ETL tools for big data analysis, classification and research, generating visual reports big data visualization technology, and sending customers information through the client software, it can provide intelligent manufacturing enterprises with timely, effective, accurate and comprehensive information service. Enterprises can monitor the running status of the product in real time by sensors embedded in the product and obtain the sales and customer data by the business platform. Through the analysis and prediction of these data, enterprises can carry out online value-added services such as fault warning, remote monitoring, quality diagnosis and remote operation and maintenance. With these online, customized and convenient value-added services, the product value space is expanded, making the business model with the product as a core turn to the "manufacturing + service" model, and finally building the big data service system of the intelligent manufacturing industry cluster.

Acknowledgements

The study is supported by Natural Science Foundation of Fujian Provincial Science and Technology Department (No.2017J01508) and Key Projects of Fujian Provincial Education Department (No.JA15388) and Quanzhou Science and Technology Project (No.2014Z133) (NO.2012Z122).

References

1. White paper on industrial big data(2017 Edition). https://wenku.baidu.com/view/66f51af69fc3d5bbfd0a79563c1ec5da51e2d670.html(2017)
2. Jay Lee. From Bigdata to Intelligent Manufacturing.(2017)
3. Daxin Zhu, Xiaodong Wang.An Efficient Dynamic Programming Algorithm for a New Generalized LCS Problem.IAENG:International Journal of Computer Science , (2016)
4. Daxin Zhu, Xiaodong Wang.A Simple Polynomial Time Algorithm for the Generalized LCS Problem with Multiple Substring Exclusive Constraints.IAENG:International Journal of Computer Science , (2015)

MATEC Web of Conferences **153**, 09001 (2018) https://doi.org/10.1051/matecconf/201815309001
ICMME 2017

Elevation Effect in Urban Water Distribution Model

Rangsan Wannapop , Thira Jearsiripongkul and Krit Jiamjiroch

Mechanical Engineering Department, Faculty of Engineering, Thammasat University, Rangsit Campus, 99 Moo 18, Paholyothin Road, Klong Nueng, Klong Luang, Pathumthani,12120, Thailand

Abstract. Metropolitan Waterworks Authority (MWA) is Thailand's national government agency responsible for the supply of water to 3 provinces Bangkok, Nonthaburi and Samut Prakan with more than 2,384.9 km^2 of service area and 2,281,058 consumers in the year 2016. Bangkok, which is both the capital and the economic center of the country, is densely populated. Consequently, there is a huge demand for water; MWA has to supply 5.914 million cubic meters of water per day. Because the metropolitan water supply area is a densely populated city, the water supply system is very complex like a spider's web. For this reason, MWA has adopted EPANET software for its water supply managing tool in the main pipeline system. There are some mistakes in the main pipe network; the elevations of the nodes are not assigned, so there are some errors. In this study, we have assigned elevations for all nodes on the pipeline network based on mean sea level (MSL). After adjusting the elevation of each node, it was found that the new pipeline network has increased the correlation between means to 0.893 from the existing model mean of that is 0.803 of accuracy up 0.09 (11.2%).

1 Introduction

Metropolitan Waterworks Authority (MWA) is Thailand's national government agency responsible for the supply of water to 3 provinces Bangkok, Nonthaburi and Samut Prakan. These 3 provinces are heavily populated. Bangkok is the capital and economic centre of Thailand, Samut Prakan is an industrial city and Nonthaburi is a growing city. There are many different types of customers, residential and public water subscribers, such as hotels, schools, factories, restaurants, cafes, bars, shopping malls, offices, parks, temples and residential houses of different types (duplex or multiplex villas with garden, apartment flats etc.) in the MWA service area. Therefore, MWA service area has high water consumption. Each day, MWA is required to supply 5.914 million m^3 water to the service area.

The water supply system of MWA is based on 2 main water sources (Chao Phraya River and Mae Klong River), 4 Water Treatment Plants and 18 pumping stations. The MWA plans to supply enough water to meet the needs of consumers with a minimum water pressure of 10 meters head for all service area.

Practically, it is unreasonable to expect to ensure the availability of sufficient quantity and pressure of good quality water to the various sections of the community in accordance with the demand 100% of the true. Many computer tools have been developed; out of all the tools available EPANET has become the most popular and convenient for the effective design of complex pipe networks [1]. EPANET is a decision tool to manage the water distribution system to meet the needs of the consumer at all times of day with proper water pressure. Water distribution system, however is related to production cost and water pumping energy.

2 MWA Main Pipeline Network

MWA main pipeline network is large and very complex containing 1,727.088 kilometres of pipes with diameters ranging between 500 mm and 1800 mm, which are supplied by 4 treatment plants and 18 pumping stations. All pumping stations are equipped with variable speed drive pumps. MWA has divided the service area to 5 zones, 18 branches and 669 District Metered Areas (DMAs). MWA main piping network consists of pipes, reservoirs, pumps and pressure reducing valves (PRV), etc. The piping elements are connected like a spider's web (Figure 1). Computation of flows and pressures in a complex network has been a great challenge and of interest for those involved with designs, construction and maintenance of public water distribution systems [1]. As a result, MWA has adopted EPANET software to help manage the main pipeline system. EPANET is a hydraulic (pressure and flow) and water quality behaviour (chlorine disinfectant) simulation software, developed by Lewis A. Rossman of United States Environmental Protection Agency (EPA), using "Gradient Method" as defined by Todini and Pilati (1987) to solve the flow continuity and head loss equation [2]. It is open source, so can be developed further.

Figure 1. MWA main piping network.

3 The basic equation of pipeline hydraulics system

3.1 Conservation of mass (continuity) [3]

$$Q = V_i A_i = V_j A_j \tag{1}$$

where Q = flow rate (volume/time)
V = mean velocity (length/time)
A = cross-section area (square length)

3.2 Conservation of energy (Bernoulli Equation)

$$\frac{p_i}{\gamma} + \frac{V_i^2}{2g} + z_i = \frac{p_j}{\gamma} + \frac{V_j^2}{2g} + z_j + h_f \tag{2}$$

where $\frac{V^2}{2g}$ =Velocity head (length)

$\frac{p}{\gamma}$ =Pressure head (length)

z =Elevation head or potential energy (length)

h_f =Head loss (length)

p =pressure (force/area)

g = acceleration (length/time^2)

i, j = index number (1...n)

In pipes of the same diameter, if both flow rate and velocity are the same ($Q_i=Q_j$, $V_i=V_j$). Equation (2) becomes.

$$\frac{p_i}{\gamma} + z_i = \frac{p_j}{\gamma} + z_j + h_f \tag{3}$$

$\frac{p}{\gamma}$ is Static Head or Pressure Head and $\frac{p}{\gamma} + z = H$ is Head Grade Line [4].

4 Metrology

MWA attempts to adjust the pipeline network model to reduce errors and calibrate with field measurements data. The calibrations were performed on the basis of pressures measured during 24 hours. Field measurements of water pressure at different points of the MWA network were conducted using flexible pressure transmitters. Data sets obtained from field measurements of water pressure were then used to calibrate the model predictions [5]. The locations of the field pressure measurement points are presented in figure 2. The majority of the main pipes of MWA (about 86%) are steel pipes and the diameters are 500-1800 mm. Units were set to metric system in EPANET and the head losses formula was set to Hazen-Williams.

Hazen-Williams roughness coefficient of the pipes was considered to be 120 for all the pipes in the model. The base demand of the main pipe network used the demand of each DMA. The average mean absolute error was approximately 1.586 m and this showed the difference between that model predictions and measurements of pressure was quite high. The calculation results tended to be in the same direction as the measured data, but there were still errors from many reasons such as the pipe friction factor, percentage of valve opening, wrong elevation of node etc. The calibration showed that the pressure measurements contained constant errors at many nodes, for example, in figure 3 which presents results obtained by the calibrated model for the measurement point U014, one can see that the differences between the measured and modelled pressures oscillate around the average pressure 3.892 m. This error was caused by wrong elevation of nodes. This value is a constant error in terms of Eq. (3).

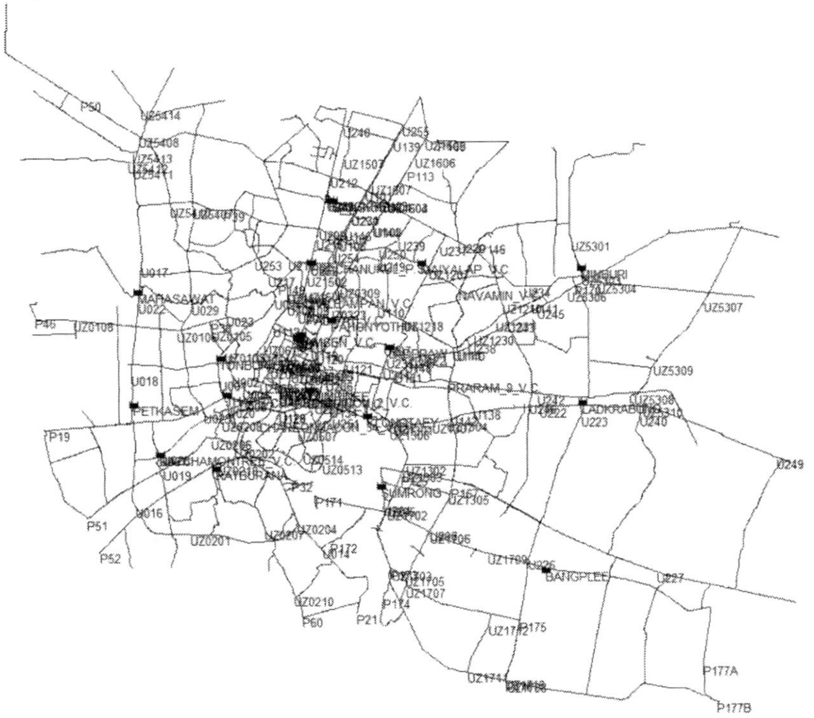

Figure 2. Field pressure measurement 134 points.

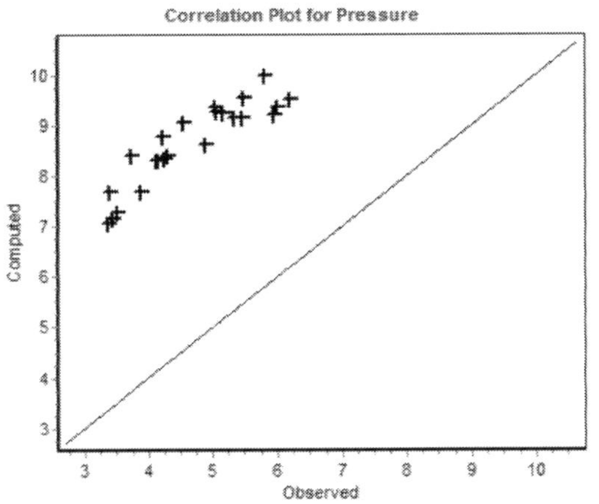

Figure 3. Correlation between computed and observed data of U014

The current pipeline network model does not assign the elevation of the node, so the elevation equals 0 for all nodes. Although MWA's service area is relatively flat, the elevations vary between 0 to 2 meters above sea level. However, some points of the MWA's pipeline are higher than the normal pipeline, especially at bridges across the many canals in MWA's service area as shown typically in figure 4.

Therefore, the pipeline at these points must be higher than the normal pipeline and these positions are often used to install a flow meter and a pressure transmitter. A pressure transmitter installed on the riser itself will indicate lower pressure than an installation in a pipe the normal level. When the water pressure from the field measurement is compared to the calculated water pressure, the pressure of the software is higher than the field measurement, as shown in figure 5.

Figure 4. Pipe laying across the canal

 https://doi.org/10.1051/matecconf/201815309001

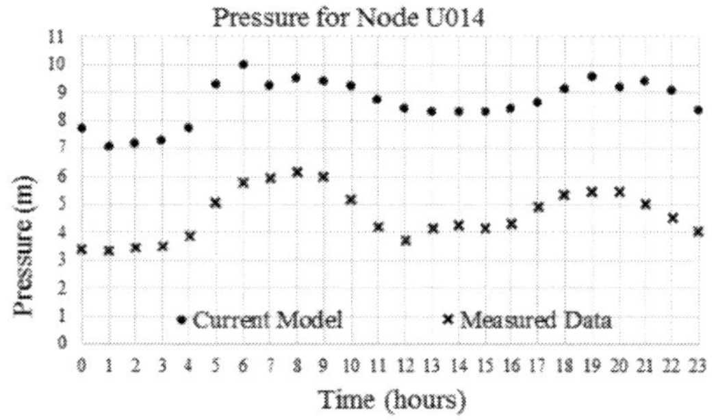

Figure 5. Computed results of current model and measured values for pressure at node U014.

Moreover, the head at each reservoir does not compensate for the elevation of each source. The head of these reservoirs is the fixed grade head, which is the source of data for calculation. When the source data is wrong, the result will be wrong. Therefore, the elevation of each node in the pipeline network model should be adjusted. Revision of node elevation can be referenced from the elevation of the roads as the main pipe is usually placed along the roads. Therefore, it is possible to estimate the elevation of the pipe line from the elevation of the roads, using data from the mean elevation map along the roads in Bangkok, 2006 – 2007 of Land Survey and Map Division, Public Works Department, Bangkok Metropolitan Administration. This is a map showing elevation values based on mean sea level (MSL) as figure 6. The elevations of roads vary between 0.09 to 1.90 meters above the sea level.

5 Result and discussion

After adjusting the elevation of nodes in the pipe network model, the model was recalibrated by comparing with the 134 field measured pressure points. It was found that the new model was more accurate than the current model. The correlation between means was up to 0.893 from 0.803. For example, at the measurement point U014 the correlation between new mean was better than the current model as shown in figure 7, where it can be seen that the results almost coincide at all times. Although overall the new model is quite good, there are shift some points that have high error such as at point number U020. It is suspected that this error may be due to the selection of the Head losses formula. Upto now, the Hazen-William formula has been used to calculate the Head losses by setting the Chw value to 120. This is the suggested value for brand-new steel pipe only. In fact, pipes of MWA have been used for a long time, so the pipe roughness has changed and should be revised in the network model. The Darcy Weisbach formula should be used for better accuracy.

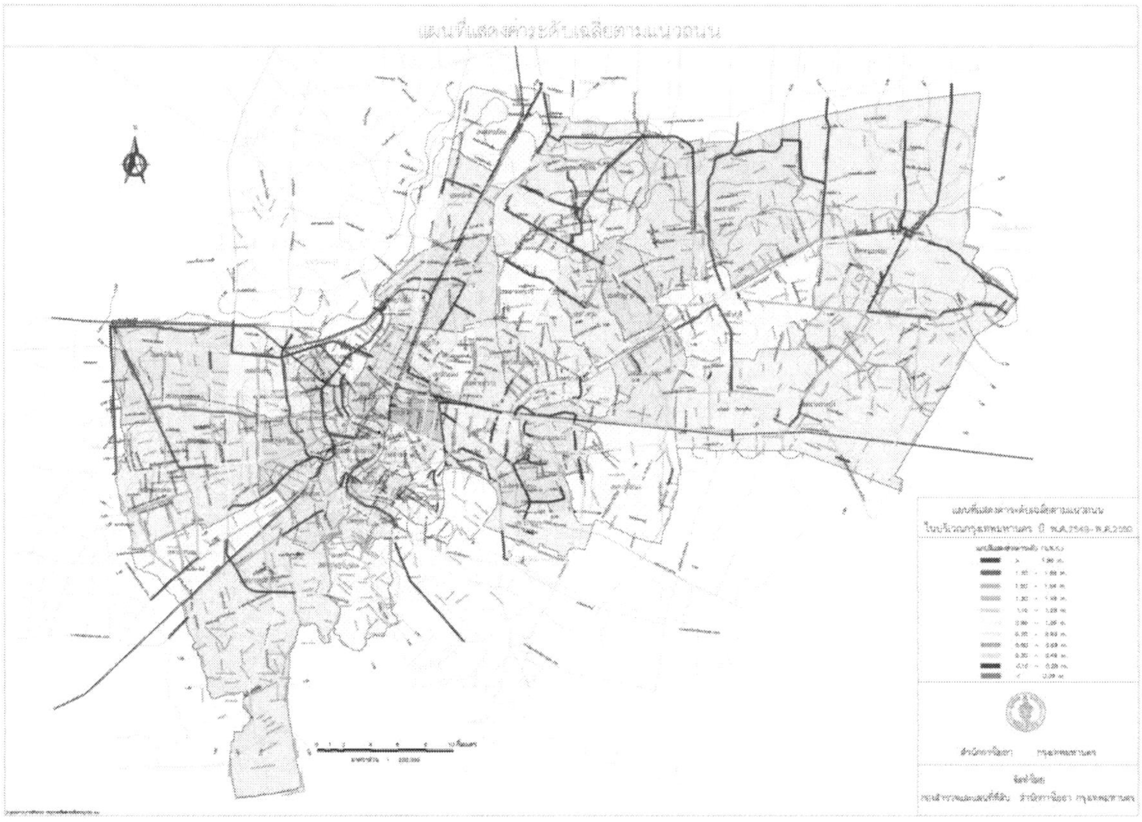

Figure 6. Bangkok roads elevation map.

https://doi.org/10.1051/matecconf/201815309001

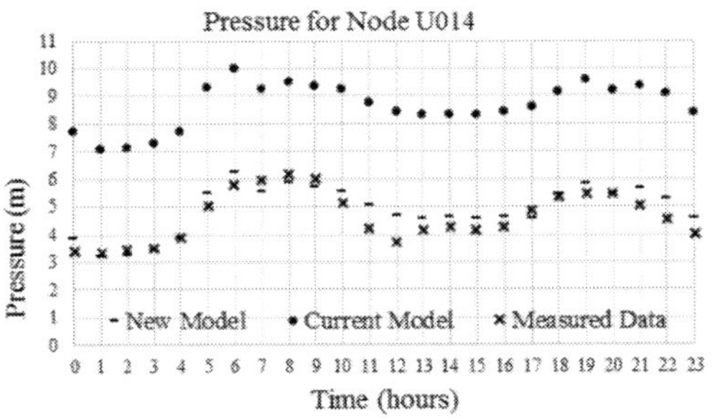

Figure 7. Computed results from the current model, new model and measured values for pressure at node U014.

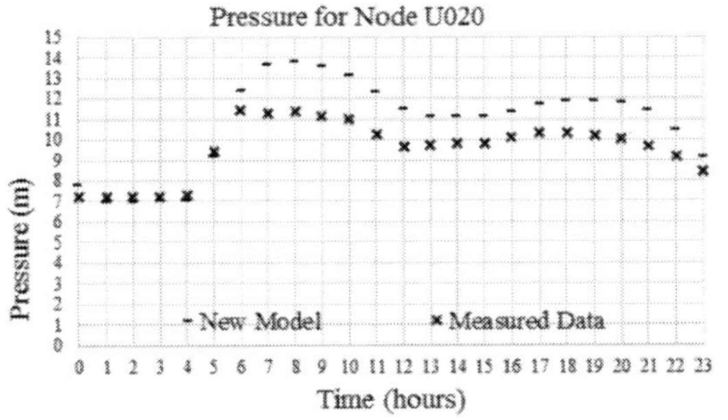

Figure 8. Computed results from the new model and measured values for pressure at node U020.

6 Conclusion

This study shows that it is necessary to assign elevation to each node in the network model, because the elevation values affect the calculation of the pressure at each node, especially at a point where a water pressure gauge is installed. These elevations were used to recalibrate the model. As a result the new network model has better accuracy.

Acknowledgement

The authors would like acknowledge Metropolitan Waterworks Authority (MWA), Thailand, for contributing important data for this study.

References

1. Dr. G. Venkata Ramanaa. Network analysis of water distribution system in rural areas using EPANET. Procedia Engineering 119 (2015) 496 – 505.
2. Lewis A. Rossman. EPANET 2 USERS MANUAL, U.S. Environmental Protection Agency (EPA). 2000
3. Bruce E. Larock. Hydraulics of pipeline systems. CRC Press LLC, 2000
4. E. Shashi Menon. Piping Calculations Manual. McGraw-Hill, 2005
5. Selami Kara. Hydraulic modeling of a water distribution network in a tourism area with highly varying characteristics. Procedia Engineering 162 (2016) 521 – 529.

https://doi.org/10.1051/matecconf/201815309002

Study on Strategic Planning of Road and Bridge Infrastructure Development in City Planning: Taking Porto-novo City of Benin Republic as Example

Dossa Didier Boko-haya [1,2], Yadong Li [1], Koffi Togbenou[1,2], Saizhi Liu [1], Changrong Yao [1] and Bin Qiang [1]

[1]*School of Civil Engineering, Southwest Jiaotong University, Chengdu 610031, China*
[2]*Polytechnic School of Abomey-Calavi (EPAC), University of Abomey-Calavi (UAC), 01 POB: 526 Cotonou, Republic of Benin*

Abstract. Concern about the townlet infrastructure construction in developing country is one of the crucial part of county town planning and development. By taking the overall planning and design in a case study of Porto-novo city at Republic of Benin, this paper analyzes the characteristics and opportunities of Porto-novo city and puts forward corresponding infrastructure construction strategy. In the end, the paper comes up with specific plan of planning and design under the background of Porto-novo's planning of development strategy.

1 Introduction

Geographically located in the east of the Benin and also situated at the southeast of Cotonou city, Porto-novo city is known as "the political capital of Benin" and is the traffic hub and material distribution center of road transport in southern of Benin and Nigeria. As such, Porto-novo needs to develop well balanced integrated multimodal transit systems, bridges and road networks by best utilizing the natural strength of each individual mode for the efficient movement of passengers and goods as well as integrate itself with the local, national and regional countries particularly with Nigeria to expand international trades through transportation networks.

Now, with Porto-novo city becoming increasingly established as Benin's political capital, large steps are being taken to alleviate these pressures through a range of improvement projects within the city and its surrounds. In this regard, the West African Urban Planning and Design (WAUPD) and National Development and Reform Committee (NDRC) of Benin preliminarily discussed the regional planning, design and development of Cotonou-Porto-novo Economic and Politic Zone [1-2]. Situated at the banded development axis of Cotonou-Porto-novo Economic and Politic Zone and according to a series of strategic thinking, the national government of Benin (NGB) has recently passed Urban Development Strategic Planning (UDSP) of Benin (hereinafter referred to as Strategic Planning (SP)). The SP made new spatial arrangement of Porto-novo city's new urban planning, design and development. As one of the three significant cities with special status of Benin Republic [3], Porto-novo city is included in the scope of new city. Thus, new thinking emerges in city infrastructure planning, design and construction.

This paper is structured in the following way. In the subsequent section, we highlight the characteristic of Porto-novo city. In section 3, we present Porto-novo city under the background of strategies planning and development. Integrated Strategies and Design of Porto-novo city's development priority needs are assessed in section 4. Section 5 finally concludes the paper and provides interpretations of where Porto-novo stands in this transition process, and how the city might be better positioned to achieve greater urban sustainability.

2 Overview and characteristics of Porto-novo city

The current study concerns the Porto-novo city which is one of the largest cities in West Africa (Figure1). Porto-novo city, 34.1 kilometers away from Cotonou, is located at the east of Cotonou and has advantageous geographical location. It spreads over a total land area of 410 km^2 and holds an estimated population of about one million.

2.1 Convenient roads network and traffic

A modern road network helps traffic move around the country more easily and is essential for any country economic growth. In this regard, Porto-novo city has convenient traffic, Porto-novo-Akpro Misserete-Igolo road cross the center of Porto-Novo, which connects Porto-novo with Cotonou and Nigeria. The national highwayRNI1[4],Hillacondji-Cotonou-Porto-novo-Akpro Misserete-Igolo expressway and Cotonou-Porto-novo railroad cross the south-east part of the country. So far, Porto-novo has basically achieved the goal that linking each village with highways. It's well developed in the east

https://doi.org/10.1051/matecconf/201815309002

and has formed network-based highway pattern as depicted in Figure 1.

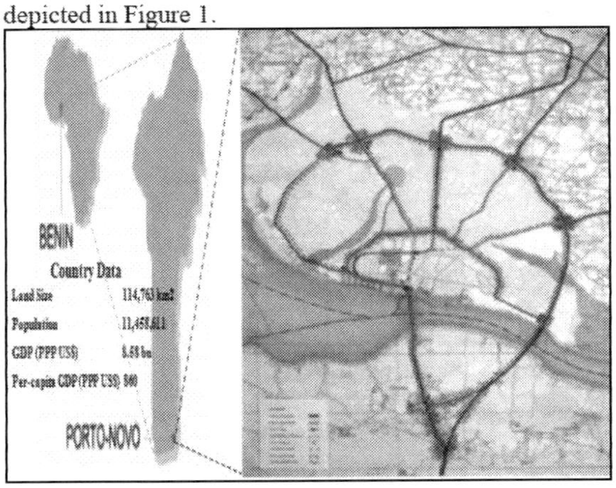

Figure 1. Country data and map of Porto-novo and it relationship to the city, Republic of Benin.

2.2 Weak economic foundation and industrial structure

Though Porto-novo has superior location advantage, its economic foundation is weak. Agricultural structure lacks of adjustment, the development of secondary industry is restricted and the tertiary industry just starts. All these obstructs Benin economic development [5-9].

2.3 Abundant tourism resource

Porto-novo city is probably the richest and most diverse heritage of any city in Benin Republic. Within 600 acres in the center of the city, history has left its footprint in the presence of successive Goun and Yoruba people, settlers and slaves, in comers from Brazil, these main four influences mingle and stand in the city creating the Porto-novo we know today. It's this past which has given the city such a diversity of place, people, culture and customs that can be found today.

In recent years, Porto-novo has undergone rapid growth, and this coupled with often severe climatic conditions has resulted in strains and pressures on its infrastructure, environment and the functioning of the city as a whole.

3 Porto-novo city's opportunities under the background of strategic planning

3.1 Opportunities brought by convenient bridge and road infrastructure development

Communities with good public transit systems are economically thriving communities and offer location advantages to businesses and individuals choosing to work or live in them [10-11]. Porto-novo has complete traffic network, which lays solid foundation for its development. National highway RNI1 crosses Porto-novo city from the south part of the country, Hillacondji-Porto-novo highway and Cotonou-Porto-novo railroad cross it administrative region. Further, the district's development will generate an increasing traffic flow. Thus, considering that the building of the "second bridge" answers to a real need of the city. In this regard, the bridge to be build will be placed at the west side of Lokpodji (an important new door for Porto-novo) at the limit between the entrance of the district and the rest of the city. The two valleys which traverse Lokpodji, on the East and West, will be recovered to facilitate the drainage of rainwater into the lagoon, without completely planning them. This approach allows the preservation of this exceptional and rich natural ecosystem. The bridges will be the connexion between the different ways of communication. Thus, developed traffic (bridge and road infrastructure) provides convenient channel for Porto-novo city's goods and population flowing to other areas; meanwhile, it can promote the development of Porto-novo city's tertiary industry such as trade and tourism (Figure 2).

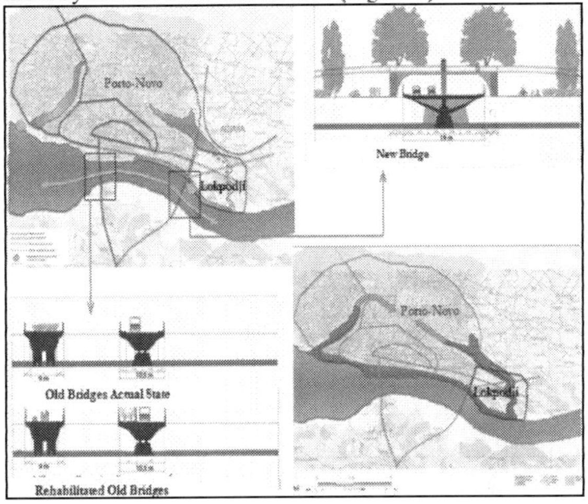

Figure 2. New bridges changing the city conception and the green belt

3.2 Opportunities brought by land use changes of city and county

The SP made a new scope demarcation of the new city (Figure 3). Then center of Porto-novo city and its villages such as Ganvie village and Lokpodji village are included in the urban development land, which have become several housing cluster in the development strategy of Benin Republic county.

In Porto-novo's space layout, Porto-novo takes its main city area as the core and forms the layout of "one center has three parts embraced by periphery lake". The current public administration building is no longer relevant in the city center, they could be located on other or less strategic areas. Though the property change of land use brings Porto-novo city with unprecedented challenges, it is a scarce opportunity for the city as whole.

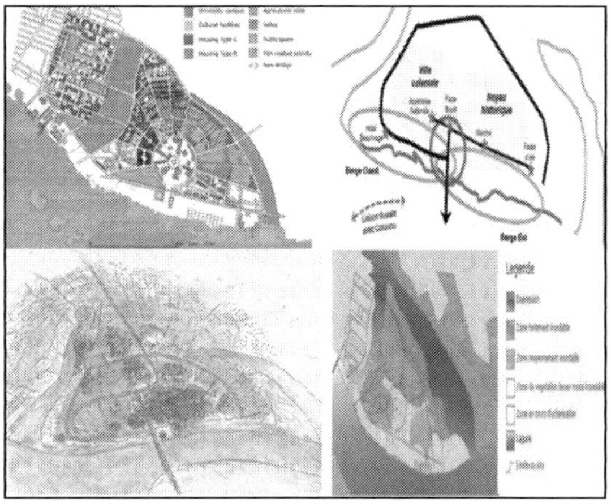

Figure 3. Strategic planning of Porto-novo city

4 Integrated strategies and design of Porto-novo city's development needs

4.1 Development strategies of Porto-novo city

4.1.1 Playing location advantage and strengthen radiation function

In the future development, Porto-novo city, situated at the central part of Cotonou and Nigeria county, should position from the overall development of Porto-novo's new towns and strengthen its connection with Cotonou, Calavi-Kpota (rossing Nokoue lake) and Nigeria to make harmonious development with Benin and Nigeria Republic and achieve mutual benefit of economic society.

As a development of new city, Porto-novo should accelerate its construction of various infrastructures and public service facility, strengthen its connection with other towns and cities. For instance, Calavi-Kpota can conveniently be connected with many destinations of Porto-novo through roadways via the lake Nokoue. Major road and bridge development activities should also focus on improving intermodal connectivity (e.g., physical integration of all feasible modes at strategic transportation hubs) and highway safety as well as maintaining roadways for quality service. Furthermore, technology should be used to improve safety and operational efficiency of all modes.

4.1.2 Optimizing the industrial structure to enhance economic strength

Industrial structure has a close relationship with economic development. As the industrial structure in Porto-novo city is relatively unitary, there is a need to concentrate on upgrading the industrial structure, changing the single structure, therefore adapting it to modern economy growth and laying a solid foundation for the long-term development of the town. In this regard, we also should spare no efforts to promote the development of the secondary or third industry, thus forming a coordinate development of economy layout. For instance, stimulate the region's economic prosperity by developing the characteristic agriculture along main line and achieve urban development strategy by improving the service level of tourism industry.

4.1.3 Highlighting historical culture and protecting the ecological environment

Porto-novo city, named for "Hogbonou Eucharistus " due to its profound culture background, was constructed in 1752. Protecting history cultural heritage is the premise of ancient city's development and improvement, so to preserve this resource, the state took the initiative with a Special Program of Rehabilitation (SPR) of the city of Porto-novo (PSRPN) in the late 1990s. Many of the richest and most diverse heritage also remained intact as well as the ancient streets and famous residential houses with Beninese traditional cultural feature. Toffa 1er, a Beninese revolution pioneer, whose native place adds more historical and herotical glory in this area. Inside the Porto-novo city has the picturesque scenery and historic celebrities.

As the indispensible scenery tourism area in Benin, Porto-novo possesses not only the unique human landscape and history background, but also uncommon ecological environment. Dynamic coordination should be used to improve transportation road network and environment. In this regard, the sustainable transportation characteristics [12] including: (i) traffic development combines with ecological environment protection and land use and (ii) traffic system not only meets the present social economic development's needs, but also provides development space for future sustainable development, would be needed.

4.2 Practical considerations of planning and design for Porto-novo city

Through the analysis of Porto-novo city's resources, socio-economic, environment conditions and super ordinate planning, Porto-novo positions itself as a living area for Benin county's development, ecotope, a backyard garden, and tourism service center of the county.

4.2.1 Porto-novo city system planning

Porto-novo city divides each village's level and scale in accordance with three levels 'City domain center, basic level village and key village', then the function system which is multi-level, multifunction and farm-out is formed. The area is framed by the Donoukin valley and serviced by a new road and bridge, connecting it to the wider city as whole (Figure 4).

Urban growth identified in this area would occur gradually, over a long period of time would be less dense in character. In this regard, the aim would be to prioritize public services assets including: (i) Supporting local business: growing wood and raphia (for construction), exploiting potential uses for water hyacinth, agriculture and market gardening; (ii) Schools and training colleges

https://doi.org/10.1051/matecconf/201815309002

(with a focus on agriculture, fishing and environmental or ecology studies); (iii) Health facilities and (iv) Sports and leisure facilities.

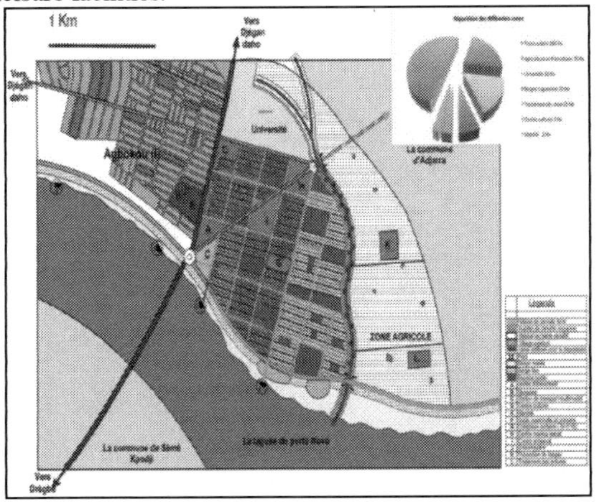

Figure 4. Porto-novo city system planning

4.2.2 The overall plan of the township

According to the new city development orientation, the distribution of construction land and long-term development trend, the new township relies on the arterial traffic and take a town built up area as the center, focusing on the development of the east, properly on the north and south development, then form a layout structure which is one center and five districts (Figure 5).

This plan naturally optimizes the space layout, keeps the integrity of the historical and cultural blocks in the old area. At the same time, Porto-novo city should develop properly and construct express way to make it more convenient to reach Benin Republic country, thus it can form a new economic growth polar.

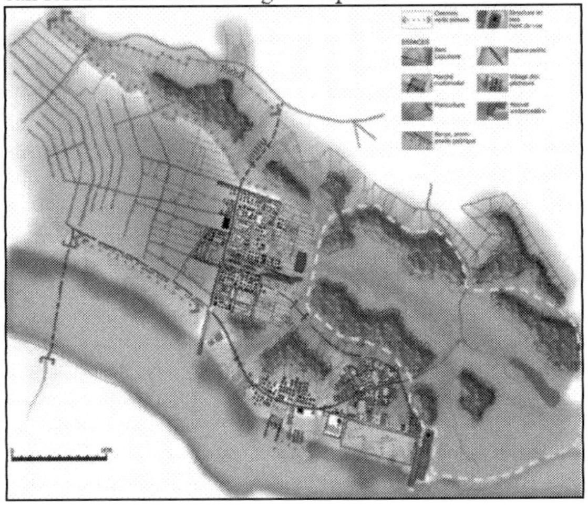

Figure 5. Overall plan of the township

4.2.3 Protection of famous historical and cultural city

The protection, inheritance and development of the city pattern plays an important role in protection of a famous city [13]. The historical blocks of Porto-novo city are concentrated, their style and features are based on eastern Cotonou's row dwellings, mixing multifarious celebrity culture. In the protection planning, the protection of ancient town's overall style and features include street space record, architectural pattern and key cultural relics. Simultaneously, the city highlights local characteristic history and culture.

By maintenance and arrangement Porto-novo exploits and utilizes its tourism resources reasonably. Taking the Lokpodji street as the center, the "ancient bridge" as the link, Porto-novo city divides its old town into two parts (south and north). Then an ancient city tourism center including Ganvie park is formed. To meet the tourism and life needs of town center Porto-novo offers necessary tourism, living, service and educational facilities.

4.3 Planning innovation key points

In order to highlight the characteristics of Porto-novo city, to create atmosphere of tourism landscape town and keep with the development of Republic of Benin's new city, this planning makes some considerable attempts.

Firstly, Porto-novo will lucubrate its historical blocks' protection, planning, design and highlight the fundamental keynote of ancient city to form tourism node. Simultaneously, Porto-novo should protrude the use of visual landscape and natural elements in the overall planning stage.

Secondly is that in the classification of land use planning, urban land classification criteria is used, that means Porto-novo city is not regarded as township land, but regarded as the city land to be planned. Thus, Porto-novo city not only meets the need of its city function, but also realise its urban group function after the development of the new city.

5 Conclusion

Porto-novo has been experiencing a very high rate of population growth and urban expansion. According to its characteristics and development of resources, location, industry and economic structure, construction and development of county townlet should adopt its good points and avoid its shortcomings to form its own unique advantages.

From the perspective of regional coordination, the overall planning of Porto-novo makes a preliminary design of urban development strategy, development tendency and urban and rural bridge and road traffic system. In order to optimize urban space, to promote the economic competitiveness and to realise common prosperity in urban and rural areas, under the circumstances that the use of city's urban and rural land changed, this planning takes the development condition and development factors as the key analyzing point.

Acknowledgment

The authors would like to thank the funding bodies of this study: the University of Southwest Jiaotong (Department of Bridge Engineering) under the Project Numbers (No. (VP13KJ1101Y14001 and 51378430).

The first and third authors would like to acknowledge and thank the Professors Cossi N. Awanou, Basile B. Kounouhéwa, Aristide B. Akpo and Clement A. Kouchade; and Dr Gabin Koto N'Gobi and Dr Ossénatou Mamadou for their continued support and encouragement.

Other important contributions of the research were made by Prof. Zhang Xun.

The conclusions in this paper are those of the authors of this paper or based on the referenced literature.

References

1. A. Anais, M. Schut. *Discussion on Strategy and urban design for the city center of Porto-novo*. pp.44-57, (2013).
2. El. Samia, B. Cukierman, A. Pringle. Discussion on Analysis of Urban Planning & Development Porto-Novo, Benin. Second Edition, pp.22-36, (2013).
3. Republic of Benin.Law n°98-005 on the organization of municipalities with special status of Benin (2001), (Accessed 18.08.2017).
4. Republic of Benin, Decree No.2001-092 of Feb. 20, 2001. Classifying economic, tourist or strategic routes of interest, pp.3-6, (2001).
5. D.D. Boko-haya, Y.D. Li, X.X. Liao, C.R. Yao, C.J. Wang and Q.Q. Xiang. Key issues and challenges in developing integrated road networks infrastructure with regional connectivity: A study case of Porto-novo in Benin Republic. MATEC Web Conf., (to be published)
6. D.D. Boko-haya, Y.D. Li, C.R. Yao and B. Qiang. Investigation on bridges connection to network carrefour in existing roads in Africa big cities: the case of cotonou in Benin Republic. Int. J. Eng. Res. App. **6**(4), pp. 53-61, (April-2016).
7. C.Dominguez-Torres,V.Foster.*Benin's infrastructure:a continental perspective.Africa infrastructure country diagnostic country report*. World Bank, Washington, DC, (2011).
8. D.D. Boko-haya, Y.D. Li, C.R. Yao, S.Z. Liu and Q.Q. Xiang. Road and Bridge Infrastructure Development Issues in Benin Republic: Analysis and Perspectives. American Journal of Civil Engineering. **5** (1), pp. 9-15, (2017).
9. D.D. Boko-haya.; Y.D. Li, S.Z. Liu, C.R. Yao, Q.Q. Xiang and B. Qiang. Assessment research on capacity strengthening in bridge and road development in the Republic of Benin. (to be published)
10. E. Zavadskas; R. Liias, Z. Turskis. Multi-attribute decision making methods for assessment of quality in bridges and road construction: State of the art surveys. Balt. J. Road Bridge Eng. **3**, pp.152-160, (2008).
11. D.D. Boko-haya, Y.D. Li, C.R. Yao, Y. Gu, B. Qiang, and Q.Q. Xiang. Development of conceptual model for overcoming the challenges of road and bridge infrastructure development: towards innovative solutions in Benin Republic, Int. J. Eng. Res. Afr. **26**, pp. 161-175, (2016). DOI: 10.4028/www.scientific.net/JERA.26.161
12. J. Yin, F. Li, X. H. Li. Research on the sustainable transportation development strategies in the developed countries. Traffic & Transportation, **7**, pp.39-42, (2009).
13. Z. J. Dong, G. Zhang, H. M. Liu. Protection and development of famous historical and cultural cities based on sustainable development-a case study of Nanjing, Study Theory, **22**, pp. 3-4, (2000).

Key Issues and Challenges in Developing Integrated Road Networks Infrastructure With Regional Connectivity: A Study Case of Porto-novo in the Republic of Benin

Dossa Didier Boko-haya [1,2], Yadong Li [1], Xiaoxian Liao [1], Changrong Yao [1], Chongjiao Wang[1], and Qiqi Xiang[1]

[1]School of Civil Engineering, Southwest Jiaotong University,Chengdu 610031, China
[2]Polytechnic School of Abomey-Calavi (EPAC), University of Abomey-Calavi (UAC), 01 POB: 526 Cotonou, Republic of Benin

Abstract. In recent years, with the rapid and continuous increase in traffic demand, a new upsurge in the construction of transport system is being made in Porto-novo. Thus, the problem of how to make modal transport system in this area play its full role and make its helpful to the development of integrated multimodal transportation networks needs to be urgently put forward. Recognizing these needs, this paper systematically addresses key issues and challenges toward developing integrated multimodal national transportation networks in the banded development axis of Abomey-Calavi-Pkota, Cotonou and Porto-novo with an emphasis to regional connectivity. Particularly, the paper addresses issues with the planning process, institutions, governance, strengths and weaknesses, funding and integrated multimodal development.

1 Introduction

In nowadays globalized and interdependent world, the development of integrated multimodal transportation systems and networks with regional connectivity is extremely vital for the socio-economic developments of any country. Located in the east part of Benin Republic and also situated at the east of Cotonou, Porto-novo is known as "the Political Capital of Benin" and is the traffic hub and material distribution center of road transport system in southeastern of Benin Republic and Federal Republic of Nigeria. As depicted in Figure 1, Porto-novo is surrounded by Abomey-Calavi and Cotonou and by Nigeria Republic from three sides (e.g., west, south and east) with a long boarder of hundreds of kilometers. As such, Porto-novo needs to develop well balanced integrated multimodal transit systems, road networks, bridges, rail, maritime and fluvial transport, by best using the natural strength of each individual mode for the efficient movement of passengers and goods as well as integrate itself with the local, national and regional countries particularly with Nigeria to expand international trades through transportation networks [1].

Mobility is the engine of economies and is extremely vital for the communication functioning of society. In this sense, being a signatory of West-African highway and Trans-Benin-Niger railway networks, Benin through Porto-novo has the obligation to develop its share of road and rail networks for seamless connection with the region and beyond. The country shares the vision and the strategic goals embraced by ECOWAS, NEPAD, African Union, and WAEMU, as well as the governance principles adopted at sub-regional and regional levels [2]. As a member country of West-African Association for Regional Cooperation (WAARC), African Union (AU), and other Regional Forums (RF), the banded development axis of Abomey-Calavi-Kpota, Cotonou and Porto-novo and other regional countries (Nigeria and Togo) cities have shown significant interest to improve transport connectivity among the regional countries for trade, tourism and commerce development purpose (Boko-haya, 2016, 2017). Specifically, Togo, Benin and Nigeria have expressed interest to utilize transport infrastructures (e.g., road, rail and ports) of Benin for bilateral trade as well as for transshipment and transit purpose. Besides, the Federal Republic of Nigeria has expressed interest to connect some cities with Benin through Porto-novo by roads and water ways.

Recognizing that Porto-novo becoming increasingly established as Benin's political capital, large steps are being conducted to alleviate these pressures through a range of improvement of infrastructure projects within the city and its surrounds area. For this purpose, the development of transportation in Porto-novo has been heavily biased towards road and with hundred kilometers of highways such as national, regional and district roads and bridges, road network becomes the dominant mode of transport with a modal share of over 93%, while the investment level in railway and inland waterway sectors are very minimal [3-4]. Being a riverine city with many rivers, the waterways could have also been developed well in this area. Nonetheless, despite a large share of investment, Porto-novo road network system capacity is considerably inadequate.

As a means of achieving sustainable and cost effective integrated multimodal transportation systems at the banded development axis of Abomey-Calavi-Kpota, Cotonou and Porto-novo and other regional countries (Nigeria and Togo) cities level for the efficient movement of people and goods as well as to serve as a regional transportation hub, Benin through Porto-novo needs to exercise a sound multimodal transportation planning process for the identification, prioritization and selection of implementable transportation infrastructure projects under its own budget constraints. The focus of the presented planning process would be to support by the NGB policies and priorities in this the area of Porto-novo. In addition, it requires strengthening institutional capacities and overcoming institutional barriers for the unbiased modal developments.

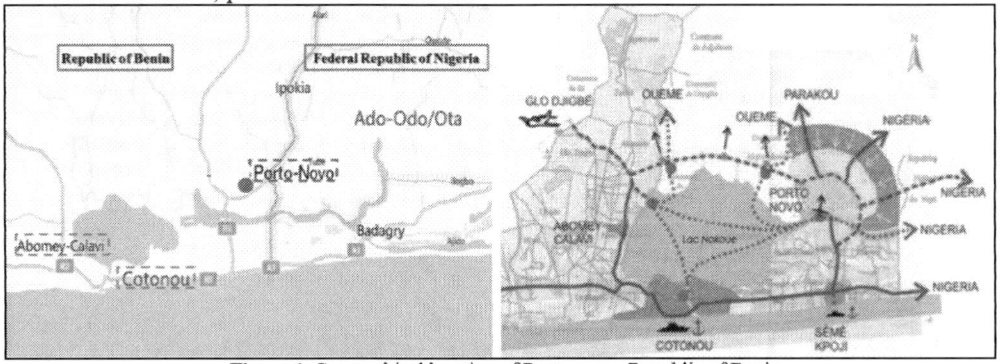

Figure 1. Geographical location of Porto-novo, Republic of Benin

2 Overview of the study area

2.1 Transport policies and plans in Porto-novo

As part of the Millennium Development Goals (MDGs), women equity, gender equity and poverty alleviation are some of the major goals of the authorities of this city where policies and programs are also formulated to achieve such goals. In this regard, the National Development and Reform Committee (NDRC), Planning Commission (PC), West African Urban Planning and Design (WAUPD) of Benin formulate long term prospective plans and short term five years plans on a rolling basis for all sectors including transportation sector [5-8]. Presently, Porto-novo has been formulating its five years plan (2016-2020). This five year plan includes the wish list of development programs and projects encompassing all sectors. However, these plans are developed without any commitments of funds for the implementation purpose. As such, many programs and projects are not implemented in a timely manner or even within the planning time frame (World Bank, 2013).

Porto-novo was stripped of its attributes of capital in the 60s and 70s with the transfer of the Presidency and the Ministries to Cotonou. With the advent of democratic renewal in 1990, the status of capital of Benin was reconfirmed and Porto-novo resurgence was emphasized with the implementation of a Special Rehabilitation Program (SRP) with the main goal to restore attributes to the city as Benin Republic capital. As part of the reform of territorial administration and decentralization [9-10], Law No. 98-005 of 15 January 1999, gives Porto-novo, as the capital, a common character with special status, 5 quarters and 86 districts. Table: 1 listed the identified weaknesses and strengths of Porto-novo city.

Table: 1 Identified weaknesses and strengths of Porto-novo city

Weaknesses	Strengths
• Difficulties in establishing itself as Benin capital near the economic capital	• Capital of Benin Republic
• Road Network, bridges and infrastructure far below the needs of the population	• Economic benefits from trade with Nigeria
• Inadequate urban services	• Environment quality, unpolluted, along the lagoon
• Weak mobilization of financial resources	• Very largely composed of Green Spaces
• Development anarchic areas not surveyed and unrenovated	• Built Heritage (African, Afro-Brazilian colonial)
• Instabilities of resources from the informal trade	• Knowledge of quality craftsmanship
• Structure of tourist reception inappropriate	• A growing structure industry
• Difficulties in collecting taxes and tax	
• Low effective project management system	
• Governance decision fail to strike a balance between long term development objectives	

The city of Porto-novo is likely the valuable and most different heritage of any city in Benin economic foundation is weak, despite its superior geographical location advantage. Adjustment is the critical deficiency in modern farming structure, the secondary industry development is restricted and the tertiary industry just starts. All these impedes Benin economic development [11-12].

2.2 Regional connectivity: convenient and practical consideration of notable infrastructure development for Porto-novo City

Communities with suitable public transit systems are economically booming communities and offer location advantages to businesses and individuals choosing to live or work in them. Thus, in recent years, Porto-novo (Benin) and Nigeria have made some good progress in the improvement of transportation connectivity between these two neighboring countries. Porto-novo authorities, through its latest Municipality Integrated Multimodal Transport Policy (MIMTP) 2016, has expressed the need of developing integrated multimodal sustainable transportation systems by efficiently inter-connecting road networks, bridges, rail, (e.g., sea ports and land) waterway transport (Boko-haya et al., 2016). As stated in the prospective plan 2015-2050, "the main objective of the perspective plan with respect to transportation is to develop an efficient, safe, sustainable and regionally balanced transportation system in which various modes complement each other, interface properly and, where possible, provide healthy competition to each other [6,8]."

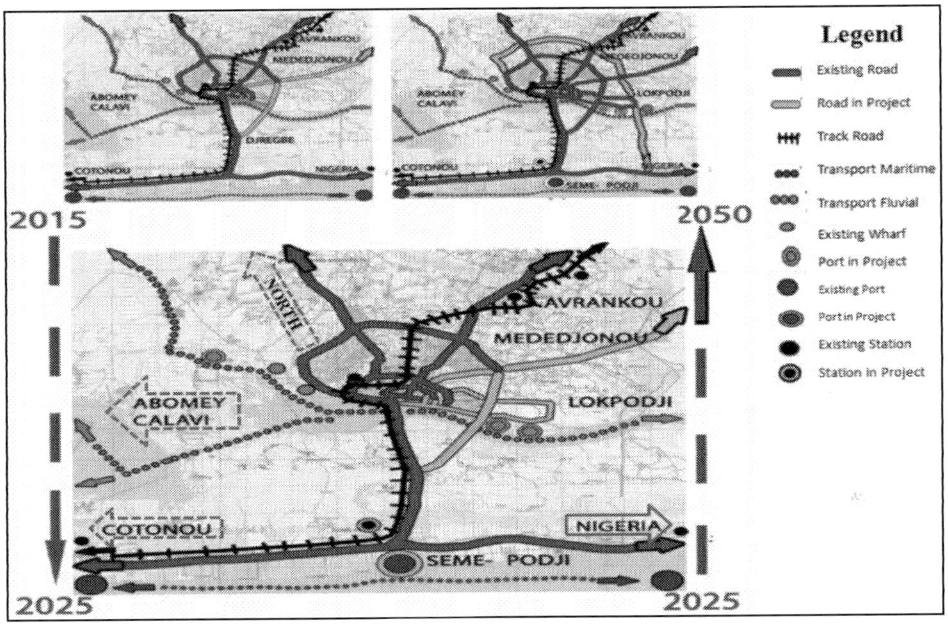

Figure 2. Evolution of the transport's map from the area of Porto- novo, Benin between 2015-2050.

- City of Abomey-Calavi Municipality (Glo-Djigbe Airport, and new bridge & road networks to be built over Nokoue lake)
- City of Cotonou (Economic Capital) Metropolitan Municipality (Airport, Sea port and Nokou lake road network to be built)
- City of Porto-novo (Political Capital) Metropolitan Municipality (Waterways, new bridge and road networks to be built)

The prospective plan further gives importance to redevelop three main cities (Porto-nono, Cotonou and Abomey-Calavi) and also two main sea-ports (e.g., Seme-Podji and Allada sea ports) and efficiently connect them with the national transportation networks for seamless movement of people and goods. Further, the plan focuses on improving rail connectivity between Cotonou and Porto-novo as well as improving roadway connectivity with the regional countries including Nigeria and Togo. It also gives importance to improve connectivity between rural roads, maritime and fluvial transport which where feasible in area of Abomey-Calavi-Kpota (crossing nokoue lake), Cotonou and Porto-novo. The construction of a road with a new bridge, transport of passengers and the construction of new debarkations as well as train (track road) rehabilitation and construction of freight's harbor in Sèmè Kpodji. Thus, freight could arrive by the harbor and then will go in all the area by the train. The Porto-novo modern road network city, helps traffic move around Cotonou, Nigeria and Abomey-Calavi more easily and is essential for Benin economic growth. Thus, Porto-novo has convenient modal traffic, it roads cross the city center, which connects then Porto-novo with Cotonou, Abomey-Calavi-Kpota and Nigeria. The national highway RNI1, 2 and 3 [9], Cotonou-Porto-novo expressway and Cotonou-Porto-novo railroad cross the south-east part of Benin. Porto-novo has basically achieved the goal that linking each village with highways, port and sea port. It's well developed in the east and has formed network-based highway pattern as shown in Figure. 2.

3 Analysis of transport's map evolution

Through the analysis of Porto-novo's environment resources, socio-economic conditions, and super ordinate planning, the city positions itself as a living area for Benin county's development, inter-connecting road networks, bridges, rail, (e.g., sea ports, and land) maritime and fluvial transport. However, the study found that there are weaknesses in the planning process and prioritization of infrastructure projects that have prevented to achieve the intended benefits from the investments. Actually, there is no comprehensive coordinated, and continuous multimodal transportation planning process in the identification, prioritization and selection of transportation infrastructure projects (e.g., rail, road and inland water) that would measure the achievements of policy goals and objectives.

Several other ministries have the responsibilities on the surface transportation development activities. For instance, Ministry of Planning, Economic Restructuring and Employment (MPREPE), Ministry of Decentralization, Local Governance, Administration and Territorial Development (MDGLAAT) coordinates the activities of the territorial administration, regional and national roads, rails, bridges and rural roads, are under the jurisdiction of Ministry of Public Works, Transport and Urban Planning, Ministry of the Environment and Sustainable Development (MCVDD) and inland waterways and ports are under the Ministry of Maritime Economy and Shipping [2]. Mostly, projects are identified, prioritized and selected severally along modal lines by the respective modally segregated ministries or subordinates agencies in association with the Ministry of Planning. In Benin, such an approach often takes place under the influence of politics and without a comprehensive project evaluation. What is more, there is no formal public participation or involvement in the planning and infrastructure project development process.

Due to lack of funds or other reasons, unnecessary projects are frequently implemented and important projects infrastructure are postponed. Despite, NGB has given much attention to develop transport infrastructure projects through Public Private Partnership (PPP) agreements in the five year plan (2014-2018), however, the intended results has not been achieved during the planning period (Boko-haya et al., 2016, 2017). The NGB fiscal budget consists of two parts including development budgets or Annual Development Program (ADP) and non-development (revenue). The funds for the revenue budget come from NGB authorities' sources. However, as the NGB has very limited financial resources for the development budget, Benin relies on the international development partners (e.g., African Development Bank, World Bank, International Monitory Fund etc) to bridge the development budget gap while compromising NGB priorities and needs. The limited ADP funds are allocated to selected sectoral programs and projects and the implementation process of funded programs and projects infrastructure begins during the same fiscal year although some projects may last for multiple years. As a central planning authority, the PC in collaboration with the Ministry of Finance plays a key role in the preparation of programs and projects to be included in the ADP for funding during each fiscal year.

4 Integrated transportation development needs

Based on the diagnostic analysis, Porto-novo is still far away from developing modally balanced integrated multimodal transportation systems. Especially, Porto-novo needs to invest heavily in the maritime, fluvial transport and rail sectors. Since Porto-novo, Cotonou and Nigeria share approximately 60 trans-boundary rivers, the collaboration of Porto-novo (Benin) and Nigeria in the joint development of waterways as well as costal shipping would be very beneficial for both the countries. Porto-novo and Abomey-Calavi-Kpota can conveniently be connected with many destinations.

With well developed and integrated multimodal transportation systems, Porto-novo can be able to use waterways for trade with Abomey-Calavi-Pkota and Nigeria and provided that the inland waterway ports are physically integrated with major roads, highways and rail transportation networks and hubs. For instance, Nokoue lake in Cotonou can be well connected with roads and even rails. Lokpodji can also be connected with Abomey-Calavi via waterways. In this view, significant investment will be needed in the development of inland waterway ports. Aids service would be needed to dredging work to maintain appropriate channel depths and night navigational. These activities will be carried/contracted out with the involvement of the private sectors or executed under various public private partnership (PPP) agreements. Other forms of models include the popular concession structure, Build-Operate-Transfer (BOT) and also Build Own Operate Transfer (BOOT), Build Transfer Operate (BTO), Design Build Operate Maintain (DBOM), Design Build Finance Operate (DBFO), Operate and Maintain (O&M), Design and Build (DB), Design Construct Manage and Finance (DCMF), Build Lease and Transfer (BLT), Design Construct, Manage and Finance (DCMF) and several other similar concession acronyms (Boko-haya et al., 2016 & Eaton and Akbiyikli, 2005).

There exists one rail gauge in Benin and Niger as well as other West African nations. Therefore, Benin should collaborate with regional countries such as Nigeria and Niger to standardize rail gauges for seamless service within and across boundaries through Porto-novo city. Then, rolling stocks further need to be standardized for seamless cross boarder service. Thus, strategic railway investment would be needed.

Regarding the investment in road and bridge sector, the major focus will be to increase the capacity of key national highways including all designated highway sections within the RN1, 2 and 3 as well as priority developments of the bridges located at Calavi-Kpota (rossing Nokoue lake) and Lokpodji (Figure.1 and 2). Major road development activities should therefore focus on improving intermodal connectivity (e.g., physical integration of all feasible modes at strategic transportation hubs) and highway safety as well as maintaining roadways for quality service. To improve safety and operational efficiency of all modes including ports (e.g., river, land and sea) and intermodal facilities/terminals, technology should be needed. In this sense, an ITS (Intelligent Transportation Systems) vision for ensuring interoperability of equipments and the systems needs to be developed and a priority base incremental deployment of ITS technologies needs to be guaranteed.

5 Conclusion and key recommendations

The presented paper reviews the programs, policies, planning, and development of surface transportation systems (e.g., highway or roadway, waterways and railways) in Porto-novo, Benin Republic. Moreover, it identifies the institutional and planning weaknesses as well as development priority needs from multimodal transportation systems and regional connectivity prospective. Presently, there are inconsistencies between the policies and how various projects are identified, selected and funded making it harder to achieve the different policies goals. Besides it's suggested that a comprehensive and continuous multimodal transportation planning process needs to be developed for the prioritization, the programs selection and infrastructure projects as well as for the analysis and evaluation of transportation policies as whole. In addition, the money borrowed from the multilateral banks and agencies and limited resources need to be used transparently justifying the policy goals and objectives.

As specific recommendations listed below, there is a lot of work to be done in a thorough going manner that lead to the: (1) Need for formal collaboration and coordination among different surface transportation agencies and ministries as well as public participations in the transportation planning and project development process for the unbiased modal developments; (2) Development of maritime and fluvial transport must be given a priority and Nigeria, Togo and Benin should collaborate in the development of cross-border waterway systems for moving freight and passengers and (3) According to the investment priorities and needs, a formula should be developed to allocate annual development program funds for the multimodal developments and modal integration aim.

Acknowledgment

The authors gratefully acknowledge the financial support received from the University of Southwest Jiaotong (Department of Bridge Engineering) under Project. (No. VP13KJ1101Y14001 and 51378430).

The first author would like to express his sincere gratitude to Prof. Yadong Li for the continuous support of my study and research, for his patience, motivation, enthusiasm, and immense knowledge. Besides, I would like to thank Professors Yongle Li, Changrong Yao and Zhang Xun questions on my work.

My sincere thanks also goes to my labmates for offering me the opportunities in their groups and leading me working on diverse exciting projects. Any opinions, findings, conclusions and recommendations expressed in this publication are those of the authors.

References

1. D.D. Boko-haya, Y.D. Li, K. Togbenou, S.Z. Liu, C.R. Yao and B. Qiang. Study on strategic planning of road and bridge infrastructure development in city planning: taking Porto-novo city of Benin Republic as example. MATEC Web Conf. (to be published)
2. D.D. Boko-haya, Y.D. Li, S.Z. Liu, C.R. Yao, Q.Q. Xiang & B. Qiang. Assessment research on capacity strengthening in bridge and road development in the Republic of Benin. (to be published)
3. C.D. Torres and V. Foster. Benin's Infrastructure: A Continental Perspective, pp.5-8, (2011).
4. D.D. Boko-haya, Y.D. Li, C.R. Yao, S.Z. Liu and Q.Q. Xiang. Road and bridge infrastructure development issues in Benin Republic: analysis and perspectives. American Journal of Civil Engineering. **5** (1), pp. 9-15, (2017).
5. D.D. Boko-haya, Y.D. Li, C.R. Yao, Y. Gu, B. Qiang, and Q.Q. Xiang. Development of conceptual model for overcoming the challenges of road and bridge infrastructure development: towards innovative solutions in Benin Republic, Int. J. Eng. Res. Afr. **26**, pp. 161-175, (2016). DOI: 10.4028/www.scientific.net/JERA.26.161
6. A. Anais, M. Schut. Discussion on strategy & urban design for the city of Porto-novo, pp.44-57, (2010).
7. République du Bénin. Etudes nationales de perspectives à long terme, Bénin 2025 alafia, stratégies de développement du Bénin, (2000).
8. El. Samia, B. Cukierman, A. Pringle. Discussion on analysis of urban planning & development Porto-novo, Benin. 2nd edition, pp.22-36, (2010).
9. République du Bénin. Programme d'action du gouvernement (2016-2021), Cotonou, (2016).
10. Republic of Benin. Law n°98-005 on the organization of municipalities with special status http://www.decentralisation-benin.org/IMG/pdf_loi_98-005.pdf , 2017.
11. Republic of Benin, Decree N°2001-092 of February 20, 2001. Classifying economic, tourist or strategic routes of interest, pp.3-6, (2001).
12. E. Zavadskas, R. Liias, Z. Turskis. Multi-attribute decision making methods for assessment of quality in bridges and road construction: State of the art surveys.Balt.J.RoadBridgeEng.**3**,pp.152-160, (2008).

AUTHOR INDEX

AUTHOR INDEX